For at least 30 years, there have been close parallels between studies of bird song development and those of the development of human language. Both song and language require species-specific stimulation at a sensitive period in development and subsequent practice through subsong and plastic song in birds and babbling in infant humans, leading to the development of characteristic vocalizations for each species.

This book illustrates how social interactions during development can shape vocal learning and extend the sensitive period beyond infancy and how social companions can induce flexibility even into adulthood. Social companions in a wide range of species, including not only birds and humans but also cetaceans and nonhuman primates, play important roles in shaping vocal production, as well as the comprehension and appropriate usage of vocal communication.

Social influences on vocal development will be required reading for students and researchers interested in animal and human communication and its development.

Social influences on vocal development

Social influences on vocal development

EDITED BY

CHARLES T. SNOWDON
John T. Emlen Professor of Psychology and Zoology,
University of Wisconsin, Madison

AND MARTINE HAUSBERGER
Directeur de Recherches, CNRS, UMR 373:
Ethologie, Evolution, Ecologie, Université de Rennes I

CAMBRIDGE
UNIVERSITY PRESS

PUBLISHED BY THE PRESS SYNDICATE OF THE UNIVERSITY OF CAMBRIDGE
The Pitt Building, Trumpington Street, Cambridge CB2 1RP, United Kingdom

CAMBRIDGE UNIVERSITY PRESS
The Edinburgh Building, Cambridge CB2 2RU, United Kingdom
40 West 20th Street, New York, NY 10011-4211, USA
10 Stamford Road, Oakleigh, Melbourne 3166, Australia

First published 1997

Printed in the United Kingdom at the University Press, Cambridge

Typeset in Monotype Ehrhardt 9/12 pt

A catalogue record for this book is available from the British Library

Library of Congress Cataloguing in Publication data
Social influences on vocal development / edited by Charles T. Snowdon
   and Martine Hausberger.
      p.   cm.
   Includes bibliographical references and index.
   ISBN 0 521 49526 1
   1. Sound production by animals.   2. Birdsongs.   3. Language
acquisition.   I. Snowdon, Charles T.   II. Hausberger, Martine.
   QL765.S63   1997
   591.59 – dc20   96-27781 CIP

ISBN 0 521 49526 1 hardback

Contents

Contributors

Luis F. Baptista, Department of Ornithology and Mammalogy, California Academy of Sciences, Golden Gate Park, San Francisco, CA 94118, USA

Eleanor D. Brown, Division of Birds, National Museum of Natural History, Smithsonian Institution Washington, DC 20560, USA

Dorothy L. Cheney, Department of Biology, University of Pennsylvania, Philadelphia, PA 19104, USA

A. Margaret Elowson, Department of Psychology, University of Wisconsin, 1202 West Johnson Street, Madison, WI, 53706-1696, USA

Susan M. Farabaugh, Department of Anatomy, School of Medicine, Auckland University, Auckland, New Zealand

Todd M. Freeberg, Department of Psychology, Indiana University, Bloomington, IN 47405, USA

Sandra L. L. Gaunt, Borror Laboratory of Bioacoustics, Department of Zoology, Ohio State University, 1735 Neil Avenue, Columbus, OH 43210-1293, USA

Susan Goldin-Meadow, Department of Psychology, University of Chicago, 5730 South Woodlawn Avenue, Chicago, IL 60637, USA

Marjorie Harness Goodwin, Department of Anthropology, University of California at Los Angeles, Los Angeles, CA 90024, USA

Martine Hausberger, Ethologie, Evolution, Ecologie, CNRS, UMR 373, Université de Rennes I, Campus de Beaulieu, F-35042 Rennes Cedex, France

Annick Jouanjean-l'Antoëne, Ethologie, Evolution, Ecologie, CNRS, UMR 373, Université de Rennes I, Campus de Beaulieu, F-35042 Rennes Cedex, France

Andrew P. King, Department of Psychology, Indiana University, Bloomington, IN 47405, USA

John L. Locke, Department of Speech Science, University of Sheffield, 20 Claremont Crescent, Sheffield S10 2TA, UK

Brenda McCowan, Marine Research Center, Marine World Foundation, Marine World Parkway, Vallejo, CA 94589, USA

Douglas A. Nelson, Borror Laboratory of Bioacoustics, Department of Zoology, Ohio State University, 1735 Neil Avenue, Columbus, OH 43210-1293, USA

Laura L. Payne, Museum of Zoology, University of Michigan, 1109 Geddes Avenue, Ann Arbor, MI 48109-1079, USA

Robert B. Payne, Museum of Zoology, University of Michigan, 1109 Geddes Avenue, Ann Arbor, MI 48109-1079, USA

Irene Maxine Pepperberg, Department of Ecology and Evolutionary Biology, University of Arizona, Tucson, AZ 85721, USA

Diana Reiss, Department of Biology, Columbia University, New York, NY 10022, USA

Rebecca S. Roush, Department of Zoology, University of Wisconsin, Madison, WI 53706, USA

Laela S. Sayigh, Department of Biological Sciences, University of North Carolina at Wilmington, Wilmington, NC 28403, USA

Robert M. Seyfarth, Department of Psychology, University of Pennsylvania, Philadelphia, PA 19104, USA

Catherine Snow, Graduate School of Education, Harvard University, Cambridge, MA 02138, USA

Charles T. Snowdon, Department of Psychology, University of Wisconsin, 1202 West Johnson Street, Madison, WI 53706-1696, USA

Peter L. Tyack, Department of Biology, Wood's Hole Oceanographic Institution, Wood's Hole, MA 02543, USA

Meredith J. West, Department of Psychology, Indiana University, Bloomington, IN 47405, USA

Richard Zann, School of Zoology, Faculty of Science and Technology, LaTrobe University, Bundoora, Victoria 3083, Australia

1 Introduction

CHARLES T. SNOWDON AND MARTINE HAUSBERGER

In the late 1960s a series of developments in linguistics, developmental psycholinguistics and animal communication led to a convergent model of vocal development in human and nonhuman species. As presented by Lenneberg (1967) and Marler (1970), the development of language and of bird song required exposure to species-specific codes during a sensitive period of development, after which subsequent learning was extremely limited. The amount of input required could be quite small, and this input could be effective regardless of social interactions. Both birds and humans needed intact hearing and an extensive time for practice (babbling for human infants, subsong and plastic song for birds) to acquire adult competence in vocal production. Subsequent to this practice crystallization occurred, and further changes in vocal structure were rare.

This paradigm has led to extremely productive research over the past 25 years, not only in the study of vocal development but also in the understanding of the neurological controls of vocal production. However, as researchers interested in the ontogeny of primate and avian vocal communication, we have become increasingly aware of the need to consider some modifications to this paradigm. Some forms of social stimulation can extend sensitive periods for song learning in birds, and songs and calls in some species can be modified throughout life, often in response to changes in social stimuli. Parrots, dolphins and great apes with exceptional training acquired codes with some similarities to human language. Yet at the same time there was little evidence of vocal plasticity in nonhuman primates, suggesting a gap in continuity of developmental processes in the evolution from birds to humans. However, at the level of pragmatics – how vocalizations are used – learning within a social environment appeared to play a critical role.

Although there was a close interaction between researchers on the ontogeny of language and of bird song 25 years ago, recently there has been little effort to cross taxonomic boundaries, and most recent volumes focus on birds, or nonhuman primates or children, with little cross-species integration.

We had several goals in the development of this book. First, we saw many parallels emerging from studies on birds, marine mammals, nonhuman primates and children, and we thought there would be considerable benefit in drawing studies on all of these species together in one volume. Second, while most of the research in the past 25 years has focused on ontogenetic factors of vocal production, understanding how organisms acquired the ability to understand and make appropriate use of vocalizations and speech seemed equally important. We sought to draw attention to the importance of considering production and pragmatics. Third, an increasing number of studies suggest that affiliative social interactions are important in shaping both vocal production and vocal usage, and we have sought to illustrate the variety of social interactions that affect development.

We have chosen contributors studying nonhuman animals who have worked in field settings or in naturalistic captive environments because, as will be seen, one may not appreciate the importance of social influences on development in more restrictive environments. Among the many scientists studying child language development we have chosen contributors who also have a strong naturalistic approach to illustrate better the parallels between human and nonhuman vocal development. We have also selected contributors with different theoretical and empirical perspectives, since our goal is to promote a productive discussion among researchers rather than to present a unified point of view.

In the second chapter *Nelson* presents a critical review of social interaction and sensitive phases, presenting evidence in support of a selective model of vocal learning, where young birds acquire the potential for producing several songs, but through territorial interactions at the time of breeding, selectively reduce the number of songs to use those that match those of territorial neighbors. Nelson is critical of those who invoke "social influences" as an explanation without specifying what aspects of social interactions influence learning.

Baptista & Gaunt review a variety of studies on both oscine and nonoscine birds, illustrating how sensitive periods for song learning can be extended through experience with social companions. They suggest that social companions can focus attention on specific songs, with the social tutor serving as both a motivator and a modeller for the young bird.

West, King & Freeberg review and revise their interpretation of research on cowbirds, obligate nest parasites where young birds do not have parental models for song learning. West *et al.* call attention to the pragmatics of communication. They report that birds with highly effective songs, will not use the songs appropriately if their social environment during development has been limited or inappropriate. Much of the complexity of vocal communication can be lost by attending only to production, and some of the most important social influences will be expressed as deficits of pragmatics rather than of production.

Payne & Payne use strong inference methods in field studies of indigo buntings and village indigo birds to argue that these birds learn new songs throughout life. Yearling birds produce song types that were rarely or never recorded in their natal territory, but which are common in the area in which they are breeding. Yearling birds acquire new songs rapidly and those birds that match songs of their territorial neighbors were more likely to retain their own territories and breed successfully.

Zann reviews the literature on song learning in zebra finches, the species that has been most extensively studied. Zann combines his own fieldwork with studies done in captivity to indicate several stages of learning. Zebra

finches can memorize the songs of their fathers heard before 35 days of age, but the songs actually produced depend on subsequent experience. Both visual and vocal stimuli are important, indicating the importance of multimodal stimulation for song learning. The highly unpredictable natural environment of zebra finches in Australia coupled with the high mortality of both young and adult birds may lead to the high degree of flexibility in vocal learning found in zebra finches.

Brown & Farabaugh stress the importance of studying species of birds that live in stable year-round groups. The species they have studied, Australian magpies, American crows and budgerigars, show extensive convergence of both songs and call types, although each individual retains unique notes or syllables. They view vocal sharing or convergence as a "social badge" that signifies membership within a social group. They argue that song sharing will be found not only between territorial rivals, as noted in other chapters, but also among affiliative partners.

Hausberger continues the theme of song sharing among affiliative partners in her studies of European starlings. She demonstrates a hierarchical structure to starling songs, with some themes being species specific, others being regionally specific, other themes specific to social groups or roosts, and still others marking individual affiliations. Two different species of starling will share themes when they belong to the same colony. When social relationships among a group of starlings are experimentally altered, birds develop a new set of themes to share with new affiliative partners. She reports song sharing among females, as well as males, especially outside of the breeding season, and also suggests that vocal convergence signals group membership.

Pepperberg presents social modelling theory as a method for exceptional vocal learning. Social modelling theory emphasizes not only production but also the functional referent of a signal and is effective in exceptional learning in emphasizing attention, motivation, and comprehension. She provides an experimental study to illustrate the effectiveness of social modelling theory in vocal learning in parrots, and she analyzes the studies of language analogs in great apes to show that the most successful of these studies are those that most closely follow social modelling as a training technique.

McCowan & Reiss show that dolphins can easily learn to imitate artificial signals relating to specific objects, and then use these imitations in spontaneous play with the same objects. They also present a naturalistic study of whistle development in dolphins, arguing for a hierarchical structure similar to that suggested by Hausberger, with species-specific, group-specific and individual specific whistle structures. As with birds and human infants, there is overproduction of variants by young dolphins, with whistles becoming less variable with increasing age.

Tyack & Sayigh, though differing in their definition of the diversity of whistle types described by McCowan & Reiss, also provide evidence of social learning in dolphins. They show that signature whistles are affected by social and acoustic factors and that dolphins can acquire new whistle forms throughout life. Adult males that form coalitions acquire and use each other's signature whistles, supporting the idea that shared vocalizations serve as an affiliative badge.

Snowdon, Elowson & Roush review vocal development in marmosets and tamarins, emphasizing the need to use modern acoustic methods to determine subtle changes in call structure. They describe three phenomena that have parallels to data from birds and humans. The trill structure of pygmy marmosets is flexible and shows evidence of convergence with changes in social companions, yet individual specific features are retained. Marmosets also show extensive "babbling," with parallels to both human babbling and the subsong and plastic song of birds. Study of the ontogeny of food-associated cells in tamarins indicates that social factors can inhibit the appropriate production and usage of calls.

Seyfarth & Cheney review their research on vocal development in Old World primates and argue strongly for a greater emphasis on studying social influences on usage and responses to calls rather than on production. They note that calls used in affiliative interactions have some flexibility in development compared with alarm calls, but note that the basic pattern for vocal production in primates is either innate vocalizations or subtle modification within constraints. Evidence of modification within constraints has been increasingly common in studies published in the last decade, suggesting that new acoustic analysis methods may be important in documenting plasticity. In an important cross-fostering study, Seyfarth and Cheney demonstrate that the greatest flexibility monkeys show may be in learning how to respond appropriately to vocalizations.

Locke & Snow discuss the importance of babbling and of linguistic input in the vocal development of children. There is evidence for vocal accommodation similar to that reported in birds, dolphins, and monkeys. Adults adjust their language to levels appropriate for children, and children produce more speech-like vocalizations in social contexts. Children who hear more words learn faster. Children whose parents neglect them acquire language more slowly than do normal children or children of abusive parents. Locke & Snow note that vocal accommodation is not only possible but necessary in adults for social-group membership and for achieving intimacy with listeners.

Goldin-Meadow presents some important studies on communication in deaf children that at first appear to contradict the results of Locke & Snow. She compares communicative skills in deaf children in the USA and China (where mothers provide more important input and more elaboration than American mothers) and finds no differences in the rate of development of gestural communication in deaf children. Deaf children consistently produce more elaborate gestures and signs than hearing parents provide as input, and a community of deaf people in Nicaragua has created its own communication system. Thus, gestural languages can develop in the absence of input, and it would appear that social influences play little role in the ontogeny of communication in deaf children. Yet, if viewed from a pragmatic perspective, the elaboration of gestural communication in deaf children may be motivated by the need to communicate with someone.

Jouanjean-l'Antoëne describes a longitudinal ethological study of language development in dizygotic twins. She focuses on the different social interactions each twin has with her parents. There are preferential interactions or relationships of each twin with one parent. Parents differentially reinforced each twin. Individual differences lead to favored social interactions within families and these favored relationships play an important role in determining from whom a child will learn and the nature of what is learned.

Goodwin concludes the book with her work on the social use of language in pre-teenaged children. She finds very different types of communication style in play groups of girls versus boys, yet there is considerable flexibility. In general, girls communicate in a non hierarchical fashion, and boys have a clear hierarchical structure to their exchanges. Yet when girls supervise play of younger siblings, play "mother," challenge strangers or interact with boys, they display the same hierarchical language structure as that of boys. This flexibility of language usage is more evidence at the pragmatic level of how language is variable in response to different social companions.

Several themes emerge from these chapters. Five main themes arise.

1. Vocal learning is not only learning to produce, but also learning to use and comprehend vocalizations

Birds, like dolphins, may use their song or whistle types differentially according to social context (Hausberger, McCowan & Reiss, Tyack & Sayigh). Monkeys use different calls according to the type of predator (Seyfarth & Cheney). Young girls may modify their style of communication according to social partners (Goodwin). Vocal comprehension and usage, like vocal production must be learned. Cowbirds may develop effective songs, but do not know how to use them if they are raised in an inadequate social environment (West *et al.*). Young tamarins learn from older conspecifics how to use food calls appropriately, as vervet monkeys learn how to use alarm calls (Snowdon *et al.*, Seyfarth & Cheney). Fully referential, functional and socially interactive inputs ensure that parrots learn not only to produce but also to comprehend allospecific vocalizations (Pepperberg), while dolphins learn to associate arbitrary sounds with particular objects and then use these sounds when playing spontaneously with these objects (McCowan & Reiss).

Different processes seem to be involved for production, comprehension and usage in terms of timing (McCowan & Reiss) or degree of fixity (Seyfarth & Cheney). In children comprehension precedes vocal production as does vocal usage (Locke & Snow, Jouanjean-l'Antoëne). Interestingly, when all three aspects are being considered, there is no "gap" between birds and humans. Vocal development in mammals such as dolphins and non-human primates is continuous with birds and humans.

2. Social inputs influence the stages of development and may delay vocal learning

One common feature of vocal development seems to be a stage of babbling (mammals) or subsong/plastic song (birds), characterized by an overproduction and variability of structure (Nelson, McCowan & Reiss, Snowdon *et al.*, Locke & Snow). This babbling or subsong is generally thought to be a "training" phase where individuals may have the opportunity to progressively develop appropriate vocal structures by comparing their own production to a model (which may be in memory). Alex the grey parrot also displays solitary "speech" (Pepperberg), while deaf children use solitary gestures (Goldin-Meadow). Plastic song stages are particularly important in the action-based learning theory, where the animals select from an array vocalizations that are socially appropriate (Nelson). Although Nelson argues that this selective process may account for reports of delayed learning in closed-end learners, several authors in this volume defend the possibility of later learning under social influences.

However, there is also evidence of new vocal learning occurring beyond the sensitive period. Baptista & Gaunt show that the sensitive period can be extended by social tutors, Payne & Payne provide compelling evidence of learning of new songs in indigo buntings during the first breeding season and beyond. Hausberger describes the development of new song themes in starling after changes in social companions and the re-emergence of themes not heard for months or years. Pepperberg shows how parrots can acquire English words at various ages. McCowan & Reiss show that dolphins can imitate novel sounds, and Tyack & Sayigh show that dolphins can acquire new signature whistles well beyond infancy.

Several of the changes in vocal production that occur outside of a sensitive period or in response to affiliative interactions are subtle changes in vocal structure that are observable only through careful spectrographic analyses. Many of the changes reported in these chapters do not represent the acquisition of completely new vocalizations, but rather subtle modifications in structure of existing vocalizations. Thus, modern spectral analysis methods are critical to the demonstration of social influences on vocal structure. The finding of plasticity at all ages is not necessarily in conflict with the age-limited learning model proposed 25 years ago. The chapters by Nelson and Zann on birds, Snowdon *et al.* and Seyfarth & Cheney on primates, and Goldin-Meadow on deaf children indicate that there are aspects of vocal production (or gestures in the deaf children) that are fixed or modifiable only at an early age. While vocal structures do become solidified, crystallization is probably too strong a metaphor to use, since some flexibility is apparent in birds, dolphins, nonhuman primates and humans at all ages.

3. The nature of social influences is complex

Nelson notes the need to specify and to understand the mechanisms of social influences, and several chapters suggest possible mechanisms. Zann shows that multimodal stimulation appears necessary for zebra finch songlearning. Pepperberg shows experimentally how social modelling leads to exceptional learning. Seyfarth & Cheney describe the importance of social reinforcement, and Hausberger and Jouanjean-l'Antoëne illustrate how individual-specific relationships shape imitation of songs or language.

Social interactors provide an attentional focus, multimodal stimulation, and reinforcement. Each of these mechanisms might be operative in social interaction. As Nelson rightly emphasizes, there is a need for defining "social influences" that are not necessarily "social interactions." Mere proximity may provide social influence without any obvious interaction (Hausberger).

Attention appears as a common and major feature in vocal learning from the tutor as well as the learner. Children respond more to negative attention from the mother than to no attention at all (Locke & Snow). This is also part of social modelling theory (Pepperberg) and is considered a primary factor in bird song learning (Baptista & Gaunt, Zann). The learner must also be viewed as an actor in its own development (Nelson, Zann, Locke & Snow, Jouanjean-l'Antoëne) and not as a passive receiver, as has often been the case.

4. Vocal learning may be a social indicator

Birds may select particular forms of songs through aggressive affiliative interactions with other birds (Nelson, Baptista & Gaunt, West et al., Payne & Payne, Zann, Brown & Farabaugh, Hausberger). Nelson, Payne & Payne, and Zann show how conflict over territory formation can lead to song sharing, but there are an impressive number of examples on the importance of affiliative interactions in vocal development (Brown & Farabaugh, Hausberger, Pepperberg for birds; McCowan & Reiss and Tyack & Sayigh for dolphins; Snowdon et al., and Seyfarth

& Cheney for nonhuman primates; Locke & Snow, Jouanjean-l'Antoëne and Goodwin for humans). Especially with birds, most studies have focused on vocalizations relating to territorial defense and mate selection, but the broad taxonomic range of species where vocal development is influenced by affiliative interactions suggests future research should focus more on affiliative processes.

Brown & Farabaugh suggest that in some species vocal learning has evolved to allow individuals to share vocalizations with a particular subset of conspecifics. The concept of "vocal accommodation" (Locke & Snow) may be extended to nonhuman species. Organisms adapt aspects of their vocal production to match or approach those of their social partners. This process is related to positive affect. Such vocal sharing may involve pairs of birds or nonhuman mammals or may extend to social groups or even to the processes of complex dialects that for some authors have a function of shared vocal identity.

Interestingly, in all of the species mentioned here as exhibiting vocal sharing, individuality is maintained through individual-specific vocalizations (Hausberger, McCowan & Reiss, Tyack & Sayigh). It appears important that the vocal system not only provides group identity but also allows, for individual identity, an "optimal vocal sharing" system.

5. Theoretical principles must be developed to account for the variation in the role of social influences on vocal development

There are socioecological correlates of the importance of social influences on vocal development. The species that appear to show the greatest capacity for learning new vocalizations are those with the greatest mobility, migratory birds and marine mammals, or those with the most unpredictable environments, zebra finches. High mobility and highly unpredictable environments would place selective pressures on vocal plasticity, since individuals would be likely to encounter other conspecifics with very different tutors or geographic dialects.

Animals living in year-round social groups with low mobility would be expected to have less advantage in broad plasticity of vocal structure. However, since individuals do disperse and the membership of social groups does change over time, the capacity to make subtle adjustments in vocal structure might be valuable to provide a social "badge" or indicator of group membership. Evidence for, or speculation about, vocal accommodation or a social badge has been raised with each of the taxonomic groups presented here.

Since individuals differ greatly in behavior, it seems sensible that there might be greater evidence for learning or plasticity in responding to the calls of others or in using signals appropriately than in the production of signals. Hence, it should not be surprising to find evidence of social influences on vocal pragmatics regardless of mobility or stability.

The chapters presented here collectively illustrate the importance of social influences on vocal development not only on production, but perhaps more importantly on usage and responses to vocalizations. These chapters show that there is a broad continuity in the processes of vocal development across all of the vertebrate taxa. It should no longer be necessary to focus on birds, marine mammals, nonhuman primates and humans as separate entities each requiring a different type of developmental process. It is our hope that the chapters in this book will lead to a new integrative study of vocal development in all its aspects that will involve multi-disciplinary, multi-species studies.

REFERENCES

Lenneberg, E. (1967). *Biological Foundations of Language*. Wiley, New York.

Marler, P. (1970). Bird song and human speech. Could there be parallels? *American Scientist* **58**: 669–74.

2 Social interaction and sensitive phases for song learning: A critical review

DOUGLAS A. NELSON

INTRODUCTION

A major focus in the study of bird song over the past three decades has been on the involvement of learning during development. At a basic level, two models of learning mechanisms have been proposed: instructive and selective (Jerne 1967; Changeux *et al.* 1984). In an instructive model, environmental stimulation adds information not already present in the behavioral repertoire. When a young bird memorizes a novel song, it is instructed. In contrast, in a selective model, learning consists of the selection of behavior(s) from a pre-existing repertoire as a function of experience. At the time of stimulation, the animal already possesses the potential or ability to perform the behavior. Therefore, the test to distinguish between the two models is to present a novel stimulus and to record whether it is learned.

Research on song learning has been guided largely by an instructive model of learning, embodied in the sensori-motor model first proposed in Konishi's (1965) study of song development in the white-crowned sparrow (*Zonotrichia leucophrys*). The sensorimotor model includes two stages: a sensory (instructive) phase in which songs are memorized, and a sensorimotor phase in which the bird compares its own song, via auditory feedback, to the memory trace acquired earlier.

One consequence of song learning is the formation of geographic "dialects" in which males at one location sing similar songs that differ from those of the same species at other locations. If vocal plasticity in birds is mediated solely by an instructive mechanism, then song matching dialects arise when males breed in the same area where they acquired their song(s). In sedentary populations, dialects will result when males settle and breed as adults in the same area where they acquired their song. When males dis-

perse long distances from the birthplace to breed, then the ability to acquire novel songs must persist until they settle upon a breeding territory later in life and acquire songs from their new neighbors (Kroodsma 1982). The observation that neighboring males in wild populations sing similar songs, coupled with information or assumptions about the degree of natal dispersal, is often taken as *prima facie* evidence of how long the instructive phase persists.

In recent years much experimental work has been devoted to identifying factors that influence the timing and expression of the sensory phase. In particular, a major focus has been on the social factors that may influence what and when a young bird is instructed. To borrow from Kroodsma (1978), what, when, and how a bird is instructed about song can potentially influence many other aspects of a species' biology, including mate choice, reproductive success, and population structure.

In this chapter I review some of the recent research on song learning. I argue that many of the experiments and field observations on song "learning" may actually have lumped together two different mechanisms of learning: the familiar instructive mechanism and the more recently described selective mechanism (Marler & Peters 1982a; Marler 1990; Nelson & Marler 1994). After describing how vocal plasticity may be mediated by a selective mechanism, I review recent experimental research on song learning within a framework incorporating selection. The possibility that song "learning" may be a heterogeneous process invites confusion. I use the terms "acquisition" or "memorization" to refer to the sensory phase wherein a male commits a novel song to memory, and "action-based learning" (Marler 1991), or "selective attrition" (Marler & Peters 1982a) to refer to the preferential retention of learned song(s) as a function of experience.

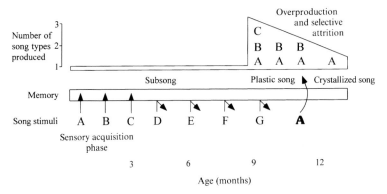

Fig. 2.1. Schematic diagram of song development in a closed-ended learner, the white-crowned sparrow (Marler 1970; Nelson & Marler 1994; Nelson *et al.* 1995). In a closed-ended learner, song stimuli are committed to memory only during a restricted interval, termed the sensitive phase. Different song stimuli (A–G) are presented at intervals throughout the first year of life. The ability to memorize song is assayed by the production of imitations. Here, song stimuli A, B, and C are memorized, while D, E, F, and G are rejected, as symbolized by the bent arrows. Males "overproduce" by imitating more (A–C) than is needed for the mature repertoire of one crystallized song type (A). Song types are gradually lost from the overproduced repertoire until the final type remains. If a stimulus that matches one type in the overproduced repertoire is presented (the bold "A" at 11 months), then song type A is crystallized. A nonmatching type presented at the same time to a different, control, bird is not memorized and reproduced (Nelson & Marler 1994). The experience-dependent plasticity of vocal behavior in males of this species at nine months of age and beyond is based on learning by selection from an over-produced repertoire, and not on learning by memorizing novel stimuli.

LEARNING BY SELECTION

A selective learning mechanism includes two stages (Changeux *et al.* 1984): (a) the production of a variety of behaviors, in this case songs or their components (notes, syllables, phrases) and their underlying neural representations and (b) the selective retention and production of a subset of this variety as a function of experience. Kroodsma (1974), Baptista (1975), and Jenkins (1977) described the second phase of the selective process, but apparently did not recognize the potential of a selective mechanism as an alternative basis of vocal plasticity operating *outside* the sensory memorization phase. Marler & Peters (1981, 1982a) clearly described the entire selective attrition process in their studies of song development in the swamp sparrow (*Melospiza georgiana*).

Song development proceeds similarly in the swamp sparrow and white-crowned sparrow (Nelson & Marler 1994; Nelson *et al.* 1995). The latter species is used to illustrate the selective attrition model in Fig. 2.1. Swamp spar-

rows and white-crowned sparrows are closed ended "learners" (or "memorizers" in the terminology of this chapter), in that males memorize songs only during a restricted time in the first few months of life. This is established by presenting males with a changing roster of different tutor stimuli throughout life, and then examining the birds' vocal production for specific imitations. After a storage interval of several months, during which males may "babble" in subsong (Marler & Peters 1982b; Snowdon *et al.* Chapter 12), imitations of specific tutors appear in plastic song.

During the overproduction phase of plastic song, males not only reproduce accurate imitations of their tutors, but also display "combinatorial improvisation" (Marler & Peters 1982b), in which novel songs are created by combining pieces from different tutor songs. This process is illustrated in Fig. 2.2. The overproduction of imitated song material and, in particular, the production of novel combinations (song patterns) by recombining elements

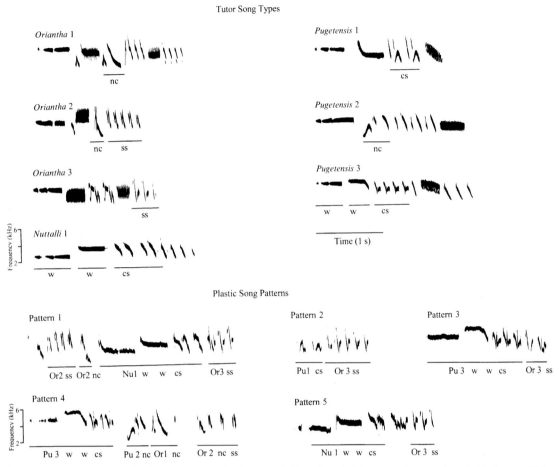

Fig. 2.2. An example of song overproduction by one male Puget Sound white-crowned sparrow in early plastic song. Males of this species sing one song type in their mature, crystallized repertoire. Shown at the top of the figure are the seven tutor song types he imitated in whole or in part. The phrases he used in generating the four song patterns on the bottom of the figure are underlined. For example, Pattern 1 consists of the simple syllables (ss) and note complex (nc) from tutor type *Oriantha* 2 (Or2), followed by the two whistles (w) and one complex syllable (cs) from tutor *Nuttalli* 1 (Nu1), and ending with three simple syllables from tutor *Oriantha* 3 (Or3). Pattern 3 was also counted as a song type, because he sang it repeatedly on at least two consecutive weeks in plastic song. Pu, *Pugetensis*. (From Nelson *et al.* 1996a.)

from different tutors provide the raw material upon which a subsequent selective stage can act. In the white-crowned sparrow, the most improvisation of novel song patterns occurs early in plastic song, followed by a gradual attrition (Fig. 2.3). Recombinatorial improvisation of song material is common to other species (northern cardinal, *Cardinalis cardinalis* (Lemon 1975); winter wren, *Troglodytes troglodytes* (Kroodsma 1981); song sparrow, *Melospiza melodia* (Marler & Peters 1987); nightingale, *Luscinia*

megarhynchos (Hultsch 1990); field sparrow, *Spizella pusilla* (Nelson 1992; indigo bunting, *Passerina cyanea* (Margoliash *et al.* 1994)), as is invention, another process that will generate diversity in song (Kroodsma & Verner 1978). Both improvisation and invention are creative processes, which will introduce greater diversity into a male's vocal production.

Marler & Peters then hypothesized the next step in a selective learning mechanism. They noted that matched

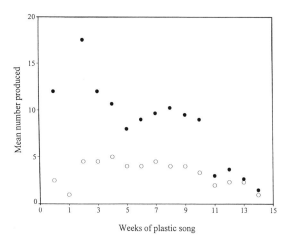

Fig. 2.3. The time course of overproduction and attrition in 11 male mountain white-crowned sparrows tutored with 32 different tutor song types between the ages of 10 and 90 days. Plotted are the mean number of different tutors imitated each week in plastic song (○) and the mean number of different song patterns produced each week (●). Plastic song week 0 began at about 270 days of age. Song patterns are unique sequences of song phrases formed by combining phrases from one or more tutors (see Fig. 2.2). Males do most of their improvisation in the first month of plastic song, followed by a decline, or attrition, in the number of different song patterns produced until they sing one stable "crystallized" song. The number of tutors imitated each week, in whole or in part, stays relatively constant throughout plastic song until just prior to crystallization. Some males crystallize a single "hybrid" song, based on two or more different tutor songs, which is why the mean number of tutors imitated in crystallized song is greater than 1.

counter-singing, the tendency of territory neighbors to exchange similar songs (Marler 1960; Kroodsma 1974; Baptista 1975; Jenkins 1977; Falls *et al.* 1982), could lead, via selective attrition of non matching song types, to local song dialects. That is, if males preferentially retained the matching song(s) from their overproduced plastic song repertoire, this would lead to song sharing among males at a locality. This is shown in Fig. 2.1, where song stimulus **A** is presented again at 11 months of age, when the male is already singing imitations of A and B. The male discards song type B from his overproduced repertoire and retains A as his sole crystallized song type. Marler & Peters also noted that, in other circumstances, males might selectively *delete* matching songs from their repertoire; it is not uncommon for neighboring males to have very different

songs (e.g., white-eyed vireo, *Vireo griseus* (Bradley 1981); nightingale (Hultsch & Todt 1981); fox sparrow, *Passerella iliaca* (Martin 1979)). Byers & Kroodsma (1992) presented some evidence that selective rejection of matching songs from an overproduced repertoire occurs in chestnut-sided warblers (*Dendroica pennsylvanica*). It is not clear why song sharing among neighbors may be favored in some species, and non sharing in others.

Direct evidence for a selective attrition process has been obtained only recently. Field sparrows sing one "simple" song type in their crystallized repertoire (Nelson & Croner 1991), but males arriving on particular territories for the first time in spring produce two to four song types (Nelson 1992). Using a quantitative measure of song similarity, I documented that males preferentially kept the one song type in their repertoire that best matched one of their neighbors, and discarded the remaining songs. In some cases, matching of neighbors' song was quite close, in others only approximate. I return to this point later. Several lines of evidence suggested that instruction was not occurring as birds established their territories

1. Several males were found singing two or more song types as the first birds to return in their respective fields; that is, there were no adult tutors within earshot. These males did not alter their existing songs after other males arrived, but merely discarded their nonmatching song(s).
2. Discarded songs did not resemble neighbors' songs more closely than was expected at random, indicating that discarded songs were not acquired from neighbors.
3. Neighbors' songs were usually not the most similar of all songs in the local population to the retained songs. Many other similar non-neighboring songs could be found locally, suggesting that instruction was not occurring from neighbors.
4. No male ever changed his song type after the attrition phase, even if new neighbors with new songs moved in. Only one male changed his territory between years.
5. Males who did not share their song with any neighbor were not uncommon, especially in large fields containing many territories.

While this observational evidence suggested that vocal plasticity in yearlings involved selection rather than instruction, an experimental approach, in which the birds' experience with song could be controlled, was needed to distinguish between instruction and selection.

For this experimental evidence we turned to the white-crowned sparrow, a species that has been the focus of much research and controversy (see e.g., Baker & Cunningham 1985; Kroodsma *et al.* 1985). Before settling on a single song type, male white-crowned sparrows also overproduce song both in the laboratory (Fig. 2.2; Nelson & Marler; 1994; Nelson *et al.* 1995) and in the wild (Baptista 1974, 1975; Baptista & Morton 1988; De Wolfe *et al.* 1989). To distinguish between instructive and selective accounts of song learning, we designed an experiment using birds with known histories of song experience (Nelson & Marler 1994). Ten hand-raised, male, white-crowned sparrows were tutored with a diversity of songs as fledglings. These males produced imitations of at least two tutor song types in their plastic song, which began at about 275 days of age, at least five months after the songs were acquired. To experimentally mimic the vocal interaction between males that occurs in nature, we played to half the males one randomly chosen tutor song that matched one of the song types in their plastic song repertoire (song type **A** shown in Fig. 2.1). The other five males each heard a different novel song type (type G in Fig. 2.1). We predicted that if song sharing results from the selective retention of songs that match those of neighbors, then males should retain the song that matches the type played to them. If instruction occurs in plastic song, then the controls should acquire the novel song type. All five experimental males crystallized the matching song type, and none of the controls acquired the novel type, demonstrating that selection, rather than instruction, occurs in the plastic stage of song development.

The possibility remained that the males hearing matching playback as yearlings were instructed by that experience, and so altered their songs to match the playback song even more closely. This possibility was suggested by a report that, in one small island population (Baptista 1975), neighboring males have very similar songs, suggestive of instruction during territory establishment. Close neighbor–neighbor sharing exceeding a chance level was not found in another *nuttalli* population (Trainer 1983). Using digital spectrogram cross-correlation (Clark *et al.* 1987), we were unable to find any evidence of instruction in plastic song. The crystallized

songs of the birds that heard matching playback in the spring did not resemble the songs of their tutors any more closely than did those in the control group, who sang from memory songs acquired five or more months previously. In this species, instructive and selective effects on vocal learning are temporally distinct processes, with the sensory phase for acquisition restricted to the first three months of life (Marler 1970; Nelson *et al.* 1995 and see below). Song dialects in the migratory *oriantha* race, and perhaps in *nuttalli* as well (DeWolfe *et al.* 1989; Nelson *et al.* 1995) appear to be maintained by the selective attrition of previously acquired songs during territory establishment.

Another example of learning by selective attrition is provided by detailed work on the brown-headed cowbird (*Molothrus ater*). West & King (1988) have shown that males are attentive to a visual display given by females, a wing stroke. Increased performance of certain songs is associated with performance of a wing stroke by females. Here, the effect of social interaction is clearly on retention of songs already acquired, and this study provides a clear example of how instructive and selective learning are different processes, since instruction in this instance occurs by a modality (audition) that differs from the modality involved in selection (vision). Female cowbirds do not sing, so it is the presence of a visual display that leads males to preferentially sing certain songs. In the white-crowned sparrow example, both instructive and selective processes are mediated by audition. It is not clear, though, whether the tuition by female cowbirds can lead to the song sharing in local populations of cowbirds described by Dufty (1985). Presumably, if females in the population have similar preferences, then males would converge on these preferred songs. Female cowbirds of different subspecies do give copulation solicitation displays preferentially to songs of their own subspecies (King *et al.* 1980). Perhaps there are more localized preferences as well.

The selective attrition model does not replace sensorimotor learning as a mechanism of vocal learning. Sensorimotor learning is the fundamental process by which a bird's own vocal production is shaped by experience. It provides the raw behavioral material that the selective model needs in order to operate, although the two processes may operate in parallel (Konishi 1985), which may make distinguishing between them difficult (see below). The selective model, however, may be responsible for vocal plasticity in some instances where the instructive, sensory acquisition phase of the sensorimotor model has been *assumed* to occur.

We have studied the learning by selection model in closed-ended learners or memorizers in which there is a clear sensitive phase. Other species, such as the European starling (*Sturnus vulgaris*), are open-ended memorizers, and can learn novel songs up to 18 months of age, and perhaps beyond (Chaiken *et al.* 1994). Chaiken *et al.* (1994) rightly emphasize that only controlled laboratory experiments provide definitive information on when birds are able to memorize novel songs. There is no logical reason why overproduction and selective attrition cannot occur in open-ended memorizers as well, but we have argued that action-based learning may be especially favored in migratory closed-ended memorizers (see below). By possibly confusing vocal plasticity resulting from memorization with that due to action-based learning, there has been a tendency in recent years to conclude that song memorization is more open-ended than it may actually be. Nowhere is this more apparent than in the debate over the effects of social tutors on song learning and sensitive phases.

SENSITIVE PHASES FOR SONG ACQUISITION

The issue of sensitive phases in song development has engendered debate, much of it centered on the effects produced by tape tutoring versus live, or "social," tutoring. Early work on song learning stressed the importance of an early sensitive phase, in the first few months of life, during which males memorized tutor songs (e.g., Marler 1970). Subsequent work, much of it centered on Nuttall's white-crowned sparrow, has had the goal of demonstrating that living, "social" tutors extend the sensory acquisition phase. The apparent differences between the two tutoring methods have led some authors to label the results of tape-tutoring experiments method specific (Petrinovich 1990), anomalous (Clayton 1987a), or "abnormal" (Slater 1989). These authors did not reveal what is "normal," knowledge of which can come only from field studies (Kroodsma 1985). I also argue below that interpretation of song learning in the field is not as easy as it might seem. By recalling songs from memory, birds may be able to change their song in response to altered (social) circumstances even after the

ability to memorize new songs is lost. In reviewing these laboratory studies I hope to establish that simpler, alternative hypotheses can explain the results attributed to "interactive, social" tutoring, or that claimed differences between interactive and non interactive tutoring are not robust.

Several studies have shown that birds will acquire songs from live tutors either after 50 days of age in Nuttall's white-crowned sparrow (Baptista & Petrinovich 1984, 1986), or in yearling male indigo buntings (Payne 1981). A feature common to these studies is that the subjects were *isolated* from adult song prior to being exposed to the living tutor. An alternative explanation for the apparent extension of the sensitive phase is that prior isolation from song, not socially interactive tutoring, prolongs sensitivity, a possibility with precedents in other systems (Konishi 1985; Baker & Cunningham 1985). Indeed, studies on these and other species have found that acquisition from tape-recordings can occur in a bird's second year *if* birds receive little or no song experience prior to that time (northern cardinal (Dittus & Lemon 1969); indigo bunting (Rice & Thompson 1968); marsh wren (Kroodsma & Pickert 1980); chaffinch, *Fringilla coelebs* (Slater & Ince 1982); mountain white-crowned sparrow (Whaling *et al.*, unpublished data)). The majority of birds that hear song early in life do not require novel songs later. The most parsimonius explanation of these data sets is that song deprivation early in life, not socially interactive tutoring later in life, extends the sensitive phase for song acquisition.

The experiments using early deprivation from song demonstrate that song acquisition can be delayed under these circumstances. It is worth while considering how likely this is to occur in the wild. Male songbirds can acquire songs from remarkably little exposure early in life (12–50 presentations on one day, blackbird, *Turdus merula* (Thielcke-Polz & Thielcke 1960); 10–20 presentations, nightingale (Hultsch & Todt 1989); 30 repetitions on one day, song sparrow, *Melospiza melodia* (Peters *et al.* 1992); 120–252 presentations over 20 days, Nuttall's white-crowned sparrow (Petrinovich 1985)). To argue that local song dialects in migratory species result when yearling males acquire songs at the time and place they settle on territories, one must assume that the majority of males are isolated from song for the first nine months or so of their lives. This assumption in necessary because the bulk of the evidence indicates that sensitive phase closure is in part dependent upon stimulus experience; that is, the sensitive phase is self-terminating (ten Cate 1989; Bolhuis 1991). White-crowned sparrows, at least, sing in every month of the year, on migration, on the wintering grounds, and during the breeding season (Baptista 1977; DeWolfe *et al.* 1989; DeWolfe & Baptista 1995). Hatching-year indigo buntings trapped on autumn migration and then held in acoustic isolation in the laboratory develop normal songs, indicating that they do hear song in their hatching year (Margoliash *et al.* 1994). I feel it is worth while to consider learning mechanisms that can produce song sharing among neighbors without requiring the restrictive assumption that males do not hear song in their hatching year.

MULTIPLE EFFECTS OF LIVE TUTORING

In many studies using "social" tutoring, one tutor, usually the last one presented, is present during both the sensitive phase and during plastic song. This tutor could exert both instructive and selective effects on learning. If, during vocal practice, subjects are housed with tutors that differ from those present during song acquisition, quite different results are obtained in that the tutor present only during vocal practice is not imitated (e.g., Nelson & Marler 1994). If the goal is to determine the relative sensitivity of birds to *acquire* song as a function of age or tutor experience, then housing subjects with a tutor during vocal practice may well confound plasticity due to acquisition with that due to selection during matched counter-singing. Since most studies do not analyze plastic song, an incomplete picture of when songs are acquired may be the result.

For example, Eales (1985) reported that when young zebra finch (*Taeniopygia guttata*) males were moved from their father to a different tutor at different ages (35, 50, 60 days), birds "learned" more from the second tutor the later they were removed from the father. Slater *et al.* (1991) exposed young zebra finches to a series of tutors during the sensitive phase, and also found a tendency to "learn" more from later tutors. Eales (1985, p. 1299) concluded that "the zebra finch male appears to be most sensitive to learning during the period when he is most rapidly developing his song output, in the latter half of the sensitive phase." Since zebra finches begin to produce songs soon after 35 days of age, these results may reflect the selective retention of the

more recent tutor(s), as much or more than they reflect differential sensitivity to memorize song (Zann, Chapter 6). Zebra finches clearly do memorize their father's song prior to 35 days. Böhner (1990) showed that male zebra finches develop accurate imitations if the father is removed at 35 days of age and the young birds are left to complete development housed by themselves. Böhner's design does not confound the effects of sensory acquisition and action-based learning. Whether zebra finches overproduce during development is not clear, so the interpretation above is speculative, but Slater *et al.* (1993) mentioned that some song elements are dropped, apparently as a function of what other siblings in the group sing.

Petrinovich & Baptista (1987) presented Nuttall's white-crowned sparrows with different living conspecific tutors before and after 50 days of age. The subjects were exposed to the second tutor until they crystallized song the next spring (200 days or more), and so were able to counter-sing matching songs with the second tutor. It is not clear that social tutoring extended the sensitive phase relative to nonsocial (taped) tutoring, because the tape tutoring in an earlier study that served as the tape control (Baptista & Petrinovich 1986) began at 50 days of age and *terminated at 90 days*, well before plastic song begins. This second tape tutor could not exert the selective effect on song retention, that Nelson & Marler (1994) demonstrated, while the second live tutor could. The experimental conditions yielded different results because the tape tutoring was optimized (unintentionally perhaps) to reveal differences in memorization, while the live tutoring optimized expression of the last song heard. Had the tape tutoring been extended into the next spring, as was done with the live tutors, or had plastic song been recorded, I predict that evidence of acquisition after 50 days from tapes would appear. In other experiments using tape tutors, we have found that the median age of song acquisition in *nuttalli* is 70 days (Nelson *et al.* 1995). It is worth stressing that, statistically, significantly fewer *nuttalli* acquired songs from a live tutor after 50 days than before (Petrinovich & Baptista 1987, p. 971), and none (of two birds) acquired songs from live tutors after 100 days. The various experiments are entirely consistent with an early sensitive phase for song acquisition in Nuttall's white-crowned sparrow that terminates in the second or third month of life (Marler 1970; Baptista *et al.* 1993; Nelson *et al.* 1995).

Byers & Kroodsma (1992) reported that in chestnut-sided warblers, learning was enhanced by a live tutor relative to tape tutoring. The birds were tutored with one set of tape-recorded songs in their first summer, and a different set, for 17 days, at an unstated time the next spring. In contrast the live-tutored group was exposed to the *same* live tutor in the summer and for 38 days during their plastic singing the next spring. The authors correctly discount the possibility that the tape tutoring in the first summer was insufficient for acquisition to occur. Here again, though, the tape-tutored birds did not have the opportunity to counter-sing matching songs in the spring, while the live-tutored group did. It is not clear that live tutoring was responsible for the treatment effects in this study.

The various experiments on marsh wrens (*Cistothorus palustris*) by Kroodsma (1978) and Kroodsma & Pickert (1984) are also difficult to interpret, for the same reasons as discussed above. Clearly, marsh wrens can acquire novel songs from live tutors as yearlings. It is possible that extended tape tutoring of yearlings during plastic song would produce the same result.

The indigo bunting is a species frequently referred to as one in which the sensitive period extends into at least the second calendar year of life (Payne 1981; Slater *et al.* 1993). As noted above, the experimental evidence was based on hand-reared birds held in isolation from song until May or June of their second calendar year (Payne 1981). Recent studies by Margoliash *et al.* (1991, 1994) demonstrated that buntings given social tutoring in the first few months of life, as normally occurs in the wild, do not acquire novel songs as yearlings. The song changes by wild buntings in their second calendar year that Payne *et al.* (1988) described may not be the result of instruction then, as has been assumed, and may be the result of selection from a previously acquired repertoire. Payne & Payne (Chapter 5) compare the songs of yearling buntings to *entire* songs they might have heard in their hatching year and to *entire* songs that were present in their first breeding season. They conclude that songs were "learned" in the first breeding season. Their analysis assumes, however, that entire songs are the units of memory, and ignores the ability of yearlings to create novel songs by recombining song figures they memorized in their hatching year (Margoliash *et al.* 1994).

Song matching appears to result from selective attri-

tion in buntings. Margoliash *et al.* (1991, 1994) described how variable, plastic song persists in yearling and older male buntings, and this song material provides the capacity for modification of song. Hatching-year buntings captured on fall migration and then housed with a live adult tutor did not acquire novel syllables from the living tutor, but rather selectively retained pre-existing syllables that most closely matched the tutors, and discarded non-matching syllables, as in white-crowned sparrows (Nelson & Marler 1994); or improvised on pre-existing syllables to effect a closer match (Margoliash *et al.* 1994). In contrast to the white-crowned sparrow, instruction does appear to play a role in yearling buntings, but this role is limited to modifying pre-existing material. Whether this is a species difference or an effect of live tutors is not known. One source of uncertainty in Margoliash's work is the fact that all the birds were first captured as migrants, so the full extent of their prior song exposure was not known. The white-crowned sparrows were hand reared from the age of five to eight days in controlled acoustic conditions (Nelson & Marler 1994). Nevertheless, plasticity in bunting song is age limited. Eleven buntings that were housed with live tutors later in their second summer did not add any new syllables, nor did any of the adult tutors.

The apparent learning in yearling chaffinches studied by Thorpe (1958) and Thielcke & Krome (1989, 1991) also involved birds that had been captured the previous fall, and so had received tuition in the wild. It is possible this late learning was, in fact, selective attrition of previously learned songs. This latter interpretation is given some credence by Thielcke & Krome's (1991) finding that there was a "gap" in the sensitive phase in mid winter, when the birds did not learn. The birds did learn when tutored in March and April of their second calendar year, when they were, apparently, in plastic song (the authors noted that only two birds learned in plastic song). In the selective attrition model, learning is tied to song performance, so learning should not occur when birds do not sing.

To summarize this section, several studies have demonstrated that male birds are capable of acquiring songs from live tutors at older ages than was previously thought to be possible. I do not think these results demonstrate that social interaction is the causal variable responsible for this effect, because subjects in the noninteractive (tape tutor) condition did not hear matching song late in development as did the live-tutored subjects. The noninteractively tutored birds may well have memorized songs at the same age as the live-tutored birds, but did not reveal this because matched tape tutoring did not continue into plastic song, and the investigators did not examine plastic song for evidence of other imitations. The fact that tape-tutored birds that do not hear tutors during vocal practice base their crystallized song on songs experienced earlier in life is strong evidence that birds are more sensitive then, and/or the song memories acquired early in life are more durable. Other studies that have revealed apparent learning of new songs used subjects captured in the wild so that the quality and quantity of early song exposure was unknown. The apparent acquisition of novel songs in the laboratory may simply reflect recall of songs memorized earlier in the wild. This may also apply to cases of song change in old birds in field studies (e.g., white-crowned sparrow (Baptista & Petrinovich 1986); indigo bunting (Payne *et al.* 1988); great tit, *Parus major* (McGregor & Krebs 1989)).

WHAT IS "SOCIAL INTERACTION"?

As discussed in the previous section, it is difficult to determine when, and how, interactive tutoring may exert its effect(s). Most studies have either explicitly, or implicitly, assumed that social interaction affects the sensory memorization phase. Interaction is defined as mutual or reciprocal action or influence. Advocates of social modelling theory tend to provide a very rich interpretation of their results, using terms such as "live interactive" tutoring or "social" tutoring to describe treatment differences (Payne 1981; Pepperberg 1985; Baptista & Petrinovich 1986). Rarely are simpler interpretations entertained. While it has been shown in several species (indigo bunting, zebra finch, white-crowned sparrow, starling) that birds preferentially learn songs when presented with a living tutor versus song alone or a tape-recording, the controls for other, simpler, interpretations are rarely used. These experiments cannot exclude the interpretation that an auditory stimulus coupled with any kind of visual stimulus, even a static one, enables learning more effectively than an auditory stimulus alone.

For example, the enhanced learning by two nightingales in a "social" context (Todt *et al.* 1979) is frequently cited as evidence of an effect of social interaction on song learning. The social tutor in this study was not another

nightingale, but rather the birds' human caregiver who held the loudspeaker broadcasting the tutor songs. In what sense is this social or interactive tutoring? The literature on avian filial imprinting (Bolhuis & van Kampen 1992; van Kampen & Bolhuis 1993) demonstrates that pairing quite arbitrary sounds with visual stimuli can enhance imprinting success. It may simply be that any stimulus combination that enhances the subject's attention or arousal (e.g., the expectation of food in young nightingales) makes the auditory stimulus with which it is paired more readily acquired, and that "social interaction" or "aggression" are not necessary to enhance song acquisition. ten Cate (1994) has also warned that explanations involving very specific effects may not necessarily be involved in the effect that "social context" has on song learning and imprinting.

Several experiments have attempted to tease apart the behavioral variables that might contribute to social interaction. Waser & Marler (1977) reared two groups of male canaries (*Serinus canarius*). One group of five males was kept with a live adult tutor, while a second brood of three heard the live tutor's singing relayed to it via a microphone–amplifier–loudspeaker audio link. Those in social contact with the tutor shared an average of 74% of their syllables with the tutor, while the yoked controls shared only 58%, a difference significant at $P=0.04$.

Eales (1989) compared the learning success of three groups of zebra finches. (1) Control males could interact visually and vocally with their live tutor in a neighboring cage. (2) Seven males in four broods could interact vocally with their live tutor but could not see him (each brood had its own tutor). (3) Eight males in three broods could only hear a live tutor through a loudspeaker, but not interact vocally with him. All three of these broods were tutored by the same male, apparently not one of the tutors of group 2. Significantly more males learned from the tutor in group 1 than in groups 2 and 3 combined, while groups 2 and 3 did not differ from one another; suggesting that vocal and visual contact are important. Interpretation is difficult because the control birds were housed in a large bird room with many other zebra finches, while the experimental groups were housed in sound-isolation boxes. It would have been preferable to house the control birds in sound isolation boxes also. In a colonial breeder like the zebra finch, this difference in general stimulation might be important in itself, a hypothesis supported by a small study reported by Slater *et al.* (1988, p. 15) in which learning after 65 days was enhanced in birds kept in the bird room relative to birds kept in isolation boxes. Two other studies using a similar design of routing a live tutor's songs via a loudspeaker to a group of noninteractive pupils failed to demonstrate differences in learning, although samples were small and the data are from pooled samples (marsh wrens (Kroodsma & Pickert 1984); white-crowned sparrows (Baptista & Petrinovich 1986); and see below).

Yearling indigo buntings learned more song figures from a live tutor they could see and hear than from a live tutor they could only hear (Payne 1981). This study is perhaps the clearest example that a visible tutor is preferred over a solely audible one. Whether *interaction* with the visible tutor is necessary is unknown. None of the studies mentioned can distinguish when or how social tutors have an effect. This issue clearly deserves further study.

Evidence that increased arousal or attention may facilitate song acquisition comes from recent experiments on zebra finches which show that males may acquire songs from tape-recordings in an operant task (Adret 1993a). Males that could control when a taped stimulus was presented learned more than males that could only hear the stimulus, although, surprisingly, the latter group did learn from tape alone. There is a suggestion from preliminary evidence that acquisition in this paradigm is facilitated by increasing the duration of song exposure (Adret 1993b; see below).

Several papers have suggested that the key aspect of social interaction that facilitates song learning is aggression, namely that young males preferentially learn from the tutor that is most aggressive (Payne 1981; Clayton 1987b). Unfortunately, neither author presented any data on aggression. To make a persuasive link between aggression and learning, one would need to know how aggression varies over the course of development (is it highest during memorization, production?). Recent studies on zebra finches failed to replicate Clayton's finding, although very different procedures were used in each case (Williams 1990; Slater *et al.* 1991; Mann & Slater 1995). In observations on one aviary containing ten breeding pairs and two unpaired zebra finches, Williams (1990) found that young males copied the two adults that provided the most parental care. There is an urgent need for more detailed work on the behavior associated with song learning.

Some of the effects attributed to interactive tutoring

(increasing acquisition success, extending the sensitive phase) can be achieved by using higher presentation frequencies of noninteractive tape tutoring. In a recent experiment, we compared song acquisition in two groups of mountain white-crowned sparrows. Males that were exposed to 32 different song types between the ages of 10 and 90 days of age memorized 283% more song phrases, and learned from 155% more tutors, than did males exposed to 16 tutor types in the same period (Nelson *et al.* 1996b). This difference is similar to one found in European starlings, which learn more material from live tutors than from taped tutors (Chaiken *et al.* 1993, 1994). Extension of the sensitive phase has been attributed to interactive tutors (Baptista & Petrinovich 1984, 1986; Kroodsma & Pickert 1984). Instead, these effects might more simply be explained by the increased frequency of exposure to tutor song.

Simple effects of exposure frequency on song acquisition occur in other taxa. Song sparrows learn more from tutors presented over six weeks than from one week's exposure (Marler & Peters 1987). Nuttall's white-crowned sparrows are more likely to learn, and learn more accurately, from tutor songs presented 250 times than presented 120 times (Petrinovich 1985). Nightingales learn the sequence of tutor songs on a tape more accurately from 50 presentations than from 25 presentations (Hultsch & Todt 1992). Increasing exposure frequency up to 25 presentations, but not beyond, enhances the probability of acquisition in this species (Catchpole & Slater 1995).

Three studies on the zebra finch have examined the effect of (live) tutor song output on acquisition success. Böhner (1983) found that relative differences in song rate between 16% and 130% did not affect preference for the father as tutor. In one case where the father sang only 10% as often as the other tutor, the other tutor's song was learned. Clayton (1987b) also reported that, using tutors with normal output, song output did not affect tutor choice, but she did not present data on song rates. Finally, Slater *et al.* (1991) reported no effect of tutor output, but the song rate differences between copied (47.2 song phrases per hour) and noncopied (44.1) tutors was very small (7%). Perhaps tape-tutoring studies have demonstrated effects of presentation frequency because they employed greater ranges of variation, and, because other factors were controlled, the effect of presentation frequency was made clearer.

In contrast to these results from the laboratory, hatching-year male song sparrows are more likely to learn a song type that is shared by two or more tutors than would be expected on the basis of the song type's proportional representation in the local population (Beecher *et al.* 1994). That is, presumed differences in presentation frequency did not affect whether a song type was acquired. Significantly, the field evidence on when song sparrows acquire their songs is perfectly consistent with laboratory experiments on the *timing* of acquisition (Marler & Peters 1987): songs were acquired in the first few months in the hatching year. In the wild, sedentary population of song sparrows studied by Beecher *et al.*, hatching-year males established "floater" territories, and acquired their songs from their adult neighbors. The claimed differences between laboratory and field studies that Beecher *et al.* (1994) remarked upon (accurate copying from one tutor, copying shared songs, possible "active" sampling of tutors) arise simply from the fact that laboratory studies have not yet been designed to address these questions (i.e., there are no data), and do not necessarily mean that laboratory experiments may be "incomplete or partially misleading" (Beecher *et al.* 1994, p. 1452).

TUTOR–PUPIL OR PUPIL–PUPIL INTERACTION?

Another concern in most studies involving "social" tutoring is whether individual pupil birds can be treated as independent replicates when they are housed together and the potential exists for them to interact *during vocal practice*. In virtually every study involving live tutors, groups of subjects were housed together with a small number of tutors. The only exceptions I am aware of are the recent study by Margoliash *et al.* (1994) in which single pupils were housed with one tutor inside sound isolation chambers, and Payne's (1981) study in which each subject was caged next to a pair of tutors (one in visual contact, one not) and out of contact, in most cases, from other buntings. In some studies, brothers reared together did not develop similar songs (Baptista & Petrinovich 1986; Eales 1987a; Slater *et al.* 1991; Mann & Slater 1995), while in other studies they did (Waser & Marler 1977; Cunningham & Baker 1983; Kroodsma & Pickert 1984; Eales 1987b; Petrinovich & Baptista 1987; Williams 1990; Byers & Kroodsma 1992; Slater *et al.* 1993; Volman & Khanna,

1995). Eales (1987a) noted that if one of the young birds came into song early, he could also influence the others.

Pooling data from nonindependent replicates may lead to rejection of null hypotheses at spuriously low probability levels (Machlis *et al.* 1985). Note that I am not objecting to the practice of housing subjects together on the grounds of poor external or ecological validity (see Campbell & Stanley 1963). It seems likely that group housing more closely approximates the learning situation in the wild than does individual housing, as long as the brood remains together and is dependent upon paternal care. This design does pose problems of internal validity, however, because all statistical methods assume independence of replicates. A conservative approach would set the sample size used in statistical hypothesis testing to be the number of *groups* reared.

CONCLUSION

In this review, I have examined the evidence that social interaction affects the sensitive phase for song learning. While numerous experiments have found a treatment difference associated with the use of live tutors, it is not clear that there is a real difference relative to noninteractive (tape) tutors, because there have been other differences between how tape and live tutors have been presented that make a straightforward comparison impossible. Specifically, tape-tutored birds have not been given the opportunity to counter-sing a matching song during the later stages of vocal development, whereas live-tutored birds have been able to.

It also is not clear whether the treatment differences reflect effects on when song is memorized. Social or interactive tutors could affect the song learning process in several ways. Behavioral differences among tutors (e.g., aggression, paternal care, etc.) might lead some to be chosen over others either simultaneously or sequentially as models for song memorization. These differences could lead to variation in the number of songs acquired and/or in the timing of acquisition. Another potential effect is that social interaction could affect choices made among a repertoire of songs that had been acquired at another time or place. These theoretical possibilities are not mutually exclusive. Since most studies do not examine plastic song, analysis of which provides the most complete picture of what the bird has acquired, and use extended periods of tutoring relative to the time course of development, it is not possible to disentangle these potentially separate effects. Plastic song may provide more information on what and when song is memorized than does analysis of crystallized song alone. Returning to Fig. 2.1, analysis of plastic song reveals that the bird memorized song from one to three months of age (songs A, B and C), while crystallized song suggests memorization was restricted to the first month (song A).

For those interested in the function or evolution of song learning, the mechanisms by which song changes as a result of experience may make little difference. I suggest that making distinctions between "instruction possible at all times" and "early instruction followed by selection" scenarios may help to explain several aspects of song learning.

First, if male birds are being instructed by their territory neighbors, then why do many males within local dialects not share their song with their neighbor, especially if there is reproductive advantage to be gained (Payne 1982; Payne *et al.* 1988; but see McGregor & Krebs 1984)? Several authors have remarked upon this failure to match (Jenkins 1977; Payne 1982; Nelson 1992). Mismatching in particular individuals might be traceable to differences in the variety of song material encountered during an early sensitive phase. If a male failed to acquire some songs early in life, then he would not be able to counter-sing a matching song if presented with those songs outside the sensitive phase. Those males with a larger repertoire of memorized songs might have a better chance of matching their neighbor's songs via selective attrition wherever they settle. This remains to be tested in the field. As discussed earlier, it may be advantageous in some circumstances or species for selective mismatching of songs to occur.

Another aspect of the developmental process that seems difficult to understand from a purely instructive perspective is the phenomenon of overproduction itself (Figs. 2.1 and 2.2). Why do males go through an extensive phase of vocal experimentation, and why do they acquire more songs than they "need" ahead of time in order to match their neighbors? We have argued that overproduction and improvisation have the function of increasing a bird's chances of matching his song(s) to those of his neighbors when the time and place of territory occupation cannot be predicted in advance (Nelson *et al.* 1996b). Just as sexual recombination may increase the odds that some

offspring will do well in an uncertain environment, or the proliferation of antibodies may meet an unknown antigenic challenge, so too the overproduction of song may enable a bird to fit into a social environment that cannot be predicted in advance. If the future breeding dialect could be predicted, or if males were routinely capable of acquiring their songs as yearlings, there would seem to be little reason for overproduction.

A selective attrition mechanism may also contribute novel perspectives on understanding the evolution of song developmental mechanisms. An emerging theme in comparative studies of vocal development is the finding that sedentary and migratory populations of the same species differ in that males in sedentary populations usually share songs more closely with neighbors than do males in migratory populations (Kroodsma 1978; Kroodsma & Canady 1985; Ewert & Kroodsma 1994; Nelson *et al.* 1995, 1996a). Sedentary Nuttall's white-crowned sparrows overproduce significantly less, spend less time in vocal rehearsal, and acquire their songs at a significantly later time, and over a broader range of ages than do migratory male mountain white-crowned sparrows, *Z. l. oriantha* (Nelson *et al.* 1995). We suggest that *nuttalli* dialects result more from *acquisition* of matching songs from territory neighbors, while dialects in *oriantha* result from the selective attrition of songs from a large overproduced repertoire acquired in the hatching year. We tested this scenario by studying development in the Puget Sound white-crowned sparrow, a migratory subspecies most closely related to *nuttalli* (Nelson *et al.* 1996a). Our hypothesized linkage between overproduction and migration was confirmed by finding that male *pugetensis* resembled *oriantha*, and differed significantly from its close relative *nuttalli* on all but one measure of vocal development. It appears that a shared migratory annual cycle is a better explanation of variation in vocal development than is recency of common ancestry.

A final consideration derives from the intense interest in the neurobiology of song production and perception. If there are effects of social interaction on song learning, we need to examine visual inputs, or, at least, auditory input contingent to that part of the song system involved in song memorization at the time when memories are acquired. If social interaction's effects can be explained simply by the amount of stimulation, as I have suggested, or by social interaction occurring during song *production*, and not necessarily during memorization, then there may be very different consequences for how research in this area should proceed.

ACKNOWLEDGEMENTS

Part of this chapter was written while I was at the Ethology Section, Leiden University. I thank Carel ten Cate for making my stay there possible.

REFERENCES

Adret, P. (1993a). Operant conditioning, song learning and imprinting to taped song in the zebra finch. *Animal Behaviour* **46**: 149–59.

Adret, P. (1993b). Vocal learning induced with operant techniques: An overview. *Netherlands Journal of Zoology* **43**: 125–42.

Baker, M. C. & Cunningham, M. A. (1985). The biology of bird song dialects. *Behavior and Brain Sciences* **8**: 85–133.

Baptista, L. F. (1974). The effects of songs of wintering white-crowned sparrows on song development in sedentary populations of the species. *Zeitschrift für Tierpsychologie* **34**: 147–71.

Baptista, L. F. (1975). Song dialects and demes in sedentary populations of the white-crowned sparrow (*Zonotrichia leucophrys nuttalli*). *University of California Publications in Zoology* **105**: 1–52.

Baptista, L. F. (1977). Geographic variation in song and dialects of the Puget Sound white-crowned sparrow. *Canadian Journal of Zoology* **63**: 1741–52.

Baptista, L. F., Bell, D. A. & Trail, P. W. (1993). Song learning and production in the white-crowned sparrow: Parallels with sexual imprinting. *Netherlands Journal of Zoology* **43**: 17–33.

Baptista, L. F. & Morton, M. L. (1988). Song learning in montane white-crowned sparrows: From whom and when. *Animal Behaviour* **36**: 1753–64.

Baptista, L. F. & Petrinovich, L. (1984). Social interaction, sensitive phases and the song template hypothesis in the white-crowned sparrow. *Animal Behaviour* **32**: 172–81.

Baptista, L. F. & Petrinovich, L. (1986). Song development in the white-crowned sparrow: Social factors and sex differences. *Animal Behaviour* **34**: 1359–71.

Beecher, M. D., Campbell, S. E. & Stoddard, P. K. (1994). Correlation of song learning and territory establishment strategies in the song sparrow. *Proceedings of the National Academy of Sciences, USA* **91**: 1450–4.

Böhner, J. (1983). Song learning in the zebra finch (*Taeniopygia guttata*): Selectivity in the choice of a tutor and accuracy of song copies. *Animal Behaviour* **31**: 231–7.

Böhner, J. (1990). Early acquisition of song in the zebra finch, *Taeniopygia guttata*. *Animal Behaviour* **39**: 369–74.

Bolhuis, J. J. (1991). Mechanisms of avian imprinting: A review. *Biological Review* **66**: 303–45.

Bolhuis, J. J. & van Kampen, H. S. (1992). An evaluation of auditory learning in filial imprinting. *Behaviour* **122**: 195–230.

Bradley, R. A. (1981). Song variation within a population of white-eyed vireos (*Vireo griseus*). *Auk* **98**: 80–7.

Byers, B. E. & Kroodsma, D. E. (1992). Development of two song categories by chestnut-sided warblers. *Animal Behaviour* **44**: 799–810.

Campbell, D. T. & Stanley, J. C. (1963). *Experimental and Quasi-Experimental Designs for Research*. Rand-McNally, Chicago.

Catchpole, C. K. & Slater, P. J. B. (1995). *Bird Song. Biological Themes and Variations*. Cambridge University Press, Cambridge.

Chaiken, M., Böhner, J. & Marler, P. (1993). Song acquisition in European starlings, *Sturnus vulgaris*: A comparison of the songs of live-tutored, untutored, and wild-caught males. *Animal Behaviour* **46**: 1079–90.

Chaiken, M., Böhner, J. & Marler, P. (1994). Repertoire turnover and the timing of song acquisition in European starlings. *Behaviour* **128**: 25–39.

Changeux, J.-P., Heidman, T. & Patte, P. (1984). Learning by selection. In *The Biology of Learning*, ed. P. Marler & H. S. Terrace, pp. 115–33. Springer-Verlag, Berlin.

Clark, C. W., Marler, P. & Beeman, K. (1987). Quantitative analysis of animal vocal phonology: An application to swamp sparrow song. *Ethology* **76**: 101–15.

Clayton, N. S. (1987a). Song learning in cross-fostered zebra finches: A re-examination of the sensitive phase. *Behaviour* **102**: 67–81.

Clayton, N. S. (1987b). Song tutor choice in zebra finches. *Animal Behaviour* **35**: 714–21.

Cunningham, M. A. & Baker, M. C. (1983). Vocal learning in white-crowned sparrows: Sensitive phase and song dialects. *Behavioral Ecology and Sociobiology* **13**: 259–69.

DeWolfe, B. B. & Baptista, L. F. (1995). Singing behavior, song types on their wintering grounds and the question of leap-frog migration in Puget Sound white-crowned sparrows. *Condor* **97**: 376–89.

DeWolfe, B. B., Baptista, L. F. & Petrinovich, L. (1989). Song development and territory establishment in Nuttall's white-crowned sparrows. *Condor* **91**: 397–407.

Dittus, W. P. J. & Lemon, R. E. (1969). Effects of song tutoring and acoustic isolation on the song repertoires of cardinals. *Animal Behaviour* **17**: 523–33.

Dufty, A. M. Jr (1985). Song sharing in the brown-headed cowbird (*Molothrus ater*). *Zeitschrift für Tierpsychologie* **69**: 177–90.

Eales, L. A. (1985). Song learning in zebra finches: Some effects of song model availability on what is learnt and when. *Animal Behaviour* **33**: 1293–300.

Eales, L. A. (1987a). Song learning in female-raised zebra finches: Another look at the sensitive phase. *Animal Behaviour* **35**: 1356–65.

Eales, L. A. (1987b). Do zebra finch males that have been raised by another species still tend to select a conspecific song tutor? *Animal Behaviour* **35**: 1347–55.

Eales, L. A. (1989). The influences of visual and vocal interaction on song learning in zebra finches. *Animal Behaviour* **37**: 507–8.

Ewert, D. N. & Kroodsma, D. E. (1994). Song sharing and repertoires among migratory and resident rufous-sided towhees. *Condor* **96**: 190–6.

Falls, J. B., Krebs, J. R. & McGregor, P. (1982). Song-matching in the great tit: The effect of song similarity and familiarity. *Animal Behaviour* **30**, 997–1009.

Hultsch, H. (1990). Recombination of acquired songs as a correlate of package formation. In *Brain–Perception–Cognition*, ed. N. Elsner & G. Roth, p. 433. Thieme Verlag, Stuttgart.

Hultsch, H. & Todt, D. (1981). Repertoire sharing and song-post distance in nightingales (*Luscinia megarhynchos* B.). *Behavioral Ecology and Sociobiology* **8**: 183–8.

Hultsch, H. & Todt, D. (1989). Memorization and reproduction of songs in nightingales (*Luscinia megarhynchos*): Evidence for package formation. *Journal of Comparative Physiology A* **165**: 197–203.

Hultsch, H. & Todt, D. (1992). The serial order effect in the song acquisition of birds: Relevance of exposure frequency to song models. *Animal Behaviour* **44**: 590–2.

Jenkins, P. F. (1977). Cultural transmission of song patterns and dialect development in a free-living bird population. *Animal Behaviour* **25**: 50–78.

Jerne, N. (1967). Antibodies and learning: Selection versus instruction. In *The Neurosciences. A Study Program*, ed. G. C. Quarton, T. Melnechuk & F. O. Schmitt, pp. 200–5. Rockefeller University Press, New York.

King, A. P., West, M. J. & Eastzer, D. H. (1980). Song structure and song development as potential contributors to reproductive isolation in cowbirds (*Molothrus ater*). *Journal of Comparative and Physiological Psychology* **94**: 1028–39.

Konishi, M. (1965). The role of auditory feedback in the control of vocalization in the white-crowned sparrow. *Zeitschrift für Tierpsychologie* **22**: 770–83.

Konishi, M. (1985). Bird song: From behavior to neuron. *Annual Review of Neuroscience* **8**: 125–70.

Kroodsma, D. E. (1974). Song learning, dialects, and dispersal in the Bewick's wren. *Zeitschrift für Tierpsychologie* **35**: 352–80.

Kroodsma, D. E. (1978). Aspects of learning in the ontogeny of bird song: Where, from whom, when, how many, which, and how accurately. In *The Development of Behavior: Comparative and Evolutionary Aspects*, ed. G. M. Burghardt & M. Bekoff, pp. 215–30. Garland STPM, New York.

Kroodsma, D. E. (1981). Winter wren singing behavior: A pinnacle of song complexity. *Condor* **82**: 357–65.

Kroodsma, D. E. (1982). Learning and ontogeny of sound signals in birds. In *Acoustic Communication in Birds* vol. 2, ed. D. E. Kroodsma & E. H. Miller, pp. 1–23. Academic Press, New York.

Kroodsma, D. E. (1985). Limited dispersal between dialects? Hypotheses testable in the field. *Behavior and Brain Sciences* **8**: 108–9.

Kroodsma, D. E., Baker, M. C., Baptista, L. F. & Petrinovich, L. (1985). Vocal "dialects" in Nuttall's white-crowned sparrow. In *Current Ornithology* vol. 2, ed. R. F. Johnston, pp. 103–33. Plenum, New York.

Kroodsma, D. E. & Canady, R. A. (1985). Differences in repertoire size, singing behavior, and associated neuroanatomy among marsh wren populations have a genetic basis. *Auk* **102**: 439–46.

Kroodsma, D. E. & Pickert, R. (1980). Environmentally dependent sensitive periods for avian vocal learning. *Nature* **288**: 477–9.

Kroodsma, D. E. & Pickert, R. (1984). Sensitive phases for song learning: Effects of social interaction and individual variation. *Animal Behaviour* **32**: 389–94.

Kroodsman, D. E. & Verner, J. (1978). Complex singing behaviors among *Cistothorus* wrens. *Auk* **95**: 703–16.

Lemon, R. E. (1975). How birds develop song dialects. *Condor* **77**: 385–406.

Machlis, L., Dodd, P. W. D. & Fentress, J. C. (1985). The pooling fallacy: Problems arising when individuals contribute more than one observation to the data set. *Zeitschrift für Tierpsychologie* **68**: 201–14.

Mann, N. I. & Slater, P. J. B. (1995). Song tutor choice by zebra finches in aviaries. *Animal Behaviour* **49**: 811–20.

Margoliash, D., Staicer, C. A. & Inoue, S. A. (1991). Stereotyped and plastic song in adult indigo buntings, *Passerina cyanea*. *Animal Behaviour* **42**: 367–88.

Margoliash, D., Staicer, C. A. & Inoue, S. A. (1994). The process of syllable acquisition in adult indigo buntings (*Passerina cyanea*). *Behaviour* **131**: 39–64.

Marler, P. (1960). Bird songs and mate selection. In *Animal*

Sounds and Communication, ed. W. E. Lanyon & W. N. Tavolga, pp. 348–67. American Institute of Biological Sciences, Washington, DC.

Marler, P. (1970). A comparative approach to vocal learning: Song development in white-crowned sparrows. *Journal of Comparative and Physiological Psychology (Monograph)* **71**: 1–25.

Marler, P. (1990). Song learning: The interface between behaviour and neuroethology. *Philosophical Transactions of the Royal Society, London B* **329**: 109–14.

Marler, P. (1991). Song learning behavior: The interface with neuroethology. *Trends in Neurosciences* **5**: 199–206.

Marler, P. & Peters, S. (1981). Sparrows learn adult song and more from memory. *Science* **213**: 780–2.

Marler, P. & Peters, S. (1982a). Developmental overproduction and selective attrition: New processes in the epigenesis of birdsong. *Developmental Psychobiology* **15**: 369–78.

Marler, P. & Peters, S. (1982b). Subsong and plastic song: Their role in the vocal learning process. In *Acoustic Communication in Birds* vol. 2, ed. D. E. Kroodsma & E. H. Miller, pp. 25–50. Academic Press, New York.

Marler, P. & Peters, S. (1987). A sensitive period for song acquisition in the song sparrow, *Melospiza melodia*: A case of age-limited learning. *Ethology* **76**: 89–100.

Martin, D. J. (1979). Songs of the fox sparrow. II. Intra- and interpopulation variation. *Condor* **81**: 173–84.

McGregor, P. K. & Krebs, J. R. (1984). Song learning and deceptive mimicry. *Animal Behaviour* **32**: 280–7.

McGregor, P. K. & Krebs, J. R. (1989). Song learning in adult great tits (*Parus major*): Effects of neighbors. *Behaviour* **108**: 139–59.

Nelson, D. A. (1992). Song overproduction and selective attrition lead to song sharing in the field sparrow (*Spizella pusilla*). *Behavioral Ecology and Sociobiology* **30**: 415–24.

Nelson, D. A. & Croner, L. J. (1991). Song categories and their functions in the field sparrow. *Auk* **108**: 42–52.

Nelson, D. A. & Marler, P. (1994). Selection-based learning in bird song development. *Proceedings of the National Academy of Sciences, USA* **91**: 10 498–501.

Nelson, D. A., Marler, P. & Morton, M. L. (1996a). Overproduction in song development: An evolutionary correlate with migration. *Animal Behaviour* **51**: 1127–40.

Nelson, D. A., Marler, P. & Palleroni, A. (1995). A comparative approach to vocal learning: Intraspecific variation in the learning process. *Animal Behaviour* **50**: 83–97.

Nelson, D. A., Whaling, C. & Marler, P. (1996b). The capacity for song memorization varies in populations of the same species. *Animal Behaviour* **52**: 379–87.

Payne, R. B. (1981). Song learning and social interaction in indigo buntings. *Animal Behaviour* **29**: 688–97.

Payne, R. B. (1982). Ecological consequences of song matching: Breeding success and intraspecific mimicry in indigo buntings. *Ecology* **63**: 401–11.

Payne, R. B., Payne, L. L. & Doehlert, S. M. (1988). Biological and cultural success of song memes in indigo buntings. *Ecology* **69**: 104–17.

Pepperberg, I. M. (1985). Social modeling theory: a possible framework for understanding avian vocal learning. *Auk* **102**: 854–64.

Peters, S., Marler, P. & Nowicki, S. (1992). Song sparrows learn from limited exposure to song models. *Condor* **94**: 1016–19.

Petrinovich, L. (1985). Factors influencing song development in the white-crowned sparrow (*Zonotrichia leucophrys*). *Journal of Comparative Psychology* **99**: 15–29.

Petrinovich, L. (1990). Avian song development: methodological and conceptual issues. In *Contemporary Issues in Comparative Psychology*, ed. D. A. Dewsbury, pp. 340–59. Sinauer, Sunderland, MA.

Petrinovich, L. & Baptista, L. F. (1987). Song development in the white-crowned sparrow: modification of learned song. *Animal Behaviour* **35**: 961–74.

Rice, J. O. & Thompson, W. L. (1968). Song development in the indigo bunting. *Animal Behaviour* **16**: 462–9.

Slater, P. J. B. (1989). Bird song learning: Causes and consequences. *Ethology, Ecology and Evolution* **1**, 19–46.

Slater, P. J. B., Eales, L. A. & Clayton, N. S. (1988). Song learning in zebra finches (*Taeniopygia guttata*): Progress and prospects. *Advances in the Study of Behavior* **18**: 1–34.

Slater, P. J. B. & Ince, S. A. (1982). Song development in chaffinches: What is learnt and when? *Ibis* **124**: 21–6.

Slater, P. J. B., Jones, A. & ten Cate, C. (1993). Can lack of experience delay the end of the sensitive phase for song learning? *Netherlands Journal of Zoology* **43**: 80–90.

Slater, P. J. B., Richards, C. & Mann, N. I. (1991). Song learning in zebra finches exposed to a series of tutors during the sensitive phase. *Ethology* **88**: 163–71.

ten Cate, C. (1989). Behavioral development: Toward understanding processes. In *Perspectives in Ethology*, vol. 8, ed. P. P. G. Bateson & P. H. Klopfer, pp. 243–69. Plenum, New York.

ten Cate, C. (1994). Perceptual mechanisms in song learning and imprinting. In *Causal Mechanisms of Behavioural Development*, ed. J. A. Hogan & J. J. Bolhuis, pp. 116–46. Cambridge University Press, Cambridge.

Thielcke, G. & Krome, M. (1989). Experimente über sensible Phasen und Gesangsvariabilität beim Buchfinken (*Fringilla coelebs*). *Journal für Ornithologie* **130**: 435–43.

Thielcke, G. & Krome, M. (1991). Chaffinches *Fringilla coelebs* do not learn song during autumn and early winter. *Bioacoustics* **3**: 207–12.

Thielcke-Polz, H. von & Thielcke, G. (1960). Akustisches Lernen verschieden alter schallisolierter Amseln (*Turdus merula* L.) und die Entwicklung erlenter Motive ohne und mit kunstlichem Einfluss von Testosterone. *Zeitschrift für Tierpsychologie* **17**: 211–44.

Thorpe, W. H. (1958). The learning of song patterns by birds, with especial reference to the song of the chaffinch *Fringilla coelebs*. *Ibis* **100**: 535–70.

Todt, D., Hultsch, H. & Heike, D. (1979). Conditions affecting song acquisition in nightingales (*Luscinia megarhynchos* L.). *Zeitschrift für Tierpsychologie* **51**: 23–35.

Trainer, J. (1983). Changes in song dialect distributions and microgeographic variation in song of white-crowned sparrows (*Zonotrichia leucophrys nuttalli*). *Condor* **100**: 568–82.

van Kampen, H. S. & Bolhuis, J. J. (1993). Interaction between auditory and visual learning during filial imprinting. *Animal Behaviour* **45**: 623–5.

Volman, S. F. & Khanna, H. (1995). Convergence of untutored song in group-reared zebra finches. *Journal of Comparative Psychology* **109**: 211–21.

Waser, M. S. & Marler, P. (1977). Song learning in canaries. *Journal of Comparative and Physiological Psychology* **91**: 1–7.

West, M. J. & King, A. P. (1988). Female visual displays affect the development of male song in the cowbird. *Nature* **334**: 244–6.

Williams, H. (1990). Models for song learning in the zebra finch: Fathers or others? *Animal Behaviour* **39**: 745–57.

3 Social interaction and vocal development in birds

LUIS F. BAPTISTA AND SANDRA L. L. GAUNT

INTRODUCTION

Experiments on song development have a long history dating from those performed by the Baron von Pernau with chaffinches (*Fringilla coelebs*), published in 1768 (Thielcke 1988). The Baron documented regional song dialects, showed that learning was restricted to a time window or "sensitive phase," and demonstrated the bird's preference for learning songs of conspecifics over those of allospecifics or "stimulus filtering." With the advent of instrumentation that allowed capture and analyses of sound, i.e., magnetic tape-recording and sound spectrographic technology, Thorpe (1958) was able to test these conclusions objectively. He played tape-recorded conspecific and allospecific songs to hand-raised, naive chaffinches reared in acoustic isolation. His experiments confirmed the existence of sensitive phases and stimulus filtering mechanisms during the song-learning process.

Thorpe's study set a standard for avian song ontogeny protocol: hand-raised naive experimental or pupil birds are isolated in sound-proof chambers and exposed to tape-recorded vocalizations of model or tutor birds. An investigator is then able to: (a) control the number of songs played to the experimental birds to determine the minimum number of songs required to effect learning (Petrinovich 1985; Hultsch & Todt 1992; Peters *et al.* 1992; Hultsch 1993); (b) demonstrate that birds may recognize conspecific song by sound alone (Konishi 1985); and (c) test the effect of sound degradation on the choice of tutors by pupils (Morton *et al.* 1986).

That social factors could influence the choice of a song tutor by a pupil was first appreciated by Nicolai (1959). His work suggested that tape-tutoring experiments, although valuable, may in some cases test only what pupils can do under the conditions imposed by the investigator and not the pupil's actual or potential capabilities in nature. The greater significance of this observation has only more recently been appreciated. Our review of studies indicates that social factors may affect (a) quality of vocalizations (songs and calls) learned, (b) choice of tutor to be learned from, (c) length of sensitive phase, and (d) rate of vocal ontogeny. These factors in turn affect dispersion of song or call types, breeding success, survival, and in some cases mate choice.

SOCIAL INTERACTION AND SONG QUALITY

Although naive individuals of some species such as chaffinches, white-crowned sparrows (*Zonotrichia leucophrys*) and swamp sparrows (*Melospiza georgiana*) produce perfect or near-perfect copies of tape-recorded tutor songs (Thorpe 1958; Marler 1970; Marler & Peters 1977; Becker 1982), some species will not learn from such a model. Among these are obligate and facultative social learners.

Obligate social learners will not acquire normal vocalizations when they are exposed only to taped acoustic stimuli. Some examples follow. Thielcke (1970, 1984) exposed naive short-toed tree-creepers (*Certhia brachydactyla*) and Eurasian tree-creepers (*C. familiaris*) of various ages to tape-recorded conspecific song. Although syllable structure, which develops independent of learning (Thielcke 1965), appeared normal in experimental birds, syntax never resembled that of the tape-recorded tutor song. One individual learned the song of an adult that it could hear giving contact calls and singing nearby. The adult's contact call stimulated the juvenile to interact socially with it, thus affecting learning. Naive juveniles

[23]

reared together developed very similar songs, again indicating that social interaction affected learning of correct syntax (Thielcke 1984).

Kroodsma & Verner (1978) exposed three naive sedge wrens (*Cistothorus platensis*), housed together, to tape-recorded adult song between 15 and 90 days of age. Although there was no evidence of learning by copying from the taped songs, the songs of the experimental birds bore some resemblance to one another. This suggests that sedge wrens also require social interaction with a live tutor before song may be acquired.

Model stimuli are acquired by **facultative** social learners in isolation but this aspect of learning is much improved if pupils are exposed to live tutors. Some examples follow. Although neighboring indigo buntings (*Passerina cyanea*) sang more similar songs than non-neighbors (Thompson 1970), naive juveniles exposed to tape-recorded conspecific song learned only some syllables from the model (Rice & Thompson 1968). Payne (1981) subsequently raised juveniles next to adults with whom they could interact and found that experimental birds produced accurate copies of the tutors' songs.

Similarly, Waser & Marler (1977) exposed one group of juvenile domestic canaries (*Serinus canarius*) to the songs of a live tutor. The tutor's songs were simultaneously picked up by microphone and piped to a second group who could thus hear but not see the tutor. The first group learned more syllable types than the second. Hand-raised common starlings (*Sturnus vulgaris*) also acquired more syllables and developed larger repertoires when tutored with live conspecifics than when tutored by tapes (Chaiken *et al.* 1993).

In his work with various species of Darwin's finches ("geospizines"), Bowman (1983) noted that a large ground-finch (*Geospiza magnirostris*) raised with parents to age 42 days produced a good copy of his father's song. Another naive large ground-finch and three sharp-beaked ground-finches (*G. difficilis*) tape tutored with the father or another male's song produced poor copies of tutor songs. These data, and field observations on other geospizines (see below), collectively indicate that fledglings exposed to their father produce good copies of his song, whereas tape-tutored fledglings learn only some elements from the recordings.

Immelmann (1969) studied song development in parent-raised zebra finches (*Taeniopygia guttata*) and in zebra finches cross-fostered under Bengalese finches (*Lonchura striata*). Both groups of pupils produced good copies of parent or foster parent songs as adults.

Price (1979) hand-raised naive zebra finches so that they imprinted on him. He then tutored them with a tape-recorder hung around his neck whenever he fed them. Only a few conspecific syllables were learned. However, if naive zebra finches can control the delivery of the tape-recorded tutor song, they produce good imitations of the tape tutor in the absence of social stimulation. This result was demonstrated by Adret (1993a,b) in operant conditioning experiments where naive birds between 35 and 65 days post-hatching could work to turn on 15 seconds of tutor tape by pressing a key, a learned directed activity. Earlier operant experiments by ten Cate (1991) using perch landing to activate the tape failed to show imitation. However, perch landing may be an inappropriate stimulus pairing as it is a spontaneous locomotor activity. Thus, no association between this activity and control of access to the song stimulus need be made. When Adret (1993b) changed the successful key-pecking control panel to a perch-activated panel, males failed to copy the tape and developed impoverished songs. He hypothesized that control of the stimulus, like interaction with a live tutor, may play a "permissive" role in learning by increasing the young birds' attention toward the tutor song.

Social interaction, then, need not be the only stimulus that facilitates learning in all avian species. Rather a pairing of certain stimuli with the auditory stimulus may enhance the pupil's attention, arousal and/or motivation such that learning is enhanced. This phenomenon is amply illustrated by acquisition of human speech by African grey parrots (*Psittacus erithacus*). Parrots interacting with human tutors learned better than those exposed to tapes or videos (Pepperberg 1993, and see Chapter 9; see also West *et al.* 1983 on starlings). In another case of human interaction affecting learning, Todt *et al.* (1979) found that, although naive common nightingales (*Luscinia megarhynchos*) will learn some syllables played from taped conspecific song, individuals allowed to observe the investigator operate the loudspeaker acquired complete songs from the tape.

CHOICE OF SONG TUTOR

The transmission of cultural traditions between and within generations has been attributed to three, not necessarily exclusive, modes (Cavalli-Sforza *et al.* 1982), and these can be applied to song traditions in birds. Vertical tradition is defined as learning from parents: genes are passed concomitantly with memes (song components). Horizontal tradition occurs when age peers learn from each other. Oblique tradition describes learning from genetically unrelated adults.

Learning from fathers

A male Eurasian bullfinch (*Pyrrhula pyrrhula*), raised by canaries, learned his foster father's allospecific song even though he could hear bullfinches nearby (Nicolai 1959). A son of this bullfinch learned the father's canary song, and his son sang his grandfather's allospecific song, ignoring neighboring bullfinches. Nicolai concluded from these observations that in nature male progeny form an "emotional" bond with the father and as a consequence learn his song.

Similarly, zebra finches raised by Bengalese finches or their own species learned their foster father's or father's song, although all birds could see and hear other conspecifics nearby (Immelmann 1969; Böhner 1983). These authors concluded that vertical tradition is the norm in zebra finches. However, Slater *et al.* (1988) separated juvenile zebra finches at various ages and exposed them to unrelated male conspecifics; they found that the juveniles could copy new syllables from these adults within a certain time window. The authors argued from these results that learning occurs after dispersal from the natal area. Zann (1990, 1993, and see Chapter 6) followed histories of free-living, color-banded son/father groups and found that learning plasticity is not necessarily realized in the wild, and that about 60% of the sons do indeed sing their father's songs as mature adults and 80% use their father's contact calls. Vertical tradition has also been demonstrated in the Bengalese finch and marsh tit (*Parus palustris*) under laboratory conditions (Immelmann 1969; Dietrich 1980; Rost 1987). These results remain to be substantiated under natural conditions.

Field observations of color-marked father/son groups of various Darwin's finches (*G. fortis*, *G. conirostris*, and *G. scandens*) have also been made (Grant 1984; Millington & Price 1985; Gibbs 1990). Again, most sons tended to sing their father's song as adults, although the number of cases of vertical tradition differed from year to year.

Millington & Price (1985) found that 92% of the sons of male medium ground-finches (*G. fortis*) sang their father's song type in 1980, whereas Gibbs (1990) found 78% when sons from 1981 to 1984 were pooled. Gibbs' (1990) study is noteworthy in that he also found that males with the rarer song types had longer lifespans and left more progeny. Over a period of seven years, the rarer song types increased in frequency whereas the commonest song type decreased. Biological and cultural evolution appeared to be coupled in these finches; however, this correlation may be spurious.

Learning from fathers has thus far been demonstrated under natural conditions only in these studies of zebra finches and *Geospiza* finches. Both groups are denizens of ephemeral habitats, and both have been selected for early maturation and may reproduce by three months of age (Immelmann 1982; Gibbs *et al.* 1984). Although it is not certain why, a premium exists to learn songs from fathers, which in turn may enable males to acquire vocal signals at a younger age, preparing them for early breeding.

Learning from age peers

Several lines of evidence suggest that, although many species learn song from adults, juvenile birds of some of these same species are also influenced by each other's vocal development. As discussed above (and for a review, see Baptista 1996), naive hand-reared young oscines deprived of contact with adult song models tend to produce abnormal vocalizations. This is also true for chaffinches and white-crowned sparrows. However, if the young of these species are isolated from adults, but not from each other, they tend to develop very similar, albeit abnormal, songs (Thorpe 1958; Marler 1970). Peer stimulation in juvenile common starlings isolated from adults not only resulted in shared song motifs but also facilitated the development of species-specific syntax in the absence of normal song models (M. Chaiken, personal communication).

Providing model song to peer-grouped zebra finches, canaries and white-crowned sparrows results in normal species-typical song, and songs within the groups are similar (Slater *et al.* 1993; Cunningham & Baker 1983; Baptista *et al.* 1993a). Group-isolated short-toed tree-creepers sing abnormal but similar motifs (Thielcke 1984). Evidence from non oscines also supports learning from age peers. Juvenile Anna's hummingbirds (*Calypte anna*) raised in groups develop very similar songs (Baptista & Schuchmann 1990).

An interesting case of maintenance of some species-specific song characteristics by grouped naive birds has been demonstrated in juncos. Dark-eyed juncos (*Junco hyemalis*) in the San Francisco Bay area tend to sing songs with single trills of one syllable type (Konishi 1964). In contrast, yellow-eyed juncos (*Junco phaeonotus*) sing songs containing several trill phrases each of several syllable types (Marler & Isaac 1961). Dark-eyed juncos raised in isolation sing complex song types resembling the song of the yellow-eyed junco, whereas isolated yellow-eyed juncos develop abnormal songs resembling dark-eyed juncos in that they consist of single trills. Group-raised dark-eyed juncos sing single trills, and group-raised yellow-eyed juncos sing multi-trilled songs (Marler *et al.* 1962; Marler 1967). Thus, although group-raised isolate juncos of both species cannot produce normal syllable types, they stimulate each other to produce songs with conspecific structure.

Although experimental studies provide considerable evidence for song learning from age peers, evidence from the field is lacking. Such evidence is difficult to gather, however, as following free-living juvenile groups is not an easy task.

Learning from unrelated adults

Under natural conditions, learning from unrelated adults has been the most widely documented trend in avian species. Examples of five such situations are given below:

1. *From nearby territorial males*: Nice (1943) reported that dispersing juvenile song sparrows in Ohio learn song from adult rivals while establishing territories. Each individual's repertoire may change as some themes are discarded or are modified. This observation was made long before the availability of sound spectrography (sonography). More recently, Beecher *et al.* (1994) followed color-marked song sparrows in a Washington population and demonstrated that each individual's repertoire contained at least one song from each nearest neighbor.

Similarly, Bewick's wrens (*Thryomanes bewickii*) when settling on territory may produce songs similar to the father's, but soon discards these for songs learned from territorial neighbors (Kroodsma 1974). By day 60, fledglings are producing song components typical of their

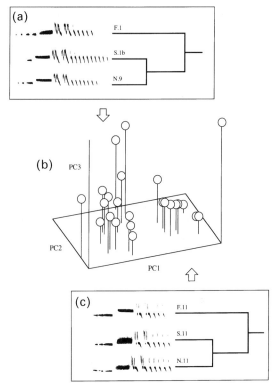

Fig. 3.1. (a) Cluster analysis of frequency and temporal components of white-crowned sparrow song, demonstrating that son (S) 1b is more similar to neighbor N.9 than to father (F) 1. (b) Principal components analyses of frequency and temporal components of songs of 24 white-crowned sparrows recorded in Golden Gate Park, San Francisco, showing two clusters comprising two song neighborhoods. This indicates that songs of nearest neighbors are more similar to each other than to those of non-neighbors. (c) Cluster analysis of a second father/son/neighbor triad, as in (a). Cluster (a) is from the song neighborhood to the left of (b) and cluster (c) is from the song neighborhood to the right. (Unpublished data from D. A. Bell *et al.*)

adult repertoire. Marsh wrens (*Cistothorus palustris*) learn not only the large repertoires of neighboring territorial adults but also the order in which they are sung (Verner 1975). Kroodsma (1979) has suggested that songs are acquired from adults during vocal duels that may represent ritualized dominance/subordinance interactions.

Neighboring Nuttall's white-crowned sparrows (*Z. l. nuttalli*) tend to have more similar introductory phrases than do non-neighbors (Baptista 1975, 1985; Fig. 3.1) thus forming small "sub dialect" clusters as described earlier by Nottebohm (1969) for the congeneric rufous-collared sparrow (*Z. capensis*). Cunningham *et al.* (1987) subjected temporal and frequency measurements of white-crowned sparrow songs to multivariate statistical analysis and found that nearest neighbors clustered together in multivariate space, although not all neighboring males had similar song. From this they argued for vertical song tradition in these sparrows, i.e., sons learned from fathers and settled nearby. D. A. Bell *et al.* (unpublished data) followed color-marked nestlings from hatching to settling sites and, using traditional sound spectrograph scanning and multivariate statistical testing, compared their songs with those of fathers and territorial neighbors. Juveniles most often matched the song of territorial neighbors rather than songs of their father, indicating that oblique tradition is the norm (Fig. 3.1). Frequency and temporal measurements of song subjected to principal components analyses revealed that songs within a neighborhood are more similar to each other than to songs from an adjacent neighborhood (Fig. 3.1). Songs of color-marked yearling montane white-crowned sparrows (*Z. l. oriantha*) seldom matched those of their fathers (Baptista & Morton 1988).

Oblique song traditions have also been documented in saddlebacks (*Philesturnus carunculatus*) of New Zealand by Jenkins (1978). Each male sings from one to four song types that may be shared by as many as 20 males. Young males move into an area and adopt the songs of the local males. However, in contrast to white-crowned sparrows, birds that seldom change song in the field after their first year (Baptista & Morton 1988), a saddleback may sing songs previously undocumented as coming from him when he was three or four years old.

A color-marked population of Eurasian linnets (*Carduelis cannabina*) breeding on the island of Mellum, North Sea, Germany, sings a dialect distinct from the mainland and the neighboring island of Wangerooge (Nicolai 1986). Breeding birds migrate out from the island in September and return the next April. First year juveniles from other populations are recruited annually in this population, bringing with them alien dialects. They gradually changed their song to match the local dialect, and some males take two breeding seasons before matching is completed.

Learning from unrelated neighbors is argued for a nonoscine genus, the violet-ear hummingbirds (*Colibri* spp.). Gaunt *et al.* (1994) studied two of the four members of this genus, the green violet-ear (*C. thalassinus*) and the sparkling violet-ear (*C. coruscans*). In both species, males form small singing assemblages of two to six birds within hearing of one another, and often out of hearing range of other assemblages. Subjective, visual scanning of spectrograms suggested that neighbors had the same song, and that their songs were distinct from other assemblages. A digital cross-correlation method was used to obtain an objective value of similarity between songs of all birds in a population. These data were subjected to statistical cluster analysis that, in all but one case, clustered neighbors on the same terminal branches that reflected the geographic distribution of song patterns (Fig. 3.2). Further support for song sharing was obtained by an evaluation for randomness of this distribution with a Mantel test. Neighboring green violet-ears have since been found, by means of DNA fingerprinting techniques, not to be first-order relatives (A. S. Gaunt *et al.* unpublished data).

House finches (*Carpodacus mexicanus*) in New York exhibit regional dialects that are shared by birds in a local population (Mundinger 1975). Neighboring house finches in California do not share song themes but share syllables (Bitterbaum & Baptista 1979). It appears that, during social interaction, individuals select elements from neighbors' songs from which to reconstruct their own themes.

Extensive studies of indigo buntings (Payne 1982, 1983) demonstrated that first-year males select breeding sites near an established adult male neighbor and may learn to match the adult's song during repeated matched countersinging. First-year males that match the songs of adult neighbors have a higher chance of mating and fledging young than non matchers (Payne *et al.* 1988), although the reason is not clear.

Brood parasitic viduine finches (*Vidua* spp.) lay their eggs in nests of various estrildid finches (Nicolai 1964, 1973; Payne 1973a). The brood parasites acquire the songs and social calls of their host species and use them during courtship (Payne 1973a,b). Because foster father and nestlings are not blood related, this instance of learning must be regarded as an example of oblique tradition.

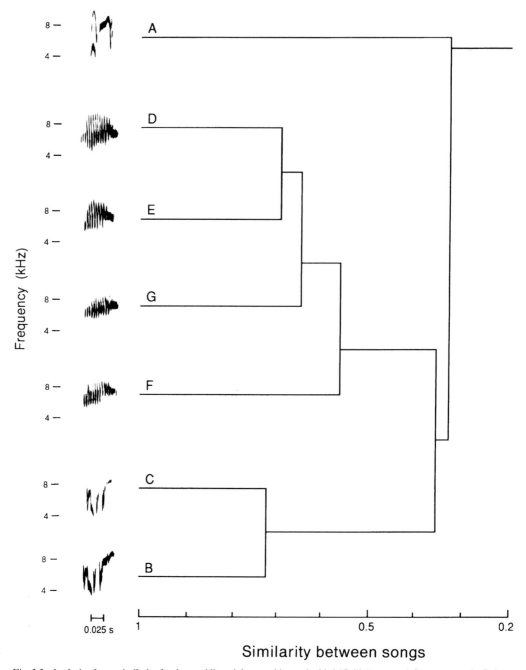

Fig. 3.2. Analysis of song similarity for the sparkling violet-eared hummingbird (*Colibri coruscans*) from one area in Quito, Ecuador. Two assemblages of neighboring birds had been identified in the field from this area. Hierarchical cluster analysis (average, UPGMA) of a matrix of sound similarity values obtained from spectral cross-correlation of these birds' songs cluster birds from each neighborhood together with one exception, bird A. (From Gaunt *et al.* 1994.)

Fledgling viduines acquire conspecific and additional allospecific songs from conspecific and allospecific adults, respectively, after dispersal (Payne, 1990).

2. *From alpha males*: Two groups of hand-raised village weaverbirds (*Ploceus cucullatus*) acquired song, not from their fathers from whom they were isolated, but from the dominant male in each group (Jacobs 1979). In addition, one bird that hatched in an aviary sang the song of the alpha male in the enclosure.

In the promiscuous village indigobird (*Vidua chalybeata*), males chose high, conspicuous perches from which to sing, and females visit these sites to seek copulations. Some males achieve more copulations than other males (Payne & Payne 1977). Neighbors share song types that change from year to year, and Payne (1985; Payne & Payne, Chapter 5) found that males tend to copy the song of the most successful male in any given breeding season.

3. *From same sex*: Bertram (1970) described a unique variation of oblique tradition in Indian hill mynas (*Gracula religiosa*). Females learn from females and males learn from males. Songs in female common starlings are unpredictable and thus difficult to record. However, Hausberger (1993; and see Chapter 8), using testosterone implants, found that under captive conditions male–male and female–female diads were formed, and, again, female starlings learned songs from females and males from males in these diads. Another example of female–female song learning has been reported for the bay wren (*Thryothorus nigricapillus*), where females share many song types with other females in the population but not with males (Levin cited by Catchpole & Slater 1995).

Although there are copious data demonstrating learning in female white-crowned sparrows in the field (Baptista *et al.* 1993b), almost all efforts to tutor females in the laboratory have met with failure (Baptista & Petrinovich 1986; Petrinovich & Battista 1987). This failure to acquire and sing new songs occurred irrespective of whether live or tape tutors were used, but only male song models were presented. Perhaps as in sturnids, female white-crowned sparrows are more likely to sing new songs if tutored by female song models.

4. *From mates*: Female bullfinches learn elements from their father's song, then add motifs from their mate's song after pair formation (Nicolai 1959). These songs are retained throughout life. Moreover, a widowed female does not learn new syllables if it pairs with a new male.

Mundinger (1970) paired different species of *Carduelis* finches and found that mates learned each other's flight calls. Males would learn from females and vice versa. Acquisition of flight calls from mates have been shown in other cardueline genera, including *Carpodacus* (Samson 1978; Mundinger 1979) and *Loxia* (Groth 1993). Although *Carduelis* species appear to learn new calls as adults (Mundinger 1970), crossbills do not appear to do so (Groth 1993).

5. *From flock members*: Mammen & Nowicki (1981) found differences between flocks of black-capped chickadees (*Parus atricapillus*) in temporal and spectral parameters of their chick-a-dee call. Three new, artificial flocks, were made by rearranging extant flocks. After one month, the duration of the "dee" syllable and total call duration changed in each flock. Flock members converged in some characteristics of their calls. Playback studies showed that individuals could recognize flock members by these flock-specific calls (Nowicki 1983).

Flocks of various *Amazona* parrot species also have flock-specific calls (Nottebohm 1970; for a review, *see* Baptista 1993). Yellow-rumped caciques (*Cacicus c. cela*) live in temporary groups or colonies and males in each colony share song types that are stable in a breeding season but change in successive breeding seasons (Feekes 1982; Trainer 1987). Shared colony-specific song then aids in group formation and maintenance. Other functions of shared colony-specific songs have been suggested, including indicators of nonaggressive behaviors.

There are benefits to be gained from learning vocalizations of mates or flock members, and they are the same as those of being members of the flock. Survival rates are higher for flock members than for birds living alone because flock members can aid group defense of territory (see Brown & Farabaugh, Chapter 7), in locating food sources, giving warnings of predators and in mobbing the latter (for a review see Parsons & Baptista 1980). The observation that black-capped chickadees can recognize flock members by their flock-specific calls and that these are used in mobbing lends credence to these ideas (Nowicki 1983). Flock calls may enable members to monitor movements of flock members and thus stay together (see also Brown & Farabaugh, Chapter 7).

Indirect influences

Adult males or females in a population may indirectly affect choice of song tutor by fledgling males. For example, two white-crowned sparrow song types occur at Fortress Hill, Alberta, Canada. Sons most often sing themes matching those of the neighbors whose songs are most similar to their fathers' (M. R. Lein, unpublished data). Thus, father's song "primes" the juvenile to select a song to be matched during settling. Böhner (1990) has described a similar priming effect in zebra finch song learning. Zebra finches may also learn preferentially from males who look like their father (Mann *et al.* 1991).

Juvenile brown-headed cowbirds (*Molothrus ater*) of two subspecies housed together selectively learned songs from males if females of that subspecies were also present. Choice of song model was influenced by subtle cues given by females (West & King 1985, 1988).

DEVELOPMENTAL STAGES

It is now well established that all or nearly all oscines acquire at least some of their species-specific song characteristics through imitative learning (Baptista 1996). Oscines may in turn be divided into "open-ended" versus "age limited" learners (Nottebohm 1993).

Open-ended learners such as canaries, common starlings, red-winged blackbirds (*Agelaius phoeniceus*) and *Vidua* species are birds that may acquire new song over a number of years or possibly throughout life (Yasukawa *et al.* 1980; Adret-Hausberger *et al.* 1990; Nottebohm 1993; Payne & Payne, Chapter 5). Age-limited learners are those species among which vocal acquisition is restricted to a small time window or "sensitive phase." This window may be a period of less than 100 days or may extend to the spring following the hatching year (e.g., Thorpe 1958; Immelmann 1969; Marler 1970; Thielcke & Krome 1989).

In age-limited learning there are actually two sensitive phases associated with the song-learning process (for a review see Nottebohm 1993). The first or "sensory phase" is that period when birds receive and store acoustical information. The second or "motor phase" is that period where stored information is vocalized (Konishi 1965). The first manifestations of the motor phase, "subsong," are usually limited to amorphous sounds bearing little resemblance to model songs. These vocalizations gradually develop into "plastic song" wherein syllables of the model

are first recognizable, and with additional practice continue to improve and develop into "crystallized" adult song (e.g. Marler 1970; DeWolfe *et al.* 1989).

In the laboratory a period of some 200 days may separate the sensory from the motor phase, the "memorization phase" (Konishi 1965; Marler & Peters 1982). During this period experimental birds do not sing. In this section we demonstrate that social interactions, in some species, may extend the sensory phase and hasten the onset of the motor phase.

The sensory phase

As adults, white-crowned sparrows readily vocalize conspecific song from tape models, if the models are played to them between the ages of 10 and 50 days (Marler 1970; Petrinovich 1985). However, experimental birds produce better copies of the model if tutored between 10 and 30 days than between 30 and 50 days (Petrinovich 1985). Birds exposed to taped song from 50 days onward will not produce recognizable copies of that song (Marler 1970; Baptista & Petrinovich 1986).

Data obtained from several species have been interpreted to indicate that social interaction with live tutors extends their sensitive phase. Hand-raised naive white-crowned sparrows exposed to *live* conspecific or red ava-davat (strawberry finch, *Amandava amandava*) tutors beyond 50 days of age produced good copies of the tutors' songs (Baptista & Petrinovich 1984, 1986). Marsh wrens acquire songs from tape models if they are presented to them before 45 days of age and not beyond. However, birds exposed to live adults the following spring acquired some of these adults' songs (Kroodsma & Pickert 1984).

Song sparrows exposed to taped songs from day 10 to 350 acquired most of their song by day 40. Some learning continued until almost day 60 (Marler & Peters 1987). In contrast, Cassidy (1993) followed marked song sparrows from hatching to breeding sites on islands in British Columbia and found that one 273-day-old individual was able to learn a new song from his territorial neighbor. These data indicate that live tutors may extend the sensory phase when songs are acquired.

The storage phase

The 200-day silent period following the sensory phase is manifest not only in white-crowned sparrows raised in sound-proof chambers (Konishi 1965; Marler 1970), but also in birds raised in cages exposed to ambient sound, including conspecific adult song in the laboratory (Baptista & Petrinovich 1986). In the field, however, fledglings of the sedentary *Z. l. nuttalli* as young as 28 days of age have been known to sing adult-like song, i.e., songs at the late plastic stage (Baptista *et al.* 1993a).

In July 1994, Baptista and colleagues encountered a fully streaked juvenile white-crowned sparrow singing a crystallized song and defending a territory. This individual responded to playback from a tape-recorder placed next to a trap by approaching, singing and displaying aggressively, and was captured and color marked. John Wingfield (personal communication) has also encountered a few juvenile song sparrows singing adult songs and defending territories.

Early song crystallization is not restricted to sedentary emberizids. DeWolfe & Baptista (1995) found that hatching-year juveniles of the migratory Puget Sound white-crowned sparrow (*Z. l. pugetensis*) may sing crystalized or almost crystallized songs upon arrival in the wintering grounds in California. These observations suggest that, in all three taxa of white-crowned sparrows, the memorization, or silent, phase described from laboratory experiments is much truncated, or perhaps absent altogether, in the field. We suggest that this is a consequence of the richer sensory environment encountered under natural conditions (DeWolfe & Baptista 1995).

Stimulus filtering

It is now well known that a number of avian species may recognize conspecific song by sound alone (for a review see Baptista 1996). For example, naive hand-reared swamp sparrows (Marler & Peters 1977) or white-crowned sparrows (Marler 1970; Konishi 1985) exposed to tapes with allospecific and conspecific songs, selected conspecific songs as models to be imitated. Moreover, white-crowned sparrows exposed to taped song sparrow songs rejected the models altogether and sang simplified isolate songs (Marler 1970).

In contrast, naive white-crowned sparrows exposed to live red avadavat, dark-eyed juncos or song sparrows acquired and sang these species' songs even though they could hear conspecific song nearby (Baptista & Petrinovich 1984, 1986). Indeed, red avadavat song was acquired beyond the 50-day critical period when acquisi-

tion ceases if taped songs were used in tutoring (Baptista & Petrinovich 1984). These data suggest that social interaction may override innate preferences or barriers to learning alien songs.

Aggression retards song development

Although we have shown in this review that social interaction enhances song learning in various ways, aggressive social interaction can retard acquisition of song. Pupils separated from live tutors by a screen will learn tutor songs (Baptista & Petrinovich 1986). However, experimental subjects allowed to interact freely in the same cage develop abnormal syllables (Casey & Baker 1993). This result could, again, be due to confined laboratory conditions because white-crowned sparrows and song sparrows acquire song during male–male aggression (Nice 1943; DeWolfe *et al.* 1989) or female–female competition (Arcese *et al.* 1988).

O'Loghlen & Rothstein (1993) found that brown-headed cowbirds do not sing local dialects until their second year post-hatching. They suggest this is due to: (a) late breeding that deprives hatching-year birds of the opportunity to hear adult conspecifics who have discontinued singing, and (b) ontogeny of song extending to the second year.

The capacity to acquire new song, even if the songs are not expressed, is present in the first year as evidenced by testosterone-treated yearlings that sing normal, nonlocal songs. We suggest that songs may be acquired on the wintering ground and early in the breeding season but are not produced because subordinate yearlings are inhibited by the aggressive behavior of local adults. O'Loghlen & Rothstein entertained but dismissed this possibility because testosterone-treated yearlings *do* sing, a situation that was interpreted as being due to yearlings overcoming inhibition. We submit that, although testosterone may increase the motivation to sing, it may not overcome the inhibition to sing *local* song.

DISCUSSION

We have reviewed literature showing that, although exposing isolated pupils to tape-recorded model songs is a valid tool in song ontogeny studies (see Introduction, above), social factors are important in affecting the quality and nature of the final product or choice of tutor (Table 3.1) and must be taken into consideration. A continuum seems

Table 3.1. *Choice of tutor in avian vocal development (song and call)*

Species	Tutor	Reference
Tree-creepers	Only live tutor	Thielcke 1970, 1984
Sedge wren	Only live tutor	Kroodsma & Verner 1978
Bullfinch	Canary foster father	Nicolai 1959
Zebra finch (lab.)	Own or Bengalese finch foster father	Immelmann 1969, Böhner 1983
Zebra finch (wild)	60% song of father	Zann 1990
	80% contact call of father	Zann 1993
Darwin's finches	Mostly father	Grant 1984; Millington & Price 1985; Gibbs 1990
Viduine finches	Estrildid host species	Nicolai 1964, 1973; Payne 1973a
Song sparrow	Adult rivals	Nice 1943
	Adult nearest neighbor	Beecher *et al.* 1994
Bewick's wren	Primarily nearest neighbor	Kroodsma 1974
Saddleback	Neighbors	Jenkins 1978
White-crowned sparrow	Neighbors at settling site	DeWolfe *et al.* 1989; D. A. Bell *et al*, unpublished data
Green violet-ear (hummingbird)	Unrelated neighbors	Gaunt *et al.* 1994, and unpublished data
Indian hill mynah	Own sex	Bertram 1970
Common starling	Own sex	Hausberger 1993
Various carduelines	Mate	Nicolai 1959; Mundinger 1979; Samson 1978; Groth 1993
Black-capped chickadee	Flock members	Mamman & Nowicki 1981
Yellow-rumped cacique	Colony members	Feekes 1982; Trainer 1987
Village weaverbird	Alpha male	Jacobs 1979
Village indigobird	Alpha male	Payne 1985

to exist in song-learning ability, with species that will not learn from tape models at one extreme (e.g., sedge wrens and tree-creeper species) to those that will vocalize good copies of taped sounds at the other end (e.g., swamp sparrow and chaffinches). Between these extremes are species that will accept some taped sounds as models, but learn better if allowed social interaction (e.g., indigo buntings and nightingales). We have also shown that the nature and mode of stimuli, be they social or other factors associated with vocal acquisition, are varied and can differ between species and/or experimental conditions.

Nelson (Chapter 2) has expressed some concerns regarding studies on social interaction and sensitive phases that are addressed below:

1. By commencing tutoring of white-crowned sparrows at 50 days of age, Nelson maintained that investigators were depriving the experimental birds and thus extending the sensitive phase. However, most birds tape tutored before 50 days produced good copies of tutor songs and no birds tape tutored after 50 days showed signs of learning, even though some birds in the latter group were

tutored for twice or three times as long (Table 3.2). Yet 17 of 26 live-tutored individuals produced their instructors' songs when exposed to them beginning at day 50. These data indicate that the live tutor is more effective than the tape beyond 50 days of age, i.e., the difference between live and tape tutoring is real.

But what exactly is the effect of live tutoring? Nelson has argued that live tutors did not instruct but only evoked stored songs. We feel that live tutors do both. In field playback studies, subsets of white-crowned sparrows and song sparrows responded to allospecific songs, indicating that allospecific songs were stored (Catchpole & Baptista 1988; Baptista & Catchpole 1989). Thus, although it has been suggested that emberizid sparrows may not learn alien songs, and that filtering is on the sensory mode (Mulligan 1966; Marler 1970; Konishi 1985), white-crowned sparrows stored but did not vocalize allospecific songs (Baptista 1990). However, white-crowned sparrows presented with live song sparrow tutors both stored and vocalized alien species' song (Baptista & Petrinovich 1986).

Table 3.2. *Summary of some experiments on tape- versus live-tutored male white-crowned sparrows*

Condition	N	Age tutored (days)	Total days tutored	No of birds[a]		Reference
				Yes	No	
Isolates	5	None	None	0	5	Baptista & Petrinovich 1986
Tape	1	8–28	20	1	0	Marler 1970
Tape	1	35–56	21	1	0	Marler 1970
Tape	5	11–113[b]	90	5	0	Petrinovich 1985
Tape	2	10–137[b]	20–102	2	0	Petrinovich 1985
Tape	1	37–76	39	0	1	Petrinovich 1985
Tape	2	50–71	21	0	2	Marler 1970
Tape	6	50–100	50	0	6	Baptista & Petrinovich 1986
Tape[c]	2	50–±200	±150	0	2	Baptista & Petrinovich 1986
Live	22	50–±200	±150	15	7	Baptista & Petrinovich 1986
Live[c]	4	50–±200	±150	2	2	Baptista & Petrinovich 1986

Notes:

N, number of individuals in each experiment.

[a] Positive (yes) response by birds signifies that the signal was not only stored but was produced when the model was presented to them at various ages.

[b] For variations between experiments, see details in Petrinovich 1985.

[c] Yoke-controlled experiment.

Similarly, although white-crowned sparrows may store taped conspecific songs presented to them after 50 days, it takes a live tutor to effect both storage and vocal performance. This finding is in keeping with Pepperberg's (Chapter 9) thesis that the tutor is both "motivator" as well as "modeller." The importance of social interaction as motivator increases with time (Table 3.2). It is noteworthy that one white-crowned sparrow, tape tutored as a yearling after onset of plastic song, modified its song (Nelson & Marler 1994, p. 10 500). Perhaps if a live tutor were used as instructor and motivator, more yearlings would have sung new songs.

2. Nelson is correct in saying that by tutoring white-crowned sparrows beginning at 50 days of age, investigators are presenting an incomplete picture of when song is actually acquired. However, the period of 50 days was chosen as the cut-off point because tape-tutoring beyond that age had proved ineffective in early experiments (Marler 1970), and it was believed that fledglings could not learn beyond that age (Cunningham & Baker 1983). Since fledglings dispersed from their parental territories on average at 50 days of age, it was believed that dispersing individuals brought with them the song or songs acquired at their birthplace. Consequently, a dialectal

area was regarded as equivalent to a deme. By showing that fledglings could learn beyond 50 days (Table 3.2; Petrinovich & Baptista 1987), investigators showed that fledglings could potentially learn new songs when settling, so songs sung on territory are not a reliable label of a bird's origin. Biochemical studies also show no genetic differences between dialectal populations (for reviews see Catchpole & Slater 1995; Brown & Farabaugh, Chapter 7). Thus, deme and dialectal area are not synonymous.

3. Nelson argued that by tape tutoring control birds aged between 50 and 90 days, Baptista & Petrinovich (1986) were not giving birds an opportunity to counter-sing a matching song with the tape stimulus, because tutoring at this age was well before the plastic song stage (which is after 200 days in the laboratory). By using live tutors until subjects had crystallized song, investigators were not demonstrating an extension of the sensitive phase, but rather the product of match counter-singing between tutor and pupil.

Nelson's observations of the onset of subsong are based on captive birds held in sound-proof boxes. Naive hand-raised birds maintained in open cages may begin subsong at between 16 and 22 days of age. One short-tailed fledgling, attended by its parents in the wild, counter-sang with the tape-recorded song using subsong. Wild fledg-lings may sing almost crystallized songs by 28 days of age (Baptista *et al.* 1993a, and see above) and may crystallize long before 90 days of age (DeWolfe *et al.* 1989; DeWolfe & Baptista 1995). By presenting subjects with songs between 50 and 90 days of age, Baptista & Petrinovich (1986) were giving subjects ample opportunity to counter-sing a matched song. Why did subjects not sing when presented with tapes?

Understanding the physiological state of birds may provide insight to this question. Konishi (1965) demon-strated the role of testosterone in advancing song development in Nuttall's white-crowned sparrows. Captive Gambel's white-crowned sparrows (*Z. l. gambelii*) have lower titers of testosterone than do wild individuals (Wingfield & Farner 1978; Wingfield & Moore 1987). Moreover, whereas wild Gambel's white-crowned spar-rows exhibit annual cycles of growth and regression of volumes of song centers in the brain, captive birds do not show these cycles unless treated with testosterone (Smith *et al.* 1996). The same results have recently been obtained with *Z. l. nuttalli* (E. A. Brenowitz *et al.*, unpublished data). If testosterone and neuron cycles are in some way involved with singing, then this may explain why labora-tory and field observations on song development differ. In the relatively sterile conditions of the laboratory, neither tape nor live tutor may evoke early onset of vocal produc-tion as observed in the wild, perhaps due to both neural substrate and/or hormonal titers being suboptimal. There may also be a higher stress level in captives, e.g., corticosterone levels are two to three times higher in cap-tives than in free-living birds (Marra *et al.* 1995).

4. Nelson argued that Baptista & Petrinovich (1986) showed effects of dosage (number of songs heard) rather than live tutoring, since control tape-tutored birds were not tutored until song had crystallized. Although samples are small, experiments testing effect of dosage have been conducted. A live tutor was placed in the middle chamber of a hexagonal test cage with outer chambers for pupils (illustration in Baptista & Petrinovich 1986, p. 1360). The tutor could potentially interact with four 50-day-old naive pupils across a screen; however, pupils could see, but not interact with, each other.

A directional microphone was suspended above the tutor such that each time he sang his voice tripped an elec-tronic switch that activated a tape-recorder. Each song sung caused one of this tutor's songs, previously recorded on tape, to be played to two naive, 50-day-old male white-crowned sparrows held singly in sound-proof chambers. In this yoke-controlled experiment, a tape rather than the actual vocalizations was used as a control because there is evidence that juveniles may cue in on the calls of adults, which stimulates the pupil to learn and sing the tutor's song (Thielcke 1970). The tape- and live-tutored subjects thus heard the same number of the same tutor's songs until their own songs were crystallized at ±200 days of age. No tape-tutored birds showed any signs of learning, whereas two out of four live-tutored birds produced excellent copies of tutors' songs. Thus, it was quality of the stimulus rather than number of songs that stimulated the live-tutored birds to store and produce the model presented by the live tutor.

We must thus conclude that the differential effects of live versus tape tutoring are real. Only the live tutor is an effective enough motivator to evoke use of song(s) stored by experimental birds after 50 days of age. But why, then, did a portion (35%, Table 3.2) of live-tutored white-crowned sparrows *not* sing new songs to which they were exposed after 50 days of age? Dominant white-crowned sparrows in captive flocks may interact with some sub-ordinates in a hierarchy but ignore others altogether (see Table 2 in Parsons & Baptista 1980). In retrospect, it is unfortunate that the hexagonal holding cage (Baptista & Petrinovich 1986; Petrinovich & Baptista 1987) was used in the experiments as the tutor might choose to interact with a pupil or ignore him altogether. If acquiring and singing new songs is a consequence of male–male inter-action (DeWolfe *et al.* 1989), then a pupil in a hexagon may not sing his tutor's song if he is ignored by him. This explanation may account for why two 100-day-old white-crowned sparrows did not sing new songs (Petrinovich & Baptista 1987). Perhaps more (or all) pupils would sing new songs if held singly in cages separated by a screen from a single tutor.

Vocal signals may be acquired by inheritance, learning or both (Baptista 1996). What then are the advantages of learning over inheritance of song? Learned signals differ from inherited signals in that learning is more imperfect in transmission from generation to generation, and cul-tural mutations may occur and become fixated in a population much faster than are genetic mutations (Jenkins 1978).

Cultural mutations give rise to variations that permit birds to distinguish between individuals by song (Miller 1979). Zann (1990) has postulated that this mechanism may function to enable birds to recognize siblings and thus avoid incest. Indeed, Grant (1984) found that females of the large cactus finch (*Geospiza conirostris*) never paired with a male singing her father's song (see Learning from fathers, p. 25). However, females of other geospizine species paired randomly with respect to song types (Millington & Price 1985).

Learning on the breeding site after dispersal from the natal area seems to be a widespread phenomenon (Mundinger 1982). A few studies indicate that song structure may be selected for maximum transmission to distant receivers in a particular habitat (Nottebohm 1985; Handford 1988; for reviews, see Baptista & Gaunt 1994; Tubaro & Segura 1994). The plasticity associated with the learning process enables young birds to acquire quality vocal signals when settling. The potential for rapid cultural mutation (Jenkins 1978) could also enable birds to colonize new habitats and in a "few generations" evolve songs that transmit best in those habitats. However, Date & Lemon (1993) played representative songs of American redstarts (*Setophaga ruticilla*) from three different habitats and found no evidence to support the hypothesis that some song types transmit better in some habitats than others.

Various authors (e.g. Payne 1982, 1983; Rothstein & Fleischer 1987) have postulated that vocalizations may function to identify a male and his quality, thus providing an honest signal. Young birds would then imitate these high quality signals. This has been demonstrated only once in the field in viduine finches (Payne & Payne 1977; Payne 1985) and once in the laboratory in village weaverbirds (Jacobs 1979).

Morton (1982 has postulated that young birds store an undegraded image of a neighbor's song, then use that image as a "standard" to measure degradation of that song over distance. If so, this behavior would enable the receiver to assess a neighbor's song and make appropriate responses, e.g., ignore, threaten, counter-sing with or attack the sender. This hypothesis is supported by some playback experiments (McGregor et al. 1983; McGregor & Falls 1984) that demonstrate greater reaction to undegraded than to degraded song. However, this response was only to familiar songs. Responses to degraded forms of an unfamiliar song were not different from responses to undegraded unfamiliar song (for a further critique of this ranging hypothesis see Catchpole & Slater 1995).

There appears to be yet another function of song sharing in lekking species. Song from a *Colibri* lek is nearly constant because males cooperate by (a) maintaining song when one or more males cease singing for other activities, and (b) singing in non interference when two or more birds are performing, i.e., each bird's songs alternates with his neighbor's in time. It is suggested (Gaunt et al. 1994) that this cooperative behavior, used to attract mates, is perceived from a distance as a single song. Thus, a signal can be detected at nearly all times in order to attract females that search for males sporadically and unpredictably throughout the year.

Although there is considerable evidence that birds prefer to learn conspecific songs over allospecific song, social interaction in some species has been demonstrated to override this preference. This behavior could have adaptive significance in local areas where feeding territories are limited and more than one species has similar requirements that result in intensified competition for resources. Experiments using playback of song show that species under these conditions will react to each other's song (for reviews, see Baptista 1990; Baptista & Gaunt 1994). Such interaction can occur between two completely unrelated species, e.g., Bewick's wren and song sparrow (Gorton 1977). The ability to store allospecific songs or acquire and vocalize those songs enable two species to recognize each other as competitors and reduce actual combat that can be energetically costly, injurious and/or fatal.

The above discussion indicates that storing and producing are two independent but important aspects of song learning. Distinguishing between the processes of storing/recall and learning anew is not always easy. For example a white-crowned sparrow in the laboratory may discard an early learned song for one it was exposed to after 50 days (Petrinovich & Baptista 1987). However, wild-trapped and laboratory-tutored juveniles may sing more than one song type during practice sessions (Baptista & Morton 1988; Nelson & Marler 1994). As proposed for saddlebacks (Jenkins 1978), fledgling sparrows may wander over large areas and store all song patterns encountered. Matched counter-singing when settling selectively evokes use of certain themes and discards others. Cassidy's

(1993) study on song sparrows is noteworthy because it points out that the existence of extended sensitive phases is still an open question.

Just as there are multiple mechanisms for sound production in birds due to differences in syringeal structure, function and control (Gaunt & Gaunt 1985), there are different patterns of vocal acquisition between bird species due to different selective pressures in a plethora of life history strategies and ecological settings. Thus, caution is needed when interpreting laboratory results, and, whenever possible, field studies should be pursued to verify conclusions derived from such studies. Developmental studies reviewed here have explored mainly questions of how vocalizations are acquired, and it is to be expected that new modes will be revealed. However, we look to the future for studies elucidating questions of why variation in modes of vocal tradition or relative contribution of learning and inheritance have been selected in various species.

ACKNOWLEDGEMENTS

We thank Martine Hausberger and Hansrudi Güttinger for inviting L. Baptista to speak at the Jacques Monod Conference at Aussois in 1992. This review is an outgrowth from that lecture and the interaction with the conference participants. Subsequently Martine Hausberger and Charles Snowdon invited us to prepare this chapter.

Barbara B. DeWolfe, Robert E. Lemon, Eugene S. Morton, Robert B. Payne, and Richard Zann read earlier versions of this manuscript and offered helpful comments. Douglas A. Bell prepared Fig. 3.1 and Kathleen Berge typed the bibliography and assisted in editing and updating various versions of the paper. Research by Baptista on song acquisition in white-crowned sparrows was supported by funding from NSF (BNS 8919453).

REFERENCES

Adret, P. (1993a). Operant conditioning, song learning and imprinting to taped song in the zebra finch. *Animal Behaviour* **46**: 149–59.

Adret, P. (1993b). Vocal learning induced with operant techniques: an overview. *Netherland Journal of Zoology* **43**: 125–42.

Adret-Hausberger, M., Güttinger, H.-R. & Merkel, F. W. (1990). Individual life history and song repertoire changes in a colony of starlings (*Sturnus vulgaris*). *Ethology* **84**: 265–80.

Arcese, P., Stoddard, P. K. & Hiebert, S. M. (1988). The form and function of song in female song sparrows. *Condor* **90**: 44–50.

Baptista, L. F. (1975). Song dialects and demes in sedentary populations of the white-crowned sparrow (*Zonotrichia leucophrys nuttalli*). *University of California Publications in Zoology* **105**: 1–52.

Baptista, L. F. (1985). The functional significance of song sharing in the white-crowned sparrow. *Canadian Journal of Zoology* **63**: 1741–52.

Baptista, L. F. (1990). Song learning in the white-crowned sparrow (*Zonotrichia leucophrys*): Sensitive phases and stimulus filtering revisited. *Current Topics in Avian Biology*, International Centennial Meeting of the Deutsche Ornithologen-Gesellschaft, Bonn, pp. 143–52. Verlag DO-G, Bonn.

Baptista, L. F. (1993). El estudio de la variacion geografica usando vocalizaciones y las bibliotecas de sonidos de aves neotropicales. In *Curación moderna de colecciones ornitológicas*, ed. P. Escalante-Pliego, Proceedings of IV International Congress of Neotropical Birds (Quito, Ecuador, 1991), pp. 15–30. American Ornithological Union, Washington, DC.

Baptista, L. F. (1996). Nature and its nurturing in avian vocal development. In *Ecology and Evolution of Acoustic Communication in Birds*, ed. D. E. Kroodsma, & E. H. Miller, pp. 39–60. Cornell University Press, Ithaca, NY.

Baptista, L. F., Bell, D. A. & Trail, P. W. (1993a). Song learning and production in the white-crowned sparrows: Parallels with sexual imprinting. *Netherlands Journal of Zoology* **43**: 17–33.

Baptista, L. F. & Catchpole, C. K. (1989). Vocal mimicry and interspecific aggression in songbirds: Experiments using white-crowned sparrow imitation of song sparrow song. *Behaviour* **109**: 247–57.

Baptista, L. F. & Gaunt, S. L. L. (1994). Advances in studies of avian sound communication. *Condor* **96**: 817–30.

Baptista, L. F. & Morton, M. L. (1988). Song learning in montane white-crowned sparrows: From whom and when. *Animal Behaviour* **36**: 1753–64.

Baptista, L. F. & Petrinovich, L. (1984). Social interaction, sensitive phases and the song template hypothesis in the white-crowned sparrow. *Animal Behaviour* **32**: 172–81.

Baptista, L. F. & Petrinovich, L. (1986). Song development in the white-crowned sparrow: Social factors and sex differences. *Animal Behaviour* **34**: 1359–71.

Baptista, L. F. & Schuchmann, K.-L. (1990). Song learning in the anna hummingbird (*Calypte anna*). *Ethology* **84**: 15–26.

Baptista, L. F., Trail, P. W., DeWolfe, B. B. & Morton, M. L. (1993b). Singing and its functions in female white-crowned sparrows, *Zonotrichia leucophrys*. *Animal Behaviour* **46**: 511–24.

Becker, P. (1982). The coding of species-specific characteristics in bird sounds. In *Acoustic Communication with Birds*, vol. 1. ed. D. E. Kroodsma & E. H. Miller, pp. 214–52. Academic Press, New York.

Beecher, M. D., Campbell, S. E. & Stoddard, P. K. (1994). Correlation of song learning and territory establishment strategies in the song sparrow. *Proceedings of the National Academy of Sciences, USA* **91**: 1450–4.

Bertram, B. (1970). The vocal behaviour of the Indian hill mynah, *Gracula religiosa*. *Animal Behaviour Monographs* **3**(2): 79–192.

Bitterbaum, E. & Baptista, L. F. (1979). Geographical variation in songs of California house finches (*Carpodacus mexicanus*). *Auk* **96**: 462–74.

Böhner, J. (1983). Song learning in the zebra finch (*Taeniopygia guttata*): Selectivity in choice of a tutor and accuracy of song copies. *Animal Behaviour* **31**: 231–7.

Böhner, J. (1990). Early acquisition of song in the zebra finch (*Taeniopygia guttata*). *Animal Behaviour* **39**: 369–74.

Bowman, R. I. (1983). The evolution of song in Darwin's finches. In *Patterns of Evolution in Galapagos Organisms*, ed. R. I. Bowman, M. Berson & A. E. Leviton, pp. 237–537. American Association for the Advancement of Science, San Francisco.

Cassidy, A. L. (1993). Song variation and learning in island populations of song sparrows. Ph.D. dissertation, University of British Columbia.

Casey, R. M. & Baker, M. C. (1993). Social tutoring of adult male white-crowned sparrows. *Condor* **95**: 718–23.

Catchpole, C. K. & Baptista, L. F. (1988). A test of the competition hypothesis of vocal mimicry using song sparrow imitation of white-crowned sparrow song. *Behaviour* **106**: 119–28.

Catchpole, C. K. & Slater, P. J. B. (1995). *Bird Song: Biological Themes and Variations*. Cambridge University Press, Cambridge.

Cavalli-Sforza, L. L., Feldman, M. W., Chen, K. H. & Dornbusch, S. M. (1982). Theory and observation in cultural transmission. *Science* **218**: 19–27.

Chaiken, M., Böhner, J. & Marler, P. (1993). Song acquisition in European starlings (*Sturnus vulgaris*): A comparison of the songs of live-tutored, tape-tutored, untutored and wild caught males. *Animal Behaviour* **46**: 1079–90.

Cunningham, M. A. & Baker, M. C. (1983). Vocal learning in white-crowned sparrows: Sensitive phase and song dialects. *Behavioural Ecology* **13**: 259–69.

Cunningham, M. A., Baker, M. C. & Boardman, T. J. (1987). Microgeographic song variation in the Nuttall's white-crowned sparrow. *Condor* **88**: 261–75.

Date, E. M. & Lemon, R. E. (1993). Sound transmission: a basis for dialects in birdsongs. *Behaviour* **124**: 291–311.

DeWolfe, B. B. & Baptista, L. F. (1995). Singing behavior, song types on their wintering grounds and the question of leap-frog migration in Puget Sound white-crowned sparrows (*Zonotrichia leucophrys pugetensis*). *Condor* **97**: 376–89.

DeWolfe, B. B., Baptista, L. F. & Petrinovich, L. (1989). Song development and territory establishment in Nuttall's white-crowned sparrows. *Condor* **91**: 397–407.

Dietrich, K. (1980). Vorbildwahl in der Gesangsentwicklung beim Japanischen Mövchen (*Lonchura striata* var. *domestica*, Estrildidae). *Zeitschrift für Tierpsychologie* **52**: 57–76.

Feekes, F. (1982). Song mimesis within colonies of *Cacicus c. cela* (Icteridae, Aves). A colonial password? *Zeitschrift für Tierpsychologie* **58**: 119–52.

Gaunt, A. S. & Gaunt, S. L. L. (1985). Syringeal function and avian phonation. In *Current Ornithology*, vol. 12, ed. R. F. Johnston, pp. 213–46. Plenum Press, New York.

Gaunt, S. L. L., Baptista, L. F., Sánchez, J. E. & Hernandez, D. (1994). Song learning as evidenced from song sharing in two hummingbird species (*Colibri coruscans* and *C. thalassinus*). *Auk* **11**: 87–103.

Gibbs, H. L. (1990). Cultural evolution of male song types in Darwin's medium ground-finches (*Geospiza fortis*). *Animal Behaviour* **39**: 253–63.

Gibbs, H. L., Grant, P. R. & Weiland, J. (1984). Breeding of Darwin's finches at an exceptionally early age in an El Niño year. *Auk* **101**: 872–4.

Gorton, R. E. Jr (1977). Territorial interactions in sympatric song sparrows and Bewick's wren populations. *Auk* **94**: 701–8.

Grant, B. R. (1984). The significance of song variation in a population of Darwin's finches. *Behaviour* **89**: 90–116.

Groth, J. G. (1993). Call matching and positive assortative mating in red crossbills. *Auk* **110**: 398–401.

Handford, P. (1988). Trill rate dialects in the rufous-collared sparrow, *Zonotrichia capensis*, in northwestern Argentina. *Canadian Journal of Zoology* **63**: 2383–8.

Hausberger, M. (1993). How studies on vocal communication in birds contribute to a comparative approach of cognition. *Etología* **3**: 171–85.

Hultsch, H. (1993). Tracing the memory mechanisms in the song acquisition of nightingales. *Netherlands Journal of Zoology* **43**: 155–71.

Hultsch, H. & Todt, D. (1992). The serial order effect in the song acquisition of birds: Relevance of exposure frequency to models. *Animal Behaviour* **44**: 590–2.

Immelmann, K. (1969). Song development in the zebra finch and other estrildid finches. In *Bird Vocalizations*, ed. R. A. Hinde, pp. 61–74. Cambridge University Press, Cambridge.

Immelmann, K. (1982). *Australian Finches*. Robertson Publishers, Sydney.

Jacobs, C. H. (1979). Factors involved in mate selection and vocal development in the village weaverbirds (*Ploceus cucullatus*). Ph.D. dissertation, University of California at Los Angeles.

Jenkins, P. F. (1978). Cultural transmission of song patterns and dialect development in a free-living bird population. *Animal Behaviour* **26**: 50–78.

Konishi, M. (1964). Song variation in a population of Oregon juncos. *Condor* **66**: 423–36.

Konishi, M. (1965). The role of auditory feedback in the control of vocalization in the white-crowned sparrow. *Zeitschrift für Tierpsychologie* **22**: 770–83.

Konishi, M. (1985). Bird song: from behavior to neuron. *Annual Review of Neuroscience* **8**: 125–70.

Kroodsma, D. E. (1974). Song learning, dialects and dispersal in the Bewick's wren. *Zeitschrift für Tierpsychologie* **35**: 352–80.

Kroodsma, D. E. (1979). Vocal dueling among marsh wrens: Evidence for ritualized expressions of dominance/subordinance. *Auk* **96**: 506–15.

Kroodsma, D. E. & Pickert, R. (1984). Sensitive phases for song learning: Effects of social interaction and individual variation. *Animal Behaviour* **32**: 389–94.

Kroodsma, D. E. & Verner, J. (1978). Complex singing behaviors among *Cistothorus* wrens. *Auk* **95**: 703–16.

Mammen, D. L. & Nowicki, S. (1981). Individual and within-flock convergence in chickadee calls. *Behavioural Ecology and Sociobiology* **9**: 179–86.

Mann, N. I., Slater, P. J. B., Eales, L. A. & Richards, C. (1991). The influence of visual stimuli on song tutor choice in the zebra finch *Taeniopygia guttata*. *Animal Behaviour* **42**: 285–93.

Marler, P. (1967). Comparative study of song development in emberizine finches. *Proceedings of the 14th International Ornithological Congress* (1966): 231–44.

Marler, P. (1970). A comparative approach to vocal learning: Song development in white-crowned sparrows. *Journal of Comparative and Physiological Psychology* **71**: 1–25.

Marler, P. & Isaac, D. (1961). Song variation in a population of Mexican juncos. *Wilson Bulletin* **73**: 193–206.

Marler, P., Kreith, M. & Tamura, M. (1962). Song development in hand-raised Oregon juncos. *Auk* **79**: 12–30.

Marler, P. & Peters, S. (1977). Seletive vocal learning in a sparrow. *Science* **198**: 519–21.

Marler, P. & Peters, S. (1982). Long-term storage of learned birdsongs prior to production. *Animal Behaviour* **30**: 479–82.

Marler, P. & Peters, S. (1987). A sensitive period for song acquisition in the song sparrow, *Melospiza melodia*: A case of age-limited learning. *Ethology* **76**: 89–100.

Marra, P. P., Lampe, K. T. & Tedford, B. L. (1995). Plasma corticosterone levels in two species of *Zonotrichia* sparrows under captive and free-living conditions. *Wilson Bulletin* **107**: 296–305.

McGregor, P. K. & Falls, J. B. (1984). The response of western meadowlarks (*Surnella neglecta*) to playback of undegraded and degraded songs. *Canadian Journal of Zoology* **62**: 2125–8.

McGregor, P. K., Krebs, J. R. & Ratcliffe, L. M. (1983). The reaction of great tits (*Parus major*) to playback of degraded and undegraded song type. *Auk* **100**: 898–906.

Miller, D. B. (1979). Long-term recognition of father's song by female zebra finches. *Nature* **280**: 389–91.

Millington, S. F. & Price, T. D. (1985). Song inheritance and mating patterns in Darwin's finches. *Auk* **102**: 342–6.

Morton, E. S. (1982). Grading, discreteness, redundancy, and motivation-structural rules. In *Acoustic Communication in Birds*, vol. 1, ed. D. E. Kroodsma & E. E. Miller, pp. 183–212. Academic Press, New York.

Morton, E. S., Gish, S. L. & van der Voort, M. (1986). On the learning of degraded and undegraded songs in the Carolina wren. *Animal Behaviour* **34**: 815–20.

Mulligan, J. A. (1966). Singing behavior and its development in the song sparrow, *Melospiza melodia*. *University of California Publications in Zoology* **81**: 1–76.

Mundinger, P. (1970). Vocal imitation and individual recognition of finch calls. *Science* **168**, 480–2.

Mundinger, P. (1975). Song dialects and colonization in the house finch, *Carpodacus mexicanus*, on the east coast. *Condor* **77**: 407–22.

Mundinger, P. (1979). Call learning in the Carduelinae: Ethological and systematic considerations. *Systematic Zoology* **28**: 270–83.

Mundinger, P. (1982). Microgeographic and macrogeographic variation in the acquired vocalizations of birds. In *Acoustic Communication in Birds*, ed. D. E. Kroodsma & E. H. Miller vol. 2., pp. 147–208. Academic Press, New York.

Nelson, D. A. & Marler, P. (1994). Selection-based learning in bird song development. *Proceedings of the National Academy of Sciences, USA* **91**: 10498–501.

Nice, M. M. (1943). Studies in the life history of the song sparrow II. The behavior of the song sparrow and other passerines. *Transactions of the Linnaean Society* **6**: 1–238.

Nicolai, J. (1959). Familientradition in der Gesangsentwicklung des Gimpels (*Pyrrhula pyrrhula* L). *Journal für Ornithologie* **100**: 39–46.

Nicolai, J. (1964). Brutparasitismus der Viduinae als ethologisches Problem. *Zeitschrift für Tierpsychologie* **21**: 129–204.

Nicolai, J. (1973). Das Lernprogramm in der Gesangsausbildung der Strohwitwe (*Tetraenura fischeri* Reichenow). *Zeitschrift für Tierpsychologie* **32**: 113–38.

Nicolai, J. (1986). The effect of age on song-learning ability in two passerines. *Acta XIX Congressus Internationalis Ornithologici, Ottawa* pp. 1098–105.

Nowicki, S. (1983). Flock-specific recognition of chickadee calls. *Behavioural Ecology and Sociobiology* **12**: 317–20.

Nottebohm, F. (1969). The song of the chingolo, *Zonotrichia capensis*, in Argentina: Description and evaluation of a system of dialects. *Condor* **71**: 299–315.

Nottebohm, F. (1970). Ontogeny of bird song. *Science* **167**: 950–6.

Nottebohm, F. (1985). Sound transmission, signal salience and song dialects. *Behavioural and Brain Science* **8**: 112–13.

Nottebohm, F. (1993). The search for neural mechanisms that define the sensitive period for song learning in birds. *Netherlands Journal of Zoology* **43**: 193–234.

O'Loghlen, A. L. & Rothstein, R. I. (1993). An extreme example of delayed vocal development: Song learning in a population of wild brown-headed cowbirds. *Animal Behaviour* **46**: 293–304.

Parsons, J. & Baptista, L. F. (1980). Crown color and dominance in the white-crowned sparrow. *Auk* **97**: 807–15.

Payne, R. B. (1973a). Behavior, mimetic songs and song dialects, and relationships of the parasitic indigobirds (*Vidua*) of Africa. *Ornithological Monographs* **11**.

Payne, R. B. (1973b). Vocal mimicry of paradise whydahs (*Vidua*) and response of male whydahs to song of their hosts (*Pytilia*) and their mimics. *Animal Behaviour* **21**: 762–71.

Payne, R. B. (1981). Song learning and social interaction in indigo buntings. *Animal Behaviour* **29**: 688–97.

Payne, R. B. (1982). Ecological consequences of song matching: Breeding success and intraspecific song mimicry in indigo buntings. *Ecology* **63**: 401–11.

Payne, R. B. (1983). The social context of song mimicry: Song-matching dialects in indigo buntings (*Passerina cyanea*). *Animal Behaviour* **31**: 788–805.

Payne, R. B. (1985). Behavioral continuity and change in local song populations of village indigobirds *Vidua chalybeata*. *Zeitschrift für Tierpsychologie* **70**: 1–44.

Payne, R. B. (1990). Song mimicry by the village indigobird (*Vidua chalybeata*) of the red-billed firefinch (*Lagonosticta senegala*). *Die Vogelwarte* **35**: 321–8.

Payne, R. B. & Payne, K. (1977). Social organization and mating success in local song populations of village indigobirds (*Vidua chalybeata*). *Zeitschrift für Tierpsychologie* **45**: 113–73.

Payne, R. B., Payne, L. L. & Doehlert, S. M. (1988). Biological and cultural success of song memes in indigo buntings. *Ecology* **69**: 104–17.

Pepperberg, I. M. (1993). A review of the effects of social interaction on vocal learning in African grey parrots (*Psittacus erithacus*). *Netherlands Journal of Zoology* **43**: 104–24.

Peters, S., Marler, P. & Nowicki, S. (1992). Song sparrows learn from limited exposure to song models. *Condor* **94**: 1016–19.

Petrinovich, L. (1985). Factors influencing song development in the white-crowned sparrow (*Zonotrichia leucophrys*). *Journal of Comparative and Physiological Psychology* **99**: 15–29.

Petrinovich, L. & Baptista, L. F. (1987). Song development in the white-crowned sparrow: Modification of learned song. *Animal Behaviour* **35**: 961–74.

Price, P. H. (1979). Developmental determinants of structure in zebra finch songs. *Journal of Comparative and Physiological Psychology* **93**: 260–77.

Rice, J. O. & Thompson, W. L. (1968). Song development in the indigo bunting. *Animal Behaviour* **16**: 462–9.

Rost, R. (1987). Entstehung, Fortbestand und funktionelle Bedeutung von Gesangsdialekten bei der Sumpfmeise *Parus palustris* – Ein Test von Modellen. Ph.D. thesis, University of Konstanz.

Rothstein, S. I. & Fleischer, R. C. (1987). Vocal dialects and their possible relation to honest status signalling in the brown-headed cowbird. *Condor* **89**: 1–23.

Samson, F. B. (1978). Vocalizations of Cassin's finch in northern Utah. *Condor* **80**: 203–10.

Slater, P. J. B., Eales, L. A. & Clayton, N. S. (1988). Song learning in zebra finches (*Taeniopygia guttata*): Progress and prospects. *Advances in the Study of Behavior* **18**: 1–34.

Slater, P. J. B., Jones, A. & ten Cate, C. (1993). Can lack of experience delay the end of the sensitive phase for song learning? *Netherlands Journal of Zoology* **43**: 80–90.

Smith, G. T., Brenowitz, E. A., Wingfield, J. C. & Baptista, L. F. (1996). Seasonal changes in song nuclei and song behavior in Gambel's white-crowned sparrows. *Journal of Neurobiology* **28**: 114–25.

ten Cate, C. (1991). Behaviour-contingent exposure to taped song and zebra finch song learning. *Animal Behaviour* **42**: 857–9.

Thielcke, G. (1965). Die Ontogenese der Bettellaute von Garten – und Waldbaumläufer (*Certhia brachydactyla* Brehm and *C. familiaris* L.). *Zoologische Anzeiger* **174**: 237–41.

Thielcke, G. (1970). Lernen von Gesang als möglicher Schrittmacher der Evolution. Z. Zool. *Systematick Evolutionsforschung* 8: 309–20.

Thielcke, G. (1984). Gesangslernen beim Gartenbaumläufer (*Certhia brachydactyla*). *Die Vogelwarte* 32: 282–97.

Thielcke, G. (1988). Neue Befunde bestätigen Baron Pernaus (1660–1731). Angaben über Lautäusserungen des Buchfinken (*Fringilla coelebs*). *Journal für Ornithologie* **129**: 55–70.

Thielcke, G. & Krome, M. (1989). Expermente über sensible Phasen und Gesangsvariabilität beim Buchfinken (*Fringilla coelebs*). *Journal für Ornithologie* **130**: 435–53.

Thompson, W. L. (1970). Song variation in a population of indigo buntings. *Auk* **87**: 58–71.

Thorpe, W. H. (1958). The learning of song patterns by birds, with especial reference to the song of the chaffinch, *Fringilla coelebs. Ibis* **100**: 535–70.

Todt, D., Hultsch, H. & Heike, D. (1979). Conditions affecting song acquisition in nightingales (*Luscinia megarhynchos*). *Zeitschrift für Tierpsychologie* **51**: 23–35.

Trainer, J. M. (1987). Behavioral associations of song types during aggressive interactions among male yellow-rumped caciques. *Condor* **89**: 731–8.

Tubaro, P. L. & Segura, E. T. (1994). Dialect differences in the song of *Zonotrichia capensis* in the southern pampas: A test of the acoustic adaptation hypothesis *Condor* **96**: 1084–7.

Verner, J. (1975). Complex song repertoire of male long-billed marsh wrens in eastern Washington. *Living Bird* **14**: 263–300.

Waser, M. S. & Marler, P. (1977). Song learning in canaries. *Journal of Comparative and Physiological Psychology* **91**: 1–7.

West, M. & King, A. P. (1985). Social guidance of vocal learning by female cowbirds: Validating its functional significance. *Zeitschrift für Tierpsychologie* **70**: 225–35.

West, M. & King, A. P. (1988). Female visual display affect the development of male song in the cowbird. *Nature* **334**: 244–6.

West, M. J., Stroud, N. & King, A. P. (1983). Mimicry of the human voice by European starlings: The role of social interactions. *Wilson Bulletin* **95**: 635–40.

Wingfield, J. C. & Farner, D. S. (1978). The annual cycle of plasma irLH and steroid hormones in feral populations of the white-crowned sparrow, *Zonotrichia leucophrys gambelii*. *Biology of Reproduction* **19**: 1046–56.

Wingfield, J. C. & Moore, M. C. (1987). Hormonal, social and environmental factors in the reproductive biology of free-living male birds. In *Psychobiology of Reproductive Behavior: An Evolutionary Perspective*, ed. D. Crews, pp. 149–75. Prentice-Hall, Englewood Cliffs, NJ.

Yasukawa, K., Blank, J. L. & Patterson, C. B. (1980). Song repertoires and sexual selection in the red-winged blackbird. *Behavioural Ecology and Sociobiology* **7**: 233–8.

Zann, R. (1990). Song and call learning in wild zebra finches in south-east Australia. *Animal Behaviour* **40**: 811–28.

Zann, R. (1993). Variation in song structure within and among populations of Australian zebra finches. *Auk* **110**: 716–26.

4 Building a social agenda for the study of bird song

MEREDITH J. WEST, ANDREW P. KING, AND TODD M. FREEBERG

INTRODUCTION

It might seem odd to suggest that the study of bird song needs a social agenda. Isn't it obvious that birds sing to attract or to repel one another? Isn't it now clear that birds need social experience to develop species-typical repertoires? Although the answer to these questions is "yes," we are convinced that only the surface structure of social influence has been uncovered – the deep structure remains to be explored. The purpose of the chapter is to defend this position by examining some of our own efforts to study social influences. We begin with an account of some of the formative experiences that shaped the directions of our research. We follow with some of the historical themes that guided us and others studying songbirds. Then, we describe some of our most recent efforts to examine new themes relevant to avian communication, themes quite familiar to those studying primates. In particular, we focus on the difference between communicative form and communicative competence. We wish to draw a greater distinction between the processes involved in developing a potentially communicative signal and the processes involved in learning how to use those signals effectively (Seyfarth & Cheney 1986 and see Chapter 13; Snowdon *et al.*, Chapter 12). We conclude that analyses restricted only to the structural nature of vocal signals are inadequate to capture the developmental processes leading to vocal communication. We must go beyond studies of songs and focus on the singers, listeners, and the contexts framing communication.

FLASHBULB MEMORIES: A BIOASSAY OF SONG, A BIOASSAY OF SINGERS

In May of 1973, we witnessed an event that led to a series of studies spanning the next two decades. We had hand raised our first brown-headed cowbird (*Molothrus ater ater*) to adulthood after many frustrating failures. That she was a female was another disappointment because we were interested in vocal development, and female cowbirds do not sing. This female, however, more than redeemed herself for her lack of vocal qualities. When we played her a tape-recording of a male cowbird's song, she responded with a copulation solicitation display (King & West 1977). It seemed an unambiguous answer to the question "Do female cowbirds, in the absence of conspecific experience, discriminate the song of their species?"

Since that day, we have performed playback experiments each year, accumulating data on over 250 different females, from eight populations, listening to the songs of over 200 different males. These experiments have been reported in a series of publications and chapters (West & King 1987; King & West 1990). Some might say we have found a methodological groove – some might consider it a rut – but it certainly became a methodological habit, one that had also been adapted by others seeking to study the functional properties of avian song (for a comprehensive review, see Searcy 1992). The logic guiding us was that we could use the females' preferences for song as a window to view the developmental variables necessary for males to develop an effective song. The fundamental assumption was that the link between perception and production would be the same in our playback procedure as in nature.

In the summer of 1992, we witnessed events that caused us to alter our view of the female playback procedure and the data obtained by using it. The observations occurred during a study of captive male cowbirds (*M. a. artemisiae*) from an ancestral part of the cowbird's extensive range, South Dakota (SD) (Freeberg *et al.* 1995) in which we combined playback measures and extensive measures of courtship. We had captured the males from juvenile flocks and taken them to the laboratory where they were housed individually with canaries (*Serinus canaria*) or female cowbirds from the same site. The housing design was almost as familiar to us as the playback protocol – we were raising the male cowbirds with female cowbirds or nonconspecific companions to explore the role of social and acoustic experience during ontogeny (King & West 1990). We then intended to play back the males' songs to SD females to relate male courtship success to the potency of the male's vocalizations. Thus, we sought to combine two familiar ways of assessing vocal development in controlled, social environments, in ways somewhat similar to ten Cate's work with sexual imprinting (ten Cate *et al.* 1993).

The events of the summer of 1992 led us to reconsider our assumptions about the linkages between production and perception. First, in a large flight cage, then in a large aviary, we watched the SD male cowbirds that had been housed with canaries chase and sing to canaries. Had the males been displaying these behaviors to female conspecifics we would have labelled it "courtship." But, in both settings, female cowbirds in breeding condition were present and were ignored by the canary-housed males. The males' persistence was striking. Even when the canaries retreated, and even when the female cowbirds approached and solicited with copulatory postures, the males directed their signals to the canaries. The males' behaviors seemed all the more surprising as they were occurring during the breeding season, at a time of year when misguided courtship might mean that a male leaves no offspring at all, given the short lifespan for the species (Lowther 1993). Even more surprising was the behavior (or lack thereof) of the SD males that had been housed with conspecific females. Although they did vocalize to female conspecifics in the flight cage tests, they showed serious deficiencies in their abilities to court female cowbirds in the aviaries. They did not pursue canaries, but they failed to direct their vocalizations to females, spending the majority of the breeding season directing vocalizations to each other and to the adult males (see Table 4.1 below).

EXPERIMENTERS AS SOCIAL ENGINEERS

We did not know what to call the behaviors we were witnessing (courtship seemed inappropriate for the canary-housed males as there was no reciprocation from the canaries), but we knew we could no longer rely on the assumption implicit in past studies: we assumed that playback females could recognize effective songs, but we had also assumed that males could recognize when to sing such songs. We were wrong. When we had carried out playback work, using a tape-recorder as a substitute for a male, it had never occurred to us that we were supplying vocal expertise for the male whose song we had recorded, i.e., knowing to whom to direct vocal signals. The SD males' failures seemed to be at so fundamental a level as to unravel any notion of a vocal safety net for the species, thereby forcing us to re-evaluate many years of thinking about the primacy of vocal signals in the species' mate recognition system.

We suspect that one of the reasons we were originally so taken by the captive female's response to song was that the idea of intrinsic connections between song perception and song production fit our preconceptions of what one ought to see when studying the components of a mating system, i.e., an inevitable coupling of signal and response. The familiar diagrams of stickleback courtship with arrows connecting the behaviors of each sex stood as visual testimony to such an implicit view (Morris 1970). Such thinking also permeated classical ethology in which the assumption was made by Lorenz and others that, even if species-typical units or components of a behavioral system were learned, the overall program for many adaptive behavior sequences such as mating were typically hard wired. Lorenz (1965, p. 35) had stated that "we cannot ever hope to understand any process of learning before we had grasped the hereditary teaching machine which controls the primary programming". Although Lorenz endorsed the use of isolation rearing only for quite simple behaviors, he did identify a special use for isolation rearing in songbirds. He speculated that those vocal signals that elicited "specific responses from conspecifics were usually innate" (Lorenz 1965, p. 76). Thus, the female cowbird's reflexive response to male song would seem to exemplify the "hereditary teaching machine" at work.

We also attribute our error in reasoning to an intuitive phase of scientific thinking in general. In the same way that naming a behavior does not explain its function (the nominal fallacy), identifying a signal in one animal in one context and a response in another context does not necessarily explain how actions become linked to consequences (a "connectionist fallacy"). All scientists have the need for a connectionist attitude at times to link what they are seeing and what they are trying to find. What we wanted to find was a compensatory mechanism of species identification. The observations of the SD males, however, revealed that the "programming" in male cowbirds could be altered by environmental sources in ways that penetrated to the "primary" level itself. That the process of linking species-typical sequences of social behaviors also involved learning was something researchers studying primates were already addressing, but not those studying song systems in passerines (Kaufman 1975).

Our mistake then was not in trying to see the connections but in overestimating the meaning of the connections we had found. We, the experimenters, were engineering links by our actions and technology, links we assumed were second nature to the subjects themselves. We chose the song, the time, and the recipient. The new data suggested that if we left those variables to the males themselves, our conclusions would differ. Thus, what we realized that summer was that the time had come to return social control to the subjects, to assess outcomes, and to study the actual connections.

Before describing more details of the experiments on cowbirds and canaries in the summer of 1992, we want to articulate some of the historical reasons that led us to abandon our role as unwitting behavioral engineers, building bridges between males and females. We believe that the arguments are significant for the broader, comparative themes of this volume because we originally conceived of the experiment carried out in 1992 as a result of long-held memories of studies of social deprivation in primates.

MOTHERLESS MONKEYS, COMPANION-LESS SONGBIRDS

Systematic studies of the ontogeny of vocal behavior in songbirds by ethologists began at about the same time that experimental psychologists were examining the development of social behavior in several primate species, but especially rhesus monkeys (*Macaca mulatta*) (Mason, 1960; Harlow & Harlow 1962; Harlow *et al.* 1963). These studies revealed the clear and often irreversible deficits in the social skills of monkeys raised without conspecifics. The abnormalities were severe enough to lead some to use the animals as models of human psychopathology (Harlow 1974). The degree to which some investigators went in

attempting to strip away any semblance of a social atmosphere (raising monkeys in a small vertical chamber for one year with no physical access to other animals) strained the concept of an animal model beyond credible limits for many scientists. Thus, it always struck us as surprising that those studying avian song development, most of whom were ethologists by training, accepted the conditions of isolation as a viable context for the study of vocal communication (Thorpe 1958, 1961). Were birds considered cognitively or socially less complex than primates and therefore less likely to suffer social deficiencies? Were they innately resistant to such social perturbations? Given that the songbirds' housing conditions were sometimes no more inviting than the monkeys' chambers, it was hard to understand. Unfortunately, we do not know what reasoning went on because we could find no reports on the social traits of isolated songbirds (but see Dufty & Wingfield 1986). We have found abundant information on the nature of the vocalizations of isolated birds but little about the birds themselves. Were the singers as abnormal as their songs?

Most avian song researchers would probably answer "yes" to such a question (based on our informal discussions over many years with those raising songbirds in isolation and the general statements of writers such as Lorenz (1965)). The simplest reason why many did not attempt to study the subsequent social behavior of naive songbirds was that there did not seem to be much to study from the point of view of understanding communication. Like the motherless monkeys, hand-reared, companion-less songbirds exhibited fear, stereotypy, and flight reactions. But, what probably diverted investigators even more was that the patterns of the isolated birds' vocalizations were the true targets of inquiry, as it was assumed that the acoustic structure held an important ontogenetic answer: what is the innate basis of song ontogeny? After the advent of the sound spectrograph and the tape-recorder, it had become possible to preserve vocal "morphology" in much the same way that curators collected "skins" for taxonomic comparisons (Thorpe 1958; Baptista & Gaunt 1994). Thus, it was the state of the isolated bird's song, not the state of the bird itself, that commanded the investigators' attention.

This was not the case for the research on isolated primates, where the emotional condition of the animals came to dominate. Ironically, both fields' initial interest in creating isolates grew out of a desire to study learning. Both groups of researchers wanted animals with controlled experimental histories to study the processes by which experience created changes. The more pronounced altricial state of young primates, a condition calling for prolonged human intervention (and invention, e.g., the surrogate mother), undoubtedly fostered the investigators' shift in interests to "affectional systems" (Harlow *et al.* 1963). The interest in learning was not lost, but became less visible.

With respect to songbirds, the interest in learning grew. Isolated birds served as a control condition for contexts in which other birds were provided with differing amounts of vocal or social stimulation (Thorpe 1958). For those interested in songbirds, the nature of the isolate himself was not a concern. As long as the vocal "morphology" could be extracted and compared, investigators could proceed. By contrast, we doubt that those studying primates were fixed on so particular a form of behavior morphology. That the birds under study were called "songbirds" suggests how conspicuous and defining the behavior is.

From an ethological point of view, given an interest in phylogeny as well as ontogeny, showing that vocal characters were as reliable as morphological characters for creating taxonomies and identifying reproductive barriers were fundamental objectives. As stated by Marler (1960, p. 350), the goal was to see whether "the *structural* [italics ours] characteristics of loud and conspicuous song play[ed] a dominant role in reproductive isolation of species." By emphasizing the songs, investigators were raising the analysis of behavior to a new level of importance from a biological point of view. The study of bird song eventually became one of the leading "textbook" models of how to analyze species-typical behavior in general (Hinde 1982). As the vocalizations became easier to record and as electronic analysis tools grew in sophistication, diverse details involved in learning, memory, perception, discrimination, filtering, and sensorimotor coordination were incorporated into descriptions of vocal ontogeny. At the same time, neurobiologists proceeded to study the interface between avian brain and behavior, mapping areas involved in integrating the many facets of vocal behavior (Nottebohm, 1984; Konishi 1985; Marler 1990). The study of human language, to which specific parallels had been drawn, was also being approached in much the same way (Marler 1970; Gleason & Weintraub 1978; Beer 1982;

Locke 1993). The creation of grammars and comparisons of syntax towered over all other approaches to the study of language (see Goldin-Meadow, Chapter 15). Linguists, like birdsong researchers, stressed the vocal record with scant attention to the vocalizers.

Thus, many forces acted on the discipline to make the song the primary focus: vocal structure essentially defined the field because so much was found to be studied. Peter Marler's goal of tracing the direct linkages between song and reproductive success remained of considerable interest (especially with respect to dialects; Baker & Thompson 1985). But, in many ways, his question was muted because of the wealth of information to be assimilated about the nature and diversity of the vocal acquisition processes themselves. That songbirds differed among themselves with respect to the nature of learning invited ecological analyses (Kroodsma 1983). That the paradigm embraced issues of nature–nurture drew in developmentalists as well (Beer 1982). That avian brain could be studied in parallel with vocal behavior afforded integrative opportunities (Nottebohm 1984). That little was known about the outcomes of the birds whose vocal histories created these facts probably seemed simply irrelevant.

EMERGENCE OF SOCIAL SKILLS IN COWBIRDS

Our attempts in 1992 to look at the behavior of birds whose rearing had been manipulated was thus motivated by a long-standing interest in early deprivation, especially given that the subjects were brood parasites, whose young were raised by other species. It had become a more urgent issue, however, because we and others had collected considerable data to suggest that cowbirds naturally experienced variation in the kinds of early, social experience available across their wide, geograpahic range (Rothstein 1986, 1988; King & West 1990). In some populations, fledglings were exposed to adults; in others, juvenile birds might only rarely encounter adults, or hear their singing, until the winter or spring (O'Loghlen & Rothstein 1993). It was also clear that differences in the nature of developmental processes involved in vocal communication occurred across the range, with males in some western populations taking a year longer to develop complete vocal repertoires than males in the east (O'Loghlen & Rothstein 1993). Moreover, females differed in at least three ways:

females from the west were less reactive to song than females in the east, females from the west were more reactive to whistles that those in the east, and females from an older, ancestral population were more open to perceptual modifiability than females from a newer population (West & King 1986). Thus, important variations were present in the development and functions of vocal signals, as well as differences in how those signals were received by females. We now needed to learn more about how different social conditions affected these populations' mate recognition systems to understand the phylogenetic differences apparent across the species' range.

Hormonal studies of cowbirds also indicated that laboratory housing had multiple effects. Housing male cowbirds in cages by themselves increased levels of corticosterone, an indicator of stress (Dufty & Wingfield 1986, 1990). Males housed in identical cages, but with female cowbirds, did not show elevated levels of corticosterone. Males housed alone, or with different male companions, also showed differences in testosterone production, with the isolated males showing the lowest levels. Thus, for many reasons, we were motivated to learn more about the social characteristics of cowbirds raised with different social companions.

Beginning in 1991, we began to address these questions from a developmental perspective. We looked first at a cowbird population collected in South Dakota, in the range of the *M. a. artemisiae* subspecies. We chose it because other investigators had found compelling evidence to suggest that birds from this subspecies (although from a different part of the range) developed vocal repertoires over two years, suggesting considerable plasticity (O'Loghlen & Rothstein 1993). Thus, we set out to study males from this population with the goal of not stopping with acoustic analyses or playback tests, but including tests of actual courtship in the summer of 1992. Because we could find no comparable studies in the avian literature as examples (the closest now is that of Williams *et al.* (1993)), we borrowed some of our observational and testing procedures from those who had studied primates. We designed a series of social transitions to look at social and vocal competence in relation to social complexity. We thought of the transitions as cases of "emergence," loosely following the language and stage concepts popularized by those who had studied separation and reunion experiences in rhesus monkeys (Harlow 1974). How would cowbirds

react to social circumstances that were increasingly more challenging in terms of the social skills of their companions?

Our protocol involved five steps, with each step increasing the social complexity and the physical size of the enclosures in which we studied the birds. First, we captured the birds in the wild when they were between 50 and 100 days of age (thus they were not hand-reared or naive). Upon being taken to the laboratory in Bloomington, Indiana, they were housed in sound-proof chambers. We housed half of the males individually with two female cowbirds (the FH group) and half individually with two adult canaries (the CH group). Second, as the breeding season approached, we recorded the males' vocal repertoires while they were still with their social companions. These vocalizations (songs and whistles) were played back to South Dakota females during the same time period to assess song potency as another measure of developmental outcome. Third, we moved the males to new quarters, placing the five CH males in one cage and the five FH males in another. They remained in the cages for several weeks so that we could document their behavioral state and to carry out the first social challenge. That test involved introducing the CH and FH males one at a time to an identically sized cage that housed two unfamiliar canaries and two unfamiliar South Dakota females. How would males react to new females or heterospecifics? Fourth, after such testing, we moved the males into indoor outdoor aviaries, where we observed the males interacting with female conspecifics from two cowbird populations (South Dakota (SD) and North Carolina (NC)), with other species, and with each other. Finally, we introduced adult, experienced SD and NC males to record their courtship behavior and to calibrate their effect, if any, on the younger, more naive males.

In order to accomplish these goals within the species' six-week breeding season and to maximize our chances of seeing how males attract mates (as opposed to maintain a paired status), we rotated males and females in and out of the testing aviary after they had either shown evidence of courtship for three consecutive days with the same bird or failed to show any courtship for four or more days. The CH and FH males were thus cycled three or four times through the aviaries. Field reports of eastern populations indicated that unpaired males quickly acquire mates in the field and are hormonally ready to act if circumstances permit: thus, the design we used had analogs to situations cowbird males may face in the wild (Dufty 1982; Dufty & Wingfield 1986).

We had studied groups of cowbirds in similar aviaries in past studies (West *et al.* 1981b; West *et al.* 1983; Eastzer *et al.* 1985). The courtship patterns formed a concordant picture with courtship in field settings, females typically being socially monogamous, and males also showing monogamy more often than polygamy; only the most dominant males typically maintain two consorts (Yokel 1986; Yokel & Rothstein 1991). In no study had we found random patterns of mating, in every case, certain males were far more successful than others. Moreover, the pairing and mating success of males could often be predicted on the basis of other measures such as song potency as tested by playback, or male dominance, as measured by perch displacements among males prior to the breeding season. The number of fertile eggs obtained from the resident females also typically fell within the normal range for the species, suggesting that they found the conditions favorable for reproduction.

PHASE I: EXPOSURE TO EACH OTHER, FEMALES AND HETEROSPECIFICS

Our first observations of the birds when they emerged from the chambers suggested that all the aviary plans might have to be shelved: the birds appeared to be in fragile social states, even though they had had companions throughout the year. The day that the males were removed from their female or canary companions even the most naive observer would have been able to pick out these males from other cowbirds. Any noise in the laboratory room or unexpected sound such as a human sneeze produced frantic flying and hovering as the birds attempted to find a perch unoccupied by another male (there were many perches to choose from and burlap cloth to hide behind). In the laboratory notes, observers report that the birds appeared frozen in one place for long periods. Even so seemingly simple an act as flying to the floor of the cage to eat involved problem-solving: the males would make several forays before actually landing and eating or drinking if another male was there. For the first five days, during focal watches each morning and afternoon, we did not see two males sharing a perch nor did birds eat or drink at the same time. The CH males appeared to be more vulnerable,

and were slower to show social interactions with each other.

Perhaps not surprisingly, given what we have said about the vocalizations in truly isolated songbirds, the cowbirds' vocal behavior resumed quickly. The first vocalizations occurred the next day and included many typical cowbird sounds as well as some odd imitations and improvisations such as the sounds of female rattles and imitations of canary calls. The most obvious difference in their vocalizations concerned the delivery of the song. First, the songs were not accompanied by the male's typical song spread display. This display begins with the male raising the features on his back and chest, lifting and spreading the wings while bowing forward, sometimes to the point that his beak reaches the perch or ground (Lowther 1993). The song co-occurs with this display, although it can also occur without it. Second, most vocalizations were not directed toward other individuals. Third, the birds often were sitting down on the perch generating vocalizations with little sign of responsiveness to events around them.

Thus, upon the birds' emergence, we saw multiple deficiencies, but many of these lessened considerably as the birds settled into their new surroundings in the two large flight cages. The males' cages were placed adjacent to cages containing canaries. We suspect that the canaries' presence facilitated the gradual emergence of more calm and species-typical behaviors by both groups, but especially the CH males. (We have used canaries to calm wild birds for years, influenced by the idea of "peer therapists" (Harlow 1974).) The canaries readily approached the cowbirds through the wire and they modelled social behaviors such as coordinated feeding and the absence of fear in the presence of humans. Within a week to ten days, the males had changed considerably, singing and displacing one another, foraging as a group, and in the case of the CH males, "guarding" the cage wall shared with the canaries from other males and singing through the wire to them.

Since that time, we have also witnessed the emergence of two SD males that had been hand reared and then housed individually with canaries for a comparable period. Thus, they had had far less species-typical contact, having been deprived for 50–70 days longer than the SD wild-caught males. The emergence of the hand-reared males was also far less successful: although they would sing to canaries, they did not sing to one another, even after weeks of social housing, nor did they sing to female cowbirds when introduced to them. Their songs were also more primitive in terms of species-typical cowbirds vocal structure. Clearly, the degree of deprivation is an issue for cowbirds as it is for primates. Truly companion-less cowbirds, a condition we did not carry out, would probably have required far greater intervention to make the transition to group living.

The first social test we put to the CH and FH males, as noted earlier, involved moving them to another cage in the same room for brief periods (15–20 minutes). Residing in the case were two adult canaries and two SD females. The transfers did not appear stressful as most males vocalized within five minutes of being introduced. But group differences were apparent: the CH males sang reliably more to the canaries and the FH males more to female conspecifics. Perhaps what struck us as most odd when watching the CH males was the imbalance between the male's attention to the two species. Even when female cowbirds approached the CH males, all but one appeared not to notice, often flying around her to approach a canary or simply to move. We had not expected to see any attention to the canaries at all, but, by this time, we began to think that perhaps it would be a passing "phase." Once they saw female cowbirds giving copulatory postures, some weakened connections would be revitalized and the causal chain between female responsiveness and male song production would snap into place. It was not that the CH males did not seem to know what to do, i.e., one did copulate with a female cowbird, it was their lack of interest in doing it. When normal males see females in copulatory postures in our aviaries or in the field, they sound and act excited: they sing more rapidly, intermixing songs and whistles as they move into position to mount the female. We heard no such sounds from the CH males. Even more puzzling was that we also did not hear them from the FH males. Although they did sing to females and occasionally mount them, they would often lapse quickly back into "unmotivated" states, sitting down on the perches, vocalizing in a manner closer to hiccuping than interacting. The contrast to normal males was impressive. At this time of the year when normal males are interacting with females, songs occur as part of energetic sequences, as males maneuver to get in front of a female before singing or run down a perch to approach her, puffing up their feathers as they get closer (a tell-tale sign that a song is coming). They even sing sometimes while flying after a potential mate. Songs of a courting male typically include the song spread wing display, producing the vivid picture of a male transforming his body from a sleek, upright posture to the point that he almost falls over his feet and wings in the act of delivering the song (the display has been called the topple-over posture by some). All of this animation was missing in the FH birds: the most spirited activity remained the efforts of the CH males to interact with their canary neighbors.

The copulatory and courtship behavior that we did see we attribute in great part to the competencies and motivated states of the females – their behavior made the males look competent. These data suggested that the cage setting was an inadequate context in that we could not separate the social "scaffolding" from other birds or from the setting itself. The close distances within the cage (1.8 m $\times$ 2.4 m $\times$ 1.8 m) meant that males and females could not meter their distances from one another in typical ways with the end result that males might have approached canaries to avoid cowbirds or vice versa. Thus, we proceeded to the aviary phase in which the number of social and physical options were vastly increased.

PHASE II: COURTSHIP IN A LESS CONFINED SETTING

We used two identically sized testing aviaries to facilitate testing more birds (the dimensions were 9.1 m $\times$ 18.3 m $\times$ 3.4 m outdoors and 9.1 m $\times$ 3.0 m $\times$ 2.7 m indoors). We observed the birds every morning from 6:45 a.m. to approximately 10 a.m. Cowbirds in the field and in captivity restrict their courtship to the morning, probably because afternoons must be spent travelling to locate nests or food. There were several other groups of birds present in the testing aviaries including: SD female cowbirds, captured from the same site as the males; NC female cowbirds (*M. a. ater*), which had resided at the laboratory for over a year; canaries; and starlings (*Sturnus vulgaris*). The two classes of cowbird female were included to look for evidence of geographic preferences, preferences we had documented in other populations. The starlings were included to provide an unfamiliar heterospecific class.

Two observers recorded vocal behavior and any copulations, seven days a week, using focal sampling of males. We scored a copulation if the female adopted a copulatory posture and the male mounted her. The males' behaviors were too variable from day to day to meet the definitions

Table 4.1. *A summary of the vocal behavior of the canary-housed (CH), female-housed (FH), and normal South Dakota (SD) and North Carolina (NC) males in the two phases occurring in the testing aviaries*

Phase	Median % of directed vocalizing to recipients			
	Canaries	SD females	NC females	Males
Phase II				
CH yearlings	50	17	20	36
	(43–87)	(0–65)	(0–100)	(12–40)
FH yearlings	0	21	21	43
	(0)	(0–100)	(0–100)	(43–94)
Phase III				
CH yearlings	34	25	0	50
	(0–63)	(0–57)	(0)	(0–67)
FH yearlings	0	20	0	100
	(0)	(0–100)	(0–100)	(80–100)
SD adults	0	100	0	72
	(0)	(50–100)	(0–20)	(3–100)
NC adults	0	21	100	47
	(0)	(0–27)	(50–100)	(23–100)

Note:
The numbers represent the median percentage of days on which males directed two or more vocalizations to a given class during a 20 minute focal sample (range in parentheses). The differences in distribution of vocalizations across classes of recipients were significantly different in both phases for all groups, as tested by Friedman analyses of variance (see Freeberg *et al.* 1995).

for consortships that we had used in previous work (Eastzer *et al.* 1985). We therefore scored their behavior in terms of the use of directed vocalizations, meaning that a male had to produce a minimum of two or more directed songs or whistles while facing and moving with 0.3 m of the individual, and he had to follow the vocalizations with a chase if the recipient departed. We hoped to capture how persistent males were in their attention to the various classes even if "true" consortships were not evident.

As we noted earlier, the males' behaviors in the aviaries continued to reveal social disabilities (from the point of view of successful reproduction). Four of the five CH males vocalized most often on the most days to canaries, vocalizing second most often to other males (Table 4.1). The FH males, which did not have the distraction/attraction of the canaries, were no more focused on female cowbirds than on male cowbirds, with only one of the five males directing vocalizations regularly to them, and coming close to meeting the criteria for a consortship (Table 4.1). Even in his case, however, he failed to stay focused on the same female from session to session – he was easily distracted by other male or female cowbirds.

Mate guarding is essential for successful reproductive outcome in cowbirds for several reasons. The entire courtship cycle consists only of egg-laying, fertilization, more egg-laying, etc., with females producing anywhere from 20 to 40 eggs in a season (Holdford & Roby 1993). Moreover, sperm storage capabilities in female cowbirds suggest that the last male to copulate has an advantage (Briskie & Montogomerie 1993). Male cowbirds must therefore be on guard throughout the entire season, as well as continuing to mate often.

PHASE III: INTRODUCTION OF EXPERIENCED MALES

After watching the FH and CH males for three weeks, we added adult SD and NC males so that at any given time the aviary contained three of the CH or FH males and three adults from both SD and NC, as well as females from each population, canaries and starlings. We wondered whether the younger males would become more attentive once they saw actual courtship. The adults readily took to courting the females, with most males devoting every day in the aviary to pursuing females (Table 4.1). The FH and CH males showed little change; if anything, both groups were less persistent, although the adults did not physically attack or interfere with them (perhaps another sign of their incompetence). Others have reported that younger male cowbirds are socially inhibited by older males in competition for mates, and our data suggested this might also be operating in the aviary (Yokel 1989; O'Loghlen & Rothstein 1993). The only increase in activity was that several of the FH males began to sing more often to one another. Singing between males is frequent in normal males during the breeding season, often while females are nearby. One of the species-typical intrasexual behaviors is counter-singing (Lowther 1993). Counter-singing involves two or more males in close proximity to one another vocalizing back and forth while bowing and spreading their wings (see p. 46). We saw almost no such instances by the FH or CH males. The adults counter-sang 15 times in the first hour in the aviary and continued to do so throughout their stay in the aviary.

Taken as a whole, these data suggested that the social skills to attract mates do not automatically come packaged with the ability to vocalize. The playback data made this point even more strongly. We had played back the vocalizations from the FH, CH, and adult SD males. The SD females responded equally often to the FH and SD vocalizations (around 33%), and significantly less to the CH males (around 6%). The latter result was hardly surprising, given that at least three CH males produced vocalizations with intermingled canary elements. But it appeared that even though the FH males had effective vocalizations, as effective as those of normal SD males, they did not use them competently. These data were not the first to indicate the cowbirds' songs were necessary, but not sufficient, to predict pairing courtship success (West et al. 1981a,b). But the data here indicated that appropriate vocal use is not an automatic or inevitable link, guaranteeing that males and females engage in courtship.

These new data suggested that the key to finding the more fundamental elements of mate recognition systems relating to communication would mean changing levels of analysis from vocal structure or syntax to vocal use or pragmatics (Smith 1977; Beer 1982; Kroodsma 1988). The latter level implicates not just the song, but the act of singing and the context surrounding the communicative event. As such, it obliges those studying song to go beyond the most objective tools of their trade, acoustic analyses, and obligates them to confront the multi dimensional nature of social influences. Songs can be recorded and analyzed and matched on objective acoustic criteria: the quality and timing of song overtures, the males' proximity to a female, his persistence in following her, his tendency to guard her from other males – all of these qualities are harder to reduce to a common set of metrics. But our data suggested that answers to questions about how mate assessment operates required new levels of analysis, a point emphasized as well by Payne & Payne (1993) in their extensive studies of social learning in indigo buntings (*Passerina cyanea*).

SOCIAL REHABILITATION IN COWBIRDS

Our belief in the need to study social influences was further reinforced by the results from studying the same males in their second year. We had decided to explore the role of adult male influence in more detail, in light of the finding that the FH males had directed most of their vocalizations to other males, although not in the typical form of counter-singing. We therefore randomly assigned CH and FH males to two conditions: half lived with females, each other, and older SD males (the "PRO" condition); and half lived only with female conspecifics and each other (the "DEP" condition). Thus, all had an opportunity to interact with both sexes, but only the PRO group was provided with opportunities to interact directly with older males and to watch older males interact with each other and with females. We tested the males, now in the second year, in the following breeding season. We found that the males given experience with older males were reliably more successful at obtaining consortships and copulated more often (Fig. 4.1).

As in the first year, we had recorded the males' songs prior to aviary testing. As in the first year also, playback potency did not correlate with aviary success: the males without exposure to adults had the more potent songs (Fig. 4.1). These data complement the data collected from NC cowbirds in a previous study, demonstrating that the males "deprived" of visual exposure to adult males produced more potent songs than did males who witnessed adult male interactions (West & King 1980). We suspected then and now that the differences in potency reflected differences in male and female cowbird's reactions to potent

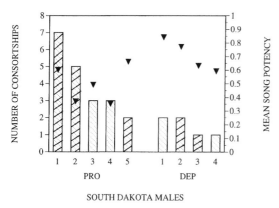

SOUTH DAKOTA MALES

Fig. 4.1. The courtship performance of the canary- and female-housed males from South Dakota in their second year is displayed, along with the potencies of the males' second-year vocalizations. The bars represent number of consortships, the triangles indicate potency. For longitudinal comparison, the following males were canary-housed in their first year: PRO males 1, 2 and 5; and DEP males 2 and 3. For explanations of PRO and DEP, see the text. The remaining males comprised the female-housed (FH) group; one of the FH males (DEP 5) had died during his second year. The differences in number of consortships and in vocal potency between groups were significant (see Freeberg *et al.*, 1995).

songs. The same song material appeared to elicit attraction in females and aggression in males. Among the most important pragmatic skills facing cowbirds is how to sing close to a female while far from a male, and with the communicative force to prevent other males from approaching and interfering. It was possible for males and females to be at great enough distances (the aviaries were 18.3 m long) from each other in the aviaries that the bioacoustics of the song would allow males to sing to females and not to be heard closely by other males. But males interact with both sexes during the breeding season, and so the dynamics of managing proximity to one sex while keeping an eye (and ear) on the other seems far from simple.

MATE RECOGNITION AND SOCIAL SKILL ACQUISITION IN INDIANA COWBIRDS

Subsequent to the experiment on SD cowbirds, we carried out one of quite similar design with Indiana (IN) cowbirds (West *et al.* 1996). These birds represent the eastern subspecies (*M. a. ater*), a group we knew more about. One of

the major differences appeared to be in the temporal properties of the development – eastern males develop repertoires equivalent to those of adults in their first year and are as successful as adults in obtaining mates in captive and field settings (Rothstein *et al.* 1986). The results of our study of IN males housed with canaries indicated intraspecific variation in certain behavioral domains, but not in others (we are in the process of studying a female-housed group now). During emergence, the CH males from Indiana did not seem quite as disturbed when placed into flight cages, although they took over a week to settle into any coordinated feeding patterns and to learn to land on a perch already containing another male, without hovering and hesitating. But the IN males housed with canaries showed much less attention to canaries in cage or aviary tests. Three of the males vocalized to them and one persisted in vocalizing to them in the aviary, but even he became less interested as time passed. Thus, species-level recognition appears to operate differently for different populations.

But, like the CH and FH males from South Dakota, the IN males were generally unsuccessful at courting females. The picture was one of variability within and across males. Their behavior was socially unorganized, resulting in low levels of courtship activity and little persistence in courting the same females. We were, however, able to use measures other than directed vocalizations because the males directed more songs to the females (they were not as distracted by the canaries). We documented consort formation, as we had in a previous study (Eastzer *et al.* 1985). A consort day was scored if a male sang at least ten songs to females in a morning, of which at least one-third were to a particular female. For a consortship to be established, three consecutive consort days with the same female had to occur. If a copulation occurred on the second day of a developing consortship, we judged that the pair of birds formed a consortship. Across the eight males, only three managed to form consortships with female cowbirds. Another male almost met the criterion for a consortship with a canary. The five males with no consortships vocalized most often to one another (producing an average rate of four songs/focal sample to females and ten songs/focal sample to males compared to rates of ten songs to females and five songs to males for the three more successful males).

No IN females copulated with the males, but copula-

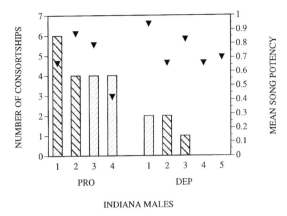

INDIANA MALES

Fig. 4.2. The courtship performance of the Indiana canary-housed males in their second year is displayed using the same metrics as Fig. 4.1. The differences in number of consortships and vocal potency were significant (M. J. West *et al.*, unpublished data). All the males had been housed with canaries in their first year; male DEP 4 had not been tested in the aviary in the first year as he had become ill but subsequently recovered; DEP 5 had no consortships; and PRO 5 died.

tions did occur between three of the males and the NC and SD females (for whom no geographically appropriate males were present). A high proportion of postures were not followed by copulation, i.e., the males often did not mount even when no males interfered and even when females reassumed postures after repositioning themselves. The canary-housed IN males' performances were then tested in relation to normal IN males. The latter competed more successfully, quickly forming consortships, as we had seen in all earlier studies of wild-reared males, including the SD and NC adults in the previous study. They also copulated frequently with the IN females, and we saw no instances of failures to mount females except in one instance where another male interfered.

The next step was to repeat the procedures comprising the second year's design used for the SD males. Thus, PRO and DEP groups were formed. As with the SD birds, rehabilitation with adult males was enormously successful at eradicating all signs of first-year incompetence in the group of males that interacted with more experienced adults over the winter (Fig. 4.2). As in the SD study, their success was not a case of modelling because the adults were not present during the testing. Nor did vocal imitation

seem the key: only two males, one in the SD study and one in the IN study, copied a song type of an adult out of the more than 30 types developed across the two groups. Finally, playback potency did not predict the IN males' performance in either their first or second year (Fig. 4.2). The second-year males housed during the winter without adult males sang reliably more potent songs according to a playback standard, but they gained no advantage in the testing aviaries as they tended to sing to males or unsystematically to the females. Looking across both studies, the data suggested that a key to initial acquisition or subsequent rehabilitation of social skills lies in social experience with adult male conspecifics, experience not contained within the acoustic loop of song copying or improvisation but gained from the broader network of social interaction within and across sexes.

To summarize, these two experiments exposed several layers of connections that now need to be studied in detail, connections we had not known could be dissociated by social manipulation. First, the CH males' orientation toward canaries made clear that species recognition is neither innately programmed nor environmentally ensured by early, juvenile experience with conspecifics. Perhaps species and mate recognition came about by an imprinting-like process. But the data from the FH males argued against imprinting, in that they should have shown strong mating affiliations towards female conspecifics and they had not. Second, the cage and aviary tests revealed that differences in social and physical setting could lead to different levels of competence (or lack thereof). Had we stopped the studies after observing the CH and FH males in the flight cage, we would have concluded that the latter showed appropriate recognition. Third, the data showed that more than the males' songs had been affected by their atypical rearing, reinforcing our view that deprived songbirds are more like deprived primates, possessing no failsafe "hereditary machinery" to guide social skill acquisition. Fourth, the failure to find correlations between vocal function and vocal use called into question the narrow, sound-centered view of communication assumed by many, including ourselves. While the data indicated a definite role of vocal behavior (we saw no consortships or copulations unaccompanied by vocalizing by males), the data suggested that we had placed too much emphasis on power and permanence of vocal properties. Not only did vocal potency fail to correlate with performance within years, it also did not correlate across years, suggesting that the social contexts exert effects after the first year. Thus, vocal structure, in and of itself, cannot withstand the burden of explanation required to connect signal production to successful reproduction. Social influence must be considered as part of initial and subsequent conditions in which learning and assessment can occur (see also Payne & Payne, Chapter 5).

VOCAL BEHAVIOR: SOMETIMES A SIGNAL, SOMETIMES A NOISE?

If vocal structure is not sufficient to capture the dynamics of avian communication, then the case for using true isolates as an experiential control seems equally untenable. Isolates' vocalizations have often been described as simpler and more repetitive than those of normal conspecifics. These sounds bear the mark of one of the most common effects of solitary rearing across a broad range of taxa, an increase in stereotyped forms of movement – the product of too much time with too little to see, hear, or do ("cage imbecility" in Lorenz's lexicon (Lorenz 1970, p. 87)). A common feature of the vocal behaviors of the CH and FH cowbirds, even during the breeding season, was the frequent occurrence of soliloquies. Acoustic analyses in other studies have revealed that such soliloquies often contain repeated cycles of the male's different song types, sung in a serial fashion. However, reactions from companions can perturb them into different patterns. We had documented precisely such effects in a previous study of FH males where we showed that even a brief flicker of a female's wing would produce behavioral and vocal effects (West & King 1988). The male immediately attempted to approach the wing-stroking female, sang at a faster pace, and tended to repeat several times the song type that had elicited the wing stroke. Most songbirds, especially in their first year, go through periods when repetition and rehearsal are conspicuous; however, most shift to different modes once the breeding season occurs. True isolates, or untutored males, such as the CH and FH males, may fail to make the shift, developing instead anomalous patterns to suit anomalous feedback. Criteria for identifying birdsong as a signal for mate attraction must therefore include empirical evidence beyond the fact of the song being sung, it must include evidence of the sequence of actions leading to the appropriate outcome.

BRINGING SOCIAL INFLUENCE TO CENTER STAGE

In our efforts to talk about different views of birdsong, we have oversimplified the history of the study of birdsong and omitted important exceptions to the classical approach of isolation rearing and/or testing. In many cases, birds were housed socially and gradually the study of social stimulation became a more central focus (Payne 1981; Slater 1983; Baptista & Petrinovich 1986). However, in reviewing studies of birdsong from the 1970s onward, we are still struck by omissions in most reports about the conditions of housing or the conditions of the birds so housed. Questions about the sizes of cages or sound-proof chambers, numbers of birds/cage, or whether other birds were present are difficult to answer. We do not believe the details are missing because investigators were attempting to withhold important information. Rather the opposite: the details are missing because the investigators (and reviewers) shared the view that the most important fact was documenting the absence or degree of reduction in species-specific vocal stimulation. We could cite any number of reports to document such omissions but choose not to as there is no way to do so without appearing to target individuals when it was a *Zeitgeist* that was responsible. Our work from that period attests to the power of that *Zeitgeist* to affect investigators' behavior. In our first efforts to study males deprived of vocal stimulation, we housed males with female cowbirds, which do not sing. In several reports, however, we referred to the birds' songs as "isolate songs" even though the birds had companions (whose effects we were later to find to be highly influential). Sometimes, we referred to the birds' state as "solitary," when in fact it was not (King *et al.* 1980; West *et al.* 1981a). Neither we nor reviewers noted such misnomers. Such is the power of a scientific paradigm.

We also recognize that we have not discussed the work of those for whom the songbirds themselves were the extensive focus of ethological study during the same period; such studies, however, typically took place in field settings, as part of investigations focusing on non developmental issues (such as song function in territorial encounters). That a gap existed between knowledge of the intrinsic neural and motor dynamics of song ontogeny under controlled conditions and the behavioral–ecological dynamics of song use was, however, recognized. Marler (1960, p. 348) was among the first to speak to the lack of any "direct information about how [songs] function". These same issues were echoed later in an overview of the field: "a single caveat looms above the bewildering array of data on vocal learning in the laboratory: what is actually happening in nature may not be adequately reflected by results of laboratory studies. The complex physical and social processes occurring in nature *cannot be duplicated in the laboratory* [italics ours]" (Kroodsma 1982, p. 16).

While we largely agree with the caveat, we would issue another caveat: few investigators have tried to duplicate the social and physical processes in laboratory settings. Social housing carries considerable scientific costs for investigators (West & King 1994). Animal enclosures must of necessity be larger and the number of animals to care for typically increases. Investigators must also be sensitive to the social dynamics to ensure that the animals treat each other in "humane" ways. But let us be clear that we do not mean that songbirds cannot be studied in the laboratory. What need to be discarded are the traditional racks and rows of small cages or small isolation chambers in which birds are confined. What need to be included are new measures and new views of what animals do when we say they are communicating. An animal's natural environment will always remain the standard against which to test knowledge gained in other settings. What we hope we have conveyed is that many of the obstacles to learning about social and cognitive capacities in songbirds have been overly limited by ingrained habits and inhospitable habitats. It is within our power to change both and put the bird and its song back together again.

FEMALE COWBIRD PLASTICITY, A NEW FRONTIER

We have focused primarily on the plasticity in male cowbirds' social skills. Thus, one might conclude that it is the female's primary programming that provides the safety net. Part of the problem in answering this point is that our attempts to modify the female's behavior have used males as potential agents of change. We had tried to probe for plasticity by exposing females to males from different geographical areas, thinking that such exposure would change their natal preferences; in most cases, it did not – it was the males who changed, modifying their songs to fit the female's preferences (King & West 1983). Now, however, we are rethinking the nature of the conditions most conducive to female perception. In Dufty & Wingfield's (1986) study of male cowbirds with and without females, it was the males with females whose corticosterone levels were lower. We wonder what the levels were for the females. Is being confined with a male stressful? A study of chimpanzees (*Pan troglodytes*) demonstrated that a female's ability to regulate a male's access to her had significant effects of copulatory behavior, with female-controlled access producing patterns more in synchrony with the female's hormonal state (Nadler *et al.* 1994). Female songbirds may also perceive forced exposure to a male as a liability, not an opportunity. We have not carried out parallel studies of female cowbirds who first live under restricted conditions and are then tested in the kinds of social contexts described here for the IN and SD males. Whether individual females' playback preferences predict actual courtship preferences awaits testing.

A study is now underway in the laboratory that comes closer to duplicating conditions under which females naturally interact with males (Freeberg 1996). The young females and young males, collected in South Dakota, were housed from the onset of the study in large indoor-outdoor aviaries, in groups containing male and female adults (numbering around 32 individuals in each condition). For two cohorts, the older generation were IN males and females and for two cohorts, the older individuals were SD adults. The results from the birds' first breeding season suggest that yearling females' pairing preferences are influenced by social housing: the females paired more often with yearling males whose experiences matched their own (although the males had obtained the experience in an aviary different from that of the females and thus were unfamiliar). Past failures to find the effects of experience may have been because the housing contexts used were not conducive to social influence in females. Females may need to witness the reactions of other young females, as well as older females, to stimulate their interest in a particular class of male. Less traditional designs may bring forth new capacities. In a study of pygmy marmosets (*Cebuella pygmaea*), Elowson & Snowdon (1994) documented evidence of vocal changes in the structure of trills when the colony conditions were changed. Thus, in certain primate species as well, changes in experimental design to retain more of the natural dynamics of social transitions may also reveal vocal plasticity, hitherto so hard to document in nonhuman primates.

These new data on females bring us back to that first female we watched respond to a recording of song in a laboratory cage. It required considerable effort to arrange to see her response. It requires much more staging to orchestrate the aviary settings we are now using so that we can look at how the males and females themselves come to organize their courtship skills. The result is that we can see the workings of the "teaching machine" Lorenz described, as we watch males and females study and test the behavioral capabilities of one another. But the "teaching" in inherent in the social settings, not inherited in the genome. Perhaps, we humans think that birds, and other organisms, "require" programming to mate appropriately because we cannot see any other way that the events could happen. We must also consider whether our methods of investigation have confined our thinking. The history of animal behavior as a science is also a history of human behavior as we learn to modify our beliefs, practices, and attitudes about other organisms. In this regard, songbirds still have much to communicate.

ACKNOWLEDGEMENTS

We thank the National Science Foundation and the Center for the Integrative Study of Animal Behavior at Indiana University for financial support. We thank the editors for advice and encouragement.

REFERENCES

Baker, M. C. & Thompson, D. B. (1985). Song dialects of white-crowned sparrows: historical process inferred from patterns of geographical variation. *Condor* **87**: 127–41.

Baptista, L. F. & Gaunt, S. L. L. (1994). Historical perspectives: Advances in studies of avian sound communication. *Condor* **96**: 817–30.

Baptista, L. F. & Petrinovich, L. (1986). Song development in the white-crowned sparrow: Social factors and sex differences. *Animal Behaviour* **34**: 1359–71.

Beer, C. G. (1982). Conceptual issues in the study of communication. In *Acoustic Communication in Birds*, vol. 2, ed. D. E. Kroodsma & E. H. Miller, pp. 279–310. Academic Press, New York.

Briskie, J. V. & Montogomerie, R. (1993). Patterns of sperm storage in relation to sperm competition in passerine birds. *Condor* **95**: 422–54.

Dufty, A. M. Jr (1982). Movements and activities of radio-tracked brown-headed cowbirds. *Auk* **99**: 319–27.

Dufty, A. M. Jr & Wingfield, J. C. (1986). The influence of social cues on the reproductive endocrinology of male brown-headed cowbirds: Field and laboratory studies. *Hormones and Behavior* **20**: 222–34.

Dufty, A. M. Jr & Wingfield, J. C. (1990). Endocrine responses of captive male brown-headed cowbirds to intrasexual social cues. *Condor* **92**: 613–20.

Eastzer, D. H., King, A. P. & West, M. J. (1985). Patterns of courtship between cowbird subspecies: Evidence for positive assortment. *Animal Behaviour* **33**: 30–9.

Elowson, A. M. & Snowdon, C. T. (1994). Pygmy marmosets, *Cebuella pygmaea*, modify vocal structure in response to changed social environment. *Animal Behaviour* **47**: 1267–77.

Freeberg, T. M. (1996). Assortative mating in captive cowbirds is predicted by social experience. *Animal Behaviour*, in press.

Freeberg, T. M., King, A. P. & West, M. J. (1995). Social malleability in cowbirds: Species and mate recognition in the first two years of life. *Journal of Comparative Psychology* **109**: 357–67.

Gleason, J. B. & Weintraub, S. (1978). Input language and the acquisition of communicative competence. In *Children's Language*, vol. 1, ed. K. E. Nelson, pp. 171–22. Gardner Press Inc., New York.

Harlow, H. F. (1974). Induction and alleviation of depressive states in monkeys. In *Ethology and Psychiatry*, ed. N. F. White, pp. 197–208. University of Toronto Press, Toronto.

Harlow, H. F. & Harlow, M. K. (1962). Social deprivation in monkeys. *Scientific American* **207**, 136–46.

Harlow, H. F., Harlow, M. K. & Hansen, E. W. (1963). The maternal affectional system of rhesus monkeys. In *Maternal Behavior in Mammals*, ed. H. L. Rheingold, pp. 254–81. Wiley, New York.

Hinde, R. A. (1982). Foreword. In *Acoustic Communication in Birds*, vol. 1, ed. D. E. Kroodsma & E. H. Miller, pp. xii–xvi. Academic Press, New York.

Holdford, K. C. & Roby, D. D. (1993). Factors limiting fecundity of captive brown-headed cowbirds. *Condor* **95**: 536–45.

Kaufman, I. C. (1975). Learning what comes naturally: The role of life experience in the establishment of species-typical behavior. *Ethos* **3**: 129–42.

King, A. P. & West, M. J. (1977). Species identification in the North American cowbird: Appropriate responses to abnormal song. *Science* **195**: 1002–4.

King, A. P. & West, M. J. (1983). Epigenesis of cowbird song: A joint endeavor of males and females. *Nature* **305**: 704–6.

King, A. P. & West, M. J. (1990). Variation in species-typical behavior: A contemporary theme for comparative psychology. In *Contemporary Issues in Comparative Psychology*, ed. D. A. Dewsbury, pp. 331–9. Sinauer, Sunderland, MA.

King, A. P., West, M. J. & Eastzer, D. H. (1980). Song structure and song development as potential contributors to reproductive isolation in cowbirds. *Journal of Comparative and Physiological Psychology* **94**: 1028–39.

Konishi, M. (1985). Birdsong: From behavior to neuron. *Annual Review of Neuroscience* **8**: 125–70.

Kroodsma, D. E. (1982). Learning and the ontogeny of sound signals in birds. In *Acoustic Communication in Birds*, vol. 2, ed. D. E. Kroodsma & E. H. Miller, pp. 1–24. Academic Press, New York.

Kroodsma, D. E. (1983). The ecology of avian vocal learning. *BioScience* **33**: 165–71.

Kroodsma, D. E. (1988). Song types and their use: Developmental flexibility of the male blue-winged warbler. *Ethology* **79**: 235–47.

Locke, J. L. (1993). *The Child's Path to Spoken Language*. Harvard University Press, Cambridge, MA.

Lorenz, K. (1965). *Evolution and Modification of Behavior*. University of Chicago Press, Chicago.

Lorenz, K. (1970). *Studies in Animal and Human Behavior*, vol. 1, trans. Robert Martin. Harvard University Press, Cambridge, MA.

Lowther, P. E. (1993). Brown-headed cowbird (*Molothrus ater*). In *The Birds of North America*, ed. A. Poole & F. Gill, pp. 1–24. Academy of Natural Sciences, Washington, DC.

Marler, P. (1960). Bird songs and mate selection. In *Animal Sounds and Communication*, ed. W. E. Lanyon & W. N. Tavolga, pp. 348–67. American Institute of Biological Sciences, Washington, DC.

Marler, P. (1970). Birdsong and speech: Could there be parallels? *American Scientist* **58**: 669–73.

Marler, P. (1990). Song learning: The interface between behaviour and neuroethology. *Philosophical Transactions of the Royal Society of London, Series B* **329**: 109–14.

Mason, W. A. (1960). The effects of social restriction on the behavior of rhesus monkeys. I. Free social behavior. *Journal of Comparative and Physiological Psychology* **53**: 582–9.

Morris, D. (1970). *Patterns of Reproductive Behavior*. McGraw-Hill Book Company, New York.

Nadler, R. D., Dahl, J. F., Collins, D. C. & Gould, K. G. (1994). Sexual behavior of chimpanzees (*Pan troglodytes*): Male versus female regulation. *Journal of Comparative Psychology* **108**: 58–67.

Nottebohm, F. (1984). Birdsong as a model in which to study brain processes as related to learning. *Condor* **86**: 227–36.

O'Loghlen, A. L. & Rothstein, S. I. (1993). An extreme example of delayed vocal development: Song learning in a population of wild brown-headed cowbirds. *Animal Behaviour* **46**: 293–304.

Payne, R. B. (1981). Song learning and social interaction in indigo buntings. *Animal Behaviour* **29**: 688–97.

Payne, R. B. & Payne, L. L. (1993). Song copying and cultural transmission in indigo buntings. *Animal Behaviour* **46**: 1045–65.

Rothstein, S. I., Yokel, D. A. & Fleischer, R. C. (1986). Social dominance, mating, and spacing systems, female fecundity and vocal dialects in captive and free-ranging brown-headed cowbirds. *Current Ornithology* **3**: 127–85.

Rothstein, S. I., Yokel, D. A. & Fleischer, R. C. (1988). The agonistic and sexual functions of vocalizations of male brown-headed cowbirds (*Molothrus ater*). *Animal Behaviour* **36**: 73–86.

Searcy, W. A. (1992). Measuring responses of female birds to male song. In *Playback and Studies of Animal Communication*, ed. P. K. McGregor, pp. 175–89. Plenum, New York.

Seyfarth, R. M. & Cheney, D. L. (1986). Vocal development in vervet monkeys. *Animal Behaviour* **34**: 1640–58.

Slater, P. J. B. (1983). Bird song learning: Theme and variations. In *Perspectives in Ornithology*, ed. A. H. Brush & G. A. Clark Jr, pp. 475–99. Cambridge University Press, Cambridge.

Smith, W. J. (1977). *The Behavior of Communicating*. Harvard University Press, Cambridge, MA.

ten Cate, C., Vos, D. R. & Mann, N. J. (1993). Sexual imprinting and song learning: Two of a kind? *Netherlands Journal of Zoology* **43**: 34–45.

Thorpe, W. H. (1958). The learning of song patterns by birds, with especial reference to the song of the chaffinch, *Fringilla coelebs*. *Ibis* **100**: 535–70.

Thorpe, W. H. (1961). *Bird Song*. Cambridge University Press, Cambridge.

West, M. J. & King, A. P. (1980). Enriching cowbird song by social deprivation. *Journal of Comparative and Physiological Psychology* **94**: 263–70.

West, M. J. & King, A. P. (1986). Ontogenetic programs underlying geographic variation in cowbird song. In *Acta XIX Congressus Internationalis Ornithologici*, vol. 2, ed. H. Ouellet, pp. 1598–605. University of Ottawa Press, Ottawa.

West, M. J. & King, A. P. (1987). Coming to terms with the everyday language of comparative psychology. In *Nebraska Symposium on Motivation*, vol. 35, ed. D. Leger, pp. 49–89. University of Nebraska Press, Lincoln, NB.

West, M. J. & King, A. P. (1988). Female visual displays affect the development of male song in the cowbird. *Nature* **334**: 244–6.

West, M. J. & King, A. P. (1994). Research habits and research habitats: Better design through social chemistry. In *Naturalistic Environments in Captivity for Animal Behavior Research*, ed. E. F. Gibbons Jr, E. J. Wyers, E. Waters & E. W. Menzel Jr, pp. 163–78. SUNY Press, New York.

West, M. J., King, A. P. & Eastzer, D. H. (1981a). The cowbird: Reflections on development from an unlikely source. *American Scientist* **69**: 57–66.

West, M. J., King, A. P. & Eastzer, D. H. (1981b). Validating the female bioassay of cowbird song: Relating differences in song potency to mating success. *Animal Behaviour* **29**: 490–501.

West, M. J., King, A. P., Eastzer, D. H. & Staddon, J. E. R. (1979). A bioassay of isolate cowbird song. *Journal of Comparative and Physiological Psychology* **93**: 124–33.

West, M. J., King, A. P. & Freeberg, T. M. (1996). Social malleability in cowbirds: New measures reveal new evidence of plasticity in the eastern subspecies (*Molothrus ater ater*). *Journal of Comparative Psychology* **110**: 15–26.

West, M. J., King, A. P. & Harrocks, T. H. (1983). Cultural transmission of cowbird song: Measuring its development and outcome. *Journal of Comparative Psychology* **97**: 327–37.

Williams, H., Kilander, K. & Sotanski, M. L. (1993). Untutored song, reproductive success and song learning. *Animal Behaviour* **45**: 695–705.

Yokel, D. A. (1986). Monogamy and brood parasitism: An unlikely pair. *Animal Behaviour* **34**: 1348–58.

Yokel, D. A. (1989). Intrasexual aggression and the mating behavior of brown-headed cowbirds: Their relation to population densities and sex ratios. *Condor* **91**: 43–51.

Yokel, D. A. & Rothstein, S. I. (1991). The basis for female choice in an avian brood parasite. *Behavioral Ecology and Sociobiology* **29**: 39–45.

5 Field observations, experimental design, and the time and place of learning bird songs

ROBERT B. PAYNE AND LAURA L. PAYNE

INTRODUCTION

Songbirds learn their songs by hearing others and then copying them, matching or improvising on the song theme (Slater 1989; Catchpole & Slater 1995). Their social behavior varies among species – they are migratory or resident, solitary or group-living, faithful partners to a single mate, polygynous or with no pair bond, and parental or nonparental in the care of their offspring (brood parasites lay in nests of other species, and their fosterers rear the young). All songbirds depend on parental care, and it has been suggested that this is the time when the young learn their songs. Later, when they are independent, the birds engage in a wider range of social interactions.

Field studies suggest that most songbirds learn their songs after the time of natal dispersal, when a bird moves from the site where it was reared to an area where it copies the song of a neighbor, rather than singing the song of its father: Bewick's wrens (*Thryomanes bewickii*), marsh wrens (*Cistothorus palustris*), saddlebacks (*Philesturnus carunculatus*), indigo buntings (*Passerina cyanea*), white-crowned sparrows (*Zonotrichia leucophrys*), and corn buntings (*Emberiza calandra*) (Kroodsma 1974; Verner 1976; Jenkins 1978; Payne *et al.* 1987; Baptista & Morton 1988; Petrinovich 1988; McGregor & Thompson 1988; McGregor *et al.* 1988). Song sparrows (*Melospiza melodia*) copy at least one song from three or four neighboring males when they settle on a territory, some time after the first four weeks of life (Nice 1943; Beecher *et al.* 1994). In two species males often copy their father (Darwin's finches (*Geospiza fortis*), zebra finches (*Taeniopygia guttata*); Millington & Price 1985; Gibbs 1990; Zann 1990), but not all individuals do this. In some other songbirds where birds were marked for individual recognition, males change their songs through the season and from season to season.

Great tits (*Parus major*) add new song types in their later adult years even when the song types were not heard in the area in previous years (McGregor & Krebs 1989). In other species the details of songs of an individual bird are apparently improvised in a gradual process of continual change (village indigobirds (*Vidua chalybeata*), purple indigobirds (*V. purpurascens*), yellow-rumped caciques (*Cacicus cela*)) (Payne 1985a,b, 1996; Trainer 1989).

When the same species or a similar species is observed in captivity, the results are consistent with the behavior of birds in the field, in that birds do not develop a normal song if they cannot hear another of their species. A bird reared in isolation does not innovate well enough to generate a normal song, as judged from song structure and from the lack of response of wild birds to the song (chaffinches (*Fringilla coelebs*), white-crowned sparrows, indigo buntings, village indigobirds); it needs to hear a song before it can perform with competence (Thorpe 1958; Marler 1970; Shiovitz 1975; Payne 1981a, 1985a). In experiments designed to test when a bird will learn, the songs presented to a young captive are those it repeated in its next season, while their performance is more likely if they hear the same song again in the next season (Marler & Nelson 1993; Nelson & Marler 1994). Learning is prolonged and songs heard later may replace songs that were heard earlier in the marsh wren. This sequential learning process in captivity may parallel the series of songs a young bird hears as it disperses from its natal area, and the number of songs to which a young bird is exposed determines the extent of the sensitive period (Kroodsma 1978; Kroodsma & Pickert 1980). The duration of the sensitive period is also modified by social experience in white-crowned sparrows, which learn at a later age when they live with a companion tutor than when tutored by a recorded song (Baptista & Petrinovich 1984, 1986). Song learning may depend on the interaction of the young with the song source. In swamp sparrows (*Melospiza georgiana*) and chaffinches a young bird copies the songs it hears from a tape-recording, and song development occurs as accurately and on nearly the same schedule as when it hears the songs of a live bird (Marler & Peters 1988; Thielcke & Krome 1989). Indigo buntings, short-toed tree-creepers (*Certhia brachydactyla*), canaries (*Serinus canarius*), white-crowned sparrows, marsh wrens, zebra finches, and European starlings (*Sturnus vulgaris*) learn more readily and more completely, when they hear the songs from a social model (a live bird with which it can interact) than from a tape-recording (Rice & Thompson 1968; Payne 1981a; Thiecke 1984; Marler & Waser 1977; Baptista & Petrinovich 1984, 1986; Kroodsma & Pickert 1984; Petrinovich & Baptista 1987; Petrinovich 1985, 1988; Clayton 1989; Chaiken *et al*. 1993).

A model of age-dependent song learning predicts that the song theme of a wild bird is one it hears in its natal year. A bird with a song not present in its natal year would be strong witness against this model. Most studies of song learning have been carried out in captivity on birds reared apart from songs they might hear after dispersal, and these studies may be misleading in attributing the time and conditions of acquisition of their song themes (McGregor & Krebs 1982, 1989; Baptista & Morton 1988; Nelson & Marler 1994).

However, a social model of song learning suggests that the song theme a bird copies depends on its interactions with others and the songs of other birds after the time of parental care. Rather than learning its song when it is with its parents and siblings, it may learn after it has become independent. This appears to be the more common circumstance of song learning in those field studies where the songs of parents, neighbors in the natal area during the season of birth, and neighbors in the area where a bird settles in its first breeding season are known. Although a young bird may learn much of its social behavior and some elements and basic structure of its song in its natal year, the details of song elements and their organization into a definitive song are not determined until after a period of social interaction in a later year.

Because the circumstances in which the captive young can learn a song may differ from the social interactions they normally encounter in the field after dispersal, estab-

lishing a territory, and chasing with neighboring conspecifics, the interpretation of captive experiments on the time and social conditions of learning is problematic. On the one hand, the design of experiments gives control on the song themes a bird hears while it is captive, but, on the other, it also introduces the absence of normal social behavior. An implication of social learning is that song learning is not a simple consequence of age but involves an assessment of the male that is copied. For example, song may be learned when one male chases another, as being chased may allow a young bird to assess its companion and provide the young bird with a criterion for song copying. Or a bird may observe a singing male as it acquires a territory or chases another male (Payne 1981b; Payne & Payne 1993b). In both contexts a bird may choose a singer, and style his own song upon the song of a neighbor, much as a female bird may choose a mate (Janetos 1980; Parker 1983; Payne & Payne 1993b). In this view, there is a strong link between social experience and song learning in birds. This view of the importance of social interactions on song development after independence and dispersal differs from one in which the set of possible songs the bird will give when it is adult is determined in an early "critical period" (Marler 1970; Marler & Peters 1982; Marler & Nelson 1993; Nelson & Marler 1994). These two models of song learning – age-dependent learning and social-dependent learning – lead to different predictions that can be tested in field studies as well as in experimental studies of behavior development.

TWO MODELS OF SONG LEARNING

Models

Two models summarize these views of song learning in songbirds. To describe the age of birds, we refer to "HY" (hatching year) for birds in their first calendar year, the year when they were born, including nestlings, fledglings, and juveniles. "Nestling" refers to dependent young birds while they are in the nest, "fledglings" to birds that have left the nest but are still fed by their parents or foster parents, and "juveniles" to young that are independent of their parents and can feed themselves and disperse within their natal area. We use "SY" (second calendar year in the northern temperature latitudes) to refer to birds in their second calendar year; this is the same as "yearlings" and "first-year males" as used in other studies.

1. Song is learned by young HY birds when they hear a song theme in an early "critical period."
2. Song is learned later in development when birds take part in social interactions with other birds, especially in their first (SY) breeding season.

Predictions

The two models suggest two sets of predictions that allow a logic of strong inference. That is, we can compare the behavior against rival predictions or logical consequences, and then determine which model is more successful in accounting for the song behavior of the birds in natural conditions (Platt 1964; Payne 1981b, 1983a). For example, we can compare the songs of birds born locally with the songs of birds born elsewhere. (1a) If the SY birds born in an area where they settle and breed ("natal returns" in migratory populations) are more likely to share the songs of their natal-year neighbors than are birds born elsewhere and move into the area ("immigrants"), then we can reject the idea that these returns learned their songs as SY or adults, because the natal returns could have recalled the songs they heard as fledglings and juveniles. (2a) However, if there is no difference between the songs of natal returns and immigrants, then we could not reject the idea that they learned after the time of dispersal. However, we can compare the songs of the immigrants with the songs heard locally in their SY year and the songs that were present in their natal year. (1b) If the immigrants have a song that was present locally in their SY year but not in their HY natal year, then we can reject the idea that they acquired the song in their natal year. (2b) If they have a song that was present in their HY year, but was not given by another local bird in their SY year, then we can reject the idea that they learned the song locally in their SY year.

The two models each lead to a prediction that allows rejection by these criteria. The model of early learning can be rejected if we find that birds are more likely to sing a song they heard during their HY year than a song they hear only in their SY year. But the model of later learning can be rejected if birds sing the songs they hear only when they are HY males, or if the natal return SY males are more likely to share these songs than are the immigrant SY males. Where birds are seasonal in dispersal and in singing,

then we can narrow the period when they can hear the song of a potential tutor to the time when they would be likely to encounter another song.

The predictions can be modified for birds that change the details of their song in their adult years. First, if older adults (born two or more breeding seasons previously) change their songs, then we can compare the songs they give when they change with the songs heard in their previous years and the songs in the adult year in which they change. If the songs differ from any songs they could have heard in their previous years, then we propose that the males learned the songs through improvisation or copying as adults. If the songs match those of other adults in the year when they change, then we suggest they copied the songs as adults.

The validity of an observational approach depends on sampling songs in a population in all years in which a bird could hear songs, on recording the songs of the bird in later years, and on observing its dispersal history. Marking individuals, determining age, and monitoring their movements and breeding are necessary to infer the normal pattern of song learning in wild birds.

TWO CASE HISTORIES

Two songbird species are of special interest for a comparative study of song development both in the field and in experimental conditions, because their life styles differ and because they are known to learn their songs from others. They differ in their social natural histories and parental behavior and these differences have important consequences for the context in which they learn their songs. However, they both disperse from their natal area and sometimes disperse from one social neighborhood to another after they reach breeding age and status, and this life history may extend the period when they can learn a song to the time when dispersal carries them into a new song area where it is important to adjust their songs to those of their new neighbors.

The indigo bunting is a common songbird in North America, a seasonal migrant, and the females rear their own young. The young can hear their fathers' songs, and, though males usually do not feed the nestlings, some males do feed their fledglings (Westneat 1988, 1990; Payne 1992). The other species, the village indigobird, is a brood parasite in Africa. The parents have no social contact with their

Table 5.1. *Major features of biology and song in two songbirds*

Features	*Passerina cyanea*	*Vidua chalybeata*
Species	Indigo bunting	Village indigobird
Region	North America	Africa
Status	Migrant	Resident
Size	15 g	13 g
Parental care	Maternal>paternal	None, brood parasite
Mating system	Social monogamy	Polygyny, no pair bond
Song repertoire/individual	1 song	24 songs
Song dialects	Yes	Yes (all 24 songs)
Song dialect size, *n* males	Mean 3–4 (range 1–22)	10–20
% adult males change song/year	5%	100%
Song matching by immigrants?	Yearlings, 80%	All ages
Do songs follow kinship lineages?	No	No?

own offspring, which are reared by another species, the red-billed firefinch (*Lagonosticta senegala*) (Payne 1973, 1983a, 1985a, 1990; Payne & Payne 1977). Both species are sexually dimorphic in plumage (blue males, brown females) and behavior (only the males sing) (Payne 1983a). In both species the birds are seasonal in their singing, and this allowed us to sample the social and song surroundings during the times when they hear other conspecifics. Both species have been observed in the field, and both have been the subjects of experimental song development (Rice & Thompson 1968; Shiovitz 1975; Payne 1981a, 1985a; Margoliash *et al.* 1991, 1994). The major features of the two species are summarized in Table 5.1.

For these two species we compare the predictions of the models of age-dependent learning and of social learning of song, and the ecological consequences of social learning.

INDIGO BUNTINGS

Populations and field observations

Indigo buntings were color banded and observed in two local populations (Niles, George Reserve) in Michigan from 1978 through 1994. The two study areas were 250 km apart. Our experimental design is to observe not just one population but two, to distinguish local effects from processes in the species' populations in general.

Buntings live in pairs; the female does most of the care of the young. The reproductive success of males was determined by finding the nests on their territories and observing the number of young buntings that fledge (Payne *et al.* 1988; Payne 1989, 1992; Payne & Payne 1989, 1993a,b). Males were aged by plumage; yearling or SY males have brown greater primary coverts, whereas older adults have all-blue greater primary coverts (Payne & Payne 1989). Males sing in spring and summer, and most young can hear the songs of many adults during their natal season, except perhaps for the latest-hatched birds. After the breeding season the buntings migrate to the tropics where they winter in flocks. We observed them in Belize for a week in February and heard no songs, even when we observed them at close range (5 m) as the birds gathered in their roosts at sunset or became active before they flew into fields to feed at dawn (Payne 1992). The other season when buntings sing is just before migration in mid-April (Johnston & Downer 1968). Because their song is seasonal, buntings might learn by hearing the song of others in spring and summer, but not in autumn and winter, so we can compare the songs of buntings and the songs they could hear as sources to learn in the breeding season.

We recorded a sample of ten or more songs of each bunting on the study area, once a week through the breeding season for the SY males. For adults we recorded a sample repeatedly through the season in the first years of the study; in the first years, we found no seasonal change in most of the adults so did not tape them as often in later years. For the buntings we used these recordings and the sightings of the color-marked birds to determine the songs that the young birds would hear from their fathers and neighbors in their natal season, and the songs of all other

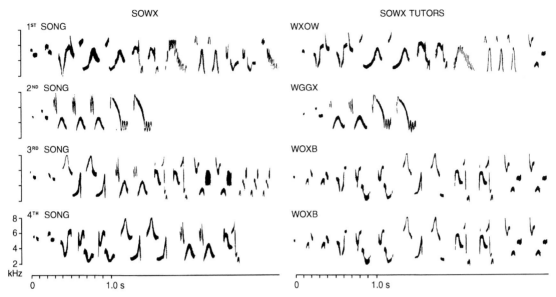

Fig. 5.1. Songs of a yearling indigo bunting SOWX before and after he moved his first breeding teritory. His earliest songs copied male WXOW then SOWX moved 500 m and settled near WGGX; then he moved again (another 900 m) and settled next to WOXB, where he copied the second part of WOXB's song, then copied the entire song. SOWX was the son of WOXB and was born on WOXB's territory.

buntings in their breeding areas in later years. We determined the songs they could hear in their natal season and compared those they hear and sing in later years; all conspecifics within several square kilometers were tape-recorded repeatedly, and we watched the buntings in the field for more than 20 000 hours.

Song matching

A bunting song is characterized by a sequence of "song figures" or "elements" delivered in pairs (Thompson 1970; Emlen 1972; Fig. 5.1). A "song theme" is a repeated pattern of three or more elements, whether shared by several males or given by only one male. When a male was recorded hundreds of times, his song usually was the same except for deletions of the terminal elements, so a male had one song theme. A few males had two songs but these were usually variations on the same theme. Song elements were scored visually from the catalog of Thompson (1970). Sequences of elements were compared with a computer program MATCH, where the number of elements in sequence were a criterion of a match between songs (Payne *et al.* 1981). We recognize the songs of two males as a "match" if the songs have at least three different elements in common and in the same sequence. In practice, when they match three they usually match five or six elements, and this is a typical song length of a bunting. In the field the buntings have a species-wide repertoire of about 100 song elements and these are geographically widespread (Shiovitz & Thompson 1970; Emlen 1971a); each bird uses only a few elements in its song and different birds use different elements. Songs rarely match between populations, and song matches within a population are usually between males on adjacent territories (Payne *et al.* 1981, 1988).

Song learning in indigo buntings, and songs as markers of the time of learning

The songs of birds reared in isolation lack the normal song elements and sequence of elements of the wild buntings. It is not enough for a bird to hear a song of its father as a nestling, or from a tape-recording; social isolate birds develop abnormal songs (Rice & Thompson 1968; Payne 1981a). The songs of social isolates do not elicit a territorial response on playback to wild birds (Shiovitz 1975). Captive buntings kept in groups with other hand-raised buntings imitate a song theme, sometimes with improvisation, when they later interact with the source of the song, either with visual contact and chases with a caged tutor, or with a wild tutor near an outdoor aviary (Payne 1981a).

In the field we saw many males arrive in spring as yearlings, give one song theme, then switch song to match the song theme of a neighboring adult (Payne 1982; Payne *et al.* 1987; Payne & Payne 1993b). We captured more than 1000 males with playback of a song. Although the buntings heard the recorded song (some males heard it hundreds of times before their capture, while others were captured after one or two songs), none matched the playback song when counter-singing with it or later in the season. Social interaction with the playback was limited, as the male responded and attacked, but the playback did not respond and interact with the male, and except for sound amplitude the playback was not contingent upon the behavior of the male.

From recording the songs of all birds in the study area over 17 years and from recordings made by Thompson in the preceding years (Thompson 1970; Payne *et al.* 1981), we found that some bunting song themes persist from year to year, when the same bird returns, or when the songs of one bird are copied by another and the copier returns in the following year. However, many songs do not persist, either because the birds (especially a yearling) switches from one song to another or because no bird with the song theme returns in a later year. Because the average number of males in a year with a common shared song theme is only between three and four (Payne *et al.* 1988; Table 5.1), and many song themes each are given by only one male, there is a chance that any particular song present in one year will be extinct in the following year. But immigrants, especially adult males in their second breeding season, introduce new songs into the local area in every spring. The turnover of song themes in a population is about 50% from year to year (Payne, 1996), and this turnover allowed us to compare the different songs from year to year, and to use the presence of different songs in the HY or SY year of a particular bunting as markers of the time when the SY male could have heard the song and learned it from another local bunting.

Song matching by yearlings and song experience in their natal season

The songs given by a SY male when he first arrives in spring allow a test of the predictions of the two models of song learning. If males learn their songs as HY fledglings and juveniles, then they should produce the songs that are given by other local males in their natal season. But if they learn later as yearlings, then they should produce the songs that older males sing then, regardless of whether the songs were there in their natal year. Also, if males learn as HY fledglings and juveniles, then the males not born in the area where a song was present in their natal year should not sing that song, but if males immigrate and learn their songs as SYs, then they should sing the local song of their yearling year.

The natal returns were banded as HY nestlings and returned to their natal area in the next season as SY males. Their songs were recorded first when they returned and also through the breeding season. For each return male we compared the first song recorded with the songs present in the current year, and in the previous year when the birds were born. At the George Reserve, three of eight males had a song that did not match any song on the study area in a previous year. The other five had a song that was there both in the previous year and in the current year. At Niles, 11 males had a song that did not match any song that we found before. The other 47 males had a song that was given both in the previous year and by a returning adult in the current year, or a song that matched an older adult that we found nearby when we extended the study area in the next year. No natal return SY had a first song like its father's song theme (Payne *et al.* 1987; Payne & Payne 1993b). Altogether, at the George Reserve 38% of the SYs had a song unlike any found in the population in the previous (natal) season, and at Niles 19% had a song unlike any in the previous season. The SY males in the natal area with songs unlike any other male within 1–2 km (an area with 50–80 males in song at the mean date when the young fledged) imply that buntings do not regularly learn their songs during their natal season, nor do they return to a site where they could have learned a song in their natal season. In the natal returns, some SY males had a song that was present in the natal area in their natal season, but none had a song that was present in their HY season only and not in their SY season.

We determined the date of fledging for each male that returned to its natal area, and the dates when all males and songs were present during its natal year and also its SY year, to find for each male whether he could have heard the song there as a HY. The breeding season is long (at the

Table 5.2. *First song of yearling indigo buntings that returned to their natal area, or were born off the study area*

Area	Origin	Was the first song that yearling gave also present locally in its natal year?[a]		Total yearling males
		No	Yes	
Reserve	Local	3	5	8
Reserve	Nonlocal	118	131	249
Niles	Local	18	40	58
Niles	Nonlocal	106	135	241

Note:
[a] For local-born buntings, "yes" includes only those songs that were given after the HY males fledged.

George Reserve, 100 days; at Niles, 108 days, from first egg to last egg laid) and the period when the SY males were fledged also is long (at the George Reserve, 24 June to 16 August, mean = 14 July; at Niles, 18 June to 29 August, mean = 30 July, for natal return males fledged from 1979 to 1985). From census one to three times per week through the breeding season we determined whether the first song was present on the study area during the time the young male was a juvenile. Most males might have heard the song in the ten days after they fledged, though none were seen in the areas where those songs occurred in their natal year.

Although 52 of 66 SY males had a first song that was like a local song of their natal year, it is unlikely that all had heard this song in their natal year. The juveniles fledged by 20 July all could have heard the song they gave in the next year as SY males. For juveniles at Niles fledged after 20 July, no males with the song theme were present after fledging date for three; for another four the last male with the song had disappeared within a week after the juvenile had fledged. As buntings remain on their natal territory for a week after fledging, 7 of the 52 were unlikely to have heard their first song (that they gave as SYs) during their natal year. Combining these cases with the males that had a first song unlike any song on their natal area in the previous year (Table 5.2), 18 of 48 (38%) males at Niles had a first song unlike any they could have heard in their natal year, the same proportion as at the George Reserve.

Most buntings that bred on the study area were born elsewhere (Payne 1991) and would not have heard the study area songs in their HY season. For males that arrived as SY immigrants and were not banded there as nestlings (nearly all local nestlings were banded before they fledged; Payne *et*

al. 1988), we compared the proportion that matched a song on the study area in the year when they were born and in their SY season. Many immigrants had a local song of the study area when first recorded, and some had a song that was present in the previous year. We compared the songs of SY males in 1982–1985, to allow for the expansion of the study area in 1979–1981 and to include nearly all buntings that were born there (Payne 1991). As with the natal returns, some immigrants with a local song may have had another song when they arrived, then switched to a new song before we recorded them, so the proportion of males with a song unlike any local song is a minimal estimate. At the George Reserve about half of the males had a first song unlike any song on the study area in the year when they were born (Table 5.2). Of the males that did not match a song that was there in their natal year, 38 (32%) matched a song that first appeared in their SY year. At Niles, 44% had a first song that did not match a previous local song. Of those, 18 (17%) matched a song that first appeared when a new adult arrived with the song in the SY male's first season.

We compared the proportion of SY natal returns and immigrants with first songs that did not match a song that we had recorded in the area in their natal year. At the George Reserve, 38% of the natal returns had a song that did not match any song of the previous year, and 32% of the immigrants had a song that did not match one in this area in their year of birth. At Niles, 38% of the natal returns, and 44% of the immigrants, had a song they could not have heard there in their natal year. Few buntings returned to their natal area at the Reserve, so we could not test whether the higher proportion of natal returns with a nonlocal song differed significantly from the proportion of

SY immigrants born of the Reserve that had nonlocal songs. Also at Niles the proportion of SY males with nonlocal first songs did not differ between local returns and birds born off the study are ($\chi^2=2.65$, ns).

The time and place where the males acquire the nonlocal songs are unknown; possibly they acquire them during spring migration, though probably not in winter, as the adults do not sing then. Juveniles sometimes remain for several weeks on their natal territory while their parents renest, and sometimes they wander to another territory; juvenile movements were not closely determined. We rarely saw unbanded juveniles on the study area, though we often saw the juveniles we had banded as nestlings. A young male that has contact with an older singing male in its juvenile season might recall the song when he returns and hears it again in his first spring. However, we have no observations of a male that associated in his natal season with a male whose song he gave in the following year. Adult males did not behave aggressively toward their own fledged brood or toward juveniles that wandered into their territory, and they did not direct their song toward the juveniles, but rather they gave alarm chips and were protective toward them, whether or not the juveniles were accompanied by an adult female.

When a SY was first recorded with a song that was on the study area in its natal year, the song theme was also given by one or more adults in his SY year as well, and he may have copied the song theme then, rather than recalling it from his HY natal year. Also, no natal returns had their father's song theme when they arrived in spring, though one eventually settled next to his father and copied the song after he had already copied the songs of two other males (Payne et al. 1987; Payne & Payne 1993b). Most immigrants had a local song theme that we recorded in the study area in their natal year, even though they were not born there, but all the immigrants that matched a local song had a song theme that was there in their SY year, so it is unlikely that they recalled a song they heard during their natal year. The observations suggest that the songs are learned in spring when the buntings are SYs, rather than in their natal season.

Song change within the yearling's spring

A prediction of the early learning model is that the later song themes of a male after he switches from first song theme are songs heard in his HY year. In contrast, a prediction of the model of later learning in social interaction is that he switches to a song that is present in his SY year, and there is no increase in the likelihood that he will switch to a song if the song was also given locally in his HY year.

About 80% of the males that remained for ten days or longer matched the song of a neighboring adult, and many of them changed from a song theme unlike that of any other local male to match the song of a neighbor during the season. The time of song copying varied among males – some copied within four days of arrival on the study area, whereas others completed one or two nestings before they switched to match a neighbor (Payne 1982; Payne et al. 1988). In males that had been banded as nestlings, their first songs did not match the song of their father (Fig. 5.1; Payne et al. 1987). A few males switched their song themes repeatedly during the season. One gave four song themes in one season, by male OBGX in 1982 (Fig. 5.2). The next year he settled in his last site of 1982 with a song unlike his songs as a SY, but like his territory neighbors in this site. He retained this last song from 1983 to 1987 and never reverted to a previous song theme.

The proportion of males that matched a local song varied with the time they remained in the study area, but the difference did not simply follow the longer tenure. We considered 28 days a minimal residence for a breeder, as this is the time required to mate and rear a brood to fledging (Payne 1992). Males that did not remain this long were more likely to have an individualist song (and not match a local song) throughout their time in the area than were males that were resident longer. In the two areas, nearly half (47%) of the SY males that remained for fewer than 28 days were recorded with individualist songs through their stay, whereas only 12% of the males that remained longer did not match a song when first recorded. The proportion of males with a first song that matched a local song was greater for the long-term residents than for the short-term residents, significantly so at the George Reserve ($n=257$, $\chi^2=6.36$, df$=1$, $P<0.02$) though not at Niles ($n=298$, $\chi^2=2.82$, df$=1$, $P>0.05$). Apparently many males that remained had learned the local songs soon after they arrived in spring. The tendency of short-term males not to match a local song, and that of long-term males to match a local song, suggest that the males that matched soon after arrival were more successful in maintaining a territory there (Payne 1982; Payne et al. 1988).

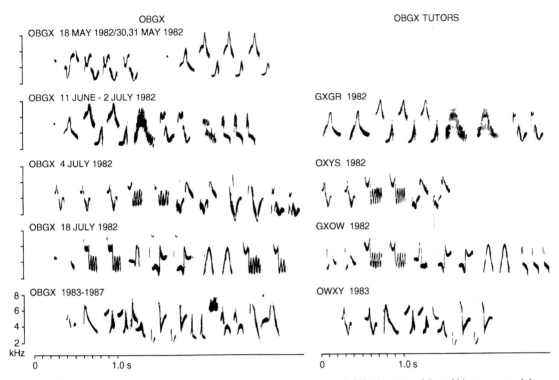

OBGX

OBGX 18 MAY 1982/30,31 MAY 1982

OBGX 11 JUNE - 2 JULY 1982

OBGX 4 JULY 1982

OBGX 18 JULY 1982

OBGX 1983-1987

OBGX TUTORS

GXGR 1982

OXYS 1982

GXOW 1982

OWXY 1983

8
6
4
2
kHz

0   1.0 s   0   1.0 s

Fig. 5.2. Repeated changes of song of a yearling male indigo bunting OBGX, his song models, and his song as an adult.

The time required to learn and perform a local song is not well known. Some SY males sang a matching song when we first found them in territories where they were not present only two days before, and some sang a matching song in early May, within a week of arrival of the earliest males in the study area. These observations suggest that SY males can acquire the song and give it within a few days of arrival in spring. Most either matched a local song when we first recorded them (most were recorded within a week of arrival, but some may have been there longer) or switched from an individualistic song and matched a local song theme in the same site.

We compared the proportion of natal returns and immigrants to test whether they matched a local song as their second song theme of the season, for the males whose first song was not on the study area in their natal year. At the George Reserve, 23 (43%) of 54 short-term immigrants and 51 (93%) of 55 long-term immigrants eventually matched a local song in 1982–1985. At Niles, 8 (29%) of 28 short-term immigrants and 45 (63%) of 71 long-term immigrants matched a local song. For comparison,

for natal returns, at the Reserve one of three with a song not there in the previous year matched by the end of the season, and at Niles one of six short-term males and three of five long-term males matched by end of the season. The proportion of natal returns and immigrants that first arrived with a new song then matched a local song was similar: natal returns were no more likely to match than were immigrants. In each case when a SY had a song that was there in the previous HY year, the song was given by other birds in the SY first spring as well. The results are consistent with the birds learning their songs in their first spring, rather than in their natal year.

We also compared whether a male with a local song switched to match another local song. At the George Reserve, 12 SY males changed from one matching song to another, and at Niles 23 changed from one matching song to another. More males changed from a nonmatching song to a matching song (George Reserve, $n=31$; Niles, $n=20$), while few changed from a matching song to a nonmatching song (Niles, $n=1$). In addition, some males changed from one song when last recorded in their SY season to another when first recorded in the next year, and some of these might have heard the new song in their SY season. At the George Reserve, 20 of the 46 males that changed between years had already changed to match a local song in their SY season; while at Niles 2 of the 36 males that changed between years had already changed to match in their SY season, while most of the others matched within their SY season but their first song may have been overlooked. These observations indicate that SY males have an open song development, even when they have already matched a local song in their SY season.

The context of song change included local breeding dispersal (Payne & Payne 1993b) when the SY males moved from one territory and settled near a new neighbor. From 1979 through 1985 at the George Reserve, 18 of 113 males (16%) that we recorded with two or more song themes moved from one territory to another, then copied the song of the new neighbor. At Niles, 14 of 76 males (18%) with two song themes moved and then copied a neighbor's song (Figs. 5.1, 5.2). Others copied a song theme and then moved, taking it to a new area, seeding a new dialect site there when they were copied by another male (Payne, 1996). At the George Reserve, 15 of 189 males (8%) copied then moved with the song from one territory and went on to use it on a new territory, and at Niles 1 of 76 males (1%) did this.

The cases where a male moves to a new territory then copies the song of a resident adult, and where a male changes his song from the first one heard on his arrival to the one of a local neighbor (and where we did not observe the male with the first song elsewhere, before it settled on its breeding territory), both stress that SY males switch from one song theme to match another local song. Also, they suggest that the social context of song change is dispersal into a territory and use of a song where its song model, rival or companion, has the same theme.

Bilingual males

A prediction of the social learning model is that SY males copy the set of songs of an adult when the model adult has more than one song theme in his repertoire. The early experience model makes no prediction of whether the male would copy one song or more than one.

Most indigo bunting males have a single song theme, but a few have two, which they alternate irregularly in a song bout. From 1980 to 1985 at the George Reserve, 5 (4%) of 136 adults and 15 (13%) of 118 SY males had two song themes, excluding males that sang one song then switched permanently to another through the season. At Niles 8 (6%) of 134 adults and 13 (8%) of 154 SY males had two song themes or distinct variants (at least three elements in sequence, different elements at the end). Bilingual adults retained their two song themes for as long as they returned to the study area.

The songs of ten males with two song themes were copied by other males (range, one to six copiers in a season). Most copiers had only one song, others had both, and a few combined the elements and element sequences of both songs into a single mixed or "hybrid" song of their own. Male GXGR was bilingual from his first adult season, with one song theme copied from a neighbor in his SY year and the other given by a neighbor then but not by GXGR until his first adult year. In his five adult seasons GXGR gave both songs; he was copied by 17 SY males and an adult. Four SY males and a new adult combined the first part of each song into a mixed song: each edited the songs differently, with cuts before or after the first three or four song elements (Fig. 5.3). Four other SYs copied for both songs. The origin of GXGR's songs from two neighbors indicates that a male can learn its song from more than one song model. The songs of his copiers indicate that males also can combine elements from different songs into songs of their own while retaining the sequence of song elements.

We found no tendency for bilingual birds to be copied more than birds with only a single song theme, and we suggest that buntings tend to copy certain males rather than certain songs.

Song changes of adults

The early experience model assumes that adults retain the same songs through their life, with each song determined in their HY year, while the social model of song learning allows some flexibility in song development according to social circumstances. Although indigo buntings usually retained the song they had at the end of their SY year and repeated the same theme as adults, a few changed, either between years or within a breeding season. The variation in whether buntings retain an old song or change to a new one suggests that song learning is not entirely dependent on age. In these adult buntings, we compared the context

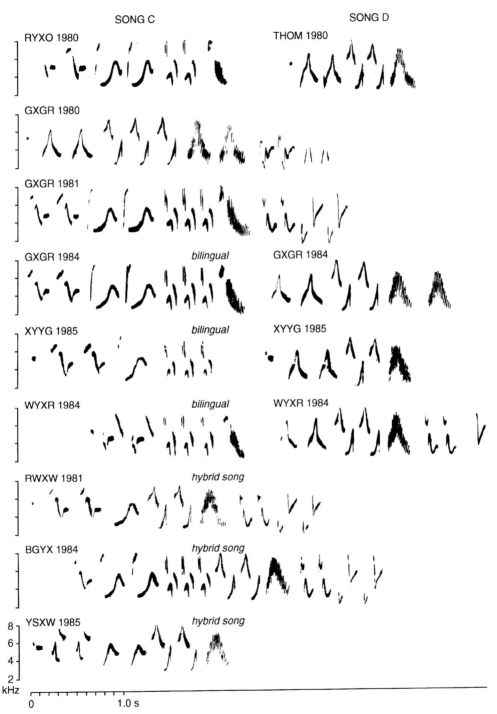

Fig. 5.3. Songs of a bilingual male indigo bunting GXGR, his song models (RYXO, THOM), and his copiers with bilingual song repertoires (XYYG, WYXR) and mixed or hybrid song (RWXW, BGYX, YSXW).

Table 5.3. *Song change between years in male indigo buntings*

Area	Age in the previous year	Song change?		G	P
		Yes	No		
George Reserve	SY	25	113	27.97	<0.0001
	Adult	6	250		
Niles	SY	18	166	14.75	<0.001
	Adult	5	27		

Table 5.4. *Association of song change between years and the factors of male age (A), reproductive success (B), song theme neighborhood size (C), and number of other males that copied the song of a male (D) in the previous year*

Area	Factors	*n* males	G	ΔG	df	P
George Reserve	A	295	33.57	—	1	<0.0001
	A+B+C+D	295	33.93	0.36	4	ns
Niles	A	359	9.92	—	1	0.0016
	A+B+C+D	359	10.65	0.73	4	ns

Note:
ns, not significant.

that precedes a change of song and the time when the songs first appeared in the study area.

Song change between years

Most buntings returned in the next year with their old song theme of the previous season, but some returned with a new one. Males returning from the SY year to their first adult year were more likely to change than were older adults (Table 5.3). Two males returned from their SY season with their old first song rather than their final song of the previous season. In both cases, other returning adults had arrived with this first song theme by the time the returning male was recorded on the study area. This switch might indicate a reversion or an independent acquisition of the song in its second breeding season. The case of GXGR suggests that a male may learn a song as a SY but not sing it until later; however, in his second season he had no neighbor within 800 m with the second song theme, though it was given elsewhere on the study area.

The adults that changed between years generally did so between their second and third breeding seasons (either they were banded as SYs in their first season, or they were captured as new adults and had a different song theme when they returned for a second adult year). In addition a few adults changed the details of their song elements at a later age, up to their seventh year.

Social factors that might affect whether a male retains a song theme or switches to a new one include his reproductive success in the previous year (that is, a male's assessment of his own success), the number of males that shared his song theme in the previous year, and number of males that copied his song (a male's assessment of his own cultural success). For returning adults where we knew age, the number of buntings fledged from all nests in the breeding season, the number of males with the song theme, and the number of SYs that copied his song (Payne *et al.* 1988; Payne 1989; Payne & Payne 1993b), we carried out a series of logistic regressions to see which factors were associated with a male's constancy or change in song theme. We took into account the effects of area (population) and age, then we tested whether the other factors accounted for more of the variance in song constancy, in terms of a significant increase in association (likelihood ratio ΔG). In the full model with all factors, male reproductive success, cultural success and neighborhood size did not explain significantly more of the variance among males than did the simple model with age alone as the independent variable (Table 5.4). The effect of age was

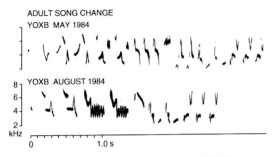

ADULT SONG CHANGE
YOXB MAY 1984

YOXB AUGUST 1984

8
6
4
2
kHz

0   1.0 s

Fig. 5.4. Song change within a season by an adult indigo bunting in his fourth year.

greater than any other factor that we tested. The results showed that age affected whether a male changed his song theme from one year to the next, and they did not reveal any other factors as significant determinants of song change.

Song change within a year

A few adult males changed their song within a year. Most of the 16 adults that changed did so with a variation on the same theme, and two kept this song when they returned for a second adult year. Four adults gave song variations early in the season that appeared similar to the variable or plastic song that some SYs give before they settle on the definitive song, then adjusted their song to one like their song of the previous year. Two males first appeared on the study area as adults and then switched within the season to match the song of a local neighbor. Another adult returned for a second year with his SY song theme that matched that of a neighbor, then added a second theme that matched that of another neighbor. One adult that was born on the study area and returned to breed there as an SY switched during this fourth breeding season from his old song to a new song. He matched a local song theme for four seasons, disappeared for three weeks halfway through the breeding season, then reappeared with a song we had not heard before (Fig. 5.4) and did not match any other local male in any previous year. We found no common circumstance in adults that changed from one song theme to another, though change was more common in males that arrived in the study area in their first adult year (age was verified in one that we banded as a nestling and next saw two years later) with a non local song, then switched to match a local song. Also, change was more common when

males were in their first adult year. So song switching was more common in the younger adults than in older adults, but we had no case where the switch was to a song that a male could have heard in his HY year, and we had several cases where the switch was to a song that was present in his first adult year, suggesting some flexibility in song development according to social circumstances.

Assessment of fitness and date of arrival

The social model as we perceive it predicts that a male will be copied if he is a success in social competition. The early experience model does not predict any relationship between whether a male is copied and his social competence or reproductive success. A copier might assess the success of a potential tutor either after comparing the success of several males, or before their success was realized in a prospective assessment. In the indigo buntings, most males match the song of a local adult in the weeks before the potential tutor has succeeded or failed in breeding. We tested whether an adult male that was a successful breeder was also a cultural success as a song model, and we also tested his success in relation to his time of arrival on the study area. Song transmission was not random among males, and the males that later would be successful were the ones whose songs were copied by the SY males, but only a small amount (<5%) of the variance among males in transmitting their song was explained by their breeding success, and there was no clear relationship between whether a male was copied and whether he had at least one fledged young bunting (Payne & Payne 1993b). Although a song was more likely to be copied in certain habitats at the George Reserve, the mean reproductive success of males did not differ in a consistent way among habitats. Cultural success did not differ among habitats at Niles. The seasonal difference in habitats explains the difference in cultural success of adults, in the same way that the season explains the copying behavior of SY males. Early males tended to settle in certain habitats, especially in thickets of black locust (*Robina pseudo-acacia*) at the Reserve, and these settlers were more likely to transmit their song to the earliest SY males (Payne *et al.* 1988; Payne & Payne 1993b). Apart from age, arrival time, and the prospective breeding success of the males, we did not find the mechanisms of social interactions that may have led to one male transmitting his song to a SY copier, though our observations hint at the SY copying the song of the male with which he most often chased.

Song learning: Do yearling males copy another male, or parts of songs of different males?

A social learning hypothesis is consistent with either a single male or a series of males being copied by a SY male. Predictions about the social circumstance and benefits of matching a song are more direct when a male establishes a social relationship by copying a certain male. If a SY male copies the songs of two or more males, then the social consequences may be more diffuse.

In the indigo buntings, most SY males copied a single neighboring male, and another 20% did not copy a local neighbor but kept the song they had when they arrived. A few (~10%) copy elements and sequences of the songs of more than one neighbor. For example, male OBGX copied several males in sequence (Fig. 5.2), and bilingual male GXGR copied two males and retained their songs as separate themes (Fig. 5.3). Also, other males copied elements from two or more neighbors, especially when the neighboring males all were SYs themselves.

A prediction of the many-males models is that SY males will develop hybrid or mixed songs with the sequences of elements of more than one neighbor, when SY males have potential song models each with a different song theme. But if they copy a single male model, then they will not combine the song elements in mixed song. In early spring, most SY males matched one neighbor, even though nearly all males had two or more song themes among their neighbors during their first ten days in spring, the time when most SY males copy the song of a neighbor (Payne & Payne 1993b). We observed some variation in detail of an element as a song was repeated in both SY and adult males. The same was observed in captive conditions, when buntings give a variable "plastic song" both in SY and in adult males (Margoliash *et al.* 1991). An implication is that the variation in songs of a SY male that matches the song theme of a neighbor may result from its improvisation upon a theme, rather than its copying parts of songs of different males.

The bilingual birds that transmitted both of their songs to SY males show that these males copied a single tutor, and did not rearrange the song elements of several males. If a SY male copies several males, then we expect much

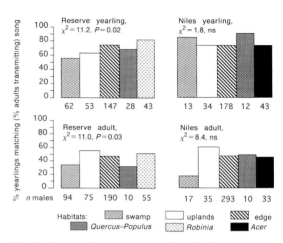

Fig. 5.5. Song copying and song matching of indigo buntings in different breeding habitats. ns, not significant.

variation in the sequence of elements among SY males with common elements in their song. In fact the sequence was fairly constant, both within a male and among his neighbors. We suggest that SY males usually copy a single neighbor, but some copy parts of the songs of more than one neighbor, and a few copy the entire song of two neighbors.

Asocial explanations of song matching

An alternative to a social determinant of song copying is that habitat differences affect which song themes are copied. Are habitat differences associated with the likelihood that a song is copied in a certain habitat because of its acoustic features (Wiley 1991)? The habitat of each territory was classified from field observations and aerial photographs. Replicate stands of the same kind of habitat supported different song themes rather than only one. Also, many song themes occurred in more than one kind of habitat (Payne *et al.* 1988). The results suggest that song themes are not habitat specific.

Were songs copied more easily in certain habitats? We compared habitats to determine the proportion of SYs that matched the song of an adult neighbor (Fig. 5.5). At the George Reserve, the proportion varied among habitats. SY males in swamps were less likely to copy, but in black locust thickets (*Robinia*) were more likely to copy. The differences were associated with the seasonal timing of the vegetation, and the mean date of settling differed among habitats. Buntings settled on arrival in spring earliest in black locust and its blackberry *Rubus* understory, where the birds nest before the leaves in the thicket canopy have grown, and latest in swamps where the nesting habitat did not develop until summer. The song differences also were consistent with a hypothesis that habitat differences affect the social interactions, as the breeding densities and the frequency of chasing and counter-singing were high in black locust, and low in the swamps. At Niles we found no differences among habitats. We suspect that the different acoustics of habitats were less important than was the attraction of buntings to those habitats that were settled early in the season and the social interactions in those habitats.

Another possibility is that song neighborhoods differ in genetic traits that bias their learning of a certain song theme (Payne 1981b). However, in the buntings the song neighborhoods are not closed genetic populations: the number of buntings in the neighborhoods is small and buntings disperse between them (Payne & Payne 1993a), buntings do not differ in body size or allozymes among song neighborhoods (Payne & Westneat 1988), males change songs from one theme to another (Payne 1982, and see p. 64), they do not mate assortatively by song theme (Payne *et al.* 1987), and there is no tendency for a male to sing a song like that of his father, or a female to mate with a male with a song like her father (Payne *et al.* 1987; Payne & Payne 1993b).

In summary, the field observations on songs of the indigo buntings suggest that songs are not learned in the HY season, but are learned and copied in the first breeding season by the SY males and sometimes later by the adults in their second breeding season. Most predictions of the social learning model were supported, especially when the predictions were counter to those of a model of learning in early life. The intense sampling of songs in the natal season and in the season when the SY males first breed allowed us to use songs as markers of the time when the buntings learn their songs. In natural field conditions, the SY males often have a song that was not present there in their natal HY season, but was present in their SY season. The proportion of natal returns and immigrant SY males with a local song was the same. These results were observed both for the first song that a SY male gave, and for the last song of the season that a SY male had after switching from his first song. Also, when the SY male

interacted with his territorial neighbors, he usually copied the song of a single male rather than parts of the songs of several males, but he often copied both songs when an adult neighbor had more than one song in a "bilingual" repertoire: the SY male copied a bird with whom it interacted, rather than a certain song.

VILLAGE INDIGOBIRDS

Populations and field observations

Village indigobirds were observed at Lochinvar National Park, Zambia, from 1972 through 1979. Two breeding populations ("Junction," "Cowpie") were studied in detail. We determined social interactions and breeding success for each male in "Junction" in 1973, and in "Cowpie" in 1976 (Payne & Payne 1977). Both "Junction" and "Cowpie" and a third, "Diptera," populations were tape-recorded across the years (Payne 1985a, 1990). The design of more than one breeding population was used to distinguish local effects from species-wide processes. Field data involved more than 2000 hours of observation of color-banded singing males. Males were also recorded as they sang in captivity, in groups of birds that were bred in captivity and in juveniles caught at Lochinvar Park and held with birds from other populations.

Individual color-banded males were tape-recorded repeatedly within and between breeding seasons. Each male sang from an arena tree called a "call-site." Each male had a large repertoire or set of song types, and he shared the set with all other males in his "social neighborhood," defined as the set of males that visit each other at their call-sites. The call-sites are usually located out of sight of other males, but sometimes are within sight; most call-sites are not within calling distance of the other sites. Males visit other males as the resident sings, perch nearby inconspicuously in trees, and apparently listen to the songs and observe the behavior of the singing resident. Then the males return to their own call-sites and sing. Males in the center of the neighborhoods are the focus of these visits, and all males that visit each other share the same set of local song types.

These brood-parasitic finches form no social pairs. Their mating success varies greatly among males as they breed in a dispersed lek. The male in the center of a neighborhood with about 20 males in 10 km² gains about half of all of the matings observed in a season. Females visit several males in rapid succession in a day, then return and mate with one, usually the male on the central call-site. A female mates repeatedly with one male, and several females mate with the same male. Males sing on their call-sites season after season, and the call-sites are used over several years, especially the call-sites where females visit often and mate (Payne & Payne 1977; Payne 1985a). The breeding season extends for four to seven months (the males that are successful early in the season continue for a longer period). In the nonbreeding season the finches live in both conspecific and mixed-species flocks and they do not sing, so singing is highly seasonal. For the indigobirds we cannot determine the songs of the father of a young male by field observations, because the breeding female indigobirds range a few kilometres beyond the territory of the male that does most of the copulations in their social neighborhood, and a young male could not hear his own father until after he is independent and joins a flock near a call-site (Payne & Payne 1977).

Song matching

Village indigobirds copy the songs of their foster species, the red-billed firefinch. Indigobirds are species-specific brood parasites and their nestlings and fledglings mimic the specific mouth patterns of the young firefinches, which are reared together with them by the foster pair. Each indigobird male has a repertoire of 20–24 distinct song types that are characterized by the shape and sequence of their song elements. Some song types mimic the songs and calls of the foster firefinch, and the others are population-specific indigobird songs that do not mimic the foster species.

Both male and female indigobirds regularly visit the singing males within the area where the males have a common set of songs, or a song dialect, and only rarely visit the singing males outside this area. The males that visit each other share the same set of 20–24 song types with each other, and we refer to the "Junction dialect" as the song types of that social neighborhood. A few males have minor variations in the form and sequence of song elements, but most males match all their mimetic and non-mimetic song types (Payne 1973, 1979, 1985a, 1990). Males in adjacent social neighborhoods have a completely different set of song types, both foster-mimetic and non-

mimetic. The other two populations studied thus had the "Cowpie" and "Diptera" song dialects.

Several morphologically distinct species of indigobird have similar behavior and they differ only in the species of foster whose songs they mimic. Their nonmimetic song elements and patterns are similar among species, and except for the presence of mimetic elements their nonmimetic songs are as similar across species as are the nonmimetic songs of different song dialect neighborhoods of *V. chalybeata*. A few males mimic the songs of an alternative foster species (about 1% of males in areas where two or more species of indigobirds occur together), so the song elements that mimic their foster species are acquired from their foster parents through learning (Payne 1973; Payne *et al.* 1993).

Experiments on song learning

Song learning in the indigobirds involves two stages. First the birds learn the general features of the calls and songs of the species that raises them. Later, the males copy the songs of adults that mimic the same kind of foster, and this is when the indigobirds acquire the mimetic songs and the nonmimetic songs that characterize the local song populations (Payne 1985a, 1990). The observations and experiments described here are on this later stage, to develop the theme of song learning in the adults.

Development of song in captive indigobirds

One male was raised by a firefinch pair in captivity and also heard his father in the aviary until 45 days of age, then was isolated until the next year. His songs lacked the complex song elements of the wild birds. When his songs were broadcast to wild indigobirds at Lochinvar Park, the wild males did not respond, though they did approach the songs of other wild males from the same area or from a remote population, so the isolate's songs were not recognized as normal by his conspecifics. Another male was captured as a juvenile at Lochinvar, then was caged next to an adult male from Tanzania. In the next season the Zambian male copied 16 nonmimetic song types of the Tanzanian male; he sang no nonmimetic Lochinvar songs. Another juvenile from Lochinvar was caged after he was about three months old with a firefinch from West Africa more than 2000 km distant; he copied the songs of the West African bird. In the following year the male indigobird transmitted his mimetic songs to another indigobird that was captured as a juvenile the year before at Lochinvar (Payne 1985a). These developmental observations on captive birds indicate that males do not sing unmodified the songs they heard as young bird in normal field conditions, but learn their songs after the age when they disperse.

Playback of songs of different song neighborhoods to wild birds

Songs were recorded from males in three neighboring song dialects and used in a playback experiment in the field at Lochinvar Park. Because the song types are used nonrandomly in certain social contexts (Payne 1979), the test songs were selected in pairs to represent the song types that were used in the corresponding context within each neighborhood (J, Junction; B, Cowpie; D, Diptera dialects). Song types J14, B9, and D18 introduced the extended song bouts on a call-site; J1, B34, and D12 were given after a male returned from chase of another male (Payne 1979, 1985a). The song types have been illustrated by Payne (1985a). The corresponding song types had a similar structure of song elements, and they were recognized as corresponding by ear (R. B. P., personal observation) between song dialects as well as distinctive markers for each song neighborhood.

Songs were broadcast to 16 males on their call-sites in each of three dialect areas. The speaker was 18–20 m from the base of a call-site in the direction of the nearest call-site. To test birds in comparable conditions, playback was begun only after a male had sung at least ten songs per minute for the previous two minutes. For two minutes before the trial, during the trial, and for 5+ minutes after playback, all observed behaviors were timed, including changes of position, flights, and the distance of the bird to the speaker, and a second tape-recorder was used to record all songs of the focal male. Playbacks were made in April and May in 1973, 1974, and 1976. For each trial a song was repeated six times in one minute. Each male had two trials, one of his own dialect and one of another dialect; one to three days were allowed between trials for a total of 16 paired tests. The song in the first trial was chosen at random; there was no significant effect of sequence of presentation, and no more than one male in a dialect had the same pairs of test songs in the same sequence. By comparing a male's response to his own song and to the corresponding song from another dialect, the design controlled for individual differences in response.

Males approached the speaker and some counter-sang with the playback songs. The following responses were compared: nth song when the male flew from the call-site, distance (meters) of closest approach, time (minutes) for male to return to site after playback, time (minutes) for male to sing after playback, time (minutes) to a rate of 10 songs/minute after playback, and time (minutes) male perched on the site after onset of playback until five minutes after playback. In no behavior did males give a significant difference in response to their own and other dialect song types (Wilcoxon matched-pairs signed-ranks tests). No male matched the playback songs in counter-song, and no male later gave the playback songs of the other dialect.

Repeated playback of a song

To test whether neighbors would transmit a new song from bird to bird, an unfamiliar song of another population 1000 km distant was broadcast near an active call-site. The call-site was chosen because the resident was seen to copulate repeatedly in 1972 and the same male was there in 1973. The design was to test whether the resident would copy the song and then other indigobirds would learn the song as it spread in a sequence corresponding to the observed social visits and mating differential of the neighborhood males. The song was broadcast to him 100 times an hour for 10 days early in the breeding season in 1973. The resident male approached the speaker during the first two days; he ignored the speaker on the other days (Payne 1985a). No male copied the song, perhaps because there was no social behavior associated with its source.

In addition to these playback tests, a common set of playback songs was used to capture most of the 85 color-marked male indigobirds at Lochinvar. No male copied the playback songs either during playback, later in the season, or in later years.

The absence of song copying in the playback tests and in capture of wild males is consistent with the view that song learning normally occurs when a male takes an active part in social interaction with another. To explore the effect of a novel song when the song was introduced together with a socially competent male, we carried out the next test.

Introduction of a novel song by transfer of males

To allow a male to establish himself on a new site and to record the spread of his songs from male to male in the area of introduction, males in two song neighborhoods were captured and released, each at the call-site of the other. The sites were 10 km apart. Before capture, all males in each neighborhood were observed to determine which male had the highest rate of female visits and copulations, then the two most successful males (orange at "mealie" site, 14 copulations in 100 hours; chartreuse at "diptera" site, 8 copulations in 100 hours) were captured and transferred on 3 April, halfway through the breeding season in 1979. Near "diptera" site, eight males contested the site on the day of capture of the resident male chartreuse; two unmarked birds eventually took up new sites nearby and both were seen to mate. Male orange returned within a few days to his old "mealie" site but was not heard to sing. The other male chartreuse disappeared after release at "mealie" and was not seen again. Neither male sang in the site where he was released, and no songs of the transferred male were recorded from the other males in the areas of introduction.

The transfer test points out a limitation of field experiments when birds can return to their source area; in some studies the birds remained where they were moved (Bensch & Hasselquist 1992; Komdeur *et al.* 1995).

Field observations of song learning and age

Males retain the same songs from one year to the next. A total of 21 color-marked males were recorded at Lochinvar Park in two or more consecutive years. Each male changed the details in all of his song types from year to year. The song variants also were shared by males, so that most males made the same changes in all of their song types. Males of all ages (birds entering their second year to birds entering their sixth year, determined from the skull of birds when they were first marked, and from recordings of marked birds across seasons) changed the details of their songs from year to year. The details of song shift both within a breeding season and from one season to the next (Payne 1985a). Song learning is a lifelong process in the indigobirds and is not restricted to early life.

Bilingual birds

A few male indigobirds sang song types of repertoire sets that were given by two, three or even four different neighborhoods. The common circumstance involved dispersal, where a bird moved from one neighborhood to

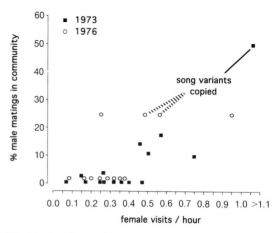

Fig. 5.6. Correlation of indigobird male's female visiting rates/copulation rates, and the males whose song variants were copied by other neighboring indigobirds.

another and copied the songs of this new neighborhood while he also gave songs of his old one, or a bird had a new neighbor that introduced a new set of songs. In one case each of these birds copied the songs of the other; the songs of the newcomer replaced the songs of the older resident, and his old song dialect became extinct. When recorded repeadly within a season or from year to year, the males with a mixed song repertoire changed the proportion of the two sets of songs; while they had a mixed repertoire they were in the process of changing from one set of songs to the other (Payne 1985a).

Song transmission by the successful breeder

Males were observed at the call-sites to determine the numbers of females that visited and mated with them. One male gained most of the matings in the song neighborhood. Males that were successful were on their call-sites for more time within a day and for a longer season, and they sang more than the unsuccessful males (Payne & Payne 1977). Most males had nearly identical songs, but a few minor variants were observed, though these did not last long as individual song markers because the variants of certain males were copied by their neighbors. To test whether the songs of certain males were copied, the success of males in attracting females and in mating were compared with their song variants and the time when the song variants appeared in the repertoire of each male (Fig. 5.6).

In 1973 the most successful male was RYRB on "junction" call-site, where he was the most successful male in the "Diptera" neighborhood in the previous year as well. He changed the details of all 16 nonmimetic song types between seasons from 1973 through 1974, and his song variants were copied by other males in 1974 (Payne 1985a). The slowest male was RBRG who sang on the margin of the neighborhood and had some songs of "Diptera" dialect and others of "Cowpie" dialect. His songs from 1972 to 1974 were unique variants of the dialect standards. In 1975, he moved 5 km and settled near "Junction" call-site where RYRB had sung; in 1975 another male took over the site. After he dispersed, RBRG changed his songs: he dropped the "Cowpie" songs and his individualistic variants of the "Junction" songs, and he had the local standards of the "Junction" songs in his fourth year (Payne 1985a).

The song variants also sometimes changed within a season. By recording each male repeatedly through the season it was possible to determine which male first changed the details of its songs, and which was copied in turn by the neighboring males in "Cowpie" dialect in 1976. "Mealie" was the most successful male in breeding early in the season and "faucet" male was more successful later. "Mealie" site was the main call-site in the first half of the season (more male visitors, more female visitors, and more copulations), the central location in the dialect area, and the longest-lived (it was seasonally active from February 1972) (Payne 1985a). Songs of "mealie" and his neighbors were copied by the other males within the season (Figs. 5.6, 5.7). "Mealie" site was near a creek and a pit well for water in Banakaila village in the early dry season in 1972 and in 1973. Its advantage shifted when a new windmill provided surface water in 1976; male "faucet" who settled there had a higher mating success than the other males late in the season as water disappeared near the other sites, but the other males had already copied the shift in song of the "mealie" male.

In both years and both populations, the males that were successful in attracting females and in copulating with them were the males whose songs were copied by the other adult males.

Song matching as social mimicry: An experimental test

The remarkable widespread sharing of song types among males that interacted with each other, and the rapid shift

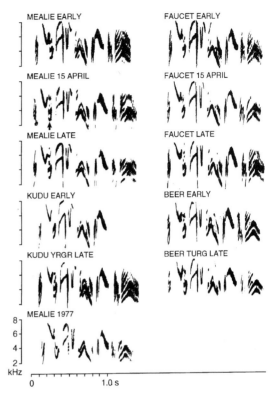

Fig. 5.7. Songs of village indigobirds in the Cowpie song neighborhood in 1976: within-season changes in song type B9. Note the change in shape of the song element indicated by the arrow. Early songs were recorded in March (February for "kudu," April for "beer"), late songs were in May (25 April for "kudu"). Male "mealie" changed first; "faucet," "kudu" and "beer" changed later. Note the continued change in this note in 1977, continuing the trend set by "mealie" in 1976.

of the details of songs from one male to others in his song neighborhood, suggest that song matching is a form of social mimicry and is related to individual recognition in these birds. Although the mimicry hypothesis is based mainly on observation, an experimental test suggests another approach.

We observed a rapid success of a previously unsuccessful male when he assumed the site of a successful male and had the songs of the previous resident. Male RYRB was the most successful breeder in 1972. After he was seen to mate, he was captured and held for a day. Later in the day his call-site was occupied by another male that sang ten song types each apparently identical with RYRB's songs. On the next day the replacement male was visited by females and he copulated at least once. RYRB was then released and after many chases he regained his site, which he kept for the season and the next two breeding seasons. In all three years RYRB had the highest mating success of all males in the population (Payne & Payne 1977). Another male used the site in 1975, was unsuccessful and disappeared. In 1976 the site was used by a male with another set of songs, and only remnants remained of RYRB's songs (Payne 1985a). RYRB's success in attracting females that mated with him was characteristic of the male, and was not simply due to location, which except for its central place in the dialect neighborhood was not obviously better endowed than other sites nearby. The structure of the call-site trees, the amount of thicket cover, the abundance of the grasses (*Echinochloa colonum*, *Setaria* spp.) on whose seeds the birds fed, the proximity of surface water, and the local density of the foster firefinches all were not correlated with the success of the resident male in attracting females or in mating (Payne & Payne 1977), so social behavior rather than habitat features appeared to explain his success.

The success of a replacement male taking over the site within a few hours after the resident was removed from the site, the swift success of the replacement male in attracting and mating with a female, and the near identity of his songs with the songs of the previous resident all lead us to suggest that a male attracts females, and repels other males, that are familiar with the songs. A similar advantage may be gained by any male in the social neighborhood area, where a female compares males as she visits on her sampling route, and males learn to discriminate the songs of their neighbors (Payne 1981b, 1983a, 1985a).

Asocial explanations of song matching

An alternative hypothesis is that habitat differences determine which song themes are used in an area. The habitats around each call-site were compared from vegetation samples and from ground and aerial photographs. The song neighborhoods of village indigobirds did not coincide with any habitat patches. Also, the habitats of the song neighborhoods did not differ in vegetation structure or in floral composition of annual and perennial grasses (the finches feed on the seeds) (Payne 1980, 1987). The lack of a vegetation difference is consistent with the view that the song dialects are the result of social interactions between

birds with these songs, rather than adaptations to the acoustic traits of their habitats.

Genetic differences are unlikely between song neighborhoods as the numbers of birds in each are small and the neighborhoods are not isolated. Dispersal of marked adults between breeding seasons was observed up to 5 km. Birds from more than one neighborhood flock together in the nonbreeding season. A few adult males sang in one neighborhood, then dispersed a few kilometers and established a territory in another one, and switched their songs to match their new neighbors. Also, some males at least three years of age on their old territories switched their songs to match the songs of their new neighbors when they immigrated from 5–10 km away (Payne 1985a,b).

DISCUSSION

Age-based models and social models of song learning

Models of the conditions under which birds learn their songs differ in their emphasis on age and social interactions. In our field studies of song development in indigo buntings and in village indigobirds, age has an effect on learning, but the details of song are learned in the first breeding season (and into later seasons in the indigobirds), and the song is not restricted to the songs heard during the first few weeks of life. In both species the predictions of the social learning model were supported, whereas the predictions of the early-age-dependent model were not. The field observations also were consistent with the experimental laboratory results in these two species.

In indigo buntings, the field observations that support a social model of song learning (and are inconsistent with the predictions of learning in the HY year) are as follows:

1. All SYs that matched a local song matched a song theme that was present in their SY year.
2. Natal return males in their first spring often have first song themes that are unlike any song themes that were present in the natal area in their natal year, or like song themes not given after they had fledged.
3. Natal return males never return with the song of their father.
4. Natal return males never have a song that was present

in their natal year but was not present in their SY year.

5. Immigrant SY males often have song themes that were present in their natal year in the area where they were not born, but never when the song was not also present in their SY year.

6. Natal return males are no more likely to have a first song that was heard locally in their HY year than were immigrant males born elsewhere.

7. SY males change their song themes when they change their social situation in dispersal from one territory to another.

8. When an adult neighbor has more than one song theme, a SY male often copies both of the neighbor's song themes, even though males generally have a repertoire of a single song theme.

All of these results were observed in both study populations, so they may be general for the species. In village indigobirds:

1. adult males sometimes change their song repertoires when they move from one neighborhood to another;

2. adult males sometimes change their song repertoires when new neighbors replace old and the new neighbors have a new set of song themes;

3. adult males modify their song themes every season to new versions that were never heard before in previous years, and these improvisations accumulate across years within individual birds;

4. all birds share songs and make the same changes in song during a season when they visit each other's territories and court the same females;

5. the songs of the individual male that attracts the most females are copied by their neighbors.

Again these results were seen in two song populations, and they may be general for the species. In both species the context provides strong evidence against an HY determination of adult song themes.

Songbirds are often viewed as learning a song in a "critical period" in early life. This view is based on experiments with isolated captive birds. Captive chaffinches did not develop normal songs when they were kept apart from conspecifics in their first summer, but did when they were kept with other chaffinches (Thorpe 1958). This study is often cited as evidence for song learning in an early sensi-

tive period, yet it also reported that "the details . . . are worked out . . . in the following spring." Later, chaffinch were found to develop a normal song when they heard a tape-recording in their natal summer or in the following spring (Slater & Ince 1982; Thielcke & Krome 1989). The view is also based on fostering experiments, where young bullfinch (*Pyrrhula pyrrhula*) were reared by canaries, and zebra finch were reared by Bengalese finch (*Lonchura striata*) (Nicolai 1959; Immelmann 1969). In a variation of this view, a young bird may hear many song themes and store them in memory, then selectively use and retain the songs it hears again in its first breeding season. It may produce a wide range of songs early in its first breeding season, then delete the songs it does not hear from other birds. In this model of "selective attrition," song learning is accompanied by "song overproduction" before a bird settles upon its definitive song theme (Marler & Peters 1982) and "action-based learning" where it compares its songs with the songs of its neighbors (Marler & Nelson 1993; Nelson & Marler 1994). In this view the songs available for performance are thought to originate in early experience. Song competence may be larger than song performance, which depends on social experience in a bird's first spring. These variations on the theme of age-dependent learning allow input by social conditions later in life, where the range of song behaviors available for a bird are determined in its natal season.

In indigo buntings, Margoliash *et al.* (1991) observed that both SY males and adults give a variable or plastic song at very low sound levels as well as a standard or "canonical" song theme, and in their bouts of variable song they sometimes include song elements not present in the standard song theme. The developmental source of these song elements is unknown. The birds they observed, including HY "immatures" captured in autumn and held into their SY season, had all heard other singing buntings in captivity when they were recorded, and their "isolated" birds were in individual cages within vocal and visual contact with other buntings. It is unknown what song elements they may have recalled from hearing the older males during the season when they were born, as the autumn birds were captured as migrants and the song elements they may have heard were unknown. Nevertheless, the difference between the normal song elements of these birds and the abnormal song elements of hand-reared birds that had only heard songs as nestlings (Rice &

Thompson 1968; Payne 1981a) suggests that the HY may learn the song elements in their natal season. The autumn-caught birds did copy the song elements and element sequence in song of older adult song models in their first spring (Margoliash *et al.* 1994). The elements can also be acquired by older buntings, because naive captives that first heard normal elements and songs recorded from wild buntings added these normal elements to their own songs after their naive songs had crystallized and the birds were 18 months of age (Rice & Thompson 1968).

The social processes involved in song learning may include an assessment of a singing male. There may be a variety of processes involved, such as being chased by a neighboring male when a male attempts to settle near a neighbor's territory, and counter-singing with a neighbor. In other cases the behavior may involve comparison of several neighbors as a male samples their singing rate or their response to his intrusion, then styles his song upon the song theme of a male that meets a criterion of assessment (Payne 1985a; Payne & Payne 1993b). It is a challenge to develop experiments where the social conditions in the field are approached in the laboratory. Birds in small cages do not learn their songs as rapidly or completely as birds in large aviaries or in the field, especially where the range of behavior of potential tutors toward their candidate copiers includes chases from nesting areas. For example, SY indigo buntings show little evidence of copying the songs of live tutors when candidates and tutors are in small cages, and HY males captured in autumn copy the songs of their first tutor social companions in their first spring, but not the songs of later tutors (Margoliash *et al.* 1994). In contrast, in the field some SY males that we observed copy a series of their neighbors as they move from one territory to another (Figs. 5.1, 5.2). In our captive buntings, all birds with two companions caged separately next to them copied the companion that they could see and chase with, rather than the one that they could only hear. Also, the captive SY male that developed a song most like the song of his social companion was in an aviary and interacted with a wild indigo bunting outside the aviary (Payne 1981a). An experimental design that includes a test of the scale of social conditions may be hepful in determining the scale that is appropriate to the normal field conditions of song learning.

Other experiments that were designed to test the importance of social behavior have not often incorporated

a range of social conditions to determine which are necessary or sufficient for song learning. In experiments with two species, social interaction appeared to be effective when recorded song was not, even when there was no visual or contact interaction with another singing bird. In short-toed tree-creepers the design allowed no visual contact, and the interaction involved counter-singing (Thielcke & Krome 1989); in nightingales (*Luscinia megarhynchos*) the song learning was facilitated by the presence of a human when the recorded song was broadcast, and the human was the "social tutor" (Hultsch & Todt 1989). Where there is no difference in experimental results, then the results might differ in another experimental design. Many observations will be necessary to determine the social conditions in learning a song, and the variety of styles of social learning. In experiments when a young starling had a live tutor, it appeared attentive, perched near the tutor and oriented towards him (Chaiken *et al.* 1993). Field observations are needed on the interactive context of song learning. We have no observations to suggest that adult songbirds attempt to gain the attention of a candidate, socially reinforce him and "teach" their songs (Payne & Payne 1993b), in the sense that behaviors of more social animals may be culturally transmitted (Whiten 1989; Boesch 1991; Caro & Hauser 1992; McGrew 1992; Wrangham *et al.* 1994).

Field and experimental observations of song learning in the indigo buntings and indigobirds are consistent with a model of song learning well after the age of independence, rather than one where learning is restricted to the songs heard early in life. Song development is more open in the indigobirds as a male continues to modify his songs through his adult life, when he copies the changing songs of the breeding male in his neighborhood, and when he disperses to a new neighborhood. Why the successful breeding male improvises upon his own songs, and why he changes the details of some song types within a season and all song types between seasons, are unknown (Payne 1985a). The rapidity with which other males track his current versions in song as they change between years might involve a greater rate of generation of song change between seasons, as most changes appear early in a breeding season. Because the songs are in a continual state of cultural evolution, the songs as they are given in one year were never heard in a previous year (Payne 1985a). This argues against the selective attrition hypothesis, which assumes that only the songs heard early in life are retained for later performance. Song development in the indigo buntings is more closely controlled by the matching of a male neighbor, and a male often changes from one song theme to another as a SY, but most males retain their song theme in later years. Also, in the buntings a male that heard a tape-recording did not copy the song in as much detail as when he heard the song of a live tutor. In both species the males learned their songs from older, more successful males, but not the songs they only heard.

A probabilistic approach to the time of song learning

Is it possible that we overlooked songs a male had heard in a previous year, and did not sing until his SY or later year, especially if he heard another male in the later year and this facilitated his performance of the "new" song theme? Possible perhaps, but not likely. In the indigo buntings, a male may have heard another male in spring migration, or outside the local study area where we tape-recorded their songs. However, it is unlikely that a male with the song of a local bird had also heard the song elsewhere. We estimate this probability based on the variation in bunting song and on the species population size. Bunting song is a series of six to eight different kinds of song element arranged in a regular sequence (Thompson 1968, 1970; Emlen 1972; Payne et al. 1981); we ignore variations in whether the elements are single, double, or multiple units. Thompson (1970) estimated that indigo bunting songs had 98 distinct song elements or "figures"; from our field work we recognize another 18 (Margoliash et al. 1991). Elements number 1 to 8 in the catalog of Thompson (1970) occur at the introduction of a song (Shiovitz 1975), and there is no regularity across song themes in the other song elements in their position and sequence in song. From these features, excluding the introductory element we estimate that a typical song has at least five different song elements in a sequence, and about 100 elements are used in the species' repertoire. The number of permutations of these elements in a song of the species is described as $_nP_r$, where P is the number of permutations, n is the number of song elements in a song, and r is the number of recognizable song elements, is described by the expression,

$$_nP_r = n(n-1)(n-2)\ldots(n-r+1).$$

For the bunting songs, $n=100$ and $r=5$, and from this we estimate the number of allowable sequences of elements to be 9.03×10^9 song themes. This value of possible song themes in buntings is much higher than the estimate of the number of individuals in the species. A species population size of 20 million buntings was estimated on the number of birds detected in continent-wide Breeding Bird Surveys, on the detectability of buntings in the surveys, and on the known density of birds in the study areas where birds were banded (Payne 1992). If the estimate of song diversity is reasonable, then it is unlikely that a bunting as a fledgling or juvenile would encounter a song theme that does not occur in its natal area, learn the song, and then repeat the song in its SY year when it heard the same song from another older immigrant into its natal area. We think it more reasonable to conclude that the buntings learned the songs as SY males, than to propose an unlikely coincidence of events in each of our 56 SY males that matched a song theme which was not present locally in their HY year and first appeared in their SY year. Also the occurrence of more than 30% of the song themes of both return and immigrant SY males did not match a local theme of their natal season, and the lack of difference between return and immigrant males, both are inconsistent with a hypothesis of song learning in the HY year.

It is also unlikely that the song changes from year to year in the village indigobirds were acquired in their HY year. The complexity of an indigobird song is similar to that of the indigo buntings (five to eight different elements in a standard sequence), and each male uses 20–24 song type themes rather than a single one. The number of elements in the species' repertoire is unknown, though one of our students counted to 460 distinct elements before terminating the exercise, so the songs are at least as variable as in the buntings. Also, when 20 000 songs were recorded from a male indigobird (for each of two adults in 1979), none of the songs matched any of the thousands of songs that were recorded and audiospectrographed there in the previous seven years. The songs were derived in a lineage of accumulated changes within song types, and these improvisations of a male indigobird were copied by his adult neighbors (Payne 1985a). Insofar as the songs were adequately sampled in the previous years, it seems more reasonable that the birds learned and changed the songs as adults than that they recalled the songs heard only in their HY year.

Ecological consequences of social learning of song

The adaptive significance of matching the song of another male may be related to individual recognition. Male indigo buntings respond differently to the songs of neighbors and of strangers, so that buntings can recognize individuals by their songs (Emlen 1971b). If a SY male styles his song upon the song of an adult neighbor, then other buntings may mistakenly identify him as the older male, and this may aid the SY in maintaining a new territory, when other males have had experience with the song theme and associate it with an aggressive interaction with the older male (Payne 1983b). Similar processes may occur in other songbirds as well (Beecher *et al.* 1994). The concept of "social mimicry" in song leads to a test of mimicry theory, which predicts that the success of an individual will vary with the ratio of mimics to models (Fisher 1958; Huheey 1988). Predictions of this ecological mimicry hypothesis were tested for the song themes. When the reproductive success of indigo buntings was compared with the number of adult "song models" and SY "song mimics" with a common theme were compared, there was no strong evidence in support of deceptive Batesian social mimicry or Müllerian social mimicry, when directed either to other males or to their mates (Payne 1983b; Payne *et al.* 1988).

Another possible advantage in matching the song of a neighboring adult is that a SY male may gain an advantage through a social relationship with the male he copies. No cooperation or coalitions were observed between singing males. However, the rates of counter-singing and chasing are greater between neighbors with matching songs than neighbors with different song themes (Payne 1983b). Playback tests showed no consistent difference in response of a bunting to his own song theme when it was shared by a neighbor and to a song of another area. Further experiments are needed to compare the response of adults before and after a neighbor has matched their songs.

The ecological consequences of song matching in the village indigobirds appears similar insofar as it aids a male to gain a breeding site through individual identification as another male. The shared songs mark the set of males that compete for females and call-sites; they visit each other but these visits are competitive rather than friendly in nature.

Intruding males are chased by the resident male, as when they stay nearby they interfere with his mating. They copy the song variants of the residents that defend the best call-sites and have the highest mating success. The more frequent intrusions at the sites of successful males than of their neighbors suggests that intruders sample the singing residents. The behaviors sampled and compared may include the proportion of times the males are on the call-site and their response to an intruder. Females may use the same cues of site attendance and visitation to assess the males in their choice of a mate. By copying a successful male, another male may establish its own site in competition with his neighbors, and then attract a female (Payne & Payne 1977; Payne 1981b, 1983a, 1985a). The changing details in the songs of the most successful males were copied by other males both within a season and between years. Also, a replacement male with the same songs as the resident male that he replaced was successful in defending a site against other males and in attracting a breeding female. As in the indigo buntings, additional field experiments are needed to test the idea that song matching has an adaptive basis in social mimicry.

Field observations, field experiments, and laboratory studies all are complementary and necessary approaches to determine the time and conditions of song learning. Nevertheless, the field observations give the more compelling evidence of behavior under natural conditions; that is, the behavior observed reveals the realized performance rather than the possible competence. We suggest that where observations of marked, free-living birds in the field and experiments with captive birds give different results, the observed field behavior has epistemic primacy. Also, in the field it is possible to test whether different song and social behaviors are associated with fitness consequences. However, the social mechanisms remain to be determined. Each species has its unique history, and it will be of interest to compare species to trace the song traits and learning style and social behavior across evolutionary lineages. We do not know whether the program of song learning in the indigo buntings and indigobirds is peculiar to their lineages or their life styles, or whether it is widespread among songbirds, because most experimental studies have not been carried out where the birds are free to interact and to choose a male to copy. The time of learning varies with social conditions, and social conditions vary with time. We suspect that other species will condition their song learning on social context as well, and field observations will suggest more realistic experimental designs on song learning.

CONCLUSIONS

We compared two models of song development with field observations and experiments in two species of small songbirds with different life styles. A model of song learning in a critical period in the first few weeks after fledging, and a model of song learning as a consequence of social interactions with other males after the age of dispersal and establishing a breeding site in the young adults, each led to a different set of predictions in a design of strong inference.

Songs were sampled intensively in color-banded populations of wild birds and were used as markers of the time and place of song learning. In both the seasonally migratory, socially monogamous and parental indigo bunting, and the resident African brood-parasitic village indigobird, the predictions of a social learning model were more successful in accounting for the song behavior than were the predictions of a model of early learning.

In the buntings, birds that return to their natal area never sing the song theme of their father, never sing the song theme that was present in their natal area in their post-fledging period unless the song theme also is present in their SY year, and often sing a song theme that first appeared in the area in their SY year and was not present in their natal HY year. Also, natal returns were no more likely than immigrants born outside the study area to sing a song theme that was present on the area in their post-fledging period. In addition, when an adult neighbor has more than one song theme, a SY male often copies both of his songs, even though the species norm is a repertoire of a single song. The social circumstances of song change include dispersal when a SY male with one song theme moves to a territory in another song neighborhood.

In the brood-parasitic indigobirds, males change their details of their entire repertoire of 20+ songs in each year, and they match the current improvisations of the most succesful breeding bird; these versions of songs were not present in previous years. The males that make the change are the males that interact repeatedly with visits to each other's call-sites and compete for the same breeding females that sample their songs on their own call-sites. Also, adult males change their entire song repertoires,

when they disperse as adults from one song-dialect neighborhood to another and copy their new neighbors' songs, and when they remain in their old neighborhood but new immigrants with other song dialects settle nearby and interact socially with them.

In both species, field experiments on song learning and observation of song development of captive birds gave results consistent with the idea that social interaction of mature birds is a usual condition of song learning in the field.

A second line of reasoning for the time and place of song learning was a probabilistic argument based on population size and the structural complexity and diversity of song in each species. Because it is highly unlikely that a male could learn a local song in a remote area, we find that the time and place of song learning usually is in the first breeding season and on the breeding area.

We suggest that when field observations and laboratory experiments give different interpretations of the time and social conditions of song learning, then additional experimental work be designed to replicate the normative field conditions.

ACKNOWLEDGEMENTS

We thank our field assistants for their observations. Amtrak and landowners allowed access to their lands at Niles, and the University of Michigan provided facilities for the studies at the E. S. George Reserve. The Department of Game, Fisheries and National Parks, Zambia, permitted studies at Lochinvar National Park. K. L. Fiala developed the program MATCH for comparing songs. For comments we thank L. F. Baptista, M. Hausberger, C. Sims Parr, C. T. Snowdon, and J. L. Woods. M. Van Bolt helped with illustrations. Field research was supported by the National Science Foundation and the National Geographic Society.

REFERENCES

Baptista, L. F. & Morton, M. L. (1988). Song learning in white-crowned sparrows: From whom and when. *Animal Behaviour* **36**: 1753–64.

Baptista, L. F. & Petrinovich, L. (1984). Social interaction, sensitive phases and the song template hypothesis in the white-crowned sparrow. *Animal Behaviour* **32**: 172–81.

Baptista, L. F. & Petrinovich, L. (1986). Song development in the white-crowned sparrow: Social factors and sex differences. *Animal Behaviour* **35**: 1359–71.

Beecher, M. D., Campbell, E. & Stoddard, P. K. (1994). Correlation of song learning and territory establishment strategies in the song sparrow. *Proceedings of the National Academy of Sciences, USA* **91**: 1450–4.

Bensch, S. & Hasselquist, D. (1992). Evidence for active female choice in a polygamous warbler. *Animal Behaviour* **44**: 301–11.

Boesch, C. (1991). Teaching among wild chimpanzees. *Animal Behaviour* **41**: 530–2.

Caro, T. M. & Hauser, M. D. (1992). Is there teaching in nonhuman animals? *Quarterly Review of Biology* **67**: 151–73.

Catchpole, C. K. & Slater, P. J. B. (1995). *Bird Song, Biological Themes and Variations*. Cambridge University Press, Cambridge.

Chaiken, M., Böhner, J. & Marler, P. (1993). Song acquisition in European starlings (*Sturnus vulgaris*): A comparison of the songs of live-tutored, untutored, and wild-caught males. *Animal Behaviour* **46**: 1079–90.

Clayton, N. S. (1989). Song, sex and sensitive phases in the behavioural development of birds. *Trends in Ecology and Evolution* **4**: 82–4.

Emlen, S. T. (1971a). Geographic variation in indigo bunting song (*Passerina cyanea*). *Animal Behaviour* **19**: 407–8.

Emlen, S. T. (1971b). The role of song in individual recognition in the indigo bunting. *Zeitschrift für Tierpsychologie* **28**: 241–6.

Emlen, S. T. (1972). An experimental analysis of the parameters of bird song eliciting species recognition. *Behaviour* **41**: 130–71.

Fisher, R. A. (1958). *The Genetical Theory of Natural Selection* 2nd edn. Dover, New York.

Gibbs, H. L. (1990). Cultural evolution of Darwin's finches: Frequency, transmisson, and demographic success of different song types in Darwin's medium ground finches (*Geospiza fortis*). *Animal Behaviour* **39**: 253–63.

Huheey, J. E. (1988). Mathematical models of mimicry. *American Naturalist* **131**: S22–S41.

Hultsch, H. & Todt, D. (1989). Song acquisition and acquisition constraints in the nightingale, *Luscinia megarhynchos*. *Naturwissenschaften* **76**: 83–5.

Immelmann, K. (1969). Song development in the zebra finch and other estrildid finches. In *Bird Vocalizations*, ed. R. A. Hinde, pp. 61–81. Cambridge University Press, Cambridge.

Janetos, A. C. (1980). Strategies of female mate choice: A theoretical analysis. *Behavioral Ecology and Sociobiology* **7**: 107–12.

Jenkins, P. F. (1978). Cultural transmission of song patterns and dialect development in a free-living bird population. *Animal Behaviour* **26**: 50–78.

Johnston, D. W. & Downer, A. C. (1968). Migratory features of the indigo bunting in Jamaica and Florida. *Bird-Banding* **34**: 277–93.

Komdeur, J., Huffstadt, A., Prast, W., Castle, G., Mileto, R & Wattel, J. (1995). Transfer experiments of Seychelles warblers to new islands: Changes in dispersal and helping behaviour. *Animal Behaviour* **49**: 695–708.

Kroodsma, D. E. (1974). Song learning, dialects, and dispersal in the Bewick's wren. *Zeitschrift für Tierpsychologie* **35**: 352–80.

Kroodsma, D. E. (1978). Aspects of learning in the ontogeny of bird song: Where, from whom, when, how many, which, and how accurately? In *Development of Behavior*, ed. G. Burghardt & M. Bekoff, pp. 215–230. Garland, New York.

Kroodsma, D. E. & Pickert, R. (1980). Environmentally dependent sensitive periods for avian vocal learning. *Nature* **288**: 477–9.

Kroodsma, D. E. & Pickert, R. (1984). Sensitive phases for song learning: Effects of social interaction and individual variation. *Animal Behaviour* **32**: 389–94.

Margoliash, D., Staicer, C. A. & Inoue, S. A. (1991). Stereotyped and plastic song in adult indigo buntings, *Passerina cyanea*. *Animal Behaviour* **42**: 367–88.

Margoliash, D., Staicer, C. & Inoue, S. A. (1994). The process of syllable acquisition in adult indigo buntings (*Passerina cyanea*). *Behaviour* **131**: 39–64.

Marler, P. (1970). A comparative approach to vocal learning: Song development in white-crowned sparrows. *Journal of Comparative and Physiological Psychology* **71**: 1–25.

Marler, P. & Nelson, D. A. (1993). Action-based learning: A new form of developmental plasticity in bird song. *Netherlands Journal of Zoology* **43**: 91–103.

Marler, P. & Peters, S. (1982). Developmental overproduction and selective attrition: New processes in the epigenesis of birdsong. *Developmental Psychobiology* **15**: 369–78.

Marler, P. & Peters, S. (1988). Sensitive periods for song acquisition from tape recordings and live tutors in the swamp sparrow, *Melospiza georgiana*. *Ethology* **77**: 76–84.

Marler, P. & Waser, M. S. (1977). The role of auditory feedback in canary song development. *Journal of Comparative and Physiological Psychology* **91**: 8–16.

McGregor, P. K. & Krebs, J. R. (1982). Song types in a population of great tits (*Parus major*): Their distribution, abundance and acquisition by individuals. *Behaviour* **79**: 126–52.

McGregor, P. K. & Krebs, J. R. (1989). Song learning in adult great tits (*Parus major*): Effects of neighbours. *Behaviour* **108**: 139–59.

McGregor, P. K. & Thompson, D. B. A. (1988). Constancy and change in local dialects of the corn bunting. *Ornis Scandinavica* **19**: 153–9.

McGregor, P. K., Walford, V. R. & Harper, D. G. C. (1988). Song inheritance and mating in a songbird with local dialects. *Bioacoustics* **1**: 107–29.

McGrew, W. C. (1992). *Chimpanzee Material Culture*. Cambridge University Press, Cambridge.

Millington, S. J. & Price, T. D. (1985). Song inheritance and mating patterns in Darwin's finches. *Auk* **102**: 342–6.

Nelson, D. A. & Marler, P. (1994). Selection-based learning in bird song development. *Proceedings of the National Academy of Science, USA* **91**: 10498–501.

Nice, M. M. (1943). Studies in the life history of the song sparrow. II. The behavior of the song sparrow and other passerines. *Transactions of the Linnaean Society of New York* **6**.

Nicolai, J. (1959). Familientradition in der Gesangentwicklung des Gimpels (*Pyrrhula pyrrhula* L.). *Journal für Ornithologie* **100**: 39–46.

Parker, G. A. (1983). Mate quality and mating decisions. In *Mate Choice*, ed. P. Bateson, pp. 141–66. Cambridge University Press, Cambridge.

Payne, R. B. (1973). Behavior, mimetic songs and song dialects, and relationships of the parasitic indigobirds (*Vidua*) of Africa. *Ornithological Monographs* **11**.

Payne, R. B. (1979). Song structure, behaviour, and sequence of song types in a population of village indigobirds, *Vidua chalybeata*. *Animal Behaviour* **27**: 997–1013.

Payne, R. B. (1980). Seasonal incidence of breeding, moult and local dispersal of red-billed firefinches *Lagonosticta senegala* in Zambia. *Ibis* **122**: 43–56.

Payne, R. B. (1981a). Song learning and social interaction in indigo buntings. *Animal Behaviour* **20**: 688–97.

Payne, R. B. (1981b). Population structure and social behavior: models for testing the ecological significance of song dialects in birds. In *Natural Selection and Social Behavior*, ed. R. D. Alexander & D. W. Tinkle, pp. 108–20. Chiron Press, New York.

Payne, R. B. (1982). Ecological consequences of song matching: Breeding success and intraspecific song mimicry in indigo buntings. *Ecology* **63**: 401–11.

Payne, R. B. (1983a). Bird songs, sexual selection, and female mating strategies. In *Social Behavior of Female Vertebrates*, ed. S. K. Wasser, pp. 55–90. Academic Press, New York.

Payne, R. B. (1983b). The social context of song mimicry: Song-matching dialects in indigo buntings. *Animal Behaviour* **31**: 788–805.

Payne, R. B. (1985a). Behavioral continuity and change in local song populations of village indigobirds *Vidua chalybeata*. *Zeitschrift für Tierpsychologie* **70**: 1–44.

Payne, R. B. (1985b). Song populations and dispersal in steelblue and purple widowfinches. *Ostrich* **56**: 135–46.

Payne, R. B. (1987). Song dialects and neighborhood habitats in the indigobirds *Vidua chalybeata* and *V. purpurascens* at Lochinvar National Park, Zambia. *Journal of Field Ornithology* **58**: 152–70.

Payne, R. B. (1989). Indigo bunting. In *Lifetime Reproduction in Birds*, ed. I. Newton, pp. 153–72. Academic Press, New York.

Payne, R. B. (1990). Song mimicry by the village indigobird (*Vidua chalybeata*) of the red-billed firefinch (*Lagonosticta senegala*). *Vogelwarte* **35**: 321–8.

Payne, R. B. (1991). Natal dispersal and population structure in a migratory songbird, the indigo bunting. *Evolution* **45**: 49–62.

Payne, R. B. (1992). Indigo bunting. In *The Birds of North America*, no. 4, ed. A. Poole, P. Stettenheim & F. Gill, pp. 1–24. American Ornithologists' Union, Philadelphia.

Payne, R. B. (1996). Song traditions in indigo buntings: Origin, improvisation, dispersal, and extinction in cultural evolution. In *Ecology and Evolution of Acoustic Communication in Birds*, ed. D. E. Kroodsma & E. H. Miller, pp. 198–220. Cornell University Press, Ithaca.

Payne, R. B. & Payne, K. (1977). Social organization and mating success in local song populations of village indigobirds, *Vidua chalybeata*. *Zeitschrift für Tierpsychologie* **45**: 113–73.

Payne, R. B. & Payne, L. L. (1989). Heritability estimates and behaviour observations: Extra-pair matings in indigo buntings. *Animal Behaviour* **38**: 457–67.

Payne, R. B. & Payne, L. L. (1993a). Breeding dispersal in indigo buntings: Circumstances and consequences for breeding success and population structure. *Condor* **95**: 1–24.

Payne, R. B. & Payne, L. L. (1993b). Song copying and cultural transmission in indigo buntings. *Animal Behaviour* **46**: 1045–65.

Payne, R. B., Payne, L. L. & Doehlert, S. M. (1987). Song, mate choice and the question of kin recognition in a migratory songbird. *Animal Behaviour* **35**, 35–47.

Payne, R. B., Payne, L. L. & Doehlert, S. M. (1988). Biological and cultural success of song memes in indigo buntings. *Ecology* **69**: 104–17.

Payne, R. B., Payne, L. L., Nhlane, M. E. D. & Hustler, K. (1993). Species status and distribution of the parasitic indigobirds *Vidua* in east and central Africa. *Proceedings of the VIIIth Pan-African Ornithological Congress, Bujumbura, 1992*, pp. 40–52.

Payne, R. B., Thompson, W. L., Fiala, K. L. & Sweany, L. L.

(1981). Local song traditions in indigo buntings: Cultural transmission of behavior patterns across generations. *Behaviour* 77: 199–221.

Payne, R. B. & Westneat, D. F. (1988). A genetic and behavioral analysis of mate choice and song neighborhoods in indigo buntings. *Evolution* 42, 935–47.

Petrinovich, L. (1985). Factors influencing song development in the white-crowned sparrow (*Zonotrichia leucophrys*). *Journal of Comparative Psychology* 99, 15–29.

Petrinovich, L. (1988). Individual stability, local variability and the cultural transmission of song in white-crowned sparrows (*Zonotrichia leucophrys nuttalli*). *Behaviour* 107: 208–40.

Petrinovich, L. & Baptista, L. F. (1987). Song development in the white-crowned sparrow: Modification of learned song. *Animal Behaviour* 35: 961–74.

Platt, J. (1964). Strong inference. *Science* 146: 347–53.

Rice, J. O. & Thompson, W. L. (1968). Song development in the indigo bunting. *Animal Behaviour* 16: 462–9.

Shiovitz, K. (1975). The process of species-specific song recognition in the indigo bunting (*Passerina cyanea*) and its relationship to the organization of avian acoustic behaviour. *Behaviour* 55: 128–79.

Shiovitz, K. A. & Thompson, W. L. (1970). Geographical variation in song composition of the indigo bunting, *Passerina cyanea*. *Animal Behaviour* 18: 151–8.

Slater, P. J. B. (1989). Bird song learning: Causes and consequences. *Ethology, Ecology and Evolution* 1: 19–46.

Slater, P. J. B. & Ince, S. A. (1982). Song development in chaffinches: What is learnt and when? *Ibis* 124: 21–6.

Thielcke, G. (1984). Gesangslernen beim Gartenbaumläufer (*Certhia brachydactyla*). *Vogelwarte* 32: 282–97.

Thielcke, G. & Krome, M. (1989). Experimente über sensible Phasen und Gesangsvariabilität beim Buchfinken (*Fringilla coelebs*). *Journal für Ornithologie* 130: 435–53.

Thompson, W. L. (1968). The songs of five species of *Passerina*. *Behaviour* 31: 261–87.

Thompson, W. L. (1970). Song variation in a population of indigo buntings. *Auk* 87: 58–71.

Thorpe, W. H. (1958). The learning of song patterns by birds, with especial reference to the song of the chaffinch *Fringilla coelebs*. *Ibis* 100: 535–70.

Trainer, J. M. (1989). Cultural evolution in song dialects of yellow-rumped caciques in Panama. *Ethology* 80: 190–204.

Verner, J. (1976). Complex song repertoire of male long-billed marsh wrens in eastern Washington. *Living Bird* 14: 263–300.

Westneat, D. F. (1988). Male parental care and extra-pair copulations in the indigo bunting. *Auk* 105: 149–60.

Westneat, D. F. (1990). Genetic parentage in the indigo bunting: A study using DNA fingerprinting. *Behavioral Ecology and Sociobiology* 27: 67–76.

Whitten, A. (1989). Transmission mechanisms in primate cultural evolution. *Trends in Ecology and Evolution* 4: 61–2.

Wiley, R. H. (1991). Associations of song properties with habitats for territorial oscine birds of eastern North America. *American Naturalist* 138: 973–93.

Wrangham, R. W., McGrew, W. C., de Waal, F. B. M. & Heltne, P. G. (eds.) (1994). *Chimpanzee Cultures*. Harvard University Press, Cambridge, MA.

Zann, R. (1990). Song and call learning in wild zebra finches in south-east Australia. *Animal Behaviour* 40: 811–28.

6 Vocal learning in wild and domesticated zebra finches: Signature cues for kin recognition or epiphenomena?

RICHARD ZANN

INTRODUCTION

In the mid-1960s Klaus Immelmann began a series of experiments with domesticated zebra finches (*Taeniopygia guttata*) that led to seminal contributions to two related fields, song learning and sexual imprinting. Immelmann (1969) manipulated the auditory and social experiences of young males in their first 100 days of life and found that those denied any contact with singing males failed to sing the normal zebra finch song at adulthood. He concluded that song in this species, like that of most songbirds, must be learned. When he isolated young from foster parents (Bengalese finches (*Lonchura striata* var. *domestica*)) at different ages he found that the sensitive phase for song acquisition began as early as 25 days of age, about a week after fledging, and ended around 80 days of age, around the onset of sexual maturity. Furthermore, young males did not learn from just any singing adult, but preferred to copy from the male with whom they formed a personal bond. In most instances this was the father or foster father and the bond was based primarily on the provisioning relationship, the most basic filial bond. Immelmann (1969) hypothesized that wild zebra finches would be likely to learn the songs of their fathers, and an early end to the sensitive phase was necessary in order to prevent learning from heterospecific estrildines.

The experimental possibilities raised by Immelmann's intriguing study of song eventually stimulated a steady series of follow-up experiments by other researchers, in particular, by P. J. B. Slater and coworkers, who used song learning in domesticated zebra finches as a model for teasing apart the subtle interactions involved in the development of behaviour (for a review, see Slater *et al.* 1988). F. Nottebohm and his group, among others, also used song in domesticated zebra finches as a model for investigating the neurobiology of learning – specifically, the nexus between brain structures and vocal acquisition, production and perception (for a review, see Nottebohm 1993).

There are a number of reasons why the Australian subspecies of zebra finch (*T. g. castanotis*) (as distinct from that inhabiting the Lesser Sundas of Indonesia, *T. g. guttata*) has become one of the pre-eminent avian models in captive studies of behavior and evolution. First, domesticated strains adapt readily to laboratory conditions. These small birds require minimum standards of aviculture and breed freely and continuously with no regard for season (Immelmann 1965). Second, their behaviors, while complex and diverse, are amazingly tractable to many research techniques. Their vocalizations, in particular, are especially convenient for study: they are stereotyped and permanent and readily elicited throughout the year. Fortunately, the song's simple structure and short duration make it relatively easy to capture, measure and characterize, yet it is sufficiently complex to reflect subtle processes of development and control. A third advantage is the short maturation interval – experimental effects on song development in zebra finches can be measured three months after hatching, whereas most song birds must be approaching 12 months of age before the full song has developed (for some exceptions, see Baptista & Gaunt, Chapter 3). Accordingly, no songbird, the canary included, has been investigated as intensively as has the zebra finch by both developmental ethologists and neurobiologists.

While laboratory workers have enthusiastically exploited the advantages of captive zebra finches, their findings on vocal development, fascinating as they are, do not tell us what happens in nature, but indicate only a restricted range of possibilities that arise under confined (and artificial) conditions. Studies of free-living zebra finches are critical to an understanding of the development and function of vocal communication in the species. However, such investigations not only are beset by the normal experimental limitations of field studies but suffer from extreme losses of experimental subjects from study populations due to exceptionally high levels of mortalities and emigration (below). Therefore, both field and laboratory approaches are necessary to elucidate the processes of vocal development in zebra finches and this is an underlying theme of this review.

In this chapter I first provide the background on the social structure and vocal system of wild Australian zebra finches. Next, I describe how learned vocalizations develop in the species in both the domesticated and wild populations, highlighting the nature and timing of the important social influences on the developmental processes. Finally I attempt to relate the developmental strategies to the social environments experienced by wild zebra finches and speculate on possible fitness consequences.

SOCIAL STRUCTURE OF WILD ZEBRA FINCHES

This highly social subspecies, which occurs throughout the arid and semi-arid parts of continental Australia, lives in more or less permanent colonies focused on nests for breeding and roosting (Zann 1994). Colonies are open social units in which mobility of members is extremely high. A continual change in membership occurs throughout the year due to the regular arrival and departure of pairs and unmated individuals as they move around a vast, drifting home range in small flocks in search of limited resources (Zann & Runciman 1994). Mobility is higher in more arid parts of the range. Sex-biased philopatry does not occur, although age of dispersal appears to vary according to available resources in males, but not in females.

Breeding seasons are long and flexible throughout the range (Zann 1996) and in central Australia occur at any time of the year, provided rains, which are erratic and aseasonal, stimulate the growth and seeding of grasses (Zann *et al.* 1995). Young are capable of breeding at 60–70 days of age and their breeding success is indistinguishable from that of experienced pairs. Precocial pairs may constitute up to 44% of pairs breeding in the second half of the

breeding season (Zann 1994). Consequently, there has been strong selection for early maturation, high fecundity and proficient parental care in this species. This life history strategy is a response to the very high rates of juvenile mortality: only 20% of fledglings survive to sexual maturity with 69% lost during the 18-day interval between fledging and nutritional independence. Adult losses range from 72–82% per annum across colonies, but mortalities are heavily confounded with dispersal. Nest predation can be as high as 66% (Zann & Runciman 1994).

Wild zebra finches live in a state of life-long social monogamy in which the pair bond is tight throughout the year. Nevertheless, re-pairing is frequent due to partner loss. Sexes do not differ in longevity, age at pair formation, or lifetime breeding success (Zann 1994). DNA fingerprinting showed that extra-pair paternity is low (2.4%), but intraspecific brood parasitism reasonably high (11% offspring, 34% of broods; Birkhead *et al.* 1990).

THE VOCAL SYSTEM

Throughout the year zebra finches are highly vocal in the wild and in captivity. In addition to the song, adult zebra finches have ten distinct vocalizations, of which the Distance Call is the most important: it is the only loud vocal signal emitted and is given in a variety of contexts (Zann 1996). The acoustic structure of the Distance Call is sex specific (Zann 1984), and males must learn it from the father in the first 40 days of life, but that of the female is not learned (Zann 1985).

The elements, or syllables, that constitute zebra finch song are organized into stereotyped phrases, or song units, that are repeated, without pause, one or more times each singing performance (Fig. 6.1). Wild birds sing phrases with a mean of 6.7 (mostly different) elements organized into a fixed sequence with a mean duration of 0.86 second, while domesticated zebra finches have phrases with a similar number of elements, but sing them at a faster rate (Zann 1993a). Fourteen distinct types of element are found among wild birds and these can be divided into two categories: call-like and noncall-like elements. The majority of elements (67%) are call-like and identical in structure to the three most common calls uttered by zebra finches, i.e., the Tet, Stack and Distance Calls, whereas noncall-like elements are unique to the song, yet are believed to be remote derivations of the Distance Call.

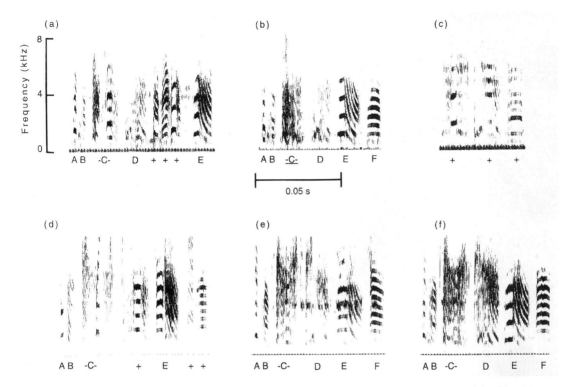

Fig. 6.1. Songs of three generations of one family of wild zebra finches. Songs of the grandfather (a), father (b) and four grandsons (c–f) are shown. Elements that match are given the same letter; underlined letters are partial matches, and elements without matches have a plus sign. Grandson "c" had no matches, having been confined to an aviary during the sensitive 35–65 day sensitive phase. Grandsons "d", "e" and "f" matched 51%, 64%, and 83% of their grandfather's elements and 50%, 100%, and 100% of their father's elements, respectively. (After Zann 1990.)

Noncall-like elements are located in the middle of the sequence, being preceded by Tets and followed by Stacks and Distance Calls, a sequence not too different from that given at take-off, suggesting the possibility that zebra finch song may have evolved from calls associated with flight intention and take-off (Zann 1993b).

The incorporation of call-notes into song is not common among songbirds. It does not occur in the Lesser Sundas zebra finch (Zann 1996), nor does it occur in closely related species (Zann 1976). However, call-notes are reported in songs of some other members of the sub-family (some waxbills and firefinches) and are also found in songs of some fringillids (tohwees) and certhiids (tree-creepers) (Baptista 1996). The warble song of budgerigars (*Melopsittacus* sp.) includes contact and alarm calls (Brown & Farabaugh, Chapter 7).

Zebra finches sing two types of song: a Directed Song, which is a signal directed at the female during the male's courtship dance (Morris 1954); and an Undirected Song which is not directed at any individual and whose function is not clear. However, the following evidence collected by Dunn (1993) suggests Undirected Song may have a role in attracting sexual partners and confining the female to the nest during egg-laying. Dunn found that if the female partner was removed during the nonbreeding season the male increased his rate of Undirected Song, but returned to baseline levels when she returned, or when he re-paired with another female. During the egg-laying period, if the male increased his rate of Undirected Song immediately the female entered the nest she stayed inside longer than when he did not sing, or did so later. Long periods of female nest attendance at this time have two advantages:

they limit opportunities for egg-dumping by other females and limit exposure to other males seeking extra-pair copulations.

Zebra finches are not territorial and the song has no aggressive intent, nor does it elicit aggression. Both song types have identical acoustic structures, but Directed is sung faster than Undirected (Sossinka & Böhner 1980) and forms longer bouts with more intense visual display (Immelmann 1968), but there are no differences in amplitude (R. Zann, unpublished observations). Directed Song is estrogen dependent whereas Undirected is more androgen dependent (Walters *et al.* 1990). Wild males sing Undirected Song throughout the year whereas Directed Song is limited to the breeding season (Dunn 1993). Domesticated females show pairing preferences towards males with high song rates (ten Cate & Mug 1984; Collins *et al.* 1994) and prefer phrases with more elements over those with fewer (Clayton & Pröve 1989) and those sung by the father over those of other males (Miller 1979a) or subspecies (Clayton 1990a). Thus, the amount of singing performed by a male and the quality of his phrase are both subject to sexual selection.

DEVELOPMENT OF VOCAL PERFORMANCE

As with most songbirds, three stages are recognizable in the motor development of song in zebra finches, but these are temporally compressed by comparison with other species: subsong begins around 25–35 days of age, plastic song begins around 50 days and full adult song around 80 days.

The male Distance Call probably originates from begging calls via the "Long Tonal Call" which emerges around 15 days of age (Zann 1985). The latter serves as a juvenile distance signal with the parents, and by day 35–40 resembles the female Distance Call. Between day 40 and day 60 it begins to change into the adult male call and a stereotyped version is first performed between 60–80 days of age.

Thus, the motor programs of the song and Distance Call reach the full "crystallized" adult versions about the same time, and while the stages of development are not exactly comparable, their neural control is shared along the way. The Long Tonal Call and the female Distance Call are under control of basic respiratory circuits located in the brain stem, but at around 40–60 days of age the song neural circuit in the forebrain becomes functional and takes over control of the male Distance Call (Simpson & Vicario 1990). Adult males will revert to the female version of the Distance Call and brain stem control if the song circuit is experimentally impaired, and females will develop song and versions of the male Distance Call if treated with estrogen at the nestling stage so that a song circuit can develop (Simpson & Vicario 1991a,b).

Clearly, motor developments of both learned vocalizations in the zebra finch overlap with the sensitive phases for model acquisition.

DIALECTS

In males the acoustic structures of Distance Calls and songs display variation at both macro- and microgeographic scales that reflect the extent of isolation between populations. Gross differences exist in acoustic structure between the subspecies (Zann 1984; Clayton *et al.* 1991), and between wild and domesticated stock of the Australian subspecies (Zann 1984; Zann 1993a). Dispersal to and from colonies prevents the formation of colony-specific versions of songs and calls, but dialects become more discrete with increasing distance, or isolation between locations. Structure of noncall-like elements is more labile across locations while that of call-like elements and macrostructural features of the phrase are more conservative (Zann 1993a).

SOCIAL INFLUENCES ON VOCAL DEVELOPMENT

After fledging (16–18 days of age), wild zebra finches gradually achieve nutritional and social independence from their parents. Complete nutritional independence is achieved soon after 35 days of age and complete roosting independence around day 48, but other types of contact with the parents are possible well after this time (Zann 1996). Investigators of song learning in zebra finches have differentiated between the period of predominantly parental and family influence and that which occurs after young have "left home." Some experimenters separated young from their parents at day 35 (e.g., Eales 1985; Slater *et al.* 1988; Zann 1990) or day 40 (Immelmann 1969; Zann 1985) by which time the nutritive bond between parent and young is almost completely severed, and young males are still in their subsong stage and have female-like Distance Calls.

Influences before nutritional independence from parents

Domesticated zebra finches are capable of fully acquiring the father's song before day 35, but no acquisition occurs before fledging (Arnold 1975). Consequently, young must acquire the model during the 15–17 day interval before nutritional independence. Evidence for learning before day 35 comes from a number of sources. For example, sons isolated from all zebra finches after day 35, and thus deprived of further opportunity for song acquisition still produce their father's song at 100 days of age just as completely as those sons confined with the father over the same interval (Böhner 1990). Young will also perform the song memorized before day 35 if social experiences with a song tutor after that age are inadequate for some reason. For example, a young male will reproduce his father's song heard before 35 days of age (a) if he cannot see the tutor, or cannot interact vocally with him (Eales 1989); (b) if he is exposed to a tutor of a species different from that of his rearing father (Clayton 1987a); (c) if he is exposed to a color morph different from that of his rearing father (Mann et al. 1991; Slater & Mann 1991); (d) if exposed to a color morph with distinctly different songs and Distance Calls from that of the father (Slater & Jones 1995) or (e) if he is exposed to two tutors of different species, either successively (Clayton 1987b) or simultaneously (Clayton 1988a).

The Distance Call of the father is also memorized before the end of the period of juvenile dependence. Specifically, a normal male call with the modulated noise component, the most sexually diagnostic feature of the call, is only reproduced at sexual maturity if the young male is exposed to zebra finch males during the first 40 days of life (Zann 1985). Sixty percent of males foster-reared to Bengalese finches gave the male zebra finch Distance Call but without the noise component. Forty percent gave versions of the Distance Calls of their Bengalese finch foster parents, either the mother's or the father's call. Obviously, the father or foster father (occasionally the foster mother) provides the auditory model for memorizing the Distance Call. Since these were the only adults to which the young were exposed in these experiments it is possible that any adult male will do, but this is not the case. Where other male models are available, copying from the father ranges from 30% among aviary-reared offspring of wild-caught birds (Zann 1985) to 72% among free-living ones (Zann 1990).

Thus, there is good evidence that the father provides a model for song and Distance Call acquisition during juvenile dependence, but how do fledglings distinguish between the two parents, and how are the preferences for the father's song and call models established? Both parents provision the offspring to an equal extent before fledging (Zann 1996), but if renesting occurs, the mother tends to attack and drive the young away from the new nest so that the bulk of the parental care falls on the father. In contrast to the captive situation, where both parents show aggression towards young during renesting (ten Cate 1982, 1984; Böhner 1983; Clayton 1987c), the father has not been observed attacking his young in the wild (Zann 1996). Parental care is not sex specific, but adult zebra finches are sexually distinct both vocally and morphologically. On the basis of experimental findings using color morphs with different degrees of sexual dimorphism, ten Cate et al. (1993) postulated that sexual imprinting upon the mother's phenotype is stronger the more sexually dimorphic the parents are, and concomitantly, the better the song learning from the father. Vos (1994) has recently found that juvenile males use the mother as the model to sexually imprint upon and use the father as a model for the opposite purpose, namely to establish a sexual aversion towards males. Nonetheless, the father simultaneously provides songs and Distance Calls that serve as the models for vocal acquisition by the young male. Conversely, juvenile females do not sexually imprint upon the father's morphological appearance, but on his behavior (Vos 1995a,b), especially vocal behaviour, which serves as a model for song recognition (Miller 1979a; Clayton 1988b; 1990b,c), and reproduction of both songs and Distance Calls if her song circuits are experimentally induced (Simpson & Vicario 1991b).

The mechanisms by which young develop sex-specific attachments to their father and mother are unknown, but probably involve general familiarization and some form of discrimination learning. Several days before fledging, young use Long Tonal Calls to communicate with approaching parents and others nearby. Several days after fledging they respond only to the calls of their own parents

(Zann 1996). This is the first manifestation of discrimination towards the parents and evidence of a filial bond. Restriction of behavior toward the parents is probably a consequence of the reinforcing events that follow the arrival of the parents, namely food for hungry young. Among wild birds, fathers are often observed singing Undirected Songs before and after bouts of food transfer, consequently young would attend to these vocal stimuli associated with primary reinforcement and would discriminate toward them. The importance of the provisioning relationship in choice of song tutor is also indicated by observations of aviary birds: occasionally males other than the father or foster father feed young (Immelmann 1969; Williams 1990). Under these circumstances song elements from both carers are incorporated into the song of the young male.

Learning from the father before day 35 post-hatching is probably the normal course of events in zebra finches, but most experimenters have concentrated on the effect of social experience and the selection of a song tutor on memorization between independence and adulthood.

Influences between independence and adulthood

Investigations of vocal learning between 35 and 80–100 days of age, the interval between nutritional independence and adulthood, have been conducted on both captive and wild zebra finches. Captive birds, in most cases domesticated stock, have been studied under two laboratory situations: (a) in highly artificial, yet rigorously controlled cage studies; and (b) in "semi-naturalistic" studies conducted in aviaries. The latter are an attempt at compromise between the advantages and disadvantages of cage and free-living studies, respectively.

Despite the young having fully memorized the father's song before 35 days of age, the song actually performed at maturity depends primarily on experiences with a suitable song model from day 35 to day 65. This was first shown by Eales (1985), who removed domesticated males from their father's cage at day 35, 50 or 65 and isolated them with a strange singer. Those isolated at 35 days learned all their song from the new tutor, while those given the new tutor at day 65 sang only the father's song and learned nothing new; those switched at day 50 gave a hybrid song. Thus, songs of tutors other than the father are principally learned between 35 and 65 days of age, a finding subsequently confirmed in other laboratory studies (Slater *et al.* 1988; Mann *et al.* 1991) and in free-living birds (Zann 1990). It is not known whether memory of the father's song is lost or "overwritten" by the subsequent memory of the new tutor's song, or whether versions of both songs exist in memory, but that of the new tutor "supplants" or "supersedes" that of the father in the production phase. Clues may exist in the plastic song, as suggested by Nelson (Chapter 2).

In recent experiments Slater & Jones (1995) found that, although males caged separately with a new tutor after day 35 tended to reproduce a song learned predominantly from the father or new tutor, there was no significant bias towards the latter. The failure of the new song model in these experiments to consistently displace that of the father was thought to be related to the high structural contrast between songs and Distance Calls of the two tutors, leading in some instances to a rejection of those provided by the new tutor. Moreover, there was a significant trend to learn both song and Distance Call from the same tutor, rather than a combination of the two. Therefore, in this experimental situation, males do not learn Distance Calls any earlier than songs; that is, the sensitive phases for both overlap extensively. When males were not given a new tutor after day 40 I found that the Distance Call of males caged in sibling groups was memorized completely by 40 days of age, and concluded that the sensitive phase terminated around this time (Zann 1985); but clearly the findings of Slater & Jones (1995) indicate the sensitive phase for the call can extend beyond this age, provided the bird is isolated with a new and acceptable tutor.

If two tutors were offered in succession in the sensitive phase there was a tendency to develop a hybrid song (Slater *et al.* 1991; ten Cate & Slater 1991), but when given a simultaneous choice of tutors the majority chose the song of one song tutor over another (Slater & Mann 1991; Slater & Jones 1995). The quality of the tutor's song, his Distance Call and appearance were significant factors, but above all, the opportunity to interact visually and vocally with the tutor was critical in the selection process. This was cleverly demonstrated by Eales (1989), who exposed males to: (a) tutors they could see and with whom they could interact vocally; (b) tutors they could not see but could interact with vocally; and (c) tutors they could hear, but not see or interact with vocally. Males in group "a" all learned from the tutor, while none learned from the tutor in group "c".

Nevertheless, in group "b" half learned from the tutor they could not see which suggests that the visual components of the song tutor are not always essential for song learning.

Frequency of singing is not an important factor in selection of tutors unless it is exceptionally low (Böhner 1983; Clayton 1987c; Mann & Slater 1995). However, quality of song is significant: when Clayton (1987c) offered two tutors, one that sang a phrase similar to that of the father and one that did not, she found a significant preference for the former. This suggests that the song of the father memorized before day 35 produces a degree of bias in choice of acceptable song tutors after this age. Moreover, there is a bias towards the song of conspecifics even though zebra finches are capable of learning songs of other species such as Bengalese finches (Immelmann 1969) or red avadavats (*Estrilda amandava*; Price 1979). For example, males foster reared by Bengalese finches could learn zebra finch elements if exposed to them after day 35, but if reared by conspecifics to day 35 and then exposed to Bengalese finches they retained all zebra finch song (Eales 1987a; Clayton 1988a). In a further experiment Clayton foster-reared males with mixed species parents of both combinations for 35 days then subsequently exposed them to tutors of both species, but at maturity they sang a greater proportion of zebra finch elements. Reasons for own-species bias are not obvious, but these findings confirm observations made originally by Immelmann (1969).

Nonetheless, visual appearance is crucial in choice of song tutors and overrides auditory biases. Clayton (1988c) demonstrated this when she simultaneously exposed zebra finches to two song tutors, a Bengalese finch that sang a zebra finch song and a zebra finch that sang a Bengalese finch song – all males learned from the conspecific. Conversely, when males were exposed to two zebra finch tutors, one that sang a normal zebra finch song and one that sang a Bengalese finch song there was no discrimination in choice of tutor. She concluded that vocal characteristics can still be important for selection of song tutors, but Distance Calls and macrostructural features of the song are more important than element morphology (Clayton 1989). Finally, Mann *et al.* (1991) offered a choice of two tutors, one the same color morph as the rearing father and a different one, and found that the pupil learned mainly from the former, again demonstrating the importance of visual appearance of the tutor in choice of a song model.

Since aggression is believed to have an important influence on sexual imprinting in the zebra finch (ten Cate 1984) it was hypothesized that the most aggressive individual may be the preferred tutor. While Clayton (1987c) found evidence in support of the hypothesis, Mann & Slater (1995) could find none. However, work in progress suggests that a male that is aggressive toward other individuals is chosen as the model rather than a male that is specifically aggressive to the pupil (P. J. B. Slater, personal communication). Finally, while unpaired males can serve as song models for captive zebra finches (Mann & Slater 1995), paired males are preferred (Mann & Slater 1994).

In cage experiments where the father is available after day 35 he is the preferred song tutor (Arnold 1975; Böhner 1983) and Mann & Slater (1994) have recently shown that the young male has at least two methods of identifying the father from other males present: they prefer (a) the male they were housed with before day 35, and (b) the male that is paired to their mother.

DELAYED SONG LEARNING – WHAT CLOSES THE SENSITIVE PHASE?

Although onset of the sensitive phase for learning is probably determined by maturation of visual and auditory perception, what closes the phase and leads to the crystallization of song? Because song learning in zebra finches normally occurs in the juvenile phase it has been generally considered an "age-limited" learner incapable of learning in adulthood ("open-ended" learner) like the canary (Nottebohm 1993; see also Baptista & Gaunt, Chapter 3). Recently Morrison & Nottebohm (1992) and Slater *et al.* (1993) found that if suitable models were withheld after day 35 the termination of the phase could be delayed. Songs of these "isolated" males (visual isolates and sound-proof chamber isolates) were quite stable yet could be modified in adulthood when models became available. The extent to which new elements were learned from the model depended on whether pupils were held in groups or singly. Morrison & Nottebohm (1992) concluded that social interaction with live tutors in natural contexts was the most important factor in closing the phase. They discounted three other possibilities, viz. the development of a stereotyped song, hearing and copying of songs of other birds, and the development of a song that conforms to a model. Social interactions were thought to crystallize the final song in two ways: by focusing learning on the song of a specific individual, and by inducing physiological changes in levels of gonadal hormones. Neither process has yet been verified. Therefore, the current hypothesis posits that song learning in zebra finches is more experience limited than age limited, and emphasizes the significant role social interactions with a singing male play in termination of the sensitive phase.

CHOICE OF SONG TUTOR IN AVIARIES

Tutor choice has been investigated in three aviary studies using small samples of domesticated zebra finches. The results are mixed. Williams (1990) and Williams *et al.* (1993) could find no significant preference for learning the father's song, whereas Mann & Slater (1995), who held their birds at lower densities, found a significant tendency to do so in one aviary (6/11 learned from the father), but not in another (1/6). Mann & Slater cautiously concluded that the father was "probably" the preferred model. One unusual aspect in William's (1990) study was that fledglings tended to form crèches where parents fed all young indiscriminately. In the wild, only parents are seen feeding their young (Zann 1993a), as was the case with the study by Mann & Slater (1995). Thus, it is likely that failure to prefer the father as a song tutor in William's studies was an artefact of confinement. Moreover, Mann & Slater (1995) have re-analysed the data of Williams *et al.* (1993), retaining unpaired birds among the potential song tutors, and found a significant tendency to prefer the father.

What social experiences link a young male with his song tutor over the 35–65 day sensitive phase? Close proximity was the only significant factor to emerge from Mann & Slater's (1995) observations. Behaviors, such as allopreening, clumping, aggression and sharing roosting nests, which might be expected to reflect the intensity of bonding between father and son, were not significant. Mann & Slater concluded that the son–father bond extends well beyond the period of nutritional dependence and that inter individual variation in bond duration was probably the main factor that determined whether the father was selected as the song tutor or not. While they hypothesized that song learning under these conditions may be no more than a "passive process" whereby the son is exposed to more frequent and louder renditions of the father's song than that of other males simply because of propinquity arising from a pre-existing social bond, Mann & Slater (1995) did not rule out the possibility of a more active and directed selection process based on important, but rarely performed behaviors.

CHOICE OF A VOCAL TUTOR IN FREE-LIVING ZEBRA FINCHES

While cage and aviary studies allow one to determine the range of vocal learning capabilities of zebra finches they do not predict what actually happens in the wild. Over three breeding seasons in the late 1980s I compared the songs and Distance Calls of 40 sons with their respective fathers ($n=20$) at the Danaher colony located in the northern part of the State of Victoria in southeastern Australia (Zann 1990). I estimated that some 61% of sons had a significant match with the song of the father and that this was not due to chance, thus concluding that these sons had copied their father's song. Similarly, I found that some 72% of sons had matches with the Distance Call from the father and presumed that he was the model, although the range of variation in just one call is more limited than that of song with its six or so elements, and so the prospects for matching by chance are correspondingly greater. When sons were compared to the fathers for both songs and Distance Calls combined there was only one male of the 40 that did not show any match at all, and in this instance, the father was not present in the colony during the 35–65 day sensitive phase, and was unable to serve as the son's tutor.

Whereas methods for determining whether one bird copied its song from another in the laboratory context are fairly straight forward, since there are only a small number of possible tutors involved, it is not so easy in the wild, where 20–40, or more, males are singing, anyone of which could potentially serve as a tutor during the 35–65 day sensitive phase. Consequently, matches between a pupil and a potential tutor may occur by chance without any copying involved, thereby leading to ambiguities or even mistakes in assigning tutors. I calculated that songs where ≥54% of elements matched were not likely to be due to chance at the 2.5% probability level, and therefore were likely to have arisen from transmission from one individual to another by copying (Zann 1990).

Considerable variation existed in copying precision.

Some sons matched their father's song in every element, whereas others, even in the same brood, showed poor matches, some not exceeding chance levels, and thus displaying no evidence of learning from the father at all (e.g. Fig. 6.1). In one of the 20 families investigated I managed to obtain recordings of song across three generations where grandfather, father and two of four grandsons showed high matches. A third grandson just failed to exceed chance matching levels and another failed to match anything at all, but he was confined to an aviary in the center of the study colony for most of his 35–65 day sensitive phase, and so excluded from effective interaction with the father (Fig. 6.1). In two families I obtained recordings of Distance Calls across four generations of males and found good matches, again suggesting high levels of transmission.

Slater & Mann (1990) argued that significant matches between free-living sons and their fathers does not necessarily mean that sons copied their songs from them, since domesticated pupils prefer to learn from a tutor that sings like the father (Clayton 1987c). Therefore, it is conceivable in the wild situation that a male will learn from a tutor that, by chance, happens to sing like his father, although there has been no cultural transmission between the two tutors. However, this event was highly unlikely at my study colony, since chance matching with the father was low (above). Low chance matching was a consequence of low philopatry because almost half of the males recorded singing in the study of song were immigrants (Zann 1993a), and even this was an underestimate because banding data showed that about 77% of breeding pairs each season were immigrants (Zann & Runciman 1994). Moreover, the probability that sons in the same broods and those in successive broods and in succesive generations would all encounter a male that happened to sing like the father at this colony must be exceptionally low. Given the fact that Mann & Slater (1995) found that 50% of aviary males learned their songs from the father it seems reasonable to conclude that 60% of wild males would do the same and that 72% would learn the father's Distance Call as well. Why 39% of wild sons, and 50% of domesticated ones fail to learn much of the father's song is a puzzle, but analysis of population structure and mating system of wild zebra finches may suggest reasons for variable strategies in vocal learning. It may simply be that the highly fluid composition of the breeding colonies causes some sons to become separated from the father during the sensitive phase. Alternatively, there could be unknown fitness consequences for learning or not learning the father's song. DNA analysis of paternity would be the ultimate test of these strategies.

KIN RECOGNITION AND VOCAL LEARNING

Cultural transmission of song and Distance Calls in the male line of zebra finch families provides cues for recognition of kin, and playback studies show that these cues are used by females to discriminate among males. Females prefer their partner's song over those of other males (Miller 1979b) and daughters prefer the songs of their fathers (Miller 1979a) or foster fathers (Clayton 1990b). The ability to memorize the father's song for the purpose of recognition, as distinct from production, first occurs between 25 and 35 days of age (Clayton 1988b). Females also learn songs of male siblings/male aviary companions that influence song preferences in adulthood (Clayton 1990b). Both sexes are also capable of discrimination learning of songs first heard in adulthood (Cynx & Nottebohm 1992).

In multiple choice arenas, domesticated zebra finches can recognize kin on first encounter (Burley et al. 1990). Kin-specific visual cues are possibly used as well as vocal ones (Burley & Bartels 1990).

Is it important for zebra finches to recognize kin? In theory, it may be important at pair formation in order to maintain an optimal level of outbreeding (Bateson 1983), but there is no strong evidence for this. At pair formation domesticated zebra finches do not discriminate between siblings and nonkin, although reproductive success is inferior in some respects (e.g., nestling survival to independence), but not others (e.g., clutch size) (Schubert et al. 1989; Fetherston & Tyler Burley 1990). Among wild zebra finches sibling–sibling pairings are probably rare because of low philopatry, but one has been detected (Zann 1996). It is conceivable that individuals that live among close kin have enhanced reproductive success. This is suggested by an aviary experiment where Williams et al. (1993) found that males from the same aviary population as their females had a higher reproductive success than those mated to females from different aviaries; moreover, males that had received song tuition from their father had

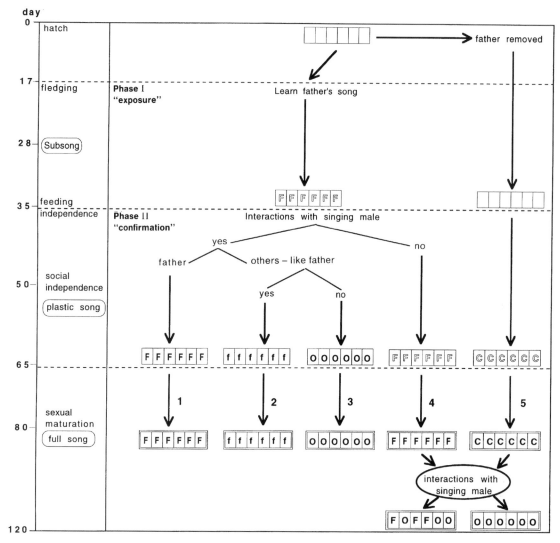

Fig. 6.2. Schematic representation of song learning options (1–5) in zebra finches. The row of boxes represents the song template: empty template, unlearned song macrostructure (Price 1979); template with letters, learned song elements (except option 5); double boxes, stable, stereotyped song. Option 1 is song learned from father, confirmed in phase II, and performed at maturity. This is the option of 60% of wild birds (Zann 1990) and 50% of aviary birds (Mann & Slater 1995). Option 2 is song learned from the father, but not confirmed by him, but by a similar singer; an experimental finding by Clayton (1987a) that is unlikely to occur in wild birds. Option 3 is song learned from the father, but "displaced" by a song learned from a new tutor in the confirmation stage. This option is found in 40% of wild birds (Zann 1990), 50% of aviary birds (Mann & Slater 1995) and in experimental birds (Eales 1985). Option 4 is song learned from the father, but experimental isolation (Morrison & Nottebohm 1992; Slater *et al.* 1993), or inadequate social encounters with other males after independence (Eales 1989; Clayton 1987a; Mann *et al.* 1991) prevents confirmation with songs of father or with others, but social encounters after day 100 adds new elements. Option 5 is no exposure to tutors during both sensitive phases produces unlearned song elements of calls and improvised sounds. Social interactions with tutors after day 100 produces learned elements (Eales 1987b; Morrison & Nottebohm 1992).

an additional advantage. While wild zebra finches appear to form subgroups within a colony, it is not known whether these are kin groups or not (Zann 1996); unfortunately, high mobility of the groups impedes investigations of this problem.

CONCLUSIONS

Current evidence for song model acquisition in the zebra finch is largely consistent with Böhner's (1990) two-stage model, where memorization of the father's song between fledging and nutritional independence is followed by a second stage in which the initial model is confirmed by social interaction with the father before adulthood (Fig. 6.2). If the father is absent, or inadequate for some reason, a new tutor is chosen based to some extent on its acoustic resemblance to the paternal model, but primarily on suitable social interactions with the pupil. While nothing is known about how the father's song is memorized during phase I, there is some evidence that the song learned and produced in phase II is not entirely an outcome of passive processes. There is a suggestion that females may exert an influence on which songs a male will learn (Jones & Slater 1993). This could be an instance of "action-based" learning (Marler & Nelson 1993), where social reinforcement leads to the selection of a particular song for production from a number that have been acquired (see Nelson, Chapter 2). Further evidence for active learning comes from Adret's (1993) successful attempt to condition young males with playback of taped song; not only did experimental males work to hear song, but they copied it more successfully than controls.

With respect to choice of song tutors – father or others – it is difficult to know if one is looking at a mixed strategy, each with its own fitness consequences comparable to that found in Galapagos finches (*Geospiza fortis*; Gibbs 1990), or simply a reflection of an inexact selection process. Possibly, seed resources affect age of dispersal of males and lead to exposure to different song models and to differential reproductive success. Similarly, with kin recognition it is difficult to know whether this is an adaptation to various physical and social environments or an epiphenomenon of vocal learning, where acquisition of models occurs in an early sensitive phase. Field experiments will be essential to answer these questions.

ACKNOWLEDGEMENTS

I thank Luis Baptista for providing references, Peter Slater for access to his unpublished studies, and the editors for suggesting improvements to the manuscript.

REFERENCES

Adret, P. (1993). Operant conditioning, song learning and imprinting to taped song in the zebra finch. *Animal Behaviour* **46**: 149–59.

Arnold, A. P. (1975). The effects of castration on song development in zebra finches (*Poëphila guttata*). *Journal of Experimental Zoology* **191**: 261–77.

Baptista, L. F. (1996). Nature and its nurturing in avian vocal development. In *Ecology and Evolution of Acoustic Communication in Birds*, ed. D. E. Kroodsma & E. H. Miller, pp. 39–60. Cornell University Press, Ithaca, NY.

Bateson, P. P. G. (1983). Optimal outbreeding. In *Mate Choice*, ed. P. P. G. Bateson, pp. 257–77. Cambridge University Press, Cambridge.

Birkhead, T. R., Burke, T., Zann, R., Hunter, F. M. & Krupa, A. P. (1990). Extra–pair paternity and intraspecific brood parasitism in wild zebra finches *Taeniopygia guttata*, revealed by DNA fingerprinting. *Behavioural Ecology and Sociobiology* **27**: 315–24.

Böhner, J. (1983). Song learning in the zebra finch (*Taeniopygia guttata*): Selectivity in choice of a tutor and accuracy of song copies. *Animal Behaviour* **31**: 231–37.

Böhner, J. (1990). Early acquisition of song in the zebra finch, *Taeniopygia guttata. Animal Behaviour* **39**: 369–74.

Burley, N. & Bartels, P. J. (1990). Phenotypic similarities of sibling zebra finches. *Animal Behaviour* **39**: 174–80.

Burley, N., Minor, C. & Strachan, C. (1990). Social preferences of zebra finches for siblings, cousins and non kin. *Animal Behaviour* **39**: 775–84.

Clayton, N. S. (1987a). Mate choice in male zebra finches: Some effects of cross-fostering. *Animal Behaviour* **35**: 596–622.

Clayton, N. S. (1987b). Song learning in cross–fostered zebra finches: A re-examination of the sensitive phase. *Behaviour* **102**: 67–81.

Clayton, N. S. (1987c). Song tutor choice in zebra finches. *Animal Behaviour* **35**: 714–21.

Clayton, N. S. (1988a). Song learning and mate choice in estrildid finches raised by two species. *Animal Behaviour* **36**: 1589–600.

Clayton, N. S. (1988b). Song discrimination learning in zebra finches. *Animal Behaviour* **36**: 1016–24.

Clayton, N. S. (1988c). Song tutor choice in zebra finches and

Bengalese finches: The relative importance of visual and vocal cues. *Behaviour* **104**: 281–99.

Clayton, N. S. (1989). The effects of cross–fostering on selective learning in estrildid finches. *Behaviour* **109**: 163–75.

Clayton, N. S. (1990a). The effects of cross–fostering on assortative mating between zebra finch subspecies. *Animal Behaviour* **40**: 1102–10.

Clayton, N. S. (1990b). Subspecies recognition and song learning in zebra finch subspecies. *Animal Behaviour* **40**: 1009–17.

Clayton, N. S. (1990c). Assortative mating in zebra finch subspecies, *Taeniopygia guttata guttata* and *T. g. castanotis*. *Philosophical Transactions of the Royal Society of London Series B* **330**: 351–70.

Clayton, N. S., Hodson, D. & Zann, R. A. (1991). Geographic variation in zebra finch subspecies. *Emu* **91**: 2–11.

Clayton, N. S. & Pröve, E. (1989). Song discrimination in female zebra finches and Bengalese finches. *Animal Behaviour* **38**: 352–62.

Collins, S. A., Hubbard, C. & Houtman, A. M. (1994). Female choice in the zebra finch – the effect of male beak color and male song. *Behavioural Ecology and Sociobiology* **35**: 21–5.

Cynx, J. & Nottebohm, F. (1992). Testosterone facilitates some conspecific song discriminations in castrated zebra finches (*Taeniopygia guttata*). *Proceedings of the National Academy of Sciences, USA* **89**: 1376–8.

Dunn, A. (1993). The song of the zebra finch: Context and possible functions. Ph.D thesis. La Trobe University.

Eales, L. A. (1985). Song learning in zebra finches: Some effects of song model availability on what is learnt and when. *Animal Behaviour* **33**: 1293–300.

Eales, L. A. (1987a). Do zebra finch males that have been raised by another species still tend to select a song tutor? *Animal Behaviour* **35**: 1347–55.

Eales, L. A. (1987b). Song learning in female-raised zebra finches: Another look at the sensitive phase. *Animal Behaviour* **35**: 1356–65.

Eales, L. A. (1989). The influences of visual and vocal interaction on song learning in zebra finches. *Animal Behaviour* **37**: 507–8.

Fetherston, I. A. & Tyler Burley, N. (1990). Do zebra finches prefer to mate with close relatives? *Behavioural Ecology and Sociobiology* **27**: 411–14.

Gibbs, H. L. (1990). Cultural evolution of male song types in Darwin's medium ground finches (*Geospiza fortis*). *Animal Behaviour* **39**: 253–63.

Immelmann, K. (1965). *Australian Finches in Bush and Aviary*. Angus & Robertson, Sydney.

Immelmann, K. (1968). Zur biologischen Bedeutung des Estrildidengesanges. *Journal für Ornithologie* **109**: 284–99.

Immelmann, K. (1969). Song development in the zebra finch and other estrildid finches. In *Bird Vocalizations*, ed. R. A. Hinde, pp. 64–74. Cambridge University Press, Cambridge.

Jones, A. E. & Slater, P. J. B. (1993). Do young zebra finches prefer to learn songs that are familiar to females with which they are housed? *Animal Behaviour* **46**: 616–17.

Mann, N. I. & Slater, P. J. B. (1994). What causes young male zebra finches, *Taeniopygia guttata*, to choose their father as a song tutor? *Animal Behaviour* **47**: 671–7.

Mann, N. I. & Slater, P. J. B. (1995). Song tutor choice by zebra finches in aviaries. *Animal Behaviour* **49**: 811–20.

Mann, N. I., Slater, P. J. B., Eales, L. A. & Richards, C. (1991). The influence of visual stimuli on song tutor choice in the zebra finch, *Taeniopygia guttata*. *Animal Behaviour* **42**: 285–93.

Marler, P. & Nelson, D. A. (1993). Action-based learning: A new form of developmental plasticity in bird song. *Netherlands Journal of Zoology* **43**: 91–103.

Miller, D. B. (1979a). Long-term recognition of father's song by female zebra finches. *Nature* **280**: 389–91.

Miller, D. B. (1979b). The acoustic basis of mate recognition by female zebra finches (*Taeniopygia guttata*). *Animal Behaviour* **27**: 376–80.

Morris, D. (1954). The reproductive behavior of the zebra finch (*Poephila guttata*) with special reference to pseudofemale behavior and displacement activities. *Behaviour* **7**: 1–31.

Morrison, R. G. & Nottebohm, F. (1992). Role of a telencephalic nucleus in the delayed song learning of socially isolated zebra finches. *Journal of Neurobiology* **24**: 1045–64.

Nottebohm, F. (1993). The search for neural mechanisms that define the sensitive period for song learning in birds. *Netherlands Journal of Zoology* **43**: 193–234.

Price, P. H. (1979). Developmental determinants of structure in zebra finch song. *Journal of Comparative Physiological Psychology* **93**: 260–77.

Schubert, C. A., Ratcliffe, L. M. & Boag, P. T. (1989). A test of inbreeding avoidance in the zebra finch. *Ethology* **82**: 265–74.

Simpson, H. B. & Vicario, D. S. (1990). Brain pathways for learned and unlearned vocalizations differ in zebra finches. *Journal of Neuroscience* **10**: 1541–56.

Simpson, H. B. & Vicario, D. S. (1991a). Early oestrogen treatment of female zebra finches masculinizes the brain pathway for learned vocalizations. *Journal of Neurobiology* **22**: 777–93.

Simpson, H. B. & Vicario, D. S. (1991b). Early estrogen treatment alone causes female zebra finches to produce learned, male-like vocalizations. *Journal of Neurobiology* **22**: 755–76.

Slater, P. J. B., Eales, L. A. & Clayton, N. S. (1988). Song learning in zebra finches (*Taeniopygia guttata*): Progress and prospects. *Advances in the Study of Behaviour* **18**: 1–34.

Slater, P. J. B. & Jones, A. E. (1995). The timing of song and distance call learning in zebra finches. *Animal Behaviour* **49**: 548–50.

Slater, P. J. B., Jones, A. E. & ten Cate, C. J. (1993). Can lack of experience delay the end of the sensitive phase for song learning? *Netherlands Journal of Zoology* **40**: 80–90.

Slater, P. J. B. & Mann, N. S. (1990). Do male zebra finches learn their father's songs? *Trends in Ecology and Evolution* **5**: 415–17.

Slater, P. J. B. & Mann, N. S. (1991). Early experience and song learning in zebra finches *Taeniopygia guttata*. *Proceedings of XXth International Ornithological Congress*, Christchurch, 1990: pp. 1074–80.

Slater, P. J. B., Richards, C. & Mann, N. I. (1991). Song learning in zebra finches exposed to a series of tutors during the sensitive phase. *Ethology* **88**: 163–71.

Sossinka, R. & Böhner, J. (1980). Song types in the zebra finch (*Poephila guttata castanotis*). *Zeitschrift für Tierpsychologie* **53**: 123–32.

ten Cate, C. J. (1982). Behavioural differences between zebra finch and Bengalese finch (foster)parents raising zebra finch offspring. *Behaviour* **81**: 152–72.

ten Cate, C. J. (1984). The influence of social relations on the development of species recognition in zebra finch males. *Behaviour* **91**: 263–85.

ten Cate, C. J. & Mug, G. (1984). The development of mate choice in zebra finches. *Behaviour* **90**: 125–50.

ten Cate, C. J. & Slater, P. J. B. (1991). Song learning in zebra finches: How are elements from two tutors integrated? *Animal Behaviour* **42**: 150–2.

ten Cate, C. J., Vos, D. R. & Mann, N. (1993). Sexual imprinting and song learning: Two of one kind? *Netherlands Journal of Zoology* **43**: 34–45.

Vos, D. R. (1994). Sex recognition in zebra finch males results from early experience. *Behaviour* **128**: 1–14.

Vos, D. R. (1995a). Sexual imprinting in zebra finch females: Do females develop a preference for males that look like their father? *Ethology* **99**: 252–62.

Vos, D. R. (1995b). The role of sexual imprinting for sex recognition in zebra finches: A difference between males and females. *Animal Behaviour* **50**: 645–53.

Walters, M. J., Collado, D. & Harding, C. F. (1990). Oestrogenic modulation in male zebra finches: Differential effects on directed and undirected songs. *Animal Behaviour* **42**: 445–52.

Williams, H. (1990). Models for song learning in the zebra finch: Fathers or others? *Animal Behaviour* **39**: 747–57.

Williams, H., Kilander, K. & Sotanski, M. L. (1993). Untutored song, reproductive success and song learning. *Animal Behaviour* **45**: 695–705.

Zann, R. (1976). Variation in the songs of three species of estridline grassfinches. *Emu* **76**: 97–108.

Zann, R. (1984). Structural variation in the zebra finch distance call. *Zeitschrift für Tierpsychologie* **66**: 328–45.

Zann, R. (1985). Ontogeny of the zebra finch distance call. I. Effects of crossfostering to Bengalese finches. *Zeitschrift für Tierpsychologie* **68**: 1–23.

Zann, R. (1990). Song and call learning in wild zebra finches in south-east Australia. *Animal Behaviour* **40**: 811–28.

Zann, R. (1993a). Variation in song structure within and among populations of Australian zebra finches. *Auk* **110**: 716–26.

Zann, R. (1993b). Structure, sequence and evolution of song elements in wild Australian zebra finches. *Auk* **110**: 702–15.

Zann, R. (1994). Reproduction in a zebra finch colony in south-eastern Australia: The significance of monogamy, precocial breeding and multiple broods in a highly mobile species. *Emu* **94**: 285–99.

Zann, R. (1996). *The Zebra Finch: A synthesis of Field and Laboratory Studies*. Oxford University Press, Oxford.

Zann, R. A., Morton, S. R., Jones, K. R. & Burley, N. T. (1995). The timing of breeding of zebra finches in relation to rainfall at Alice Springs, central Australia. *Emu* **95**: 208–22.

Zann, R. & Runciman, D. (1994). Survivorship, dispersal and sex ratios of zebra finches *Taeniopygia guttata* in southeast Australia. *Ibis* **136**: 136–46.

7 What birds with complex social relationships can tell us about vocal learning: Vocal sharing in avian groups

ELEANOR D. BROWN AND SUSAN M. FARABAUGH

INTRODUCTION

Vocal learning in birds has evolved independently in several different avian orders, but is common in only two groups, oscine songbirds and parrots. The diversity and complexity of vocal repertoire structure among the species in these two groups is enormous. However, much scientific attention in avian song learning has focused on a small group of songbirds, north temperate migrant species, in which song is restricted mainly to males and the occurrence of song and territoriality is seasonal. In addition to these birds there is also a vast number of laboratory studies on the development and neural control of male song in the zebra finch (*Taenopygia guttata*), an Australian species with an unusually compressed developmental period. This large body of research has resulted in general models of song and vocal learning drawn from only a small subset of the world's birds.

While these many studies have broadened our understanding of learned vocal communication, it is our contention that there is much to be gained by study of the vocal behavior of avian species with more complex social relationships. Such species exhibit long-term associations between well-acquainted individuals, and, for many, long-term associations are related to permanent residence in an area. Another important characteristic is the tendency to live in stable groups for at least part of the year. These groups could be, for example, a winter foraging flock of chickadees, a breeding colony of caciques, a foraging flock of cockatoos, or a permanently territorial pair of tropical wrens.

By consideration of common features of disparate groups that represent more fully the portion of the world's birds that learn vocally, we can approach a more truly universal model of vocal learning. Data on a permanently ter-ritorial Australian songbird, an atypical north temperate songbird, and a nomadic Australian parrot can contribute to a more cosmopolitan view of vocal learning in birds. The Australian magpie (*Gymnorhina tibicen*), the American crow (*Corvus brachyrhynchos*), and the budgerigar (*Melopsittacus undulatus*) differ in ecology, spacing behavior, and phylogeny. But these three focal species are similar in that they all have long-term relationships and stable social groups. In addition, all three species also share characteristics that indicate vocal learning is important in their societies. They have a striking capacity for heterospecific mimicry; crows and parrots are well known as "talking" birds in captivity. This learning ability persists into adulthood. Much of the evidence for vocal mimicry by parrots and corvids comes from hand-reared animals, which accept humans as conspecific social companions (Lorenz 1970a; see West *et al.* Chapter 4). This acceptance strongly implies a social factor in vocal learning – an influence so powerful that it overrides normal preferences for species-specific sounds. Therefore we might predict that in normal social interactions with conspecifics, the group – be it pair, family, cooperative group or flock – is one of the central source of social influences on vocal learning. Learned vocalizations should reflect particular social relationships, resulting in shared sounds used in communication both within the group (communication with collaborators) and among groups (communication with rivals).

In this chapter, we first present general background information on our three focal species, a description of their vocalizations, and data relating to the social influences on their vocal learning. Keeping in mind their similarities and differences, we discuss the evolution of vocal learning. We argue that vocal learning has evolved to allow individuals to share vocalizations with a particular subset of conspecifics, such as, territorial rivals or flock-mates, rather than with any conspecific. We then discuss vocal sharing among social rivals and within social groups, learning of songs and calls, and vocal learning by both males and females. Finally, we discuss the social influences on vocal sharing.

SHARED SONG IN THE AUSTRALIAN MAGPIE

Species background

The Australian magpie is not a true magpie at all. Although it bears a superficial resemblance to northern hemisphere magpies of the corvid genus *Pica*, this large black-and white bird is a member of the endemic Australo-Papuan songbird family Cracticidae (currawongs, butcherbirds, and Australian magpies). The Australian magpies are considered by Amadon (1962) to be members of a single species that occurs widely across Australia and southern New Guinea, and has been introduced to New Zealand. Adult Australian magpies are sexually dimor-phic. First-year birds are strikingly different from adults in beak, eye, and plumage coloration.

The central focus of the magpie's way of life is owner-ship of land with open spaces and trees, the essential pre-requisite for breeding (Carrick 1963, 1972). The magpie is a long-lived species (25 years; Carrick 1963, 1972) and is a facultative cooperative breeder exhibiting year-round territoriality. The pair bond has been thought to be monog-amous, but in some populations multiple nests on territo-ries, and feeding of nestlings by any, or all, group members, make monogamy questionable (Hughes *et al.* 1995). DNA fingerprinting studies now in progress (Hughes *et al.* 1995) should eventually resolve issues of parentage. Social organization and group size vary geographically, with a north/south gradient (Robinson 1956; Shurcliff & Shurcliff 1973; Hughes & Mather 1991). In our New Zealand study population and in Carrick's (1963, 1972), cooperative groups consist of two to seven birds. In other southern populations (Hughes *et al.* 1995) and western populations (Robinson 1956) larger groups of 15–20 birds are common. Each group communally defends a perma-nent year-round territory. There are constant boundary skirmishes as groups vie to improve their holdings. The social system is further complicated by large contingents of landless flock birds, who strive to attain membership in a group holding a high-quality territory. Cooperative groups originate when young birds (kin) remain on the natal terri-tory, or when a group of unrelated birds (nonkin) coalesces within the flock and fights its way onto a new territory (Carrick 1963; E. D. Brown *et al.* 1988). Young remain on the natal territory until between one and three years of age. Thus, kin and nonkin territorial groups, pairs, and flocks coexist in a single population. As its fortunes change, during its long life a magpie may find itself in any or all of these social milieux (Carrick 1963, 1972).

Within the cooperative group, activities are to some degree partitioned according to sex and age. Females brood the young. Males usually feed females on the nest and share duties of caring for nestlings and fledglings. Auxiliary birds aid in the reproductive effort by feeding the brooding female, nestlings and fledglings, or by helping to build the rest (Brown & Farabaugh 1991). In their southern study population with large cooperative groups, Hughes *et al.* (1995) found tremendous variability in reproductive effort and participation: a cooperative group could have one nest, attended by all group members; a group could have multiple nests, each nest attended by a different male/female pair; or a group could have multiple nests, with all nestlings at all nests fed by all group members. All group members including immature and adult birds of both sexes aid in cooperative defense of the territory, although young juveniles make a lesser contribution (Farabaugh *et al.* 1992b). Group song by birds of both sexes and all ages is integral to territorial aggression and defense. Territory-holding Australian magpies, in contrast, do not join in any communal activ-ities, such as roosting or mobbing predators, with birds outside their group. A territory-owning adult Australian magpie rarely leaves its territory.

Social behavior of the Australian magpie is specialized for year-round territorial life. Much social behavior is ori-ented toward territorial rivals belonging to other neigh-boring groups, and the magpie has many behavior patterns specialized for long distance territorial advertisement (Brown & Veltman 1987). There are frequent, often daily, aggressive encounters between territorial groups. Within the group, low-level aggressive interactions over a period of months may eventually lead to the eviction of an indi-vidual and its replacement as a breeder. This probably

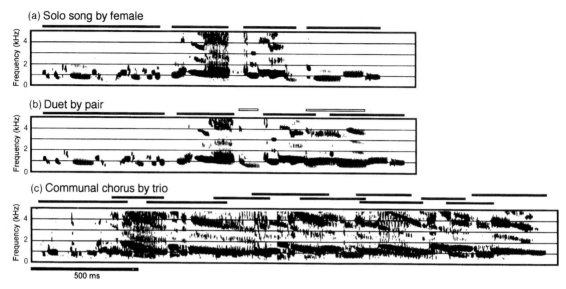

Fig. 7.1. Duet and chorus song. (a) The female's song part, sung solo, of the Australian magpie duet in (b); the female's introductory warble and three carol syllables are clearly visible. (b) Duet by a pair of magpies in Queensland. The complete song is composed of an introductory warble phrase followed by carol duet composed of five carol syllables, three by the female and two by the male. Bars above the spectrogram indicate when each bird is singing (black, female; white, male). (c) Communal song by a trio of magpies in New Zealand. Bars above the spectrogram give only a general idea of where syllables begin and end. This chorus song is composed of a brief warble phrase sung by one of the trio, followed by a 10-carol syllable chorus in which all three birds participated. (Adapted from Brown & Farabaugh 1991.)

reflects the within-group competition for breeding opportunities (or territory inheritance) that is a feature of many cooperative breeding systems (Stacey & Koenig 1990). Within the group, activities such as foraging, resting, and preening tend to be synchronized. There is no allopreening and little affiliative contact, except play with juveniles (Pellis 1981a,b).

Vocalizations

With its melodious singing, called carolling, the Australian magpie rivals the most famous singers of the northern hemisphere. Birds of both sexes and all ages sing. Magpie song can be divided roughly into two general categories, warbles and carols. This terminology has been used historically in most descriptions of Australian magpie song (e.g., Gould 1865) as well as in colloquial language. (Warble is a term that has also found its way into the literature describing continuous songs of variable length in several unrelated species, such as the European starling (*Sturnus vulgaris*; Hausberger, Chapter 8 and the budgerigar see later)). Australian magpie warbles are composed of soft, short, liquid sounds, whereas carols are extremely loud, long sounds for long distance broadcasting (E. D. Brown *et al.* 1988; Brown & Farabaugh 1991). A typical magpie song is composed of warble syllables followed by carol syllables (Fig. 7.1). The warble segment is sung solo, whereas the carol segment is sung as a duet or chorus, with all group members participating. Both types of syllable are highly individual specific; carol syllables and songs are sex specific (Farabaugh *et al.* 1988; Brown & Farabaugh 1991). Songs can consist of warble syllables alone. Unlike typical territorial songbirds of the north temperate zone, the magpie has no species-typical syllable or song repertoire. Fixed sets of syllables that form recognizable song "types" repeated on different occasions can be identified individual by individual, but not on a geographic basis, and therefore no dialects exist. Occasionally, heterospecific mimicry of other species is incorporated into a warble song (E. D. Brown, unpublished data) and can be transmitted to other group members (Robinson 1956).

Context

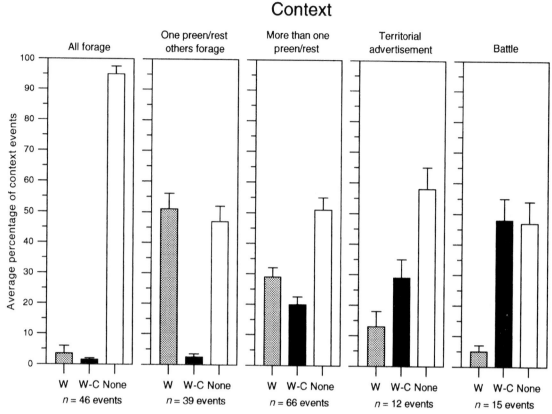

Fig. 7.2. Song context. Histograms showing the relative occurrence of different modes of Australian magpie song, or no song, in contexts of five different types. Modes of singing are W, warble (gray), and W-C, warble followed by carols (black). None means no song (white). These data were collected by means of one/zero sampling with a 15-second interval from six territorial groups for song in typical contexts. Because groupmate's activities were not independent, singing behavior and context were defined with reference to group performance. Because group activities were not always perfectly synchronous, contexts were sometimes combinations of activities. Each set of sequential intervals with the same context was considered a separate, independent context event. *N* was the number of context events. For any particular context event, the percentage of total intervals spent in each mutually exclusive song mode (or no song) was calculated. The percentages from all context events of the same type were then averaged. Error bars are standard errors. (From Brown & Farabaugh 1991.)

Song in Australian magpies functions in territorial advertisement and defense; in addition it functions in intragroup cohesion and communication. Magpie song expresses individual identity, group identity, sex, and number of group members. Full communal carol songs are heard most during territorial advertisement and during lengthy battles at territorial boundaries (Fig. 7.2), when all groupmates join in highly synchronized cooperative defense, and in these contexts probably functions both as an aggressive proclamation directed at alien rivals and also as a rallying cry or war chant to incite and coordinate group mates (Brown & Veltman 1987; Brown & Farabaugh 1991; Farabaugh et al. 1992b). In nonterritorial contexts, group membership is delineated by participation in the communal vocal display. Warble songs without carols are heard most in nonaggressive situations (Fig. 7.2), when all group members are preening or resting together, during nest-building and feeding young (E. D. Brown et al. 1988).

Vocal sharing in Australian magpie populations

To consider the extent to which patterns of vocal sharing at a particular point in time reflect social interactions among Australian magpies, we tested sharing by magpies in the same social group compared to magpies in different groups, and by magpies in kin compared to magpies in nonkin groups. We conducted fieldwork in two magpie populations, one in New Zealand and one in Queensland, Australia (E. D. Brown *et al.* 1988; Farabaugh *et al.* 1988; Brown & Farabaugh 1991). The New Zealand population contained many multiple-adult groups within which we investigated sharing of warble songs. The Queensland population contained many adult pairs, which made possible a detailed study of sex-specificity of communal carol songs (Fig. 7.1a and b). We also studied captive magpies in order to observe basic processes of vocal learning. The studies of patterns of vocal sharing could be thought of as "snapshots" in time, from which we determined mechanisms *a posteriori*, whereas study of captive birds allowed us to observe ongoing processes through time.

Sharing of warbles within and among territorial groups

At our New Zealand study site, which comprised two localities (adjacent farms), we tape-recorded color-banded magpies in six territorial groups and the local flock over an eight-month period. After extensive sound spectrographic analysis of vocal repertoire samples of 23 magpies, we determined warble song types and classified 893 warble syllable types. We then determined which warble syllables occurred in the repertoire of each bird, and calculated the average percentage of the syllables shared by every pairwise combination of the 23 birds (253 pairwise comparisons).

Warble syllable sharing patterns suggested that vocal imitation was the mechanism of sharing, and that an Australian magpie's choice of song tutor was affected by social relationship and sex. Not all syllables were shared: of the 893 syllable types found in our population of 23 birds, only 33% were shared, and the remainder were individually unique. Although syllable structure was extremely variable and complex, when sharing occurred exact copies were sung by the sharers (Fig. 7.3b). Far from being random, warble syllable sharing was related to cooperative group membership; magpies in the same group

shared significantly more syllables than did magpies in different groups (E. D. Brown *et al.* 1988). Kin groups shared no more than nonkin groups (E. D. Brown *et al.* 1988). A detailed analysis of shared warble syllable types revealed more clearly the sources of vocal imitation (Farabaugh *et al.* 1988). By dividing our pairwise comparisons of repertoire overlap into categories based on locality, we made statistical comparisons of warble syllable sharing by birds in the same territorial group, in neighboring territories, and in different local areas (Fig. 7.4c and d). Magpies shared the most syllables with their groupmates and immediate neighbors (Fig. 7.4c and d); these shared syllables were very complex in physical structure (e.g., Fig. 7.3b). They shared an intermediate amount with birds from the same locality that were not neighbors. They shared the least with birds from another locality (Fig. 7.4a and b), and these between-locality shared syllables were also simple in structure. We also tested warble syllable sharing within and between the sexes (Fig. 7.4e and f). Warble syllable types were not sex specific for the population as a whole (Fig. 7.4g and h); thus, a magpie was not more likely to share syllables with distant birds of its own sex than with distant birds of the opposite sex. But, for birds in the immediate vicinity, both groupmates and neighbors, a magpie shared significantly more syllables with birds of its own sex (Fig. 7.4 right). For whole warble songs, existing patterns of song sharing were more complicated, and involved fragments, variants, and recombinations. The birds that shared songs often belonged to the same group and were always part of the same subpopulation: no song types were common to both localities.

In contrast to shared sounds, individual-specific syllables and songs suggested the importance of other processes of vocal learning such as invention and improvisation. Through practice, invention and improvisation can lead to individually unique sounds and the proliferation of individual-specific syllables and songs in the repertoire. Warble syllable repertoire size of individual birds varied enormously, and ranged from 30 to 188. A majority of warble syllable types were unique, and a large portion of each magpie's repertoire (typically 20–40%) was composed of these unique, individual-specific syllable types. Figure 7.3a shows a series of individually unique syllables from the repertoire of Red White Yellow, a subadult member of a cooperative group. The structural similarity of the syllables suggests that they are a set of

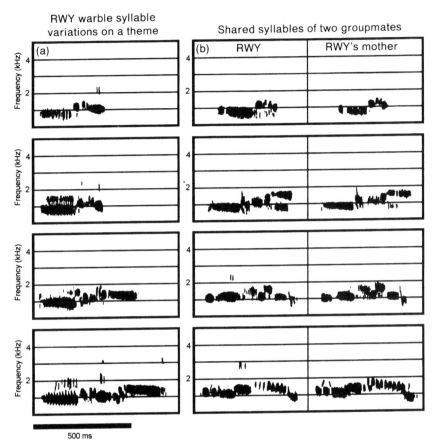

Fig. 7.3. Improvised and shared syllables from warble song of Australian magpies. (a) Column of unique individual-specific "variations on a theme" by Red White Yellow (RWY), a subadult member of a territorial kin group. The structural similarity of the syllables, and the incremental complexity of each syllable in the series suggests improvisation. (b) Syllables shared by Red White Yellow (left column) and its mother (right column). These syllables were sung by no others birds. The complexity of the sounds, and the existence of exact "copies" in each bird's repertoire, suggest that vocal imitation was the mechanism of sharing. Extreme complexity was typical of syllables shared with either groupmates or neighbors. (From E. D. Brown *et al.* 1988.)

improvisations or variations on a theme. An analogous example of improvisational whistle variations from the simpler repertoire of the tame magpie Zoe is shown in Fig. 7.5c. When we compared the whole warble songs (i.e., discrete sequences of syllables) we also found extreme variability in warble song repertoires (E. D. Brown *et al.* 1988). Song repertoire size varied widely (from five to 27), and many variations on a theme could occur in the same individual's repertoire (Fig. 7.3a), suggesting improvisation at the level of songs as well as syllables. Almost all warble song types were individual specific: of the few that were shared, none were sung by more than five of the 23 magpies in the study population.

Processes of vocal learning are influenced by social factors. The social setting of imitation certainly involves members of the same social group: the higher percentage of syllable sharing by magpies in the same group compared to magpies in different groups reflect vocal imitation of groupmates. Coincidental sharing would be an unlikely explanation, because syllables shared with groupmates were significantly more complex in structure, whereas simple syllables were shared among more distant birds

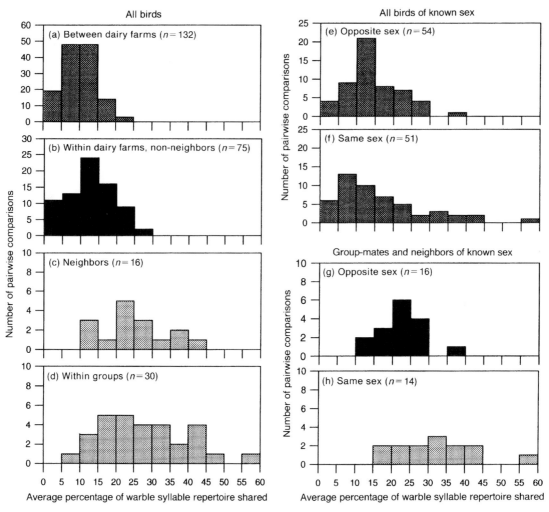

Fig. 7.4. Syllable repertoire sharing by Australian magpies on the basis of social relationship and sex. Left panel, histograms of the average percentage of the repertoire shared by birds (a) in different localities, (b) on the same farm, (c) on neighboring territories, and (d) in the same cooperative group. Repertoire sharing was determined by comparing the syllable repertoire of each bird and calculating the average percentage of repertoire sharing, pairwise, with every other bird (n = the number of pairwise comparisons). Each of the four distributions was compared to each of the others with Mann–Whitney rank sum tests: (a), (b) and (c) were all significantly different ($P < 0.001$); (c) and (d) were not significantly different ($P > 0.3$). Right panel, histograms of the average percentage of the repertoire shared by all pairwise possibilities of the opposite sex or the same sex, (e) to (f) for all birds in the population, and (g) and (h) for groupmates and neighbors. Average percentage of repertoire shared was calculated and tested statistically as above. Distributions of (e) versus (f) were not significantly different ($P > 0.79$). Distributions of (g) versus (h) were significantly different ($P < 0.008$). (From Farabaugh et al. 1988.)

(Farabaugh *et al.* 1988). Shared song could be a badge for group recognition (Treisman 1978) and unshared song could provide the basis for individual recognition. Magpies are selective about what and from whom they imitate, according to the tutor's sex and social relationship. Shared warble syllable repertoires may allow magpies to match songs of territorial rivals, especially same-sex rivals during border conflicts. Counter-singing may reduce the incidence of physical aggression, which in the magpie carries great personal risk (Carrick 1972; Farabaugh *et al.* 1988, 1992b). Within the group, use of shared song as greeting or contact among group members, and participation in communal song, can reflect affiliative and collaborative relationships, as in the case of the tame magpie Zoe (Fig. 7.5b and d).

The process of vocal imitation and the social group

To test whether syllable sharing could result from vocal imitation of groupmates, we tried an experiment with a three-year-old adult female magpie, Zoe, who was hand reared and accepted humans as conspecifics (E. D. Brown *et al.* 1988). Apart from being directed at human recipients, Zoe's behavior patterns and social responses were normal. Zoe played with, and tolerated in her aviary, four to five people, approximately the number of a typical magpie group. She attacked outsiders fiercely and often "countersang" with them (Fig. 7.5b). Her vocal repertoire was composed mainly of human whistle sounds (Fig. 7.5). Brown, one of Zoe's "groupmates," chose a simple model tutor whistle that Zoe had never been heard to utter, and whistled it repeatedly during daily social interactions in chorus songs and when greeting Zoe from a distance. Within two weeks Zoe had incorporated an exact copy into her repertoire and used it frequently (Fig. 7.5d). This shows that vocal imitation results in shared sounds, and that at least some vocal imitation is related to group membership and probably also to affiliative social interactions. Vocal imitation of other human whistles, especially those used in "counter-singing" (Fig. 7.5b) with human territorial rivals, may have been related to aggressive interactions.

Group, individual and sex specificity of communal carol songs

From fieldwork with multiple-adult groups in New Zealand, we found that Australian magpies sang group-specific carol songs. Songs contained a wide range of carol syllables that could be grouped easily into general classes, and these classes were sex specific. To study song roles in greater detail we tape-recorded Australian magpies in a Queensland population that had a high proportion of territory-holding adult pairs, because of the technical difficulty of identifying each singer's vocal role in communal chorus songs when several group members participated (Fig. 7.1c). At our Queensland study site, we color-banded 36 magpies, mapped all territories, and tape-recorded magpies on 12 territories for four months each year during two consecutive years (Brown & Farabaugh 1991). After extensive spectrographic analysis of vocal repertoire samples of 24 birds (12 adult male/female pairs), we determined carol song types and classified 204 carol syllable types in 11 general classes based on sound structure. We then determined carol syllable types and song types that occurred in the repertoire of each bird and each pair, and considered these carol repertoires in terms of the singer's sex, group membership, and territorial neighbors.

All carol songs we analyzed were group specific and unique to whatever particular territorial group sang them. Carol songs are communal. They are made up of the combined melodies of all the participating singers, like motets and madrigals, which are also composed of combined melodies rather than melody and harmony. In our analysis of pairs at the Queensland site, communal carol songs were duets comprising carol song parts sung by males and carol song parts sung by females (Fig. 7.1a and b). Group specificity and "sharing" of carol song types is not achieved by sharing of the same song parts by more than one bird in a group, but by combination of each bird's individually distinctive carol song part into a whole, the communal song shared by the group (or pair). The combination of song parts was not random; a particular carol song part of one bird was regularly combined with a particular carol song part of its mate, and was repeated the same way each time a given duet was sung. Variation in song repertoire size occurred among pairs, between mates within pairs, among males in general, and among females in general (Fig. 7.6). Females had carol song repertoires significantly larger than those of males.

When we examined carol syllables, we found major differences between and within the sexes. Most of our 11 general classes of carol syllables were sex specific (Fig. 7.7). Calculation of pairwise repertoire overlap showed

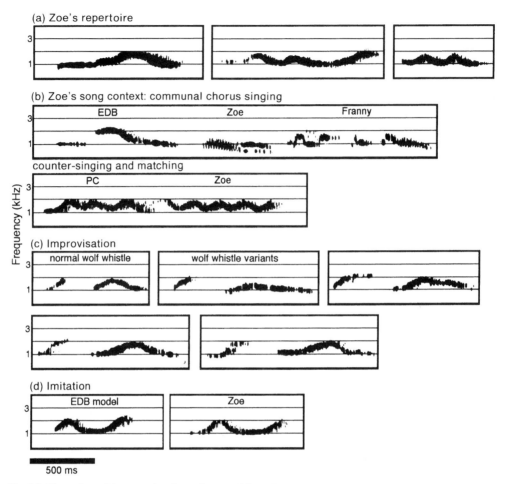

Fig. 7.5. Illustrations of the repertoire of a captive-reared Australian magpie, Zoe. (a) Representative whistles from Zoe's simple repertoire. (b) Zoe's sounds were used in normal contexts; top, communal singing by author EDB, Zoe, and juvenile wild-reared magpie Franny; bottom, counter-singing with a human territorial rival, P. Cary. (c) Normal wolf whistle, and improvised variants of the wolf whistle during rambling songs, differing in duration, temporal patterning, and timing of the frequency peak, etc. (d) Vocal imitation of social companions. Left, Model tutor whistle by EDB. Right, Zoe's imitation. (From E. D. Brown *et al.* 1988.)

that females shared a high percentage of their general syllable classes with other females, and a very small amount (6%) at the class level with males. Males shared highly with other males. However, at the level of particular syllable types, no carol syllables were shared by birds of the opposite sex. Carol syllable types were shared only by same-sex birds, but, even then, no syllable type was shared by all same-sex birds in the study population. Most (78%) carol syllable types were individual specific, and particular composition of the carol repertoire varied

from bird to bird, because magpies, especially females, sing themes with many variations. Although exact copies of some carol syllable types were shared among a few females or a few males, the social correlates of this sharing remained obscure. There was no consistent pattern of carol syllable sharing with respect to territory proximity: sharing was not significantly higher between females (or males) who were neighbors versus those who were not neighbors.

These results show that carols can express individual

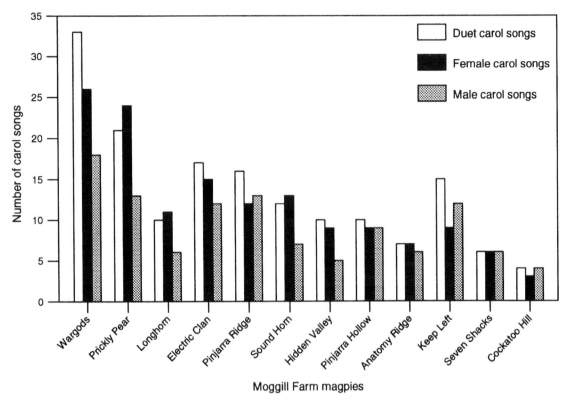

Fig. 7.6. Histograms showing the number of duets (white), females' song parts (black), and males' song parts (gray) of Australian magpie pairs at the study site at Moggill Farm in Queensland. There were relatively few duet song types actually sung by each pair, compared to the number of possible duets if song parts were combined at random. Female magpies had significantly larger song repertoires than their mates ($P < 0.05$, Wilcox on matched-pairs signed-ranks test). (From Brown & Farabaugh 1991.)

identity, sex, number of group members, and group composition. Carolling occurs especially in contexts of intergroup rivalry (Fig. 7.2): Carolling proclaims competition with rivals. Carolling may well aid in recognition and coordination during the mêlée of an intergroup battle, which can involve several territorial groups simultaneously and can last for up to 30 minutes (Brown & Veltman 1987). The contribution of females to synchronized cooperative defense is equal to that of males (Farabaugh *et al.* 1992b), and the active role of the female is correlated with a carol repertoire that is as large or larger than that of the male. Sex-specificity of syllable types may relate to male-on-male and female-on-female fighting during territorial battles. By participating in carolling, a magpie participates in a communal display with all its group mates, and this

participation can delineate an collaborative relationship (e.g., as sung by the tame magpie Zoe and her "group-mates" Fig. 7.5b) in nonterritorial as well as territorial contexts. Carolling also signals a magpie's readiness to act in concert with its group mates.

Magpie song summary

Both warble and carol components of Australian magpie song can represent shared, group-specific vocalizations. Both warbles and carols have affiliative as well as aggressive functions, and warble and carol development involves both collaborative interactions among group members as well as aggressive interactions between neighboring territorial groups or occasionally between competing group members of the same sex.

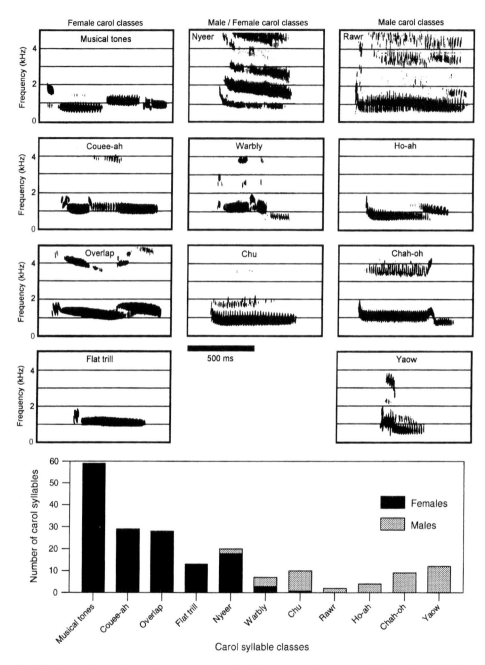

Fig. 7.7. Sex-specificity of Australian magpie carols. (Top) illustrations of the 11 general classes of carol syllables at Moggill Farm: left, female-specific carol classes; center, classes sung by both sexes; right, male-specific classes. Carol syllables sung by females were more complex structurally than those of males. (Bottom) Stacked histogram showing number of syllable types in each carol class at Moggill Farm. Black bars indicate syllable types from females' syllable repertoires, and gray bars indicate syllable types from males' syllable repertoires. Syllable classes sung exclusively by females had more different carol syllables than all but one class sung exclusively by males. (From Brown & Farabaugh 1991.)

SONG SHARING IN THE AMERICAN CROW

Species background

The American crow is a long-lived oscine songbird belonging to the family Corvidae (ravens, crows, magpies, jays, nutcrackers). This crow species has a broad geographical range through most of North America and occurs in a variety of habitats (Bent 1946). The sexes are monomorphic, both shiny black with a dark brown iris. First-year birds have duller, slightly more brownish plumage.

The crow's way of life revolves around the family. The American crow is a facultative cooperative breeder (Caffrey 1992). The pair bond is permanent and monogamous (Good 1952; Goodwin 1976). Females incubate (Good 1952); both mates share nest-building and parental duties, and remain together throughout the year. They may be aided at the nest by offspring from previous years (Kilham 1984; Caffrey 1992); cooperative groups consist of one adult breeding pair and their young from previous broods (Caffrey 1992), unlike the various combinations described previously for Australian magpies. In some populations approximately one-third of the breeding groups contain auxiliaries (Caffrey 1992), but in other regions breeding pairs may have no helpers (Ignatiuk & Clark 1991; Sullivan & Dinsmore 1992). In addition to maintaining close ties with parents and new siblings, second-year birds spend some time feeding in flocks with other unmated birds in their home area (Good 1952; Caffrey 1992). Juveniles remain with parents during the summer and fall, when the whole family begins to join a large communal roost at night (Good 1952). There is partial migration in northern populations of American crows (Bent 1946), unlike Australian magpies, none of which migrate.

Establishment and defense of a territory or exclusive area is variable among crow populations and even within a local population. Territoriality may be exhibited during part or all of the year (Kilham 1985) but in some populations defense of exclusive areas does not occur (e.g., Good 1952; Caffrey 1992). In the Maryland population observed by Brown (1979), crows nesting in coniferous forest bordering cornfields did not establish strictly demarcated exclusive areas, whereas, nearby, territories consisting of a combination of meadow and woods were assiduously defended. During the winter, most or all crows spend time in a flock, even if some continue to visit their territories each day, as does the hooded crow (*Corvus cornix*) of Europe (Loman 1985).

Social behavior of the American crow is oriented toward close, well-known individuals, not territorial rivals. Unlike Australian magpies, crows have almost no specialized territorial social behavior (Brown & Veltman 1987). Crows have a large number of close-range social behavior patterns; crow communication is further specialized for very close contact by complicated patterns of facial feather erection by which extremely subtle gradations of expression are visible (Brown 1979; common ravens (*Corvus corax*), see Gwinner 1964). Within the social group, allopreening is frequent.

Vocalizations

Although the vocalizations usually associated with the American crow are the numerous loud caws used in long distance communication and territorial advertisement, crows often sing in nonterritorial contexts. Both sexes sing. Crow song includes an extremely diverse array of sounds ranging from clicking and rattling noises to caws and melodious coos. Unlike the songs of typical north temperate territorial passerines, or of European starlings (Hausberger, Chapter 8) and Australian magpies, crow song has no stereotyped or fixed sets of syllables that form recognizable song "types" repeated on different occasions. Instead a "song sequence" is simply a discrete sequence of syllables occurring in close temporal connection, which can last from one to several seconds.

Song in corvids functions in intragroup bonding rather than territorial advertisement or defense (Brown 1985). Young crows in their first year sing together for hours at a time. Townsend (1927) and Good (1952) observed adult American crows singing during courtship and nest-building, and Gwinner & Kneutgen (1962) found that songs were shared by mated adult ravens. Lorenz's (1970b) remarks about the song of the jackdaw (*Corvus monedula*) also apply to crow song: it expresses the jackdaw's "mood," and contains imitated sounds.

Social interactions and vocal imitation in song

To investigate the relationship between vocal imitation in crow song and social interactions, Brown (1985) analyzed

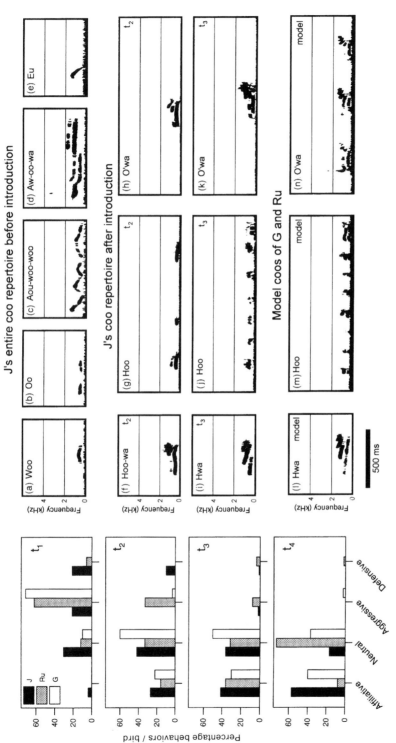

Fig. 7.8. Concurrent vocal imitation and increased affiliative interactions among American crows. Left panel, social interactions of J (black), Ru (gray), and G (white) over four sampling periods. Each bar represents the percentage of that bird's behaviors during social interactions. Behaviors were classified as affiliative (e.g., alloopreen, share food, "fluffed" posture in close proximity), neutral (e.g., watch from distance, ignore another's presence), aggressive (e.g., peck, jump at, threat posture, aggressive vocalization), or defensive (e.g., cringe, jerk or jump away). For each period (t_1 through t_4) there were approximately 50 behaviors per bird. t_1 is the time of introduction of J to Ru and G; t_2, t_3, and t_4 are 2, 8 and 10 weeks after introduction. Right panel, imitation of G and Ru "model" coos by J. Top ((a) to (e)): Complete repertoire of J's coo elements before introduction to Ru and G. J's imitations of three new coos at t_2. The three imitations at t_2 are unlike the original coos of J, but are not yet exact copies of the harmonic structure and temporal patterning of the "model" coos ((l) to (n)) sung by G and Ru. ((l) to (k)) J's imitation of three new coos at t_3. The three imitations at t_3 match the models in harmonic structure and emphasis, and temporal patterning. (Adapted from Brown 1985.)

vocal sharing by American crows in the same social group, as opposed to outsiders. Patterns of vocal sharing at a particular point in time represented a "snapshot" from which the mechanisms of sharing were inferred. A subsequent introduction experiment allowed observation of the ongoing processes involved in vocal sharing and concurrent social interactions. Crows were studied in rural Maryland.

Group specificity of song repertoires

In rural Maryland, four sibling crows were taken from the nest at between two and three weeks of age. Crows fledge at five weeks, and begin to sing at about five to six weeks of age. The four sibling crows (two females, one male, one unknown but probably female) were hand-reared together as a group, and at ten weeks of age they were housed in a large outdoor flight cage set among cornfields and woodlots in prime crow habitat. Two American crow pairs nested within 0.5 km; crows often flew in and landed on the aviary, and extremely frequent vocal interactions with local crows took place every day. Two other captive crows, J and N, one a hand-reared female of the same age as the four siblings, were housed in a second aviary 20 m away, within sight and earshot. Behavior of captive crows was documented by detailed notes and videotapes; vocalizations were tape-recorded and videotaped. A local family of American crows was also observed and tape-recorded intensively over a three-month period. Vocalizations on audiotapes and videotapes were then spectrographed; song syllable types were defined and categorized into five major classes (caws, coos, rattles, harmonic rattles, and harsh sounds). Repertoires of syllable types were determined for three social groups: the four siblings, the dyad (J and N), and the wild family.

Syllable sharing varied according to social relationship. Within each social group members shared some syllable types that were group specific, whereas other syllable types were sung by all groups, including the wild family. For example, of 11 different coo syllable types sung by the four siblings, 10 were group specific; of 5 coo syllable types sung by J, 4 were unique to her. Within a social group, the four siblings, syllable sharing also varied with social interactions. All siblings' repertoires of syllables were sampled at age four months and ten months. Appearance, disappearance, and switching of syllable types occurred in the repertoires of all four. The sibling dyads who allopreened

significantly more also shared the most song syllables. These findings suggest that learned, shared song is related to affiliative social bonds between particular individuals who know each other well.

Affiliative interactions and vocal imitation

If shared sounds are important in intragroup communication, vocal imitation should reflect social interactions. In an introduction experiment, Brown (1985) examined concurrent vocal and social processes. As adults at the age of 2.5 years, two of the siblings, G and RU (females), were moved to a new aviary immediately adjacent to that of J (female) and N. The aviaries were separated only by a wire partition with mesh large enough for the birds to put their heads through. Before introduction, they had also been able to see and hear each other clearly, but only from a distance. Previously it had been established that J had a number of unique syllable types (Fig. 7.8a–e). At each of four sampling periods during the first three months after introduction, social interactions between G, J, and RU, most of which took place at the wire partition, were documented and songs tape-recorded. During this time the character of J's social interactions with G and RU (Fig. 7.8, left panel) and her song repertoire (Fig. 7.8, right panel) changed simultaneously. The percentage of aggressive and defensive behaviors such as pecks and agonistic postures decreased, whereas the percentage of affiliative behaviors such as food-sharing and touching increased, culminating in mutual allopreening. Over the same period J imitated the coo syllables of G and RU, and progressively perfected her imitations of the "model" coos (Fig. 7.8l–n). She used the new coos exclusively in place of her original coo syllable types, during social interactions. Vocal imitation was therefore the mechanism that resulted in shared song, and shared song was related to affiliative interactions and the formation of social bonds.

CALL AND SONG SHARING IN THE BUDGERIGAR

Species background

The budgerigar is a small flock-living parakeet native to central Australia. Unlike the two passeriform songbirds of the two preceding sections, it belongs to a different order, the Psittaciformes, along with other parrots. Adult male and female budgerigars differ in the color of the cere above the bill, and juveniles differ from adults in that the forehead feathers are striped and the iris is dark.

The most important feature of the budgerigar's social milieu is the flock. The arid interior of Australia has shaped the budgerigar's nomadic existence. Extreme variability in annual rainfall, prolonged drought, and occasional plentiful but unpredictable plant growth characterize this environment. To survive the dry years, many Australian parrot species, including the budgerigar, escape through nomadic movement to better areas when conditions deteriorate (Rowley & Chapman 1991). Budgerigars live their entire lives as members of a flock, and gain advantages in avoiding predators, finding scattered resources, and acquiring the knowledge of more experienced companions. Budgerigar flocks roam great distances across the continent rather than over defined home areas, and breed opportunistically to take advantage of transient abundance of food. Once thought to be random (see e.g., Serventy 1971; Rowley 1974), geographic movements of budgerigars have an underlying seasonal pattern, and some evidence suggests that flocks return to traditional sites where they had bred successfully (Wyndham 1983). Flock size and flock dynamics are not as well understood in budgerigars (Wyndham 1980a,b, 1981, 1983) as opposed to the more sedentary well-studied Australian parrot species such as the galah (*Eolophus roseicapillus*; Rowley 1980, 1983a, 1990), Major Mitchell's cockatoo (*Cacatua leadbeateri*; Rowley & Chapman 1986, 1991) and Carnaby's cockatoos (*Calyptorhynchus latirostris*; Saunders 1974, 1982, 1983, 1986), which all form stable flocks of known individuals. Budgerigars may maintain stable flocks for a period of time, but the duration of these associations and the rate that birds join or leave flocks is unknown (Farabaugh & Dooling 1996).

Within the flock, budgerigars probably exhibit the obligate monogamy typical of almost all parrot species (e.g., for a review, see Forshaw 1989). This is because, in most parrot species, the efforts of both mates are required for successful breeding: the male not only helps to feed the young but also feeds the female during incubation and brooding. In addition, the budgerigar breeds at unpredictable times when conditions suddenly become favorable. Birds that are already paired are thought to be able to initiate breeding most rapidly; established pairs also coordinate breeding most efficiently (Rowley 1983b).

Budgerigar pairs fledge their young directly into a juvenile flock or crèche. Because hatching is asynchronous, the parents may still have nestlings to care for, and visit the juvenile crèche only intermittently to feed their fledglings for no more than two weeks (Wyndham 1981). Although each subsequent nestling is fledged into the crèche, there may be little association among the siblings, as in the galah (Rowley 1980). After two weeks, juvenile budgerigars join flocks composed of juvenile and adult birds and leave the nesting area. The young birds thus gain the advantages of improved predator avoidance and knowledge of more experienced birds within the flock. In addition they develop social bonds and eventually find a mate.

Vocalizations

Budgerigars have an elaborate vocal repertoire consisting of a number of distinct calls (e.g., contact and alarm calls) and a long complex warbling song with which they coordinate social and reproductive behavior (Brockway 1964a,b, 1969; Wyndham 1980b; Farabaugh *et al.* 1992a, 1994; Farabaugh & Dooling 1996). Vocal development occurs early in life. Budgerigar nestlings develop an individually specific food-begging call in the week before fledging, and the essential repeated pattern of frequency modulation of this call becomes the pattern of the bird's first contact call (Powell 1993; Farabaugh & Dooling 1996). This contact call is used for the first month and then young budgerigars begin imitating the contact calls of their social companions in the flock. At this same age budgerigars have a nearly complete adult vocal repertoire, which includes the two most important learned vocalizations, warble song and contact calls. (Similarities between song learning and call learning are discussed later in a separate section.).

Song

Budgerigars sing a long complex song composed of melodious warbling in which louder chirps and squawks are interjected. Like crow song, budgerigar warble song has no set length and varies greatly in structure both within and among individuals (Farabaugh *et al.* 1992a). Many functionally distinct call types are embedded in warble, especially contact and alarm calls, but other syllable classes are found only in warble. Heterospecific mimicry is also incorporated into warble, for example mimicry of human speech in warble songs of domestic budgerigars (Gramza 1970) and mimicry of other native Australian bird species in warble songs of wild-caught budgerigars (S. M. Farabaugh, unpublished data). Both males and females sing, although males warble daily at a much higher rate than do females. Warble develops through learning; males in the same social group share a significantly greater portion of their warble syllable repertoires with each other than with males in a different social group (Farabaugh et al. 1992a). We know social interaction guides what is learned, because a young male preferentially imitated the abnormal syllables and temporal patterning of the warble of his isolate-reared cagemate, rather than the normal warble, clearly audible, of birds in adjoining cages (Farabaugh et al. 1992a). Warble song is important for synchronization of breeding in mated pairs: Brockway (1965, 1967, 1969) found that learning and/or performing warble song promoted full gonadal activity in both sexes. Warble song occurs in two main contexts: directed at another bird, and accompanied by active social behavior such as beak-touching, rapid sidling, or sometimes courtship feeding and copulation; and during inactive periods of relaxation when other flock mates are warbling (Farabaugh et al. 1992a).

Contact calls

Contact calls are the sounds most frequently uttered by budgerigars (Brockway 1969; Wyndham 1980b). Contact calls are strongly frequency modulated, in the range of 2 to 4 kHz, and last about 100–300 ms (e.g., Fig. 7.9). The patterns of frequency modulation are extremely varied, but an individual bird can repeat a particular pattern with great precision from one occasion to the next. At any given time, a budgerigar has a repertoire of one to several patterns, or contact call types; usually one or two types predominate, and account for 95–100% of all contact calls uttered (Farabaugh et al. 1994). The contact call is the vocalization that budgerigars and other parrots use to maintain contact with their mate and flock (see e.g., Brockway 1964a; Wyndham 1980b; Pidgeon 1981; Saunders 1983). Contact calls are produced repeatedly when budgerigars are in flight, separated from the flock, preparing for evening roosting, or reunited with the mate after separation (Wyndham 1980b). Budgerigars have highly developed perceptual abilities for detecting and cat-

egorizing contact calls (e.g., Dooling 1986; Park & Dooling 1985, 1986; Okanoya & Dooling 1991). In the laboratory, budgerigars call more readily in response to their mate's call (Ali et al. 1993) and can learn to discriminate among different individuals' renditions of the same shared contact call type (S. D. Brown et al. 1988).

Social interactions and the process of call learning

If recognition of pairmates and flockmates are important functions of shared contact calls, social interactions among flockmates should influence call learning. Farabaugh et al. (1994) investigated this connection in a recent laboratory study. They obtained six adult male budgerigars from six different breeders. Contact calls were tape-recorded from all these birds for several days to establish that each bird had a distinctive repertoire of contact call types (Fig. 7.9, left panel). Three birds per cage were put into two adjoining cages separated only by a thin cloth partition so that birds on each side could hear but not see birds on the other side. Over the next eight weeks, contact calls were tape-recorded from each bird every three to five days. Within one week, shared contact calls appeared in the repertoires of birds kept on the same side of the partition. Sharing began as imitation of another bird's call type, but the birds quickly began to develop new types composed of pieces of the call types of various birds. Once convergence was achieved through imitation and recombination, the particular structure of the call continued to change from week to week, and these changes were evident in the calls of all of the birds that shared the call type (Fig. 7.10). A similar pattern of slow simultaneous change was found in the shared colony-specific songs of yellow-rumped caciques (Cacicus cela; Trainer 1989). At the end of eight weeks, the three birds on one side of the partition shared the same dominant contact call type, and the three birds on the other side of the partition shared other dominant contact call types (Fig. 7.9, right panel). There was only slight evidence of sharing across the partition: thus, aural contact without social contact significantly reduces learning of contact calls.

This study shows that social interactions play a large, and perhaps critical, role in the learning of contact calls by this species. Socially directed contact call learning closely corresponds with Pepperberg's (1985, and see Chapter 9)

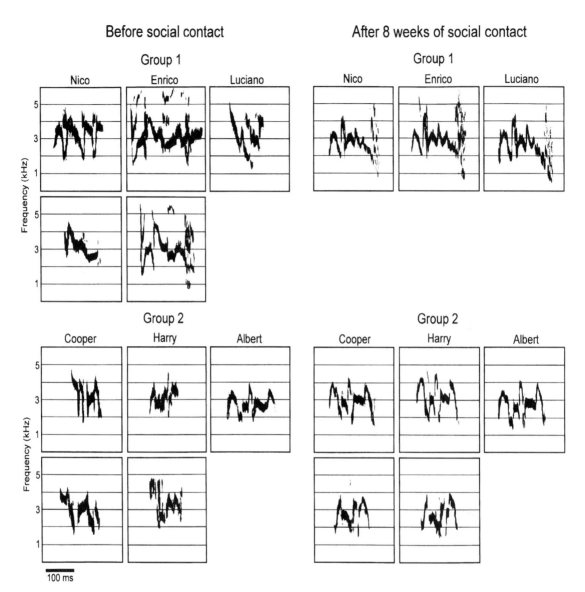

Fig. 7.9. Social interactions play a critical role in the development of flock-specific contact calls of the budgerigar. Left panel: Spectrograms of the dominant contact calls of six male budgerigars before coming into social contact. Two birds (Luciano and Albert) each had a single dominant contact call type that accounted for 97–99% of each bird's recorded contact calls. Four birds (Nico, Enrico, Cooper, and Harry) each had two dominant contact call types that together accounted for 91–96% of their recorded contact calls. No contact call types were shared by the six birds. Right panel: The dominant contact calls of the same six birds after being housed for eight weeks in two groups of three birds, each group in an adjoining cage. Birds could see, hear, and physically interact with their groupmates but could only hear the birds in the other group. Each group of three birds now shared a dominant contact call type. Two of the birds in group 2 (Cooper and Harry) each had two dominant contact call types, and they shared both. Dominant contact call types were different in the two groups. (After Farabaugh et al. 1994. Copyright © 1994 by the American Psychological Association. Adapted with permission.)

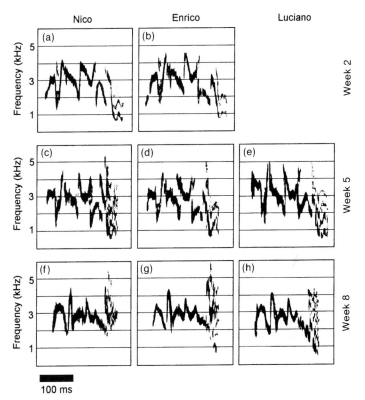

Fig. 7.10. Synchronous changes in a shared contact call type of a budgerigar group. Representative spectrograms of the shared contact call type from three different times during the study from Nico (left; (a), (c) and (f)), Enrico (center; (b), (d) and (g)) and Luciano (right; (e) and (h)). Luciano did not add this call to his repertoire until week 5. This call type, which was not a dominant type at week 2, eventually became the dominant shared call type of the three groupmates (From Farabaugh *et al.* 1994. Copyright © 1994 by the American Psychological Association. Adapted with permission.)

application of social modelling theory to avian learning. In this model, a living tutor demonstrates the sound and makes adjustments according to the behavior of the observing bird. The more intense the contact between the observing bird and the tutor, the more likely the observer is to learn the sound. One can easily imagine how this process of learning can result in a shared flock call type, and thus the shared contact call could be a badge of flock membership (Treisman 1978), and aid in synchronization and coordination of the movements and activities of flock members.

Species from diverse taxa that live in stable groups for all or part of the year have a functionally similar learned vocalization that is equivalent to the budgerigar contact call. Chickadees (Mammen & Nowicki 1981; Nowicki 1983, 1989), cardueline finches (Mundinger 1970, 1979), and even marmoset monkeys (Elowson & Snowdon 1994) have been shown to have group-specific calls, referred to variously as contact, distance, or flight calls, which are used when the group is about to move or fly off, when individuals are separated from the group, and when groupmates or mated pairs greet one another after separation. In parrots and flocking songbirds these stable flocks are usually quite mobile, and flock-specific contact calls may aid in the coordination of synchronous group movements when the flock temporarily joins larger amalgamations of similar flocks. In these diverse species, the contact call is one of the vocalizations whose major function may be formation, maintenance, and recognition of social bonds.

VOCAL LEARNING IN THREE AVIAN SOCIETIES

Although our three study species exhibit many differences in ecology, spacing behavior, and taxonomy, they all form complex social relationships with long-term associations among well-acquainted individuals, and they are similar in their tendencies to learn vocalizations from their social companions throughout their lives. By understanding similarities and differences among many groups of birds, and identifying common features, we can clarify the role of social interactions in vocal learning and approach a more general model of vocal learning. What is similar about vocal learning in parrots and songbirds? What is different about the vocalizations of species with and without vocal learning? How do learned shared vocalizations aid social relationships and interactions among individuals? Do males and females always differ in learning ability? And how do social interactions affect what is learned?

Evolution of vocal learning

Vocal learning is thought to occur in at least four orders of bird (Nottebohm 1972; Kroodsma 1982a): Passeriformes (oscine songbirds; for a review, see Kroodsma 1982a), Psittaciformes (parrots; Gramza 1970; Nottebohm 1970; Rowley & Chapman 1986; for a review, see Farabaugh & Dooling 1996); Apodiformes (Trochilidae, humming-birds; Snow 1968; Wiley 1971; Baptista & Schuchmann 1990), and Piciformes (Rhamphastidae, toucanets; Wagner 1944). Vocal learning is the norm in two of these taxonomic groups, songbirds and parrots. Although it was once thought that the neural circuitry and specialized nuclei involved in vocal learning were homologous in parrots and songbirds (Paton *et al.* 1981; De Voogd 1986), recent neuroanatomical and neurochemical studies suggest that the similarities are superficial (Striedter 1994; Ball 1994). In comparison with songbirds, the vocal control nuclei of parrots are in a different anatomical position (Paton *et al.* 1981), receive auditory input from a different auditory pathway (Brauth *et al.* 1987, 1994; Brauth & McHale 1988; Striedter 1994), and lack some of the steroid receptors that are abundant in the songbird vocal control nuclei (Gahr *et al.* 1993; Ball 1994). This lack of homology at the neurological level strengthens the argument that vocal learning has arisen independently in these

two groups, and thus makes the similarities in the function of learned vocalizations between the two groups all the more intriguing.

Because songbirds account for nearly half of all extant bird species, vocal learning is considered a common feature of birds in general, yet in the majority of avian taxa, individuals develop their vocalizations without reference to hearing conspecifics' or even their own vocalizations. These species can have proficient vocal communication without vocal learning. They can use vocalizations to communicate species and sexual identity, and, because there are consistent differences among individuals' vocalizations, they can also communicate individual identity. Domestic chickens and red jungle fowl have been shown to have different alarm calls for different classes of predator (Collias 1987; Evans *et al.* 1993), and they call and respond differently depending on their social situation (see e.g., Evans & Marler 1991). Learning is a risky way to obtain species-specific vocalizations; a bird may not learn the "correct" vocalizations. For example, galahs naturally cross-fostered after nest-robbing by Major Mitchell's cockatoos learn the calls of their foster parents and associate with their foster species, to the exclusion of their own, throughout their lives (Rowley & Chapman 1986).

For vocal learning to have evolved, it must allow some special quality that is available only with learning. One distinctive quality that vocal learning provides is a mechanism for individuals to share sounds with *particular* conspecifics, be they neighbor, kin, mate, flock, or territorial group, rather than with *any* conspecific (i.e., species-specific vocal characteristics shared with all conspecifics). By sharing vocalizations with a subset of conspecifics, a bird's vocalization can denote species, sex, individual and *group* identity. To understand why vocal learning evolved we must understand why it is advantageous for a subset of conspecifics to share sounds unique to that subset. The advantages of shared sounds may differ depending on the nature of the subset, i.e., rivals versus collaborators.

Sharing sounds with territorial rivals: Song dialects and song matching

The song dialects of territorial songbirds may be the most familiar example of learned vocalizations shared by a subset of conspecific birds. Song dialects are geographic variants of a species song. It has been proposed that song dialects are a result of young birds learning song only during an early sensitive phase and eventually settling in their birth locale; dialects were thus thought to function in assortative mating (Marler & Tamura 1962) and to correspond to genetic differences among neighboring populations or demes (see e.g., Baker 1975). This hypothesis has remained controversial (Zink & Barrowclough 1984; Zink 1985; Hafner & Petersen 1985; Baker *et al.* 1985; Rothstein & Fleischer 1987), but recent molecular evidence (Payne & Westneat 1988; Loughheed *et al.* 1993) provided no evidence of genetic differences among dialect groups. Possibly of more importance, many studies have shown that dialects are a consequence of birds learning songs from neighbors *after* settlement (see e.g., Jenkins 1978; Kroodsma 1979; Payne 1979; Payne *et al.* 1981; McGregor & Krebs 1982; Payne & Payne, Chapter 5; for a review, see Baptista 1985) or through selective attrition of previously learned songs that are at least similar to neighbors' songs (Nelson 1992, and see Chapter 2). Recent studies also show that established territory-holders continue to learn new songs: great tits (*Parus major*) learn the songs of newly arrived neighbors (McGregor & Krebs 1989); and wild European starlings that are at least four years old continue to learn new song themes (Mountjoy & Lemon 1995). Several species show change of song repertoires from year to year, which results in a greater similarity of neighbors' songs (e.g., tufted titmice (*Parus bicolor*), Schroeder & Wiley 1983; thrush nightingales (*Luscinia luscinia*), Sorjonen 1987).

It now appears that shared song, as in dialects, functions in communication, recognition, and competition between territorial rivals. The most common example of communication with shared song is song matching during counter-singing at territorial boundaries. Song matching has been shown to be associated with a high probability of attack or increasing threat in several species such as great tits (Krebs *et al.* 1981), song sparrows (*Melospiza melodia*) Kramer *et al.* 1985), and Carolina wrens (*Thryothorus ludovicianus*; Simpson 1985)). In some species, matching is part of an elaborate vocal duel where birds share large song repertoires and sing them in the same sequence, as in marsh wrens (*Cistothorus palustri*; Kroodsma 1979) and European blackbirds (*Turdus merula*; Todt 1981). By exposing European blackbirds to different playback regimes, Todt (1981) showed that blackbirds would stop singing from perches where they lost song-matching

battles, i.e., where they were matched and overlapped. Very large shared song repertoires are not the norm; it is more common for territorial songbirds to have repertoires of several song types, and they may share different songs with different neighbors, as do great tits (McGregor & Krebs 1989). Song sharing could also be advantageous because it may inhibit individual recognition in some situations. A young newly settled bird could be mistaken for an established adult territory holder if it sings the same song (Payne 1981), and an operant study involving shared songs showed that song sparrows grouped songs by song type rather than voice (Beecher *et al.* 1994). However, it is unlikely that established territory-holding adult great tits learning the songs of new neighbors accrue an advantage by being mistaken for a young newly settled bird. Nor are these adults merely increasing the size of their song repertoires, since they discard other songs (McGregor & Krebs 1989). Rather, it appears that it is important for great tits to share at least one song with each neighbor, which can be used during matched counter-singing with that neighbor (McGregor & Krebs 1989).

Territorial song matching may be aggressive partly because shared songs provide clear communication of the singer's location. Songs degrade over distance due to the differential propagation of low versus high frequencies. Birds can use this degradation to assess the location of the singer; clear songs indicate a close singer, and degraded songs indicate a distant singer (Richards 1981). Birds appear to be better at distinguishing degraded from undegraded songs when they have the song type in their own repertoire (McGregor & Krebs 1984) or are familiar with the song type (i.e., a neighbor's song that they do not share but have regularly heard (McGregor *et al.* 1983). In song sparrows, when a neighbor sings a song that the listening bird does not share, the listener often answers the neighbor by singing a song that it does share with that neighbor (Stoddard *et al.* 1991a,b). By approaching the common border and answering with a shared song the singer is unequivocally letting the neighbor know that it is nearby and ready to defend. Song matching represents a special communication between neighbors. It is a way of saying "I hear you, I recognize you, I know where you are, and I am letting you know where I am." A newly settled bird may be less able to distinguish the level of threat in its neighbor's songs because it cannot accurately assess its neighbor's location. Territorial neighbors are indeed a serious threat;

song sparrows are as likely to lose their territories to a neighbor as to a floater (Arcese 1989).

In Australian magpies, aggressive counter-singing may be one of the reasons for high warble syllable sharing among neighbors and groupmates of the same sex. Under unusual circumstances counter-singing may even be used within territorial groups: when an intruding female arrived and attempted to supplant the territorial female already in residence in a three-bird group, the two females counter-sang with matching warble songs repeatedly during the first several days of the takeover attempt. There were also violent fights between the females, which were watched with interest by the resident males (E. D. Brown, unpublished data). Counter-singing may explain why Australian magpies also exhibit even higher sharing of carol syllable classes, and males have high sharing of the particular carol syllable types. Carol songs composed of shared carol syllable classes are used in the daily counter-singing interactions between neighboring groups at territorial boundaries.

Sharing sounds with group mates: Duets, choruses, and flock-specific calls

Shared group-specific vocalizations are thought to be both passwords and proclamations of group membership (Feekes 1977, 1982; Treisman 1978). For example, yellow-rumped caciques in the same breeding colony share a repertoire of songs that they use in competition with one another for mates (Trainer 1987, 1988); caciques in the colony collaborate to chase away intruding males that lack the colony-specific song (Feekes 1977, 1982). In a similar way, American crows may recognize kin by shared family-specific song. Superb lyrebirds (*Menura novaehollandiae*), which breed in a dispersed lek, share group-specific song, including particular heterospecific mimicry, which may be used in competition for mates as well as contact with other lek members (Robinson 1991). Young lyrebirds learn at least some of the heterospecific mimetic song from other lyrebirds. This is shown by the lyrebird population introduced to Tasmania from Victoria 60 years ago, in which calls of mainland species absent from Tasmania persist in lyrebird song after many generations (Robinson 1991). Species from diverse taxa that live in stable groups for all or part of the year have a functionally similar learned call that is specific to that group (Mundinger 1970, 1979;

Mammen & Nowicki 1981; Nowicki 1983, 1989; Farabaugh *et al.* 1994; Elowson & Snowdon 1994; McCowan & Reiss, Chapter 10). Bottlenose dolphins (*Tursiops truncatus*) often imitate signature whistles of their social companions (Tyack 1986; Smolker 1993; Tyack & Sayigh, Chapter 11). Group-specific contact calls may aid in the coordination of synchronous group movements when the group or flock temporarily joins with similar flocks. Learned shared contact calls allow group as well as individual recognition for members of the group. Using perceptual testing procedures, S. D. Brown *et al.* (1988) demonstrated that budgerigars can easily discriminate among different contact call types and that birds that shared a given contact call type could also discriminate between different birds' versions of that contact call type. This perceptual ability is learned; budgerigars that did not have this contact call type in their repertoires could not separate the calls by individual bird. Thus, like the perceptual advantage that songbirds gain in assessing the distance of a singer by the degradation of its shared song (Richards 1981; McGregor & Krebs 1984), a group-living bird may gain a perceptual advantage of individual recognition of its groupmates by means of the shared call.

Group-specific vocalizations can be used in aggressive interactions between groups, and these can be distinctly different in structure from group-specific vocalizations used in affiliative interactions between groupmates. Rather than singing the same song parts, group members combined their unshared song parts into a unified, stereotyped, group-specific territorial song. The end result is still group specificity, even though the method of achieving it is different from sharing "copies" of the same songs. This formation of a group- or pair-specific unified song from combined unshared song parts is a common feature of many duetting species, where males and females defend the territory together, as in slate-colored bou-bou shrikes (*Laniarius funebris*; Wickler 1972), *Thryothorus* wrens (Farabaugh 1983), and Australian magpies (Brown & Farabaugh 1991; for a review see Farabaugh 1982). Individuals do not combine song parts at random, rather each bird sings each of its song parts with only a restricted number of song parts in its groupmate's repertoire. Pair-specific duets may aid in mate recognition and coordination during a pair-versus-pair battle at a boundary. Group-living Australian magpies synchronize their efforts during lengthy territorial battles that involve aerial and ground attacks (Bro wn & Veltman 1987; Farabaugh *et al.* 1992b). Presumably, group-specific chorus song may aid magpies' recognition of groupmates versus intruders during the mêlée of a territorial battle, as well as aiding in synchronizing their aggressive activities (Brown & Farabaugh 1991). Group-specific song may be important as an advertisement to neighbors that the territory is still defended by a particular collaborating group with whom the neighbors have had previous interactions of known outcomes. For the Australian magpie, participation in the group's combined song announces a magpie's willingness to collaborate with its groupmates, and thus the communal territorial song can also function in group cohesion.

Characteristics common to our three focal species support Thorpe's (1961; Thorpe & North 1965, 1966) contention that shared learned vocalizations function affiliatively within the group in the formation of social bonds. In the societies of our three species, long-term associations between mutually well-known individuals are the norm, yet an individual's social environment can change as it moves from one social group to another over the course of its life. In all three species, individuals share vocalizations that are used in affiliative contexts: magpies in the same group exhibit more sharing of the less aggressive warble song, crows share nonterritorial song, and budgerigars share contact calls. Vocal learning is related to affiliative social interactions: magpies share most with groupmates and neighbors, crows show increased vocal imitation with increased affiliative and decreased aggressive interactions, and budgerigars preferentially develop shared cells with birds with whom they can interact as flockmates. Both sexes are integral members of the social groups in all three species, and both sexes exhibit vocal sharing. Both sexes exhibit vocal learning of song, and both male and female budgerigars learn contact calls.

Learning of songs and calls

The division of bird vocalizations into two distinct categories, song and calls, is germane to the discussion of vocal learning in birds. On the basis of structure, ontogeny, and function, bird sounds have historically been classified as either songs or calls (see e.g., Thorpe 1961). The traditional distinction between songs that are complex learned vocalizations used in territory advertisement and mate attraction, and calls that are simple innate vocalizations, is

now seen as inaccurate and artificial. Many species that do not learn have songs (e.g., suboscines such as flycatchers; Kroodsma & Konishi 1991). Some species, such as the Corvidae, use calls for territorial advertisement and song for intragroup communication; some species have songs that are simpler in structure than their calls (e.g., black-capped chickadee (*Parus atricapillus*; Ficken *et al.* 1978)). Many species that learn song also learn calls (Mundinger 1970, 1979; Güttinger 1974; Marler & Mundinger 1975; Nicolai 1959; Poulsen 1959; Nowicki 1989; for a review, see Farabaugh *et al.* 1994). Evidence that some calls are learned has been available for many years. Sick (1939) described dialects in the rain call of the chaffinch (*Fringilla coelebs*), and such geographic variation is thought to arise through vocal learning. The ontogeny and neurologic foundations of call and song learning are similar. Male zebra finches, both wild and domestic, preferentially learn their distance calls and song from their fathers (Zann 1990, and see Chapter 6). The neural control of learned calls and song is identical in zebra finches, and differs from the neural control of unlearned vocalizations (Simpson & Vicario 1990). In budgerigars, lesions of vocal control nuclei disrupt both learned song and learned calls (Hall *et al.* 1994; Brauth *et al.* 1994).

Vocal learning by females and males

Despite the prevalence of duetting species among the world's birds (see e.g., Payne 1971; Farabaugh 1982), singing by females has received scant attention in the literature on avian song learning and function, probably because it is less common in the north temperate songbird species that have been traditionally studied. It has been known for a long time that females of some north temperate species sing (Erickson 1938; Nice 1943). Not only does the young male bulfinch (*Pyrrhula pyrrhula*) learn to sing the song of his father, the female bulfinch learns and sings the song of her mate (Nicolai 1959). In the north temperate species studied, song by females is usually less complex or more variable than song by males. Female song functions in intra pair and family group cohesion in some species, such as northern cardinals (*Cardinalis cardinalis*; Ritchison 1986), black-headed grosbeaks (*Pheuticus melanocephalus*; Ritchison 1983, 1985), and Acadian flycatchers (*Empidonax virescens*; Kellner & Ritchison 1988). In red-winged blackbirds (*Agelaius phoeniceus*;

Beletsky 1983), song sparrows (Arcese *et al.* 1988), and white-crowned sparrows (Baptista *et al.* 1993), female song is used in aggressive territorial encounters and advertisement.

Vocal learning and use of shared vocalizations by collaborating group members are by no means exclusive to one sex alone, nor do they have different functions for males and females. Crows of both sexes sing; the three crows in Brown's study whose vocal imitation of song syllables corresponding to increased affiliative behavior all happened to be females. Budgerigars of both sexes develop and use the shared flock-specific contact calls. Calls that are used to synchronize pair or flock activities are learned by male and female chickadees (e.g., Nowicki 1983, 1989; for a review, see Farabaugh *et al.* 1994). Not only male but female budgerigars produce warble song, which is important for breeding synchrony (Brockway 1965, 1967, 1969). In the Australian magpie, the two portions of song, warble and carol, are sung by females as well as males. Carol syllable and song repertoires of female magpies are more complex, and significantly larger, than those of their mates. The elaborate song repertoire of females is correlated with singing and fighting alongside males in defense of their territory, and reflects the territorial function of duet and chorus song in group territorial defense. In many tropical duetting species (for a review, see Farabaugh 1982), the song of the females is equivalent to that of the males. Farabaugh (1982) argued that this situation arises because defense by females becomes profitable when females live on permanent year-round territories, and that the coordination of the female's defense with that of the mate is the best defense against other territorial pairs and groups.

Greater female participation in singing is correlated with larger female song repertoires, and this in turn is correlated with the size of the specialized forebrain nuclei associated with vocal learning. In species with little or no female song these nuclei are much smaller in females, but in duetting species where males and females both sing frequently and have equally large song repertoires (Farabaugh 1982, 1983), these nuclei are of similar size (Brenowitz & Arnold 1985; Brenowitz *et al.* 1985). In budgerigars, where both sexes learn calls and warble song, there is also little difference in the size of the brain nuclei in their nonhomologous vocal learning neural circuitry, but detailed measurements have yet to be published.

The impact of considering female song means reassessment of some long-held ideas about song function, structure, and development. For example, although the functional significance of song repertoire size in birds has received much attention (for reviews, see e.g., Krebs & Kroodsma 1980; Kroodsma 1982b; Payne 1983; Read & Weary 1992), few authors, if any, have ever considered female song repertoires in their analysis. Usually only the song of north temperate male songbirds is considered (e.g., Read & Weary 1992). Many have discussed the role of sexual selection on song repertoire size (Kroodsma 1983, 1988; Searcy & Andersson 1986), while several studies have suggested that repertoires may play a role in male/male competition for the acquisition and maintenance of territories (e.g., great tits; McGregor *et al.* 1981; Hiebert *et al.* 1989). For species such as the Australian magpie, any hypothesis that explains the function and evolution of song repertoires must account simultaneously for the size of male, female, and combined duet repertoires. Female Australian magpies have larger song and syllable repertoires than their mates. Sexual selection is unlikely to account for magpie song repertoires, but hypotheses relating to territorial competition may well apply.

Vocal learning and social interactions

The nature of the social relationship between individuals is related to the type of social interaction that influences learning. For social rivals, vocal learning is guided by aggressive interactions. Both wild and domesticated male zebra finches usually imitate their fathers (Zann 1990, and see Chapter 6), and aggression by fathers towards sons may often occur, especially when the next brood hatches. Young zebra finches that were reared by their mother alone and then given a choice of tutors preferentially learned from the tutor who was most aggressive towards them (Clayton 1987). Aggressive territorial interactions may often account for shared song: interspecific territorial interactions lead to interspecific vocal mimicry in sparrows (Catchpole & Baptista 1988), and first-year lazuli buntings (*Passerina amoena*) preferentially copy neighbors that they counter-sing and interact with the most (Laws 1994). In the Australian magpie, aggressive interactions are involved in vocal learning, certainly with the use of group-specific carol song during territorial fights, and probably during low-level counter-singing and matching of warble song between territorial neighbors. Aggressive interactions can also be related to shared vocalizations in parrots. A tame female eclectus parrot (*Eclectus roratus*) mimicked the voice quality and words of her human social rival, and sometimes repeated this imitation immediately before aggressive attacks on her rival. In other affiliative contexts this bird imitated the voice and words of her human social companion (E. D. Brown, unpublished data).

For social collaborators, vocal learning is guided by affiliative interactions. Our studies add to the evidence that affiliative interactions can be as important as aggression in determining vocal learning. Affiliative rather than aggressive social interactions are clearly associated with learning and the use of shared vocalizations in all three of our focal species. This is shown by: the tame magpie Zoe, who mimicked her human companion's "model" whistle; the association of magpie warble with nonaggressive contexts; the concurrence of vocal imitation and affiliative interactions among crows; and the development of shared contact calls by budgerigars. Other examples of species in which affiliative interactions are associated with vocal learning are African grey parrots (*Psittacus erithacus*; see e.g., Pepperberg 1985, 1994); it is also implicit in Todt's (1975) study of competitive vocal imitation that grey parrots learned many sounds during affiliative interactions with their social companions. West *et al.* (1983) showed that European starlings learned heterospecific mimicry in conjunction with affiliative interactions. West & King (1985, 1988) also showed that brown-headed cowbirds (*Molothrus ater*) changed their vocalizations by trial-and-error learning according to the preferences shown by non-vocalizing females, and this social learning was based at least in part on affiliative interactions between the sexes.

If vocal learning has evolved so that an individual will share vocalizations with a subset of conspecifics that it interacts with on a daily basis, it is not surprising that social factors affect what, when, and from whom the individual learns. Even in species that do not produce learned vocalizations, social interactions can affect the performance and perception of vocalizations. For example, Gottlieb (1991) showed that social rearing affected ducklings' preference for sounds heard during rearing, and Evans (1991) showed that ducklings' perceptual sensitivity to conspecific calls was enhanced if the calls were presented in an interactive manner. Hultsch & Todt (1989) showed that songbirds memorize context in addition to sound, thus indicating how learned vocalizations come to be associated with particular social factors. Our studies of three avian societies with complex social relationships indicate that vocal sharing occurs both among rivals and social companions, and functions in communication and recognition between individuals that have long-term social relationships. In our species, these long-term associations are subject to change over an individual's lifetime, and thus the ability to learn vocalizations continues throughout life.

ACKNOWLEDGEMENTS

These studies were supported by National Science Foundation grants BNS–841263 and BNS–8711837 to E.D.B. and S.M.F., National Geographic Society grants to E.D.B. and S.M.F., and National Institutes of Health National Research Service Award DC00052–01 to S.M.F. We gratefully acknowledge the help of Jane Hughes, Wolfgang Schleidt, Bob Dooling, and Ed Minot. Space and facilities were provided by: the Departments of Zoology and Psychology, University of Maryland, College Park; the Department of Botany and Zoology, Massey University, Palmerston North, New Zealand; and the School of Australian Environmental Studies, Griffith University, Brisbane, Australia.

REFERENCES

Ali, N. J., Farabaugh, S. M. & Dooling, R. J. (1993). Recognition of contact calls by the budgerigar (*Melopsittacus undulatus*). *Bulletin of the Psychonomic Society* **31**: 468–70.

Amadon, D. (1962). The Cracticidae. In *Check-list of Birds of the World*, vol. 15, ed. E. Mayr & J. C. Greenway, pp. 166–172. Museum of Comparative Zoology, Cambridge, MA.

Arcese, P. A. (1989). Territory acquisition and loss in male song sparrows. *Animal Behaviour* **37**: 45–55.

Arcese, P., Stoddard, P. K. & Hiebert, S. M. (1988). The form and function of song in female song sparrows. *Condor* **90**: 44–50.

Baker, M. C. (1975). Song dialects and genetic differences in white-crowned sparrows (*Zonotrichia leucophyrs*). *Evolution* **29**: 226–41.

Baker, M. C., Tomback, D. F., Thompson, D. B. & Cunningham, M. A. (1985). Reply to Hafner and Petersen. *Evolution* **39**: 1177–9.

Ball, G. F. (1994). Neurochemical specializations associated with

vocal learning and production in songbirds and
budgerigars. *Brain Behavior and Evolution* **44**: 234–46.

Baptista, L. F. (1985). The functional significance of song
sharing in the white-crowned sparrow. *Canadian Journal of
Zoology* **63**: 1741–52.

Baptista, L. F. & Schuchmann, K. L. (1990). Song learning in
the Anna hummingbird (*Calypte anna*). *Ethology* **84**: 15–26.

Baptista, L. F., Trail, P. W., DeWolfe, B. B. & Morton, M. L.
(1993). Singing and its functions in female white-crowned
sparrows. *Animal Behaviour* **46**: 511–24.

Beecher, M. D., Campbell, S. E. & Burt, J. M. (1994). Song
perception in the song sparrow: Birds classify by song type
but not by singer. *Animal Behaviour* **47**: 1343–51.

Beletsky, L. D. (1983). Aggressive and pair-bond maintenance
songs of female red-winged blackbirds (*Agelaius
phoeniceus*). *Zeitschrift für Tierpsychologie* **62**: 47–54.

Bent, A. C. (1946). Life histories of North American crows,
jays, and titmice. *United States National Museum Bulletin*
191.

Brauth, S. E., Heaton, J. T., Durand, S. E., Liang, W. & Hall,
W. S. (1994). Functional anatomy of forebrain auditory
pathways in the budgerigar, *Melopsittacus undulatus*. *Brain
Behavior and Evolution* **44**: 210–33.

Brauth, S. E. & McHale, C. M. (1988). Auditory pathways in
the budgerigar. II. Intratelencephalic pathways. *Brain
Behavior and Evolution* **32**: 193–207.

Brauth, S. E., McHale, C. M., Brasher, C. A. & Dooling, R. J.
(1987). Auditory pathways in the budgerigar. I. Thalamo-
telencephalic projections. *Brain Behavior and Evolution* **30**:
174–99.

Brenowitz, E. A. & Arnold, A. P. (1985). Lack of sexual
dimorphism in steroid accumulation in vocal control
regions of duetting song birds. *Brain Research* **344**: 172–5.

Brenowitz, E. A., Arnold, A. P. & Levin, R. N. (1985). Neural
correlates of female song in tropical duetting birds. *Brain
Research* **343**: 104–12.

Brockway, B. F. (1964a). Ethological studies of the budgerigar
(*Melopsittacus undulatus*): Non-reproductive behavior.
Behaviour **22**; 193–222.

Brockway, B. F. (1964b). Ethological studies of the budgerigar
(*Melopsittacus undulatus*): Reproductive behavior. *Behaviour*
23: 294–324.

Brockway, B. F. (1965). Stimulation of ovarian development and
egg laying by male courtship vocalization in budgerigars
(*Melopsittacus undulatus*). *Animal Behaviour* **12**: 493–501.

Brockway, B. F. (1967). The influence of vocal behavior on the
performer's testicular activity in budgerigars (*Melopsittacus
undulatus*). *Wilson Bulletin* **79**: 328–34.

Brockway, B. F. (1969). Roles of budgerigar vocalization in the
integration of breeding behaviour. In *Bird Vocalizations*, ed.

R. A. Hinde, pp. 131–58. Cambridge University Press,
Cambridge.

Brown, E. D. (1979). The song of the common crow (*Corvus
brachyrhynchos*). M.Sc. thesis, University of Maryland,
College Park.

Brown, E. D. (1985). The role of song and vocal imitation
among common crows (*Corvus brachyrhynchos*). *Zeitschrift
für Tierpsychologie* **68**: 115–36.

Brown, E. D. & Farabaugh, S. M. (1991). Song sharing in a
group-living songbird, the Australian magpie, *Gymnorhina
tibicen*. III. Sex specificity and individual specificity of
vocal parts in communal chorus and duet songs. *Behaviour*
118: 244–74.

Brown, E. D., Farabaugh, S. M. & Veltman, C. J. (1988). Song
sharing in a group-living songbird, the Australian magpie,
Gymnorhina tibicen. I. Vocal sharing within and among
groups. *Behaviour* **104**: 1–28.

Brown, E. D. & Veltman, C. J. (1987). Ethogram of the
Australian magpie (*Gymnorhina tibicen*) in comparison to
other Cracticidae and *Corvus* species. *Ethology* **76**: 309–33.

Brown, S. D., Dooling, R. J. & O'Grady, K. (1988). Perceptual
organization of acoustic stimuli by budgerigars
(*Melopsittacus undulatus*): III. Contact calls. *Journal of
Comparative Psychology* **102**: 236–47.

Caffrey, C. (1992). Female-biased delayed dispersal and helping
in American crows. *Auk* **109**: 609–19.

Carrick, R. (1963). Ecological significance of territory in the
Australian magpie. *Proceedings of the International
Ornithological Congress* **13**: 740–53.

Carrick, R. (1972). Population ecology of the Australian black-
backed magpie, royal penguin, and silver gull. *United States
Department of International Wildlife Research Report* **2**:
41–99.

Catchpole, C. K. & Baptista, L. F. (1988). A test of the
competition hypothesis of vocal mimicry, using song
sparrow imitation of white-crowned sparrow song.
Behaviour **106**: 117–28.

Clayton, N. S. (1987). Song tutor choice in zebra finches.
Animal Behaviour **35**: 714–21.

Collias, N. E. (1987). The vocal repertoire of the red jungle
fowl: a spectrographic classification and the code of
communication. *Condor* **89**: 510–24.

De Voogd, T. J. (1986). Steroid interactions with structure and
function of avian song control regions. *Journal of
Neurobiology* **17**: 177–201.

Dooling, R. J. (1986). Perception of vocal signals by the
budgerigar (*Melopsittacus undulatus*). *Experimental Biology*
45: 195–218.

Elowson, A. M. & Snowdon, C. T. (1994). Pygmy marmosets,
Cebuella pygmaea, modify vocal structure in response to a

changed social environment. *Animal Behaviour* **47**: 1267–77.

Erickson, M. M. (1938). Territory, annual cycle and numbers in a population of wrentits (*Chamaea fasciata*). *University of California Publications in Zoology* **42**: 247–333.

Evans, C. S. (1991). Of ducklings and Turing machines: interactive playbacks enhance subsequent responsiveness to conspecific calls. *Ethology* **89**: 125–34.

Evans, C. S., Evans, L. & Marler, P. (1993). On the meaning of alarm calls: Functional reference in an avian vocal system. *Animal Behaviour* **46**: 23–38.

Evans, C. S. & Marler, P. (1991). On the use of video images as social stimuli in birds: Audience effects on alarm calling. *Animal Behaviour* **41**: 17–26.

Farabaugh, S. M. (1982). The ecological and social significance of duetting. In *Acoustic Communication in Birds*, vol. 2, ed. D. E. Kroodsma & E. H. Miller, pp. 85–124. Academic Press, New York.

Farabaugh, S. M. (1983). A comparative study of duet song in tropical *Thryothorus* wrens. Ph.D. dissertation, University of Maryland, College Park.

Farabaugh, S. M., Brown, E. D. & Dooling, R. J. (1992a). Analysis of warble song of the budgerigar, *Melopsittacus undulatus*. *Bioacoustics* **4**: 111–30.

Farabaugh, S. M., Brown, E. D. & Hughes, J. M. (1992b). Cooperative territorial defense in the Australian magpie, *Gymnorhina tibicen* (Passeriformes, Cracticidae), a group-living songbird. *Ethology* **92**: 283–92.

Farabaugh, S. M., Brown, E. D. & Veltman, C. J. (1988). Song sharing in a group-living songbird, the Australian magpie. II. Vocal sharing between territorial neighbors, within and between geographic regions, and between sexes. *Behaviour* **104**: 105–25.

Farabaugh, S. M. & Dooling, R. J. (1996). Acoustic communication in parrots: Laboratory and field studies of budgerigars, *Melopsittacus undulatus*. In *Ecology and Evolution of Acoustic Communication in Birds*, ed. D. E. Kroodsma & E. H. Miller, pp. 97–117. Cornell University Press, Ithaca.

Farabaugh, S. M., Linzenbold, A. & Dooling, R. J. (1994). Vocal plasticity in budgerigars (*Melopsittacus undulatus*): evidence for social factors in the learning of contact calls. *Journal of Comparative Psychology* **108**: 81–92.

Feekes, F. (1977). Colony-specific song in *Cacicus cela* (Icteridae, Aves): The password hypothesis. *Ardea* **3**: 197–202.

Feekes, F. (1982). Sound mimesis within colonies of *Cacicus c. cela*. A colonial password? *Zeitschrift für Tierpsychologie* **58**: 119–52.

Ficken, M. S., Ficken, R. W. & Witkin, S. R. (1978). Vocal repertoires of the black-capped chickadee. *Auk* **95**: 34–48.

Forshaw, J. W. (1989). *Parrots of the World*. Landsdowne Editions, Willoughby, New South Wales.

Gahr, M., Güttinger, H. R. & Kroodsma, D. E. (1993). Estrogen receptors in the avian brain: Survey reveals general distribution and forebrain areas unique to songbirds. *Journal of Comparative Neurology* **327**: 112–22.

Good. E. E. (1952). The life history of the American crow – *Corvus brachyrhynchos* Brehm. Ph.D. dissertation, Ohio State University, Columbus.

Goodwin, D. (1976). *Crows of the World*. Cornell University Press, Ithaca, NY.

Gottlieb, G. (1991). Social induction of malleability in ducklings. *Animal Behaviour* **41**: 953–62.

Gould, J. (1865). *Handbook to the Birds of Australia*. John Gould, London.

Gramza, A. F. (1970). Vocal mimicry in captive budgerigars (*Melopsittacus undulatus*). *Zeitschrift für Tierpsychologie* **27**: 971–98.

Güttinger, H. R. (1974). Gesang des Grünlings (*Chloris chloris*) Lakale Unterschiede und Entwicklung bei Schallisolation. *Journal für Ornithologie* **115**: 321–37.

Gwinner, E. (1964). Untersuchungen über das Ausdrucks- und Sozialverhalten des Kolkraben (*Corvus corax* L.). *Zeitschrift für Tierpsychologie* **21**: 657–748.

Gwinner, E. & Kneutgen, J. (1962). Über die biologishe Bedeutung der "zweckdienlichen" Anwendung erlernter Laute bei Vögeln. *Zeitschrift für Tierpsychologie* **19**: 692–6.

Hafner, D. J. & Petersen, K. E. (1985). Song dialects and gene flow in the white-crowned sparrow, *Zonotrichia leucophrys nuttalli*. *Evolution* **39**: 687–94.

Hall, W. S., Brauth, S. E. & Heaton, J. T. (1994). Comparisons of the effects of lesions in nucleus basalis and field "L" on vocal learning and performance in the budgerigar (*Melopsittacus undulatus*). *Brain Behavior and Evolution* **44**: 133–48.

Hiebert, S. M., Stoddard, P. K. & Arcese, P. (1989). Repertoire size, territory acquisition and reproductive success in the song sparrow. *Animal Behaviour* **37**: 266–73.

Hughes, J. M. & Mather, P. B. (1991). Variation in group size of territorial groups in the Australian magpie, *Gymnorhina tibicen*. *Proceedings of the Royal Society of Queensland* **101**: 13–19.

Hughes, J. M., Robinson, A. & Mather, P. B. (1995). Co-operative breeding in a population of the Australian magpie *Gymnorhina tibicen*. Paper presented at Australiasian Society for the Study of Animal Behaviour national meeting, Brisbane, Australia.

Hultsch, H. & Todt, D. (1989). Context memorization in the song-learning of birds. *Naturwissenschaften* **76**: 584–6.

Ignatiuk, J. B. & Clark, R. G. (1991). Breeding biology of American crows in Saskatchewan parkland habitat. *Canadian Journal of Zoology* **69**: 168–75.

Jenkins, P. F. (1978). Cultural transmission of song patterns and dialect development in a free-living bird population. *Animal Behaviour* **26**: 50–78.

Kellner, C. J. & Ritchison, G. (1988). Possible functions of singing by female Acadian flycatchers (*Empidonax virescens*). *Journal of Field Ornithology* **59**: 55–9.

Kilham, L. (1984). Cooperative breeding of American crows. *Journal of Field Ornithology* **55**: 349–56.

Kilham, L. (1985). Territorial behavior of American crows. *Wilson Bulletin* **97**: 389–90.

Kramer, J. R., Lemon, R. E. & Morris, M. J. (1985). Song switching and agonistic stimulation in the song sparrows (*Melospiza melodia*): five tests. *Animal Behaviour* **33**: 135–49.

Krebs, J. R., Ashcroft, R. & van Orsdol, K. (1981). Song matching in the great tit. *Animal Behaviour* **29**: 918–23.

Krebs, J. R. & Kroodsma, D. E. (1980). Repertoires and geographic variation in bird song. *Advances in Studies in Behavior* **11**: 143–77.

Kroodsma, D. E. (1979). Vocal dueling among male marsh wrens: evidence for ritualized expressions of dominance/subordinance. *Auk* **96**: 506–15.

Kroodsma, D. E. (1982a). Learning and the ontogeny of sound signals in birds. In *Acoustic Communication in Birds*, vol. 2, ed. D. E. Kroodsma & E. H. Miller, pp. 1–24. Academic Press, New York.

Kroodsma, D. E. (1982b). Song repertoires: Problems in their definition and use. In *Acoustic Communication in Birds*, vol. 2, ed. D. E. Kroodsma & E. H. Miller, pp. 125–145. Academic Press, New York.

Kroodsma, D. E. (1983). The ecology of vocal learning. *Bioscience* **33**: 165–71.

Kroodsma, D. E. (1988). Contrasting styles of song development and their consequences. In *Evolution and Learning*, ed. R. Bolles & M. D. Beecher, pp. 157–84. Erlbaum, Hillsdale, NJ.

Kroodsma, D. E. & Konishi, M. (1991). A suboscine bird (Eastern phoebe, *Sayornis phoebe*) develops normal song without auditory feedback. *Animal Behaviour* **42**: 477–87.

Laws, J. M. (1994). The effect of social interaction on song development in yearling lazuli buntings (*Passerina amoena*). Paper presented at the American Ornithologists' Union national meeting, Missoula, MT.

Loman, J. (1985). Social organization in a population of the hooded crow. *Ardea* **73**: 61–75.

Lorenz, K. (1970a). Companions as factors in the bird's environment. In *Studies in Animal and Human Behavior*, vol. 1, pp. 101–258. Harvard University Press, Cambridge, MA.

Lorenz, K. (1970b). Contributions to the study of the ethology of social Corvidae. In *Studies in Animal and Human Behavior*, vol. 1, pp. 1–100. Harvard University Press, Cambridge, MA.

Lougheed, S. C., Handford, P. & Baker, A. J. (1993). Mitochondrial DNA hyperdiversity and vocal dialects in a subspecies transition of the rufous-collared sparrow. *Condor* **95**: 889–95.

Mammen, D. L. & Nowicki, S. (1981). Individual differences and within-flock convergence in chickadee calls. *Behavioural Ecology and Sociobiology* **9**: 179–86.

Marler, P. & Mundinger, P. C. (1975). Vocalizations, social organization, and breeding biology of the twite, *Acanthis flavirostris*. *Ibis* **117**: 1–17.

Marler, P. & Tamura, M. (1962). Song "dialects" in three populations of white-crowned sparrows. *Science* **146**: 1483–6.

McGregor, P. K. & Krebs, J. R. (1982). Song types in a population of great tits (*Parus major*): their distribution, abundance and acquisition by individuals. *Behaviour* **79**: 126–52.

McGregor, P. K. & Krebs, J. R. (1984). Sound degradation as a distance cue in great tit (*Parus major*) song. *Behavioural Ecology and Sociobiology* **16**: 49–56.

McGregor, P. K. & Krebs, J. R. (1989). Song learning in adult great tits (*Parus major*): Effects of neighbours. *Behaviour* **108**: 137–59.

McGregor, P. K., Krebs, J. R. & Perrins, C. M. (1981). Song repertoires and lifetime reproductive success in the great tit (*Parus major*). *American Naturalist* **118**: 149–59.

McGregor, P. K., Krebs, J. R. & Ratcliffe, L. M. (1983). The reaction of great tits (*Parus major*) to playback of degraded and undegraded songs: the effect of familiarity with stimulus song type. *Auk* **100**: 898–906.

Mountjoy, D. J. & Lemon, R. E. (1995). Extended song learning in wild European starlings. *Animal Behaviour* **49**: 357–66.

Mundinger, P. C. (1970). Vocal imitation and individual recognition of finch calls. *Science* **168**: 480–2.

Mundinger, P. C. (1979). Call learning in the Carduelinae: Ethological and systematic considerations. *Systematic Zoology* **28**: 270–83.

Nelson, D. A. (1992). Song overproduction and selective attrition lead to song sharing in the field sparrow (*Spizella pusilla*). *Behavioural Ecology and Sociobiology* **30**: 415–24.

Nice, M. M. (1943). Studies in the life history of the song sparrow. II. The behavior of the song sparrows and other passerines. *Transactions of the Linnaean Society, New York* **6**: 1–238.

Nicolai, J. (1959). Familientradition in der Gesangsentwicklung des Gimpels (*Pyrrhula pyrrhula*). *Journal für Ornithologie* **100**: 39–46.

Nottebohm, F. (1970). Ontogeny of bird song. *Science* **167**: 950–6.

Nottebohm, F. (1972). The origins of vocal learning. *American Naturalist* **106**: 116–40.

Nowicki, S. (1983). Flock-specific recognition of chickadee calls. *Behavioural Ecology and Sociobiology* **12**: 317–20.

Nowicki, S. (1989). Vocal plasticity in captive black-capped chickadees: The acoustic basis and rate of call convergence. *Animal Behaviour* **37**: 64–73.

Okanoya, K. & Dooling, R. J. (1991). Detection of species-specific calls in noise by zebra finches *Poephila guttata* and budgerigars *Melopsittacus undulatus*: time or frequency domain? *Bioacoustics* **3**: 163–72.

Park, T. J. & Dooling, R. J. (1985). Perception of species-specific contact calls by the budgerigar (*Melopsittacus undulatus*). *Journal of Comparative Psychology* **99**: 391–402.

Park, T. J. & Dooling, R. J. (1986). Perception of degraded vocalizations by budgerigars (*Melopsittacus undulatus*). *Animal Learning and Behavior* **14**: 359–64.

Paton, J. A., Manogue, K. R. & Nottebohm, F. (1981). Bilateral organization of the vocal control pathway in the budgerigar (*Melopsittacus undulatus*). *Journal of Neuroscience* **1**: 1279–88.

Payne, R. B. (1971). Duetting and chorus singing in African birds. *Ostrich Supplement* **9**: 125–46.

Payne, R. B. (1979). Song structure, behavior, and sequence of song types in a population of village indigobirds, *Vidua chalybeata*. *Animal Behaviour* **27**: 997–1013.

Payne, R. B. (1981). Song learning and social interaction in indigo buntings. *Animal Behaviour* **29**: 688–97.

Payne, R. B. (1983). Bird songs, sexual selection, and female mating strategies. In *Social Behavior of Female Vertebrates*, ed. S. Wasser, pp. 55–90. Academic Press, New York.

Payne, R. B., Thompson, W. L., Fiala, K. L. & Sweany, L. L. (1981). Local song traditions in idigo buntings: Cultural transmission of behavior patterns across generations. *Behaviour* **77**: 199–221.

Payne, R. B. & Westneat, D. F. (1988). A genetic and behavioral analysis of mate choice and song neighborhoods in indigo buntings. *Evolution* **42**: 935–47.

Pellis, S. M. (1981a). Exploration and play in the behavioural development of the Australian magpie, *Gymnorhina tibicen*. *Bird Behaviour* **3**: 37–49.

Pellis, S. M. (1981b). A description of the social play by the Australian magpie, *Gymnorhina tibicen*, based on Eshkol–Wachman notation. *Bird Behaviour* **3**: 61–79.

Pepperberg, I. M. (1985). Social modeling theory: A possible framework for understanding avian vocal learning. *Auk* **102**: 854–64.

Pepperberg, I. M. (1994). Vocal learning in grey parrots (*Psittacus erithacus*): Effects of social interaction, reference, and context. *Auk* **111**: 300–13.

Pidgeon, R. (1981). Calls of the galah *Cacatua roseicapilla* and some comparisons with four other species of Australian parrot. *Emu* **81**: 158–68.

Poulsen, H. (1959). Song learning in the domestic canary. *Zeitschrift für Tierpsychologie* **16**: 461–92.

Powell, E. F. (1993). Perception of developing vocalizations in the budgerigar (*Melopsittacus undulatus*). M.Sc. thesis, University of Maryland, College Park.

Read, A. F. & Weary, D. M. (1992). The evolution of birdsong: Comparative analyses. *Philosophical Transactions of the Royal Society* **338**: 165–87.

Richards, D. G. (1981). Estimation of distance of singing conspecifics by the Carolina wren. *Auk* **98**: 127–33.

Ritchison, G. (1983). The function of singing in female black-headed grosbeaks (*Pheucticus melanocephalus*): Family-group maintenance. *Auk* **100**: 105–16.

Richison, G. (1985). Variation in the songs of female black-headed grosbeaks. *Wilson Bulletin* **97**: 47–56.

Ritchison, G. (1986). The singing behavior of female northern cardinals. *Condor* **88**: 156–9.

Robinson, A. (1956). The annual reproductive cycle of the magpie, *Gymnorhina dorsalis* Campbell, in south-western Australia. *Emu* **56**: 233–6.

Robinson, F. N. (1991). Phatic communication in bird song. *Emu* **91**: 61–3.

Rothstein, S. I. & Fleischer, R. C. (1987). Vocal dialects and their possible relation to honest status signalling in the brown-headed cowbird. *Condor* **89**: 1–23.

Rowley, I. (1974). *Bird Life*. Collins, Sydney.

Rowley, I. (1980). Parent–offspring recognition in a cockatoo, the galah, *Cacatua roseicapilla*. *Australian Journal of Zoology* **28**: 445–56.

Rowley, I. (1983a). Mortality and dispersal of juvenile galahs, *Cacatua roseicapilla*, in the western Australian wheatbelt. *Australian Wildlife Research* **10**: 329–42.

Rowley, I. (1983b). Re-mating in birds. In *Mate Choice*, ed. P. G. Bateson, pp, 331–60. Cambridge University Press, Cambridge.

Rowley, I. (1990). *Behavioural Ecology of the Galah, Eolophus roseicapillus*. Surrey Beatty & Sons, Chipping Norton, New South Wales.

Rowley, I. & Chapman, G. (1986). Cross-fostering, imprinting and learning in two sympatric species of cockatoo. *Behaviour* **96**: 1–16.

Rowley, I. & Chapman, G. (1991). The breeding biology, food, social organization, demography and conservation of the Major Mitchell or pink cockatoo, *Cacatua leadbeateri*, on the margin of the western Australian wheatbelt. *Australian Journal of Zoology* **39**: 211–61.

Saunders, D. A. (1974). The function of displays in the breeding biology of the white-tailed black cockatoo. *Emu* **74**: 43–6.

Saunders, D. A. (1982). The breeding behaviour and biology of the short-billed form of the white-tailed black cockatoo *Calyptorhynchus funereus*. *Ibis* **124**: 422–55.

Saunders, D. A. (1983). Vocal repertoire and individual vocal recognition in the short-billed white-tailed black cockatoo, *Calyptorhynchus funereus latirostris* Carnaby. *Australian Wildlife Research* **10**: 527–36.

Saunders, D. A. (1986). Breeding season, nesting success, and nestling growth in Carnaby's cockatoo, *Calyptorhynchus funereus latirostris*, over 16 years at Coomallo Creek, and a method for assessing the viability of populations in other areas. *Australian Wildlife Research* **13**: 261–73.

Schroeder, D. J. & Wiley, R. H. (1983). Communication with shared song themes in tufted titmice. *Auk* **100**: 414–24.

Searcy, W. A. & Andersson, M. (1986). Sexual selection and the evolution of song. *Annual Review of Ecology and Systematics* **17**: 507–33.

Serventy, D. L. (1971). Biology of desert birds. In *Avian Biology*, vol. 1, ed. D. S. Farner & J. R. King, pp. 287–331. Academic Press, New York.

Shurcliff, A. & Shurcliff, K. (1973). Territory in the Australian magpie (*Gymnorhina tibicen*): An analysis of its size and change. *South Australian Ornithologist* **26**: 127–32.

Sick, H. (1939). Über die Dialektbildung beim "Regenruf" des buchfinken. *Journal für Ornithologie* **87**: 568–92.

Simpson, B. S. (1985). Effects of location in territory and distance from neighbors on the use of song repertoires by Carolina wrens. *Animal Behaviour* **33**: 793–804.

Simpson, H. B. & Vicario, D. S. (1990). Brain pathways for learned and unlearned vocalizations differ in zebra finches. *Journal of Neuroscience* **10**: 1541–56.

Smolker, R. A. (1993). Acoustic communication in bottlenose dolphins. Ph.D. dissertation, University of Michigan, Ann Arbor.

Snow, D. W. (1968). Singing assemblies of little hermits. *Living Bird* **7**: 47–55.

Sorjonen, J. (1987). Temporal and spatial differences in traditions and repertoires in the song of the thrush nightingale (*Luscinia luscinia*). *Behaviour* **102**: 196–212.

Stacey, P. B. & Koenig, W. D. (1990). *Cooperative Breeding in Birds*. Cambridge University Press, Cambridge.

Stoddard, P. K., Beecher, M. D., Horning, C. L. & Campbell, S. E. (1991a). Recognition of individual neighbors by song in the song sparrow, a species with song repertoires. *Behavioural Ecology and Sociobiology* **29**: 211–15.

Stoddard, P. K., Beecher, M. D., Horning, C. L. & Willis, M. S. (1991b). Strong neighbor–stranger discrimination in song sparrows. *Condor* **92**: 1051–6.

Striedter, G. F. (1994). The vocal control pathways in budgerigars differ from those in songbirds. *Journal of Comparative Neurology* **343**: 35–56.

Sullivan, B. D. & Dinsmore, J. J. (1992). Home range and foraging habitat of American crows *Corvus brachyrhynchos* in a waterfowl breeding area in Manitoba. *Canadian Field Naturalist* **106**: 181–4.

Thorpe, W. H. (1961). *Bird Song*. Cambridge University Press, Cambridge.

Thorpe, W. H. & North, M. E. W. (1965). Origin and significance of the power of vocal imitation: With especial reference to the antiphonal singing of birds. *Nature* **208**: 219–22.

Thorpe, W. H. & North, M. E. W. (1966). Vocal imitation in the tropical bou-bou shrike, *Laniarius aethiopicus major*, as a means of establishing and maintaining social bonds. *Ibis* **108**: 432–5.

Todt, D. (1975). Social learning of vocal patterns and modes of their application in grey parrots (*Psittacus erithacus*). *Zeitschrift für Tierpsychologie* **39**: 178–88.

Todt, D. (1981). On functions of vocal matching: Effect of counter-replies on song post choice and singing. *Zeitschrift für Tierpsychologie* **57**: 73–93.

Townsend, C. W. (1927). Notes on the courtship of the lesser scaup, Everglade kite, crow, and boat-tailed and great-tailed grackles. *Auk* **44**: 549–54.

Trainer, J. M. (1987). Behavioral associations of song types during aggressive interactions among male yellow-rumped caciques. *Condor* **89**: 731–8.

Trainer, J. M. (1988). Singing organization during aggressive interactions among male yellow-rumped caciques. *Condor* **90**: 681–8.

Trainer, J. M. (1989). Cultural evolution of song dialects of yellow-rumped caciques in Panama. *Ethology* **80**: 190–204.

Treisman, M. (1978). Bird song dialects, repertoire size, and kin association. *Animal Behaviour* **26**: 814–17.

Tyack, P. (1986). Whistle repertoires of two bottlenosed dolphins, *Tursiops truncatus*: Mimicry of signature whistles? *Behavioural Ecology and Sociobiology* **18**: 251–7.

Wagner, H. O. (1944). Notes on the history of the emerald toucanet. *Wilson Bulletin* **56**: 65–76.

West, M. J. & King, A. P. (1985). Social guidance of vocal learning by female cowbirds: Validating its functional significance. *Ethology* **70**: 225–35.

West, M. J. & King, A. P. (1988). Female visual displays affect the development of male song in the cowbird. *Nature* **334**: 244–6.

West, M. J., Stroud, A. N. & King, A. P. (1983). Mimicry of the human voice by European starlings: The role of social interactions. *Wilson Bulletin* **95**: 635–40.

Wickler, W. (1972). Aufbau und paarspezifitat des Gesangsduettes von *Laniarius funebris*. *Zeitschift für Tierpsychologie* **30**: 464–76.

Wiley, R. H. (1971). Song groups in a singing assembly of little hermits. *Condor* **73**: 28–35.

Wyndham, E. (1980a). Environment and food of the budgerigar *Melopsittacus undulatus*. *Australian Journal of Ecology* **5**: 47–61.

Wyndham, E. (1980b). Diurnal cycle, behaviour and social organization of the budgerigar *Melopsittacus undulatus*. *Emu* **80**: 25–33.

Wyndham, E. (1981). Breeding and mortality of budgerigars *Melopsittacus undulatus*. *Emu* **81**: 240–3.

Wyndham, E. (1983). Movements and breeding seasons of the budgerigar. *Emu* **82**: 276–82.

Zann, R. (1990). Song and call learning in wild zebra finches in south-east Australia. *Animal Behaviour* **40**: 811–28.

Zink, R. M. (1985). Genetical population structure and song dialects in birds. *Behavioural and Brain Science* **8**: 118–19.

Zink, R. M. & Barrowclough, G. F. (1984). Allozymes and song dialects: A reassessment. *Evolution* **38**: 444–8.

8   Social influences on song acquisition and sharing in the European starling (*Sturnus vulgaris*)

MARTINE HAUSBERGER

INTRODUCTION

Vocal learning is a very widespread characteristic of songbirds, and a large variety of these learning processes has been described over the last decades. Learning can lead to different types of variation and results in song sharing that can be geographically localized and is then considered as "dialects." The distribution of these variations can be limited to a few birds and/or cover large areas. Experimental studies have given precise information about the mechanisms involved, in particular in terms of "timing." However, naturalistic validations may be necessary for us to fully understand the functional significance of vocal learning (see discussion of the sensitive periods, in the literature). There is a need for integrative studies and it is important to consider communication in its own context, which is that of social interaction (for language as a social act, see Goodwin 1990). A social organization needs an adapted communicative system, and comparative studies can give us hints about the evolutionary bases of vocal learning. Comparison can be made between species or phylogenetic groups but also between populations of a same species.

Likely candidates to help us to understand these processes are highly social animals that can adapt to different social environments. Starlings clearly correspond to this definition and here I examine through experimental and naturalistic studies the possible relation between song acquisition, song sharing and social organization.

The European starling (*Sturnus vulgaris*) occupies the largest geographic range of all the species of *Sturnus* (Feare 1984). It is present in Asia and in Europe from Scandinavia to Spain and Italy, and even in North Africa. It has also been introduced successfully in North America, New Zealand, and South Africa. This is a highly adaptable species that is confronted over its breeding range with a variety of environmental conditions. The general features of the European starling's life correspond to those of most *Sturnus* species: it feeds in flocks, nests in holes, breeds in nest colonies (when possible) and sleeps in communal roosts where several hundred starlings may perch for the night in the same site (Feare 1984).

THE STARLING AS A SOCIAL SPECIES

The availability of suitable nest holes is generally considered as a limiting factor in starling breeding: the starlings do not excavate and therefore are dependent on existing cavities. Starlings preferentially choose nest sites that are oriented eastwards, are high and have a surrounding open space (e.g. lawns) (Verheyen 1969; Coleman 1974; Feare 1984). However, Motis (1994) has shown that starlings actively look for sites where several nests are available: indeed, when possible, starlings preferentially nest in colonies (mostly of between 3 and 15 nests). They are often dependent on the human habitat for nest sites; many nests are under the roofs of houses. In most introduced populations, starlings can only nest individually because the structure of the houses allows room for only one nest. But starlings also form colonies as soon as several nest sites are available in these populations (Flux & Flux 1981; Adret-Hausberger 1988). Choosing a nest site is important not only for breeding but also for social interactions. In sedentary populations, the nest site appears as a center point for diurnal activity: Clergeau (1993), following radioequipped adults, found that they spent 38% of their time within 150 m of the nest and 58% between 150 and 2000 m during winter time! Sedentary males visit their nest every day all year round, with a decrease only in summer (molt). These visits are the occasions for numerous vocal interactions between neighbors. Migratory birds come to the breeding area between April and May, and stay in the colony until June, after having spent their time in breeding activity (Pinxten & Eens 1990) (Fig. 8.1).

Sedentary starlings are also faithful to their feeding sites: according to Caccamise & Morrison (1988), diurnal activities cover about 1 km². The size of the feeding flocks increases during the day, reaching a peak at the night roost. Site fidelity suggests that to some extent, the same birds "meet" every day.

In roosts, up to several million birds can mix, including both sedentary and migratory birds in the winter quarters. Birds arrive from different directions as flocks and start making synchronized flights before "settling" into the roost site. They then start singing intensively while re-adjusting on the branches. Wynne-Edwards (1962) and Clergeau (1993) hypothesized that there may be an internal social structure in roosts, whereas Summers *et al.* (1987) found more adult males in the center and more young females at the periphery of the roost.

After chemical destruction of a roost, Such (cited by Clergeau 1991) found that ringed migratory birds were grouped according to their geographic origin. The idea that migratory populations separate into given winter areas has been proposed by different authors (Goodacre 1959; Gromadski & Kania 1976).

Sedentary starlings seem to be more faithful to their nest site (Coleman 1974; Clergeau 1993) than are migratory birds, which seem to change nest site from one year to the next (Verheyen 1980). Nest fidelity tends to increase with the age of the colony (Merkel 1980). The breeding pattern tends to differ between sedentary and migratory populations (Fig. 8.1).

Little is known about what happens outside the nest colony. A segregation of sexes seems to occur in different contexts: males and females come back separately from the winter quarters (above), from the roost (F. W. Merkel personal communication), and tend to use different feeding sites (Peach & Fowler 1989). This may, however, concern only unpaired birds (Clergeau 1993).

Starlings are attracted to stuffed starlings in a feeding position, especially if they are similar to themselves (i.e., have similar bill color; Clergeau 1991). Bill color reflects the hormonal state, and sedentary starlings, especially the males, acquire the yellow color as early as December, in contrast to females (February) and migratory birds (after arrival on the breeding ground) (Bullough 1942). Moreover male testosterone level has been shown to be higher in dense colonies compared to those in dispersed nest sites, and testosterone is in all cases higher in the field birds than in captive animals (Ball & Wingfield 1987).

In conclusion, starlings are highly social animals (Fig. 8.1). Migratory birds spend about a fourth of their life concentrating on breeding activities. Even sedentary starlings who pair early spend two-thirds of their time in flocks and the rest interacting socially in the nest colony. Moreover an important part of the population (non-breeding birds) spends its time in flocks. What really happens in flocks and night roosts remains to be discovered.

The important social life of starlings is best described by Geroudet (1957): "la plupart du temps, il faut parler de lui au pluriel" (most of the time, one must speak of the starling using the plural).

STARLING SONG

The vocal talents of starlings have been known since antiquity, when Pliny considered their ability to mimic human speech noteworthy (West & King 1989). Ornithologists know that this species possesses a rich repertoire of calls and songs, composed of whistles, clicks, snarls, and screeches (Feare 1984). In addition, starlings are well known for their ability to mimic the sounds of other animals or even mechanical noises. Descriptions of starling song in the past reflect the difficulty of describing all the variety of sounds included. Witherby *et al.* (cited by Chaiken *et al.* 1993) mentioned a "lively rambling melody of throaty warbling, chirping, clicking and gurgling notes interspersed with musical whistles and pervaded by a peculiar creaking quality."

This complexity explains why detailed studies of starling song have been delayed long after the arrival of the sound spectrograph. As mentioned by West & King (1989), "the problem with starlings is that they vocalize too much, too often and in too great numbers, sometimes in choruses numbering in the thousands. Even the seemingly elementary step of creating an accurate catalogue of the vocal repertoire of wild starlings is an intimidating task because of the variety of their sounds."

(a)

Sunrise ⟶ Sunset

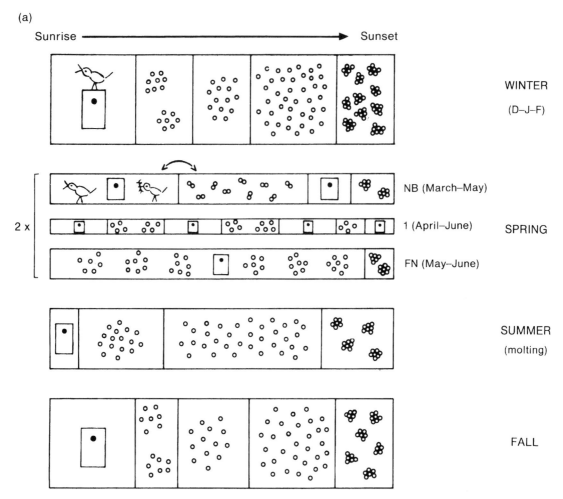

WINTER

(D–J–F)

NB (March–May)

2 x   I (April–June)   SPRING

FN (May–June)

SUMMER

(molting)

FALL

Fig. 8.1. An illustration of the starling's social life, with the average proportion of time spent in the different social contexts according to the time of day and time of year: (a) sedentary birds; (b) migratory birds. NB, nest-building; I, incubation, FN, feeding nestlings.

When in flocks, the birds alternate feeding bouts with perching bouts, when all birds are on high perches and sing together (Adret-Hausberger 1982; Feare 1984). No territorial behavior appears on the feeding sites.

In the pre-roosts, large flocks gather in high perches, where they sing intensively, in alternation with feeding bouts. They then fly to the night roost site.

In migratory populations, the males arrive first and choose a nest site. Breeding starts intensively when the females arrive. Polygyny can occur, with males attracting up to five females (Merkel 1980; Pinxten & Eens 1990).

In sedentary birds, the pairs are generally formed in the fall, and the mates visit their nest together through winter (Merkel 1979; Clergeau 1993). In a British Starling colony, Wright & Cuthill (1992) found that only 5% of the males were polygynous, and also observed that the female size is negatively correlated with the laying date as well as with the female's later investment in feeding the brood.

(b)

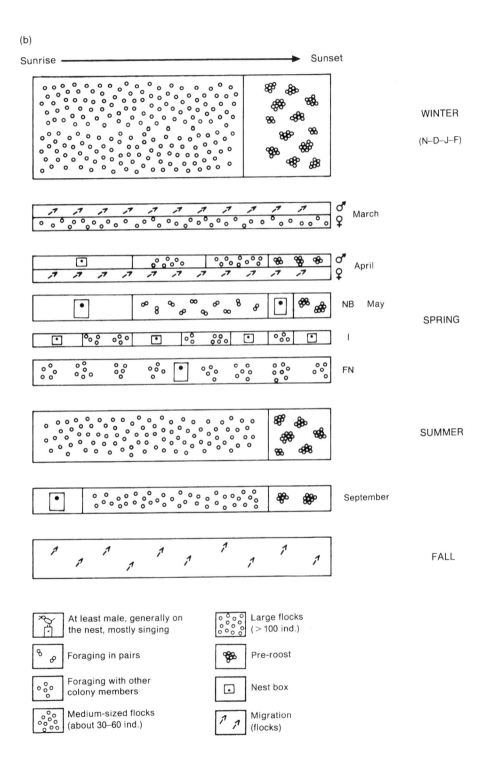

The song repertoire

Recordings of songs of more than 300 birds on three continents and from 10 populations led Hausberger *et al.* to distinguish two main categories of songs: (a) the long continuous warbling song, and (b) the loud discontinuous whistles (Hausberger & Guyomarc'h 1981; Adret-Hausberger & 1988; Hausberger *et al.* 1991). The mostly pure tone structure and loudness of the whistles is in opposition to the complexity and softness of the warbling (see Figs. 8.2 and 8.3), as was noticed by an American ornithologist: "the loud clear musical call penetrated closed windows. The call is appropriate to a bird of its strength and compels attention; the song, on the contrary is of gentle singing quality" (Bready 1929). Most authors working in field colonies have noticed that whistles are most often sung separately and are involved in interactions between males: Hartby 1969 (glissandos); Hausberger & Guyomarc'h 1981 (calls); Adret-Hausberger 1982 (whistled songs); Feare 1984, Wright & Cuthill 1992, Motis 1994, L. Henry, unpublished data (whistles). Adret-Hausberger (1982) obtained song matching responses using playback of whistles in the field but none using warbling song (M. Hausberger, unpublished data). Similarly, Eens *et al.* (1989) observed that "the whistles clearly stand out from the other categories because they are also sung separately between song bouts. Song matching between neighbours occurs only with the whistled songs, suggesting that these could be important for social relationships between males."

Heterospecific imitations can be found in both song categories without any evidence that they have any special function outside that of the song in which they are included. There is no evidence of mimicry functioning as a predator repellent for example (Hindmarsh 1984, 1986; Hausberger *et al.* 1991).

Definitions, however, differ somewhat between authors, depending on the extent to which both song categories are mixed. Warbling song sequences (=song bouts: *sensu* Eens) are defined as continuous songs with less than 1-second silent interval (Hausberger *et al.*), 1.5 seconds (Eens *et al.*) or 5 seconds (Chaiken *et al.*). Successive whistles are separated by intervals of at least 1 and up to 10 second (Hausberger 1991) and therefore they are excluded in the first case, a few are included in the second case and full sequences of whistles can be integrated in the last case. All authors agree that no silent interval of more than 0.5

second can be found in the rest of the song. Such differences lead to differences in the evaluations of repertoire sizes that are not simply due to sample sizes (compare e.g., Eens *et al.* 1991 and Hausberger *et al.* 1995b).

The studies of Eens *et al.* (1989, 1991, 1992) and Chaiken *et al.* (1993, 1994) have concentrated on the warbling song and all authors agree on the general organization of this vocalization (Fig. 8.2). Evaluations of the repertoire sizes of motif types for different males both in the field and in captivity give ranges of 17 to 55 (France, Germany, New Zealand: Adret-Hausberger & Jenkins 1988; Adret-Hausberger *et al.* 1990; Hausberger *et al.* 1995b); 21 to 67 (Belgium, New Zealand: Eens *et al.* 1991, 1992); 24 to 40 (USA: Henry unpublished data) and 60 to to 80 (USA: Chaiken *et al.* 1993). Since individual repertoires of whistles can reach 23 (Henry, unpublished data) whether or not whistles are included might account for the differences in repertoire sizes mentioned.[1]

It is possible to distinguish further between categories of whistles (Fig. 8.3). Each male has a repertoire of whistles, some of which are shared by the males of most populations (the species-specific themes) and others are individual specific within the colonies (including the heterospecific imitations sung separately). The species-specific themes show the same characteristics and variation ranges in all populations. However, within each population, each species-specific theme can be subdivided into a certain number of variants (Fig. 8.3). Playback experiments have shown us that the birds classify their songs the way we do: they reply to the whistle broadcast by matching the broadcast theme even if the particular variant played is different

---

[1] Only two whistles can be considered as an integral part of the warbling song: they are never produced separately and are immediately followed by motifs. One is a long descending uniform whistle (see also Eens *et al.* 1991) and the other is a particular form of the rhythmic theme (Adret-Hausberger 1989b).

Eens *et al.* (1991) considered the separate bouts of whistles as negligible, evaluating them to be 5% of the time spent singing. This low rate of whistling may be due to two aspects: (a) their birds, coming back from the winter quarters, are very much involved in pairing, and probably do whistle less (see Henry *et al.* 1994), (b) the colonies are not stable (at least across years), which may have an effect on social interactions. In any case, total duration may not be the best criterion to judge the importance of a song: loudness may be more important for long distance communication (see observations of Bready, p. 132). Adret-Hausberger (1984a) observed that two males can produce up to 500 whistles in two hours in their colony, while West *et al.* (1983) reported starlings repeating a whistled sound 200 times in half an hour! Whistles also appear later in the ontogeny compared to warbling, which develops slowly from subsong (Adret-Hausberger, 1989b; Henry 1994).

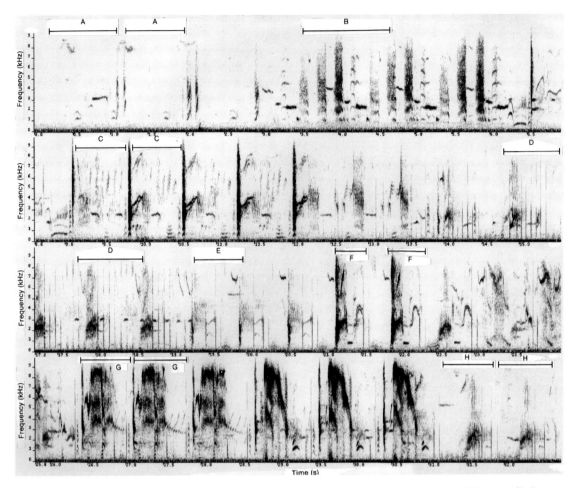

Fig. 8.2. Example of a warbling song (Adret-Hausberger & Jenkins 1988) (=song bout; Eens *et al.* 1991); song (Chaiken *et al.* 1993)). Chaiken *et al.*'s (1993) description summarizes the findings of the different authors: each song lasts up to one minute and consists of a succession of motifs (A, B, C . . .). A motif is a fixed combination of acoustic elements, with a duration of about 0.5 to 1 s. Most motifs are repeated before the next one is produced (Adret-Hausberger & Jenkins 1988; Eens *et al.* 1991). All starlings (warbling) song (bouts) conform to a basic pattern:

1. Part I (A–C) can begin with one or several whistles separated by 0.5 to 5 s (separated in Hausberger's descriptions), which are followed by a number of complex motifs (variable song types of Eens) (A→C). Heterospecific imitations are most abundant in this part. Much contrast is observed.
2. Part II (D and E) is characterized by the presence of click trains (rattle song types of Eens). All motifs are composed of a series of rapid click like beats delivered at a rate of about 16 s. Most of these motifs include broad-band buzzy elements and/or brief high frequency whistles superimposed.
3. Part III (G) consists of a series of high frequency whistles or trills (high frequency song types).

The song increases in both tempo and frequency as it progresses from beginning to end. The different frequency ranges of motifs in parts I, II and III result in a general progression from lower to higher frequencies. Within the sections II and III, Adret-Hausberger & Jenkins (1988) recognized two motifs of II (chicks) and 1 of III (high pitched trills) as universally shared. The motif, called the high pitched trills, are reminiscent of the threat screams described by Hartby (1969) and Van der Mueren (1980) and is in both contexts often associated with a wing-waving display. Mimicries can also be observed in the C section.

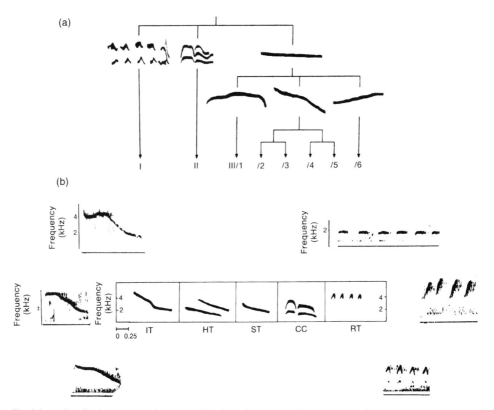

Fig. 8.3. (a) Key for the categorization of whistles. According to general characteristics in their structure, whistles can be separated into several main categories. Each of these general classes contains vocalizations that have identical combinations of various characteristics but differ in details. These classes are categorized on the basis of acoustic properties easily discerned by the human ear. First, three main categories are distinguished:

– Category I contains short notes that are regularly repeated several times. It corresponds to one specific theme: the rhythmic theme.
– Category II is characterized by two (or sometimes more) different whistle-like units. It corresponds to the composed theme.
– Category III corresponds to unitary whistles with a regular pattern of frequency modulation. It can be separated into subcategories by the specific pattern of frequency modulation.

(b) The Class I themes:

– IT: inflection theme is characterized by an inflection followed by a descending slope starts with 4–5 kHz, ends around 2 kHz. Variations occur in the frequency modulation.
– HT: "harmonic" theme: two superimposed notes (generally not harmonically related). Variations in frequency modulation.
– ST: very simple descending modulation (but see analyses). Starts at around 3 kHz ends around 2. Variations occur in the duration.
– CC: composed theme: two or more different notes associated.
– RT: repetition of a given note: the shape of note, rhythm and frequences can vary, leading to a large number of variants.

The use of an index of similarity enabled us to have a quantitative comparison of all spectrograms of a same theme. Variants appeared for each theme. Here are three examples of variants of IT on the left, three examples of RT on the right. The variants are no more different between than within populations. The most variability was observed for RT and the least for ST, which varies only in its duration.

The responses of the birds confirm the previous classification (see the text).

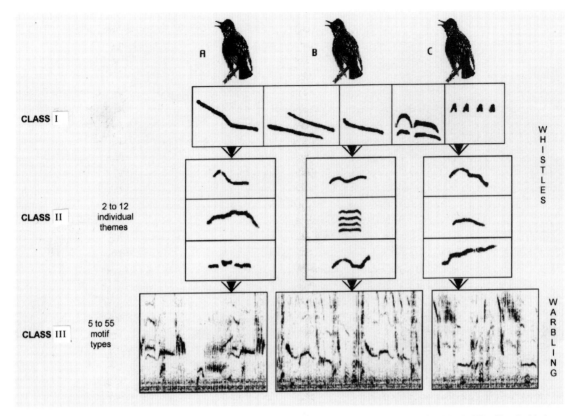

Fig. 8.4. The main song types of male European starlings from field observations of above 300 birds. The Class I whistle types are present in the repertoire of all or most males with the same characteristics and variation ranges. From the top left: inflection theme (IT), theme with harmonics (HT), simple theme (ST), composed theme (CC), rhythmic theme (RT). The individual whistle types (Class II themes) characterize one bird in its colony, and can be interspecific mimicries. At the bottom, short parts of the warbling song (Class III song) of these three birds including two motifs each time.

The occurrence of the different song categories varies according to context (see the text). The number of Class II themes in the repertoires of adult males varies from 2 to 7 in France (Hausberger & Guyomarc'h 1981), 2 to 11 in Belgium (only before warbling; Eens *et al.* 1991), 2 to 12 in Germany (Adret-Hausberger *et al.* 1990), 1 to 12 in Spain (Motis 1994), and 7 to 23 in America (L. Henry, unpublished data).

Heterospecific imitations were evaluated to about 10–20% of the warbling song in duration (Hindmarsh 1984). Other evaluations show that 30–50% of the individual whistle types and about 25% of the warbling motifs included such imitations. Their structure corresponds to that of the song category in which they are included (Hausberger *et al.* 1991).

from their own (Adret-Hausberger, 1982). Song matching responses are a good way of evaluating the classification of their song types by the birds themselves (Falls *et al.* 1982). Each bird generally produces only one variant of each theme and always replies to the broadcast whistle with its own variant of the particular theme.

The individual themes are more likely to precede warbling than the species-specific themes. Individual themes are thought to be a way of attracting the attention of a listener from a distance before starting the quieter warbling (Hausberger 1991).

In conclusion, a separation can be made between *universally shared songs* (species-specific whistles and some warbling motifs) and more *individual structures* (individual whistle themes and most warbling motifs). Therefore I suggest separating starling songs into three classes: *Class I songs*, corresponding to the "species-specific whistled themes" shared by the males of most populations; *Class II songs*, corresponding to the more individual whistles; and *Class III song*, corresponding to the warbling (Fig. 8.4).

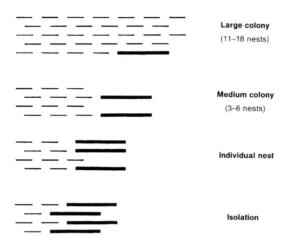

Large colony
(11–18 nests)

Medium colony
(3–6 nests)

Individual nest

Isolation

Fig. 8.5. Singing style according to social environment around the nest. Proportions of whistles (——— 1 whistle) sung in succession whether or not followed by a warbling sequence (▬▬)). These are examples only from sedentary populations at the beginning of the nest-building stage. The pattern may be different in migratory birds. Captive birds in isolation always alternate whistles and warbling, as they also mostly do when nesting individually (see also Chaiken *et al*. 1993). Actual data from M. Hausberger *et al*. (unpublished).

The effect of social context on singing behavior

Singing style

The different attitudes of authors toward the different categories of starling songs are better understood if the context of recordings is considered. Chaiken *et al*. (1993) have worked mostly on captive animals in a restricted social environment, moreover on an introduced population where starlings nest mostly individually (see also Mountjoy & Lemon 1991, 1995). Eens *et al*. (1991) study starlings that belong to rather socially unstable colonies.

Comparisons of the singing styles of starlings have revealed that they vary according to the social environment: *starlings in isolation* (captivity) or *nesting isolately* tend to *sing one or two whistles before warbling* and rarely separate both song categories, whereas *birds in large colonies* (e.g. 10 nests) tend to produce *long separate sequences of whistles* (Fig. 8.5, M. Hausberger *et al*., unpublished data). This reflects also the proportion of the two song categories in these different situations (see below)

and reinforces the idea of a particular social importance of whistles.

Repertoire use

Whistles versus warbling As mentioned before, the singing style as well as the proportion of whistles compared to warbling (in number of utterances) varies largely according to the social environment at the nest. In fact the number and proportion of whistles is positively correlated with the number of neighbors (M. Hausberger *et al*., unpublished data). Only the whistles are involved in vocal interactions between males and used in response to playback. Such observations have led us to consider the whistles and the warbling as two functionally different songs (Adret-Hausberger & Jenkins 1988; Hausberger *et al*. 1991).

In a field experiment where females were taken from the nest, we could observe that *pairing status* also played a role in the repertoire use. Males tended to produce more warbling and more Class II whistles when they were unpaired and more of the Class I whistles when they were paired (Henry *et al*. 1994). These results were interpreted as a further indicator of the more probable role of these whistles in social interactions between males. Eens *et al*. (1993), in an aviary experiment, introduced females to unpaired males and observed that both warbling and whistles increased. These results do not contradict the previous findings, since we observed this increase also for the Class II whistles, which are almost the only whistles sung in captivity (see below).

Whistles As mentioned before, *vocal interactions* between males are frequent in all contexts. Only the Class I themes tend to be used and song matching is the most frequent type of interaction observed. Therefore, vocal interactions influence the repertoire use.

Other factors, such as the *time of the year* or *nest proximity* have been shown to have an effect on the relative frequency of production of the different whistle types (Hausberger & Guyomarc'h 1981; Adret-Hausberger 1984a). For example, the simple theme is more frequently sung in the colony in the fall, and when the birds are close to their nest.

The *social context outside the colony* is another important parameter: the relative frequency of occurrence of the different Class I themes vary according to flock size or whether a roost is considered (see also p. 151). The Class II themes tend to be used less in these contexts.

Clear *differences* appear also between *captive* and *free-living* birds. L. Henry (unpublished data) observed that wild birds used a higher proportion of the Class I themes than did captive birds living in an outdoor aviary in the same area. The interactions between captive birds did not use these themes, which overall were almost never produced. Young birds raised in a social group with adults in this aviary never produced them later although they could hear them produced by the wild birds around them. Similar observations have been made by Hausberger *et al.* (1995b) and P. Marler (personal communication). *Captive birds tend to not use the Class I themes.*

Finally, differences can be observed *between populations*: the rhythmic theme was almost never recorded in Australia (Adret-Hausberger 1988), in America (Henry 1994; M. J. West, personal communication) or New Zealand (P. F. Jenkins, personal communication), whereas it is frequent in France, Germany or Britain (Adret-Hausberger 1989b) (see pp. 145, 151 for further discussion). The harmonic theme is also rarely heard in Germany. Adret-Hausberger (1988) found differences in the repertoire use of whistles between isolately nesting birds and colonial birds in Australia.

Repertoire turnover in young and adult birds

Few field studies have reported repertoire changes in adult songbirds. Adret-Hausberger *et al.* (1990) observed over two years a small colony of individually ringed starlings whose life history was known and found song repertoire changes even in older birds (up to seven years old). All classes of song did not change in the same way: the Class I shared songs remained very stable from one breeding season to the next, whereas large changes were observed in all birds for the Class II songs. Some of the males changed completely their repertoire of individual whistles. There was no consistent increase with age in the whistle repertoire size. Finally the older males only slightly increased their repertoire of Class III song (warbling) whereas a young individual modified most of his repertoire and increased it largely (see also Eens *et al.* 1992). More recently, Mountjoy & Lemon (1995) also found repertoire changes in the Class III songs of adult males. They found an increase of repertoire size with age but also a turnover in the repertoire composition.

Similar changes in the song repertoire composition could be observed in adult captive birds as a result of changes in the social group (Hausberger *et al.* 1995a). *Both males and females appeared to show an important repertoire turnover in response to social changes* (Fig. 8.6 and pp. 146–8) whereas stable social groups did not exhibit such drastic turnovers (L. Henry, unpublished data). These changes could be observed at any time of the year. Interestingly, some whistles or motifs that had been dropped at one stage could reappear a year later.

Young birds also exhibit a high rate of turnover in their first 18 months of life. The changes seemed to occur mainly after a new tutoring session began in experimental animals (Chaiken *et al.* 1994, see also pp. 138, 147).

Female song

Until recently, song in female starlings (as in many songbird species) has received very little attention. There have been reports of occasional observations of females singing (Witschi & Miller 1938; Bullough 1942; Feare 1984). Such cases were thought to correspond to older females with high testosterone levels. More recently, we observed a polygynous trio near the nest in spring where the two females were observed interacting vocally together or with the male, in song matching sequences. Only four whistle types were recorded, but they were all shared by both females, and three of them were shared with the male. Most female whistles were produced while the females were sitting close to each other and the interactions were always started by the older female. Only very short bouts of warbling were observed (Hausberger & Black 1991).

This report led to interesting questions about the singing and learning abilities of females. Two detailed studies are now available: one concentrates on a description of female song after testosterone treatment (Hausberger *et al.* 1995a); the other describes spontaneous singing in captive females (L. Henry unpublished data).

Both studies revealed the ability of females to sing complex songs and no qualitative difference was found in song composition between testosterone-induced or spontaneous songs. The individual repertoires ranged from 6 to 16 and around 7, respectively, for the whistles; and from 8 to 34 and around 17, respectively, for warbling motif types.

Individual differences in repertoire sizes were not related to age. As mentioned before a repertoire turnover could be observed according to social context (see

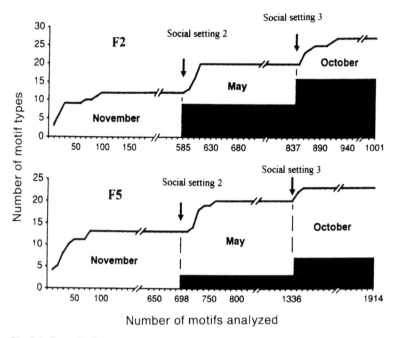

Fig. 8.6. Example of changes in song repertoire size and composition over time for two females. Cumulative curves showing how repertoires reached an asymptote at each recording session but new structures were obtained at the next session. Dark area corresponds to the number of motif types that were not recorded any more. Note repertoire turnover (from Hausberger *et al.* 1995b).

pp. 146–8). Testosterone-treated females produced more song and longer warbling sequences than did the untreated birds. However, even under testosterone treatment the birds did not produce all their potential repertoire: some song types that did not appear in two successive recording sessions separated by several months (assumed to have been dropped) reappeared again later in particular in the context of social interactions. Therefore it appears that *testosterone injections do not necessarily inform us about the "whole repertoire"* of a bird. *Social interactions may be more important* for the production of particular song types.

It is interesting that the shared structures of the males (Class I songs, clicks and high pitched trills) are mostly absent from the female repertoires in all studies. Only the "harmonic theme" was present in several females in both studies.

Spontaneous singing in females occurs mainly between November and January, and stops very rapidly if nestboxes are introduced (L. Henry, unpublished data). Females

placed in an aviary where nestboxes were present were almost never observed singing, even outside the breeding season (Hausberger *et al.* 1995a). It is therefore clear that *spontaneous singing in females is related mainly to a non-breeding situation*. This explains for a good part why female song has been neglected for so long: most studies have concentrated on the nest site and the breeding season. It is difficult to know what happens outside the colony, in flocks and roosts where singing is important.

Conclusion

Descriptive studies of starling song reveal several characteristic features:

A large complexity of structures based on both discontinuous and continuous songs.

The presence of more or less universally shared and individual structures that can be grouped into three classes.

The importance of female song in a nonbreeding context. Females do not produce the Class I songs that correspond to the male universally shared structures.

The difficulty of assessing a repertoire: repertoire use as well as singing style vary with social context.

Moreover an important repertoire turnover can be observed in both adult and young birds, apparently in response to social changes. Even testosterone does not allow us to assess the whole repertoire of females.

Starling song is a good example of the difficulties encountered in the definition of a song repertoire: discontinuous versus continuous song, recognition of units, biases related to context-dependent use of the "repertoire" (Kroodsma 1982). The data actually even bring us to question the concept of repertoire itself.

There are more and more examples of differential use of repertoire according to context (see e.g., Kroodsma *et al.* 1989). Derrickson (1987) found that the evaluation of repertoire size varies according to the way in which it is evaluated and the social situations of the observations in northern mockingbirds (*Mimus polyglottus*). What can be the significance of repertoire size when it varies according to the situation?

Female starlings sing only small parts of their repertoire when interacting: they obviously select the appropriate elements to respond to social partners. Males produce only some parts of their repertoire in a given context. Moreover, the repertoire turnovers are frequent. What is then the "repertoire"? Is it only the elements produced or also those dropped? Our experiments show that, when they were again in the appropriate social context, the females at least can reuse elements that had been dropped previously.

Finally, if observations had been made only on birds in captivity (and especially in a restricted social environment), none of us would have guessed the existence of the shared Class I themes or at least their social importance (see below).

The observations suggest that starlings select from their repertoire appropriate social elements in a way that reminds of the "action-based" learning or selective learning hypotheses (Slater 1989; Marler & Nelson 1993; Nelson, Chapter 2): the birds learn to produce more than they actually emit in a given social situation. However, it is not clear what kind of learning is involved: there is certainly an aspect of *learning to selectively use* elements of the repertoire (see also Pepperberg, Chapter 9; Snowdon *et al.*, Chapter 12; Seyfarth & Cheney, Chapter 13; Locke & Snow, Chapter 14). Our captive young starlings never sang the Class I themes: did they simply not learn them or did they learn that the aviary situation was not the appropriate environment to produce them? Therefore, I would suggest that the selective learning hypothesis (Nelson, Chapter 2) may merely reflect a general ability of songbirds to selectively use their repertoires: maybe the deleted songs of the sparrows studied would reappear in another social context?

Similarly, female singing behavior is generally underestimated (Smith 1991). In starlings, it seems that context is again involved: most studies have been done near the nest and in a breeding context. For the same reason, female song may have been underestimated in different species (see also Baptista *et al.* 1993).

Finally, our data emphasize the importance of a good field descriptive basis before or at the same time as precise experimentation (see Kroodsma 1982; Kroodsma *et al.* 1989; Payne & Payne, Chapter 5).

SOCIAL INFLUENCES ON VOCAL DEVELOPMENT: EXPERIMENTAL EVIDENCE

The effect of song tutoring

Young starlings start producing some subsong early in life and tend to have a more "crystallized" song around the age of nine months, when testosterone levels increase (Chaiken *et al.* 1993) and the first whistles appear (Adret-Hausberger 1989b; Henry 1994).

Chaiken *et al.* (1993) have compared the songs of young males raised in different social conditions: either with a wild-caught adult song tutor, individually housed but tape-tutored by a tape-recording or raised in total isolation. All birds had been taken from the nest at an early age (8–10 days) and were hand raised. Untutored birds produced mostly an abnormal song, where even the basic organization of song was missing. In contrast, both tape- and live-tutored birds developed songs with a normal basic organization, but with some syntactical abnormalities for the tape-tutored birds. Tape-tutored birds had repertoires half as large as those of live-tutored birds. Large differences occurred between both groups of birds in their degree of imitation. Tape-tutored birds imitated 10% as many motifs as did live-tutored birds. Live-tutored birds' copies constituted 69% of their repertoire of motifs, whereas imitations of the tape constituted 18% of the repertoires of tape-tutored birds, leaving more room for improvisations. Chaiken et al. (1994) also found that adult male starlings (at least 18 months old) were still able to acquire, i.e., memorize, and produce new structures. Starlings can also produce songs that they had heard 18 months earlier but had never produced. An important turnover can be observed in the repertoire (55–92% of the motifs) especially for the learned motifs. The turnovers seem to occur mostly after a change of tutoring session.

Böhner et al. (1990) supported the conclusions of this study: starlings can learn at any time, whatever their hormonal status is (photosensitive or photorefractory).

Groups of young, untutored starlings give additional information: like Chaiken's untutored birds, most of the species-specific features of the songs are missing (apart from the organization in motifs), repertoires are small and whistles are almost always absent. Interestingly, however, birds from a same group show some amount of sharing: even though song development is abnormal, the tendency to share with members of the social group is present, although the sex-specificity observed in other conditions (see below) does not seem to exist in these birds (Adret-Hausberger 1989a, and unpublished data), which may be due to the artefact of the isolation (from adults) situation (see West et al., Chapter 4).

Social contact and vocal sharing with humans

Reports of hand-raised starlings maintained in a human environment emphasize still further the ability of starlings for late learning and the importance of social contacts in vocal development (e.g., Suthers 1982). West et al. (1983) have compared the vocalizations of such animals placed in different conditions. Birds kept as a group in auditory contact with human sounds do not mimic human sounds. Birds kept in a cage with a cowbird as their only companion mimic the cowbird as well as mechanical sounds but no human speech. Finally, birds kept in strong social contact with humans (free flying, intensive care) clearly mimic human speech and whistled songs. This is an illustration of the importance of the social bond on vocal sharing.

In this study, starlings mimicked social vocalizations as well as arbitrary sounds in the environment. This also obviously reflects the natural situation, where we have found that starlings can mimic heterospecific mimicries of other starlings (Eens et al. 1992; Hausberger et al. 1995b), but also have a repertoire that reflects the acoustic environment in which they live, at least in the warbling part (Feare 1984; Hindmarsh 1986; Hausberger et al. 1991).

Conclusion

Experiments made on hand-raised starlings bring some important information:

Starlings, like most songbirds, produce abnormal songs when untutored.
A social tutor elicits much more imitation than a tape, and larger repertoires are observed in live-tutored birds.
Starlings can acquire new songs as adults and do not seem to depend on hormonal levels for learning.
Young starlings can learn from memory.
Young raised as a group imitate each other.

Starlings, like many of the songbirds tested until now, not only require hearing normal song in order to produce it but also learn much better with a social tutor (e.g., Baptista & Gaunt, Chapter 3).

It is not clear from these experiments what in the social contact is important for song acquisition. As Pepperberg (1983) mentioned, few studies have concentrated on which aspects are necessary. It is difficult at this stage to say whether it is social interaction (reciprocal action) or the mere presence of given stimuli that is important (Nelson, Chapter 2). This does not in any case exclude a social influence: sociogenesis does not require an interactive component in order to occur (Gottlieb 1993). Intermodal effects are certainly involved and enhancing the subject's attention or arousal is already a social influence: e.g., Locke & Snow (Chapter 14) emphasize the role of attention in human language acquisition; zebra finches seem to be "attentive" to singing males (ten Cate 1987).

Starlings' song acquisition does not appear to be related to agonistic interactions. The experiments with human caregivers rather suggest an importance of affiliative behavior. Obviously, starlings acquire elements of human language only if there is an active social bond and not by the mere presence of the human. Intense social interactions may be necessary to override the inhibition to learn these strange sounds (Pepperberg 1985, and see Chapter 9), whereas contact might be sufficient to acquire the more "natural" vocalizations (Bandura 1977). The importance of social input is clear in groups of hand-raised starlings that share many song elements, as was observed in other species (Becker 1978). Both the instructive and selective modalities (Nelson, Chapter 2) seem to be involved, although, as discussed before, it is not clear whether this selective phenomenon results at all in definitive attrition.

SOCIAL INFLUENCES ON SONG SHARING PATTERNS: A NATURALISTIC APPROACH

First description

In a first attempt to describe in detail starlings' whistled songs, Hausberger & Guyomarc'h (1981) found that, apart from a number of Class II themes, males from neighboring colonies shared the same variant of most Class I themes that differed from the variants sung in a more distant colony (Fig. 8.7). The structure of the rhythmic theme however, differed even between nearby colonies: each variant seemed to characterize each colony. This finding can be verified in small stable colonies (see also Adret-Hausberger & Güttinger 1984), although it might be different in large unstable colonies (M. Eens, personal communication).

We were interested in determining whether there was variation in themes between more distant sites. Therefore recordings were made regularly in an area 50×40 km when birds were singing near their nest in the morning. The songs of 150 resident adult males, recognizable by their yellow bill in winter (Bullough 1942), were recorded between December and March.[2]

Starling nests were regularly distributed over this area as they follow the human habitat, which is dispersed in this region. Starlings nested mainly under the roofs of old farms where numerous nest holes were available.

[2] Successive whistles were accumulated until a total of five minutes of whistles (independent of silent intervals) were recorded. This duration has been shown to allow the recording of all Class I and most Class II themes of an individual (Adret-Hausberger 1983; M. Hausberger et al. unpublished data). Two hours of recordings were generally needed until such criteria were met but this was the only way of being sure that all whistles recorded belonged to the same bird.

The distribution of the variants of each theme was not random, and a complex pattern of variation appeared. Each Class I theme appeared to have a particular level of variation that differed from each of the other themes. Dialectal areas covered about 5 to 10 km² for the inflection theme, 150 km² for the harmonic theme and at least several hundred square kilometers for the simple theme (Fig. 8.8a). Dialectal variations appeared to have several levels superimposed. Boundaries were in most cases sharp, with birds at the boundary singing both variants, but some areas were found that showed gradual variation in structure from the central variant.

Factors influencing the dialectal pattern

To test the generality of this dialect phenomenon, and to see what factors might have determined the emergence of such a complex pattern, further studies were made in different populations:

In the Rennes area again three years later to evaluate *variations over time* (Adret-Hausberger 1986).

In Paimpont (Brittany), where the distribution of colonies was similar but large unexploitable areas were present, limiting the possibility *of social exchanges* between birds from different colonies (Adret-Hausberger 1984b).

In Germany (Palatinate) to evaluate a *partially migratory* population where colonies have a *patchy distribution* (Adret-Hausberger & Güttinger 1984).

In Australia (New South Wales) to evaluate a *recently introduced* (1862) sedentary population characterized by *isolated nests* (restricted nest availability on houses) distributed within towns and rare small colonies (Adret-Hausberger 1988).

In Spain, where there are *dense colonies* with either a patchy distribution of monospecific or mixed with the spotless starling (Motis, 1994).

The complexity of the dialectal pattern is persistent at the temporal level but some changes were observed after three years: the boundaries had moved only slightly for the simple and composed themes, whereas boundaries appeared to have changed clearly for the harmonic and inflection themes. In both the last cases, we observed either a persistence of the old variants or an apparent drift, suggesting that the new variants were derived from the old

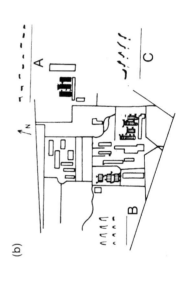

Fig. 8.7. (a) Comparison of Class I themes (inflection, harmonic and simple themes) between colonies either on same site (200 m apart) or further apart (15 km). Note similarity between A and B. (b) A map of site 1 (Rennes University Campus). The rectangles correspond to buildings where the nests are represented as arrows. The hatched or dark rectangles correspond to the buildings where the colonies were observed. The spectrograms correspond to the variant of the rhythmic theme sung by all members of the colony considered.

ones (Fig. 8.9). Moreover, structurally similar variants were close spatially, a trend that had already been observed in the previous study. Changes in dialects in this case seem to be the result of a drift. The changes are much more drastic for the rhythmic theme: the more "isolated" colonies have kept their variant unchanged, whereas two other colonies now socially linked by additional nests are sharing a new variant. This new variant corresponds to an additional form sung by one male three years earlier (a "leader"?).

The pattern of variation found in the different populations was close to that first described in Brittany (Fig. 8.8b–d). However, the social structure, in terms of nest distribution, influenced the precise distribution of the variants as well as the complexity of the system:

Large unexploitable areas (even for foraging) appear as "*natural barriers*" to the transmission of dialects: birds nesting on other sides of a forest or heath area tend to show different variants.

The distribution of colonies in Palatinate *was reflected in the variations* of the inflection theme: each variant includes the different colonies within a given village but varied from one village to the next. The absence of nests between villages seems to have hindered the transmission of dialects. Moreover the harmonic theme is almost never sung near the nest.

In Australia, where starlings mostly *nest individually*, the pattern of *variation is much simpler*: the rhythmic and composed themes were absent in most repertoires and the simple theme showed no local variations (identical everywhere). The inflection theme showed microgeographic variations within towns, where groups of males appear that share the same variant and the harmonic theme showed variations on a larger scale. The rhythmic theme was recorded in a few dispersed birds, which suggests that the birds are able to produce them but may lack the adequate stimulation to do so.

The dialectal variations found in Spain were very similar to those observed in other European populations: as in Germany, the inflection theme varied from one village to the next (same patchy distribution). However, the complexity was increased in that Motis (1994) found that some Class II themes also were shared between colony members. However the most interesting result in this study is that in sympatric areas *the spotless and European starlings shared the same dialectal variations* and the same whistle types. The two species were frequently heard in song-matching interactions. *Two species shared songs apparently as a result of social contact.* When in the same colony all starlings shared all the same dialectal variants.

Conclusion

The existence of *a complex dialectal pattern* seems to be a *general feature* of the starling. Both sedentary and at least partially migratory populations as well as introduced populations present it. In all cases the dialects are based mainly on structural changes in the same Class I themes rather than on presence/absence of song types. Observations of the same area after some years gives us some hints about how dialects may differentiate: in most cases slow drifts in the structure seem to occur. The *social structure* seems to play an important role in that the *dialectal pattern* appeared to be much simpler in a population where colonies are almost absent and to follow the distribution of nests. The rhythmic theme is interesting: it can change very quickly as a result of social changes in the colony and is precisely absent in populations where the birds nest in isolation.

Song sharing as an indicator of social bond

Eens *et al.* (1992), observing a group of adult and young starlings in an outdoor aviary, found that: (a) several whistle types (no warbling) were shared by two or more males and were copied by the young males; (b) the repertoire of adult males did not change from one year to the next apart from a slight increase; (c) the yearlings increased their repertoire; (d) the young males copied about 60% of their repertoire from one or several males living in the same aviary (including the father); (e) part of the repertoire of the young birds had an unknown origin; and (f) song learning did not seem to involve agonistic interactions.

We can conclude from these findings that young birds learn preferentially from adults with whom they are in close contact, that this learning is not related to aggressive behavior, and that adults do not change repertoire in a stable situation.

A few questions arise from the above-mentioned observations and also from the descriptions of dialects:

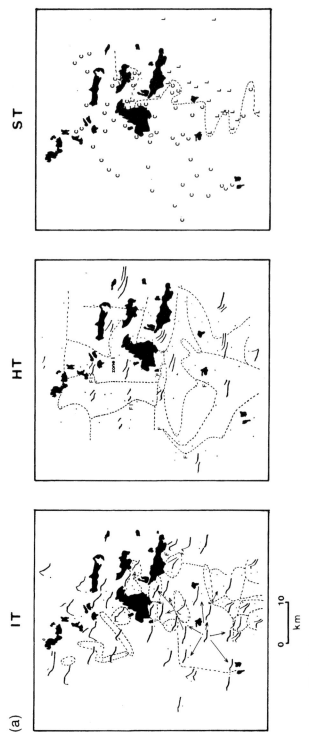

Fig. 8.8. Dialectal variations in different populations for three Class I songs: IT, inflection theme; HT, harmonic theme; ST, simple theme. (a) The distribution of dialectal variants in the Rennes area. Dialectal boundaries correspond to dotted lines. The hatched areas correspond to forests. C, L, short vs long variant of ST. Note for IT an area where dialects are not clearly differentiated, with progressive changes with distance (arrows). Part (b):

(A) Dialectal variations in Paimpont (France). Dark areas, forests. The letters correspond to different variants. C, L, short and long variants of ST. (From Adret-Hausberger 1984b.)

(B) Variations in Palatinate (Germany). The open circles correspond to villages. Note that each variant of the IT corresponds to a different village (a few examples: a–g here). L, TL, long and very long variants of ST. (From Adret-Hausberger & Güttinger 1984.)

(C) Variations of IT in Australia within the town of Armidale (left) and along the road outside the town (middle and right). On the right, variations of HT in the same area. Each cross corresponds to a recording point. (From Adret-Hausberger 1988.)

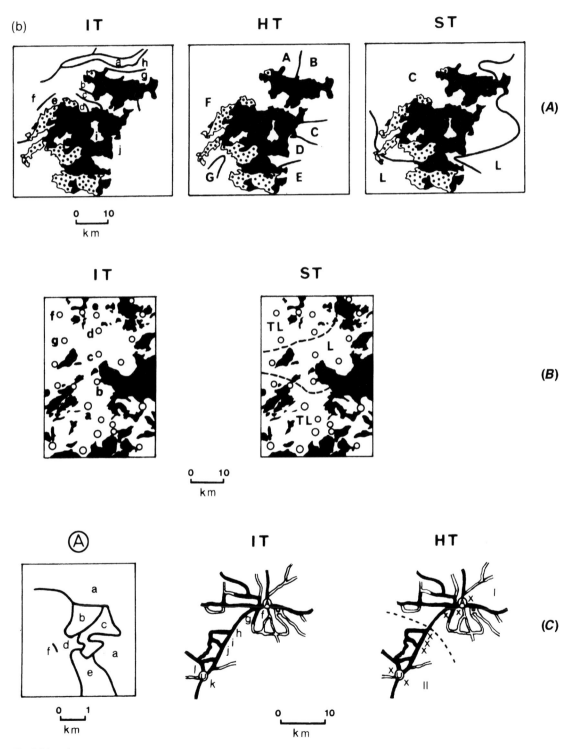

Fig. 8.8 (*cont.*)

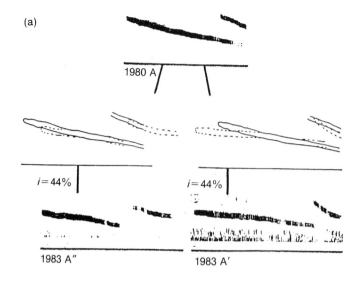

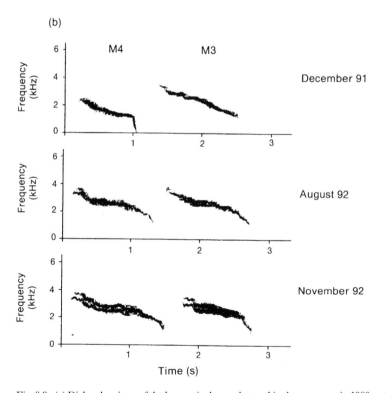

Fig. 8.9. (a) Dialectal variants of the harmonic theme observed in the same area in 1980 and 1983. The variants A′ and A″ correspond to different dialects but seem to both derive from the same original A variant. *I* index of similarity. (From Adret-Hausberger 1986.)

(b) Changes in the structure of the inflection theme of two males housed in the same aviary. Both males (M4 and M3) arrived with different variants in December 1991 but the structures they produced progressively converge over time.

(a) what determines who learns from whom, (b) what does song sharing reflect in terms of social bonds, and (c) what determines changes in the song repertoires of adults.

Two parallel studies were undertaken on captive groups of starlings: one consisted of following a group of adult and young males and females over eight months, without any period of social change; the other consisted in following a group of adult birds over a year and submitting them to social changes (L. Henry & M. Hausberger, unpublished data; Hausberger et al. 1995b).

In both studies, the activities of individuals and the closest neighbor were noted using a scan sampling method, whereas all types of interactions were noted *ad libitum*. In the second study, song acquisition by females was investigated by removing them, and injecting them with testosterone before the observations, after pairing and again at the end of the study.

Both studies gave strikingly similar results: the social organization was best defined in terms of proximities and was based, in the nonbreeding season, on within-sex groups or pairs with a few intersexual pairs (Fig. 8.10). Females, in particular tended to form social pairs that spent most of their time in close proximity for all activities. *The pattern of song sharing clearly reflected the social organization* (Fig. 8.10b). Members of the social pairs of females shared most of their songs, whereas males shared songs with other males to an extent that depended on their degree of association. Song sharing was restricted mostly to birds of the same sex. When young birds were present, they tended to associate mostly with birds of the same sex and age class. They learned from adults of the same sex with whom they had a preferential bond (in particular, father or mother). The song repertoires of adults of both sexes remained fairly stable over time in the first group, whereas very large changes were observed in the repertoire size and composition of all adult males and females after each social change (see p. 138). This corresponds to the production of new songs by newly associated birds. Some were first produced by one bird and later produced by the social partner, whereas other new songs appeared suddenly in both birds' repertoires.

Both warbling motifs and whistle types were involved in song sharing in both studies. The Class I themes, which differed between males when they arrived, showed slow drifts in their structure until all males had a similar variant (Fig. 8.9), whereas Class II themes were also shared in this context and appeared suddenly in the repertoire of one of both newly associated birds. The pattern of sharing did not change in the breeding season and very few themes or motifs were shared between males and females.

It is interesting that some males who were socially peripheral to the group showed a low degree of sharing (Hausberger et al. 1995b). Another important point is that a certain amount of individuality in repertoire always remained, even when there was much sharing.

Conclusion

Song sharing clearly reflects affiliative interactions and *sex-specificity*. The high plasticity of adult starling abilities mentioned before obviously enables them to adapt to new social situations. Within these social groups a still higher degree of sharing is observed than in most field situations, since even Class II themes are shared. This seems to be related to density, as Motis (1994) also found this type of sharing in her field colonies. These observations help us to understand the differentiation of dialects. Two different learning mechanisms tend to be involved: subtle progressive changes in the structure, leading to a convergence for the more universal Class I themes; and production of new types for the Class II themes.

Starlings, like many songbirds, show local variations. As in sparrows, the dialects are based on subtle differences in general song categories. However, these variations are of an almost unique type compared to the dialects found in most other bird species. To my knowledge, no other such complex system with several levels superimposed has been described in birds, apart from some mention of "subdialects" in white-crowned sparrows (*Zonotrichia leucophrys*; Baptista 1975). The other originality is that differences between localities are differences in the precise structure of the same song type (variants) rather than differences in the actual song types produced. Whereas all these characteristics make the dialectal system of starlings unique amongst birds, they are reminiscent of the descriptions of dialects in human:

"each word has its own story" (Petyt 1980) (each song
    type has its own level of variation).
Intelligibility of words between dialects (starlings recog-
    nize song types despite the variations): variant ways
    of saying the same thing (Fasold 1985).

(a)

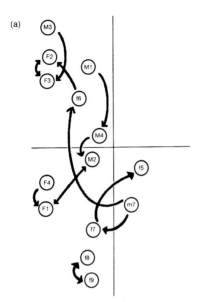

Whistle types	Fall			
	F1	F4	F3	F2
A	+	+		
P	+	+		
Q	+	+	+	
R	+			
S				
T			+	+
U			+	+
V			+	+
W			+	+
X			+	+
Y			+	+
F				
No. of nonshared types	3	3	1	0

(b)

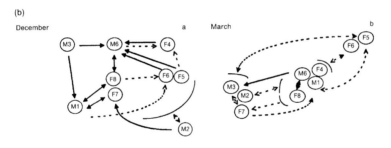

	November									May								
Whistle types	F4	F3	F2	F1	F5	F6	F8	F7	F9	F4	F3	F2	F1	F5	F6	F8	F7	F9
A	+	+								+	+							
B	+				+									+	+			
C		+		+	+									+	+			
D				+	+								+	+				
E		+		+	+								+	+	+			
F		+		+	+							+	+	+				
G																+	+	
H															+	+		
I										+	+							
J										+	+							
No. of nonshared types	6	12	11	2	2	7	2	0	6	7	2	9	3	8	6	2	6	2

Fig. 8.10. Social organization and song sharing in captive groups of starlings: M, males; F, females, m, f, juvenile birds.
(a) Social organization in an American captive group in the fall: the arrows correspond to preferences (in terms of proximity). The spatial proximity on the figure corresponds to the degree of affiliation in terms of frequencies as nearest neighbor. The table shows the pattern of sharing and the number of individual-specific structures in the adult females (spontaneous song).

(b) Social organization in captive groups of starlings: note change in group composition between both time periods. The song patterns of the females are shown in the table: recordings were made during testosterone treatment (from Hausberger *et al.* 1995b).

Variations in structure more than in word use.
Both adjacent variants at boundaries or continua with
different dialects at each end.
Progressive changes in structure over time, leading to
convergence.

Comparisons certainly cannot go further than this
descriptive level, but these similarities raise questions
about the evolutionary processes involved.

As Baker & Cunningham (1985) mentioned, dialects
are an indirect proof of learning. Our studies of captive
birds shed some light on the mechanisms that may be
involved in starlings. The three classes of song seem to be
acquired through different processes. The Class I songs,
which are male specific, are those involved in dialects and
the sharing of a same variant seems to be the result of convergence, with progressive changes in structure: such a
process is reflected in the variations over time observed in
the dialectal system and may explain the continua
observed in some areas. A similar convergence phenomenon can be observed in flocks of chickadees (Nowicki
1989), colonies of caciques (Trainer 1989), and also in male
dolphins (Tyack & Sayigh, Chapter 11). Changes in
precise structure of calls as a result of social changes are
also mentioned in pygmy marmosets (Snowdon *et al.*,
Chapter 12).

The Class I songs show a high stability in stable
colonies (see e.g., Adret-Hausberger *et al.* 1990). The
Class II and III songs show a very different pattern of variation: although they characterize each individual in its
colony in the field, some Class II whistle types and warbling motifs are shared in our social groups. This sharing
clearly characterizes social affiliations and in particular the
social pairs of females. Vocal sharing as indicator of social
bond is mentioned for various oscines (see e.g., Feekes
1982; Nowicki 1989; Brown & Farabaugh, Chapter 7) but
also in cetaceans (see McCowan & Reiss, Chapter 10;
Tyack & Sayigh, Chapter 11). In humans, Milroy (1980)
found that the degree of sharing is correlated to the social
commitment of the speaker to the local group. Deaf people
placed in a community develop a unique set of shared signs
(Kegl *et al.* 1994 by Goldin-Meadow, Chapter 15).

Locke & Snow (Chapter 14) refer to vocal accommodation as a tendency to adapt aspects of one's vocal production to match those of the partner. This phenomenon is
related to a "positive affect." In our starlings, song sharing

appeared between associated birds (in terms of proximity)
that were not aggressive towards each other. Affiliative
context is mentioned to be involved in vocal sharing of
Australian magpies, American crows and budgerigars
(Brown & Farabaugh, Chapter 7) as well as in tutor choice
in young zebra finches (Williams 1990; Mann & Slater
1995 cited by Zann, Chapter 9). In starlings, these observations confirm the experimental data on song learning,
showing the importance of the social context or social
bond in a nonagonistic relation (see pp. 139–41).

If vocal sharing is a result of social affiliations, we can
suggest that the dialects we observe in starlings have a
social basis. Interestingly, Catchpole & Rowell (1993)
observed that two groups of wrens separated by a lake have
different dialects, and the authors infer that the lake is a
social barrier. This corresponds very well with our conclusions that unexploitable areas were, for starlings, social
barriers to the transmission of dialects.

More field data are needed to understand the learning
mechanisms involved but also to be able to understand the
captive data in a functional way. The sex-specificity
observed in particular probably reflects a general disposi-
tion of the species. Song learning along sexual lines seems
to be another interesting social influence on vocal learning
(Baptista & Gaunt, Chapter 3) (see also mynahs (Bertram
1970); slate-colored shrikes (Wickler & Sonnenschein
1989); Australian magpies (Brown & Farabaugh, Chapter
7); cetaceans (Tyack & Sayigh, Chapter 11).

STARLING SONG AS A SOCIAL SIGNAL

There is both contextual and experimental evidence
showing that bird song is an important signal for all sorts
of social interactions, in addition to its role in breeding.
These observations give us some hints about the possible
social functions of songs in general and shared song in particular.

Song production in a nonbreeding context

Song does not always repel males
Mountjoy & Lemon (1991) have broadcast some starling
song near nestboxes. The birds that visited the nestboxes
(with song or "silent") were automatically caught. They
found that both males and females came to visit the nest-

boxes where song was broadcast, both sexes significantly more often than when there was no song. The design of this experiment does not allow a definitive conclusion on whether it was song or a vocalization in general that was being attractive (Kroodsma *et al.* 1989), but it shows at least one important fact: male song does not repel males.

This is not completely surprising, since it has been shown that European starlings, when possible, choose nests where other starlings are already present. It is interesting that this seems to happen also in American starlings, where for generations, most of them had to nest individually.

Song as a within-sex signal

Not only does the song not repel males, but L. Henry (unpublished data) even showed in her captive group that the males spent as much time singing whether the closest neighbor was a male or a female in a nonbreeding situation. Only at the beginning of breeding did they sing more toward females (Fig. 8.11). This feature is still more striking in females who, in a nonbreeding situation, highly preferred to sing near another female, which recalls Hausberger & Black's (1991) observations of a polygynous trio. These observations were similar in adult and yearling birds.

Moreover, socially associated males tend to sing preferentially when they sit close to each other rather than when another male is their closest neighbor (Hausberger *et al.* 1995b). Singing in proximity never led to aggression and might be the sign of a close social bond, reinforced by vocal sharing. We also observed that the song of a male directed toward a female in winter could be a good predictor of later pairing choices. Close singing again appears as an indicator of a strong bond.

Song in social nonbreeding contexts

Starlings sing abundantly outside the colony, in the different social contexts they encounter during the day (see pp. 130–1). During the day, foraging bouts are interspersed with perching and singing of the whole flock. But observers are impressed mainly by the chorus that accompanies night roosting – to the point that Delvingt (1961) suggested that it "all happens as if communal singing were necessary to prepare for sleep." There is a peak of singing at the time when the birds have just entered the roost site and fly from branch to branch before settling on a perch for the night (Adret-Hausberger 1982). Roost singing also

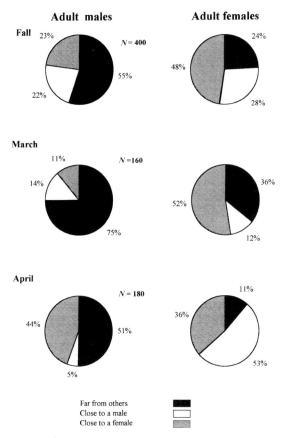

Fig. 8.11. Time spent singing activity according to social proximity within a captive social group: no close neighbor (>50 cm from any other birds), nearest neighbor is a male, or nearest neighbor is a female. *N*, number of records (scan sampling). (Henry, unpublished data.)

occurs in the morning, shortly before departure. A good description was given by Feare (1984): "roost singing in the morning could conceivably have a role in synchronising departure because the volume of singing gradually increases until eventually a sudden silence precedes the exodus wave. Singing then increases until another cessation precedes the departure of the next wave about three minutes after the earlier exodus."

Shared songs as passwords?

Social bonds

In order to test the discrimination of shared and non-shared song types by females, playback experiments were

made on our captive females (see pp. 147–8) placed in isolation. Their reactions were tested to different whistled
songs: songs of familiar nonassociated females, unique
versus shared songs of the social mate, and unique and
shared own songs. No vocal response to playback was
observed in this nonsocial situation, but videorecordings
revealed that *attention, in terms of head movement*, differed
according to the song category the birds were hearing.
Females did not show any particular reaction to the songs
of the familiar nonassociated females, nor to their or their
mate's unique own songs. However, they showed an
increase of their head movements after hearing *their social
mate's shared songs* but even more movements after hearing
their *own shared songs* (M. Hausberger *et al.* unpublished
data). This was especially interesting because the social
mates had been separated for more than eight months.
These results suggest that this special attention toward
shared themes reflects a long-term memory of a particular
social value.

Dialectal groups

Whistled songs are particularly abundant in flocks and
roosts and numerous vocal interactions can be heard. The
Class I whistles tend to be the most frequently sung,
although Class II whistles can be heard in small flocks of
neighbors near the colony in the morning and in some
roosts. The different Class I themes tend to be used differentially according to the social context: the inflection
theme is used more in small flocks, the simple theme in
large flocks, and mostly the rhythmic theme and some
Class II themes in the roosts (Adret-Hausberger 1982; E.
Bigot, unpublished data).

Starlings, like all songbirds tested, show a dialect discrimination: they respond less often and after a longer time
to the playback of a strange variant (Adret-Hausberger
1982). This may be important in the flocks and roosts
where different dialectal variants can be heard. Whether or
not this leads to differential social interactions (Payne
1981) remains to be investigated precisely.

Adret-Hausberger (1982), observing that the rhythmic
theme was a theme frequently heard in roosts, suggested
that it may help the birds from a same colony to stay close
to each other, since the variants characterized the colonies
(see pp. 141–2). A detailed study of the whistled songs produced in roosts shows now that different variants of a given
theme (mainly the rhythmic theme) are present at different places within the roost, suggesting that birds from a

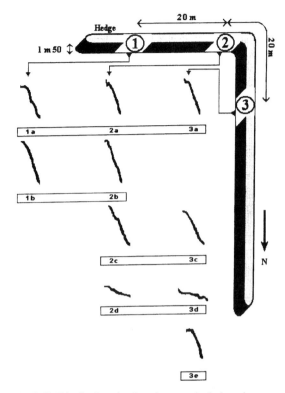

Fig. 8.12. Distribution of variants in a roost in the introductory
note of the rhythmic theme within a roost. About 100 starlings
were present. Microphones (1, 2, 3) were placed in three different sites within the roost. One variant was found in the three
sites, one was found only in site 3, and three others were
recorded from neighboring microphones. (From Bigot *et al.*
1994; E. Bigot, unpublished data.)

same area tend to be in the same part of the roost (Fig. 8.12;
E. Bigot, unpublished data).

Song sharing and social density

One intriguing result when field data are compared to
captive data is that in most field observations, only the
Class I "universal" whistles were shared (plus the clicks
and trills in warbling), whereas the "field" highly individual Class II themes and the warbling are also shared in
captivity. In fact, comparisons of populations can give us a
hint. In Australia (and apparently also in New Zealand and
America), where social density is very low, even some Class
I themes are missing: in Germany, the harmonic theme is
missing; the other extreme is in Spain, where the colony
densities are high and where some Class II themes are

FIELD OBSERVATIONS + CAPTIVE STUDIES

SPECIES: { Class I songs (whistles): *only males*: same characteristics and variation ranges in different populations

Class III song (warbling): *in males*: basic organization, ends of sequences

POPULATION: *in males*: complex local variations (see isoglosses in human dialects) with different levels of variation (Class I songs)

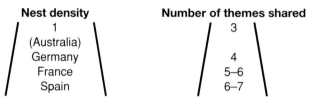

Nest density	Number of themes shared
1	3
(Australia)	
Germany	4
France	5–6
Spain	6–7

COLONY: all Class I themes shared (males)

SOCIAL GROUPS: structures shared in Class I, II and III songs
(captivity, field?)   Males: song sharing correlates with social bond
Females: social pairs share the same structures

PAIRS (♂/♀): a few individual whistles and motifs shared

INDIVIDUAL: individual whistle types + warbling: (♂ + ♀) but degree of individuality depends on social context: very high in "natural" colonies, very low in some captive conditions

Fig. 8.13. The relation between starling song variations and social structure as revealed by field and captive studies. (Modified from Hausberger 1993.)

shared in addition to the Class I themes (Fig. 8.13). The degree of song sharing is related to nest density. In none of the studies, whether in captivity or in the field, did the numerous song-matching interactions lead to aggression.

Conclusion

Starling song can be heard in all social situations and conveys different pieces of social information (Fig. 8.14): species, population, social group, individual. There are few examples of such complex information coded in birds, but a very similar situation is observed in cetaceans (Tyack & Sayigh, Chapter 11): bottlenose dolphins show individual-, group-, and species-specific whistles (McCowan & Reiss, Chapter 10).

The question is whether such variations have a functional significance. Playback experiments clearly show that starlings not only discriminate dialects but also react differently to the songs shared with a particular social partner. Starling songs do not appear to be related to ago-

nistic situations, and our observations suggest that they may rather serve a social identity function rather than merely be used in territorial/breeding encounters.

Song has been suggested to be a "password" or a "badge" in different species and Brown & Farabaugh (Chapter 7) suggest that "vocal learning has evolved to allow individuals to share vocalizations with a particular subset of conspecifics." Dialects in starlings may be used in roosts, for example, in order for colony members to find each other. Social partners may use shared songs for identifying each other. In all cases, sharing seems to be rather indicative of tolerance. If this is the case, we can understand why the degree of sharing increases with social density: the more social constraints, the more shared signals.

In humans, dialects have been said to be badges of deme or social group membership (Labov 1972). Locke & Snow (Chapter 14) mention that sound change is a necessary means of signalling group membership and achieving intimacy, and Fasold (1985) considered that dialects have a social identity function separable from the communicative function. Chambers (1985) observes that antagonisms can occur along dialectal boundaries: "strangers" are less well accepted.

Dialects in general do not seem to be very important for mating decisions in humans (Chambers 1985) nor in birds (McGregor *et al.* 1988). Female starlings do not produce the Class I themes that are involved in dialects. Shared songs occur more between social partners than between mates.

Although characteristics of the warbling song make it a likely candidate for a process of sexual selection (Eens *et al.* 1991; Hausberger *et al.* 1991), we observed in our aviaries that even silent males could pair, not even later than the other males. Familiarity appears as another important factor in mating decisions (Hausberger *et al.* 1995b). The social function of song may be as important as its reproductive function.

GENERAL CONCLUSIONS

Starling song appears to be a good example where vocal learning may have evolved as a result of social pressures. Starlings are highly social, mobile and have different types of social encounter at different times of the day or of their life. Their song system is complex, with different levels of variation that seem to play a social identity function. Field and captive studies bring us important information on social influences on song development in starlings:

Starlings learn better from a live tutor than from a tape. They can acquire new songs as adults.
Learning processes involve both selective use and song acquisition.
Song sharing is dependent upon social bond and leads to complex dialects.
Song learning occurs along sexual lines, reflecting the social organization of groups.
Shared songs are differentiated from nonshared songs by the birds themselves and may play a role of password.
The different levels of variations give information about the species, population, social group and individuals.

All these features are shared with other social species mentioned in this volume (Baptista & Gaunt, Chapter 3; Brown & Farabaugh, Chapter 7; Pepperberg, Chapter 9), particularly cetaceans (McCowan & Reiss, Chapter 10;

Tyack & Sayigh, Chapter 11), but also humans (Locke & Snow, Chapter 14; Goodwin, Chapter 17). This plasticity in vocal learning and dependency upon social bonds may be a response to rapidly changing social environments as a result of high mobility. Few of these changes are observed in stable situations, which may explain that plasticity may have been underestimated in the more stable nonhuman primate groups (Snowdon *et al.* Chapter 12).

Starlings, like cetaceans, might share with humans "processes of personal identity and group identity formation that require rapid learning and considerable flexibility even at later stages" (Locke & Snow, Chapter 14).

There are also questions that the starling studies raise about the ways vocal development can be broached:

Studies on "repertoires" have to take into account possible changes in use according to context. This is also necessary for understanding the learning process.
The nonbreeding situation is very important for studying song, in particular in females.
Captive studies have to have field validations when possible: some song types did not occur even in outdoor aviaries.
Subtle changes in structure are often underestimated compared to changes of song type. They do, however, involve learning.
There is no reason why dialects in birds should not be considered as such when several song types are involved (see data on humans in the literature).
Naturalistic studies should not be neglected in favor of solely experimental studies: only the combination of both can give us all the information about song tutor choice and the relation with social organization. As developed by West *et al.* (Chapter 4), only studies of birds raised in a group can really inform us about social influences.
Social organization has to be studied in parallel with that of vocal development: "who likes to be with whom" may be more important than actual "interactions" (see also Mason 1978; Goodwin 1990).

The neglect of the social importance of song, in particular outside the breeding context, comes from earlier assumption that song is "a social utterance, long or short, simple or complex and species-specific which is given by either sex or both and which functions primarily to repel males of the same species, to attract a mate or both" (Smith

1959). It is also generally easier to study song in this context. However, there is no reason why song may not have more functions (Smith 1991). Robinson (1990) suggested that birds also may have a need for sociability for its own sake, and therefore develop a "phatic" communication that expresses sociability. This may be the case for other phylogenetic groups as well.

ACKNOWLEDGEMENTS

I am very grateful to all colleagues with whom fruitful discussions have occurred along the years: J. C. Guyomarc'h, who initiated our studies on starling song, and also H. R. Güttinger, P. F. Jenkins, P. Clergeau, A. Motis, E. Bigot, and L. Henry. I am very grateful to H. Schuelke-Grillou and L. Henry for their help in preparing this manuscript and to M. Forasté for her help in the research process.

REFERENCES

Adret-Hausberger, M. (1982). Social influences on the whistled songs of starlings. *Behavioral Ecology and Sociobiology* **11**: 241–6.

Adret-Hausberger, M. (1983). Variations dialectales des sifflements de l'étourneau sansonnet sédentaire en Bretagne. *Zeitschrift für Tierpsychologie* **62**: 55–71.

Adret-Hausberger, M. (1984a). Seasonal variation in the whistles of starling *Sturnus vulgaris*. *Ibis* **126**: 372–8.

Adret-Hausberger, M. (1984b). Variations dialectales et "barrières naturelles" chez l'étourneau sansonnet (*Sturnus vulgaris*). *Biology of Behaviour* **9**: 213–25.

Adret-Hausberger, M. (1986). Temporal dynamics of dialects in the whistled songs of starlings. *Ethology* **71**: 140–52.

Adret-Hausberger, M. (1988). Song differentiation and population structure: The example of the whistled songs in an introduced population of European starlings *Sturnus vulgaris* in Australia. *Ethology* **79**: 104–15.

Adret-Hausberger, M. (1989a). Song ontogenesis in starlings *Sturnus vulgaris*: Are song and subsong continuous? *Bird Behaviour* **8**: 8–13.

Adret-Hausberger, M. (1989b). The species repertoire of whistled songs in the European starling: Species-specific characteristics and variability. *Bioacoustics* **2**: 137–62.

Adret-Hausberger, M. & Güttinger, H. R. (1984). Constancy of basic pattern in the song of two populations of starlings: A comparison of song variation between sedentary and migratory populations. *Zeitschrift für Tierpsychologie* **66**: 309–27.

Adret-Hausberger, M., Güttinger, H. R. & Merkel, F. W. (1990). Individual life history and song repertoire changes in a colony of starlings. *Ethology* **84**: 265–80.

Adret-Hausberger, M. & Jenkins, P. F. (1988). Complex organization of the warbling song in the European starling *Sturnus vulgaris*. *Behaviour* **107**: 138–56.

Baker, M. C. & Cunningham, M. A. (1985). The biology of bird song dialects. *Behavioral and Brain Sciences* **8**: 85–133.

Ball, G. F. & Wingfield, J. C. (1987). Changes in plasma levels of luteinizing hormone and sex steroid hormones in relation to multiple-broodedness and nest-site density in male starlings. *Physiological Zoology* **60**: 191–9.

Bandura, A. (1977). *Social Learning Theory*. Prentice Hall, Englewood Cliffs, NJ.

Baptista, L. F. (1975). Song dialects and demes in sedentary populations of the white crowned sparrow. *University of California Publications in Zoology* **105**: 1–52.

Baptista, L. F., Trail, P. W., De Wolfe, B. B. & Morton, M. L. (1993). Singing and its functions in female white-crowned sparrows. *Animal Behaviour* **46**: 511–24.

Becker, P. H. (1978). Der Einfluß des Lernens auf einfache und komplexe Gesangsstrophen der Sumpfmeise. *Journal für Ornithologie* **119**: 388–411.

Bertram, B. (1970). The vocal behaviour of the Indian Hill mynah *Gracula religiosa*. *Animal Behavior Monographs* **3**: 79–192.

Bigot, E., Hausberger, M. & Clergeau, P. (1994). Dialects and social organization within roosts in starlings. *Journal für Ornithologie* **135**: 157.

Böhner, J., Chaiken, M., Ball, G. F. & Marler, P. (1990). Song acquisition in photosensitive and photorefractory male European starlings. *Hormones and Behavior* **24**: 582–94.

Bready, M. B. (1929). *The European starling on his westward way*. The Knickerbocker Press.

Bullough, W. S. (1942). On the external morphology of the British and Continental races of the starling (*Sturnus vulgaris*). *Ibis* **6**: 225–39.

Caccamise, D. F. & Morrison, D. W. (1988). Avian communal roosting: A test of the "patch-sitting" hypothesis. *Condor* **90**: 453–8.

Catchpole, D. K. & Rowell, A. (1993). Song sharing and local dialects in a population of the European wren *Troglodytes troglodytes*. *Behaviour* **125**: 67–78.

Chaiken, M., Böhner, J. & Marler, P. (1993). Song acquisition in European starlings, *Sturnus vulgaris*: A comparison of the songs of live-tutored, tape-tutored, untutored and wild-caught males. *Animal Behaviour* **46**: 1079–90.

Chaiken, M., Böhner, J. & Marler, P. (1994). Repertoire turnover and the timing of song acquisition in European starlings. *Behaviour* **128**: 25–39

Chambers, J. K. (1985). Social adaptiveness in human and songbird dialects. *Behavioral and Brain Sciences* **8**: 102–4.

Clergeau, P. (1991). Nouvelles hypothèses de structure de dortoirs d'étourneaux et conséquence sur la gestion de l'espèce. *Bulletin de la Société Zoologique de France* **116**: 289–95.

Clergeau, P. (1993). Fonctions des dortoirs d'étourneaux: Hypothèses historiques et apport du modèle étourneau. *Oiseau et la Revue Française d'Onithologie* **2**: 88–105.

Coleman, J. D. (1974). The use of artificial nest sites erected for starlings in Canterbury, New Zealand. *New Zealand Journal of Zoology* **1**: 349–54.

Delvingt, W. (1961). Les dortoirs d'étourneaux en Belgique en 1959–60. *Gerfaut* **51**: 1–27.

Derrickson, K. C. (1987). Yearly and situational changes in the estimate of repertoire size in northern mockingbirds (*Mimus polyglottus*). *Auk* **104**: 198–207.

Eens, M., Pinxten, R. & Verheyen, R. F. (1989). Temporal and sequential organization of song bouts in the European starling. *Ardea* **77**: 75–86.

Eens, M., Pinxten, R. & Verheyen, R. F. (1991). Organization of song in the European starling: Species-specificity and individual differences. *Belgian Journal of Zoology* **2**: 257–78.

Eens, M., Pinxten, R. & Verheyen, R. F. (1992). Song learning in captive European starlings. *Animal Behaviour* **44**: 1131–43.

Eens, M. Pinxten, R. & Verheyen, R. F. (1993). Function of the song and song repertoire in the European starling (*Sturnus vulgaris*): An aviary experiment. *Behaviour* **125**: 51–66.

Falls, J. B., Krebs, J. R. & McGregor, P. H. (1982). Song matching in the great tit: The effect of similarity and familiarity. *Animal Behaviour* **30**: 997–1009.

Fasold, R. W. (1985). Bird-song dialects and human-language dialects: A common basis? *Behavioral and Brain Sciences* **8**: 104–5.

Feare, C. J. (1984). *The Starling*. Oxford University Press, Oxford.

Feekes, F. (1982). Song mimesis within colonies of *Cacicus cela*: A colonial password? *Zeitschrift für Tierpsychologie* **58**: 119–52.

Flux, J. E. C. & Flux, M. M. (1981). Population dynamics and age structure of starlings (*Sturnus vulgaris*) in New Zealand. *New Zealand Journal of Ecology* **4**: 65–72.

Geroudet, P. (1957). *Les passereaux*, vol. 3, pp. 127–41. Delachaux et Niestlé, Neuchatel.

Goodacre, M. J. (1959). The origin of winter visitors to the British Isles. 4 – Starling (*Sturnus vulgaris*). *Bird Study* **6**: 180–92.

Goodwin, M. H. (1990). *He Said – She Said*. Indiana University Press, Bloomington, IN.

Gottlieb, G. (1993). Social induction of malleability in ducklings: Sensory basis and psychological mechanism. *Animal Behaviour* **45**: 707–19.

Gromadski, M. & Kania, W. (1976). Bird-ringing results in Poland. Migrations of the starling *Sturnus vulgaris*. *Acta Ornithologica* **15**: 280–312.

Hartby, E. (1969). The calls of the starling. *Dansk Ornithologisk Forenings Tidsskrift* **62**: 205–30.

Hausberger, M. (1991). The organization of whistle sequences in starlings. *Bird Behaviour* **9**: 81–8.

Hausberger, M. (1993). How studies on vocal communication in birds contribute to a comparative approach of cognition. *Etología* **3**: 171–85.

Hausberger, M. & Black, J. M. (1991). Female song in European starlings: The case of noncompetitive song matching. *Ethology, Ecology and Evolution* **3**: 337–44.

Hausberger, M. & Cousillas, H. (1996). Categorization in birdsong: From behavioural to neuronal responses. *Behavioural Processes* **35**: 83–91.

Hausberger, M. & Guyomarc'h, J. C. (1981). Contribution à l'étude des vocalisations territoriales sifflées chez l'étourneau sansonnet *Sturnus vulgaris* en Bretagne. *Biology of Behaviour* **6**: 79–98.

Hausberger, M., Henry, L. & Richard, M. A. (1995a). Testosterone-induced singing in female starlings. *Ethology* **99**: 193–208.

Hausberger, M., Jenkins, P. F. & Keene, J. (1991). Species-specificity and mimicry in bird song: Are they paradoxes? A reevaluation of song mimicry in the European starling. *Behaviour* **117**: 53–81.

Hausberger, M., Richard, M. A., Henry, L., Lepage, L. & Schmidt, I. (1995b). Song sharing reflects the social organization in a captive group of European starlings (*Sturnus vulgaris*). *Journal of Comparative Psychology* **109**: 222–41.

Henry, L. (1994). Influences du contexte sur le comportement vocal et socio-sexuel de la femelle étourneau (*Sturnus vulgaris*). Thèse, Université de Rennes.

Henry, L., Hausberger, M. & Jenkins, P. F. (1994). The use of repertoire changes with pairing status in male European starlings. *Bioacoustics* **5**: 261–6.

Hindmarsh, A. M. (1984). Vocal mimicry in starlings. *Behaviour* **90**: 302–24.

Hindmarsh, A. M. (1986). The functional significance of vocal mimicry in song. *Behaviour* **99**: 87–100.

Kroodsma, D. E. (1982). Song repertoires: Problems in their definition and use. In *Acoustic Communication in Birds*, ed. D. E. Kroodsma & E. H. Miller, vol. 2, pp. 125–46. Academic Press, New York.

Kroodsma, D. E., Bereson, B. C., Byers, B. E. & Minear, E.

(1989). Use of song types by the chestnut-sided warbler: Evidence for both intra- and inter-sexual functions. *Canadian Journal of Zoology* **67**: 447–56.

Labov, W. (1972). *Sociolinguistic patterns*. University of Pennsylvania Press, Philadelphia.

Marler, P. & Nelson, D. A. (1993). Action-based learning: A new form of developmental plasticity in bird song. *Netherlands Journal of Zoology* **43**: 91–103.

Mason, W. A. (1978). Ontogeny of social systems. In *Recent Advances in Primatology*, vol. 1, *Behaviour*, ed. D. E. Chivers & J. Herbert, pp. 5–14. Academic Press, London.

McGregor, P. K., Walford, V. R. & Harper, D. G. C. (1988). Song inheritance and mating in a songbird with local dialects. *Bioacoustics* **1**: 107–29.

Merkel, F. W. (1979). Lebenslauf eines Starenweibchens. *Natur und Museum* **109**: 348–52.

Merkel, F. W. (1980). Sozialverhalten von individuell markierten Staren in einer kleinen Nistkastenkolonie. Die Rolle der Polygynie. *Luscinia* **44**: 133–58.

Milroy, L. (1980). *Language and Social Networks*. Blackwell, Oxford.

Motis, A. (1994). Territorialitat interspecifica de *Sturnus vulgaris* I *Sturnus unicolor*, dues aloespecies en contacte secundari: Habitat de cria, conducta agonistica I cants territorialis. Doctoral thesis, University of Barcelona.

Mountjoy, D. J. & Lemon, R. E. (1991). Song as an attractant for male and female European starlings, and the influence of song complexity on their response. *Behavioral Ecology and Sociobiology* **28**: 97–100.

Mountjoy, D. J. & Lemon, R. E. (1995). Extended song learning in wild European starlings. *Animal Behaviour* **49**: 357–66.

Nowicki, S. (1989). Vocal plasticity in captive black-capped chickadees: The acoustic basis and rate of call convergence. *Animal Behaviour* **37**: 64–73.

Payne, R. B. (1981). Population structure and social behavior: Models for testing the ecological significance of song dialects in birds. In *Natural Selection and Social Behavior*, ed. R. D. Alexander & D. W. Tinkle, pp. 108–20. Chiron Press, New York.

Peach, W. J. & Fowler, J. A. (1989). Movements of wing-tagged starlings from an urban communal roost in winter. *Bird Study* **36**: 16–22.

Pepperberg, I. M. (1985). Social modeling theory: A possible framework for understanding avian vocal learning. *Auk* **102**: 854–64.

Pepperberg, I. M. (1993). A review of the effects of social interaction on vocal learning in African grey parrots (*Psittacus erithacus*). *Netherlands Journal of Zoology* **43**: 104–24.

Petyt, K. M. (1980). *The Study of Dialect: An Introduction to Dialectology*. A. Deutsch, London.

Pinxten, R. & Eens, M. (1990). Polygyny in the European starling: Effect on female reproductive success. *Animal Behaviour* **40**: 1035–47.

Robinson, F. N. (1990). Phatic communication in bird song. *Emu* **91**: 61–3.

Slater, P. J. B. (1989). Bird song learning: Causes and consequences. *Ethology Ecology and Evolution* **1**: 19–46.

Smith, R. L. (1959). Songs of the grasshopper sparrow. *Wilson Bulletin* **71**: 141–52.

Smith, W. J. (1991). Singing is based on two markedly different kinds of signalling. *Journal of Theoretical Biology* **152**: 241–53.

Summers, R. W., Westlake, G. E. & Feare, C. J. (1987). Differences in the ages, sexes and physical condition of starlings at the centre and periphery of roosts. *Ibis* **129**: 96–102.

Suthers, H. B. (1982). Starlings mimic human speech. *Bird Watcher's Digest*, 37–9.

ten Cate, C. (1987). Listening behaviour and song learning in zebra finches. *Animal Behaviour* **34**: 1267–8.

Trainer, J. M. (1989). Cultural evolution in song dialects of yellow-rumped caciques in Panama. *Ethology* **80**: 190–204.

Van der Mueren, E. (1980). Intraspecific aggression in a group of caged starlings. *Gerfaut* **70**: 455–70.

Verheyen, R. F. (1969). Le choix du nichoir chez l'étourneau. *Gerfaut* **59**: 239–59.

Verheyen, R. F. (1980). Breeding strategies of the starling. In *Bird Problems in Agriculture*, ed. E. N. Wright, I. R. Inglis & C. J. Feare, pp. 69–82. British Crop Protection Council, London.

West, M. J. & King, A. P. (1989). Mozart's starling. *American Scientist* **78**: 106–14.

West, M. J., Stroud, A. N. & King, A. P. (1983). Mimicry of the human voice by European starlings (*Sturnus vulgaris*): The role of social interaction. *Wilson Bulletin* **95**: 635–40.

Wickler, W. & Sonnenschein, E. (1989). Ontogeny of song in captive duet-singing slate-coloured boubous (*Laniarius funebris*). A study in birdsong epigenesis. *Behaviour* **111**: 220–33.

Williams, H. (1990). Models for song learning in the zebra finch: Father or others. *Animal Behaviour* **39**: 745–57.

Witschi, E. & Miller, R. A. (1938). Ambisexuality in the female starling. *Journal of Experimental Zoology* **79**: 475–87.

Wright, J. & Cuthill, I. (1992). Monogamy in the European starling. *Behaviour* **120**: 262–85.

Wynne-Edwards, V. C. (1962). *Animal Dispersion in Relation to Social Behaviour*. Oliver & Boyd, Edinburgh.

9 Social influences on the acquisition of human-based codes in parrots and nonhuman primates

IRENE MAXINE PEPPERBERG

INTRODUCTION

Most studies of the effects of social interaction on the ontogeny of vocal communication in birds and primates concentrate on the normal course of development of species-specific codes: how birds learn conspecific song, how nonhuman primates develop their natural repertoire of calls, and how human infants develop language. The effects of social interaction, however, are probably even more important during *exceptional* learning (Pepperberg 1985): learning that is unlikely to occur in the normal course of events. Such learning, defined and described below, has been documented for a number of species, including humans. I have been particularly interested in examining how social interaction can influence a specific type of exceptional learning – the development of inter-species communication between humans and birds. My research on the effects of social interaction on the acquisition of a vocal, English-based code by grey parrots (*Psittacus erithacus*) clearly demonstrates how social and environmental input[1] can engender learning that would not otherwise occur (e.g., Pepperberg 1990a). Interestingly, an analysis of research on ape language also demonstrates how social interaction may be a particularly effective means of teaching nonvocal human-based communication codes to nonhuman primates.

Although characterizing the effects of social and environmental influences on exceptional learning is not a simple task, my work has shown that a conceptual framework, social modelling theory, can be used (a) to character-

ize how social input influences learning and (b) to delineate the critical features of input necessary for exceptional learning. Hence, in this chapter I focus on four issues concerning social input and the development of communication. (a) I review the principles of human social modelling theory that are most relevant for understanding how birds learn to communicate; (b) I delineate the specific aspects of input that influence the development of communication; (c) I describe experiments – my own with birds and those of other researchers with nonhuman primates – that demonstrate the effects of different types of input on learning; and (d) I discuss how the principles of social modelling might be used to predict learning outcomes.

A POSSIBLE FRAMEWORK FOR STUDYING THE EFFECTS OF SOCIAL INTERACTION ON VOCAL LEARNING

Few researchers dispute that environmental interactions are involved in the development of an organism's ability to communicate, but significant disagreement exists as to the *extent* to which social interaction, input from live models, and observation during and after explicit communicative acts – or a combination thereof – affect this development (e.g., see discussions in Mowrer 1954, 1958; Macphail 1982, 1987; Gleitman *et al.* 1984; Pepperberg 1985, 1988a,b, 1991, 1993; Furrow & Nelson 1986; Kuczaj 1986; Tomasello & Farrar 1986; Bedrosian *et al.* 1988; Cazden 1988; Marler & Peters 1988a; Petrinovich 1988; Veneziano 1988; Lock 1991; Goldfield 1993; Harris 1993). Menyuk *et al.* 1995 also described the effects of varying pragmatic, discourse, and conversational aspects of caregiver input with a child's stage of development; note other chapters in this volume, e.g., for children, Goldin-Meadow Chapter

---

[1] Social input can be subsumed under environmental input, but I mention these aspects separately to emphasize the role of social interaction. The environment may, for example, provide an opportunity for observational learning that involves no interaction.

15; for birds, Baptista & Gaunt, Chapter 3). If an isolated white-crowned sparrow (*Zonotricia leucophrys*) can learn its conspecific song by hearing a limited number of tape renditions (12 songs/day for 21 days; Petrinovich 1985), why should researchers consider social interaction of much importance? At least two reasons come to mind.

First, whether a bird so tutored understands how to use such a learned vocalization is unclear. Birds must learn not only what to sing but also when to sing, and the appropriate context for song (King & West 1983, 1989; Kroodsma 1988); such learning occurs through social interaction. Researchers have demonstrated that wild birds respond less vigorously to songs of laboratory-reared conspecific isolates than to songs of wild conspecifics (e.g., Thielcke 1973; Shiovitz 1975; Searcy *et al.* 1985), but comparisons have not been made with songs sung by birds that learned from conspecific audiotapes in a laboratory.

Second, during conspecific song learning social interaction may not significantly alter but rather act to facilitate or modify the course of development. For example, some positive correlation between high model status and efficacy of acquisition (rate and amount) of modelled behavior was found in humans (Mischel & Liebert 1967; Bandura 1977) and birds (Payne 1978, 1981, 1982, 1983; Mundinger 1979; Snow & Snow 1983; Baptista & Morton 1988). The actions of boys who were regarded as having high status by their peers were imitated far more often than were the actions of other group members at a summer camp (Lippitt *et al.* 1952); with respect to song acquisition, birds may react in similar ways (see Payne & Groschupf 1984). In cases where no single model is presented as "superior" and the modelled patterns are diverse, both humans (Bandura 1977) and birds (e.g., Laws 1994) engage in behavioral innovation: they *combine* elements of the modelled behaviors to develop their repertoire. And sometimes a tutor may simply be a more effective stimulus: certain birds are likely to learn more from live tutors than from tapes and some birds choose to learn songs of live tutors with whom they can interact rather than songs presented by tape or by noninteractive tutors (Waser & Marler 1977; Kroodsma 1978; Todt *et al.* 1979; Payne 1981; Kroodsma & Pickert 1984b; see also Brown & Farabaugh, Chapter 7). Data showing, for example, that juvenile Bewick's wrens (*Thryomanes bewickii*) will, ostensibly before song crystallization, counter-sing with adults using "nearly perfect renditions of the adult songs"

(Kroodsma 1974, p. 360), suggest that the effect of social interaction on song acquisition is not a laboratory artefact (for comparable data on white-crowned sparrows, see Baptista 1983).

Thus, studies to determine how organisms learn to communicate must delineate the exact role (or roles) played by social input and its context, and search for the specific mechanisms through which such input can influence learning. The task of determining how learning is affected by different types of input can be made easier if there already exists a conceptual framework that identifies the critical factors to be studied and the experiments that examine these factors. I have suggested (Pepperberg 1985, 1986a,b, 1988b, 1991) that human social psychology, in the form of social modelling theory (e.g., Bandura 1971, 1977), provides just such a framework for studies of communication.

Social modelling theory was derived initially from an analysis of procedures used by psychologists to enable humans to overcome strong inhibitions or phobias (Bandura 1971); researchers subsequently realized that the theory was also relevant for teaching certain aspects of communication (e.g., Brown 1976; Snow & Hoefnagel-Höhle 1978; Pepperberg 1981, 1985; Fey 1986). The theory systematically identifies individual contributions to the learning process that cannot otherwise always be easily distinguished. The theory encompasses several levels of interaction. One level involves separating learning situations into three functionally distinct categories: two types of active input – social modelling and social interaction – and the passive phenomenon of observational learning. A second level involves determining the optimal form of input for a given type of learning. For the purposes of this chapter, the modelling situation provides the best context for explaining the theory.

Modelling, in this theory, considers the learner as an active force in the learning process and the trainer as a motivator as well as a modeller. Unlike a subject in an operant paradigm, in which the learner responds to a limited number of simple stimuli (e.g., a ball, a block) with predictable actions learned by trial and error (e.g., to point to one item rather than to the other) and is rewarded with some unrelated item (e.g., a piece of food or a generic "That's right"), a learner in a modelling paradigm is supposed to use the modelled act as a directive. Optimally, the modelled act signals which aspects of the environment should be noted, emphasizes the common attributes – and thus the possible underlying rules – of diverse acts, provides a contextual understanding for the reasons for the acts, and demonstrates the consequences of the acts. During training, for example, a dysfunctional human student (e.g., an autistic child) for whom a communicative act is intentionally being modelled would be shown a tray of objects and would observe the following interaction between two trainers:

Trainer 1: (points to a ball) "What's this?"
Trainer 2: "Ball."
Trainer 1: "Yes, that's a ball. Here's the ball."

The ball is transferred, the receiver plays for a few moments, and the roles of questioner and receiver are then exchanged. This interaction provides an example of appropriate speaker–listener–respondent relationships, in which both an action and the consequences of the action are clearly demonstrated by an interactive tutor and model.

Social modelling theory thus emphasizes the roles of attention, comprehension, and motivation for learning. It consists of a set of principles – the second level referred to above – that delineates the optimal form of social input for a particular type of learning (Bandura 1971, 1977). Because some of its principles presume the existence of cognitive factors (e.g., symbolic coding, cognitive organization, mental rehearsal) that cannot be unquestionably attributed to all nonhuman subjects (cf. Griffin 1985), not all of the principles may be relevant for animal studies. I have, however, previously suggested how a subset of four of these principles are clearly applicable to the study of avian vocal learning (Pepperberg 1986a,b, 1988b, 1991).

One of these principles states that the level of competence of the student must be taken into account (note Jouanjean-L'Antoëne, Chapter 16). Human children, for example, are most likely to imitate actions (both vocal and physical) that are just slightly beyond their level of competence (Piaget 1954; Ryan 1973; Scollon 1976; K. E. Nelson 1978, cited by Kuczaj 1983; Krashen 1980, 1982; Kuczaj 1982a,b, 1983; Masur 1988); thus an interaction that models a new behavior that differs only slightly from an existing behavior or that encodes only slightly novel information is most easily learned. Some evidence exists that birds respond similarly.

(1) White-crowned sparrows tutored by live strawberry finches (*Amandava amadava*) or dark-eyed juncos (*Junco hyemalis*) produced extremely close renditions of the strawberry finch song, which is both simple and similar to conspecific isolate song, but copied only some syllables of the juncos, who produce multiple songs that are more complex than that of the finch (Baptista 1985; Baptista & Petrinovich 1986)

(2) Song sparrows (*Melospiza melodia*) are more likely to acquire swamp sparrow (*Melospiza georgiana*) syllables and notes from tapes altered to present the alien material in syntax patterns that closely approach conspecific song than from tapes that have not been altered (Marler & Peters 1988b)

(3) A grey parrot, already having "gr" (from "green"), "[ā]" (from p[ā]per), "k" (from "key"), and "arrot" (from "parrot") produced the novel vocalizations "gray" and "carrot" after minimal exposure to these labels, whereas months of training were required to produce novel labels such as "color" and "shape," for which antecedents did not exist (Pepperberg 1983; Pepperberg *et al.* 1991).

A corollary is that for the learning process to continue, the tutor/model must constantly adjust the demonstration to take into account – and continue to challenge – the increasing knowledge of the student. For humans, considerable evidence supports the contention that tutors who actively adjust their input positively effect the output of their students (e.g., Tomasello 1992; Dunham *et al.* 1993). For birds, similar evidence exists for the case of human tutoring of a grey parrot (Pepperberg 1981, 1990c), but is only circumstantial for the case of avian–avian tutoring. A female brown-headed cowbird (*Molothrus ater ater*) does indeed actively and continually direct the song of a male by a wing stroke display (West & King 1988), but the extent to which she may adjust her direction is not easily analyzed (see Caro & Hauser 1992). An English-speaking adult grey parrot that works with two humans to train a juvenile grey to label will often repeatedly identify a single object multiple times in a session, speaking particularly clearly and distinctly; such repetitive behavior with clarification is observed much less frequently outside of these sessions (I. M. Pepperberg, unpublished data). The intentionality of the behavior, however, cannot be proved. For birds such as zebra finches (*Taeniopygia guttata*), the effects of different types of adult–juvenile interaction and the models presented by adult male–female pairs during the juveniles' song acquisition period remain a matter for study (see e.g., Williams 1990; Williams *et al.* 1993; Mann & Slater 1994, 1995; see also Zann, Chapter 6).[2]

A second principle states that the modelling process must help the student understand how the new material relates to present circumstances and what advantage is conferred by learning the new material. Thus training is most effective when two conditions are met: (a) the student sees and then practices the targeted behavior under conditions similar to those in its regular environment, and (b) the appropriate use and consequence of the behavior are explicitly demonstrated (Bandura 1971, 1977; Brown 1976; Harris *et al.* 1986; note Schwartz & Terrell 1983; Baptista & Petrinovich 1984, 1986; Payne *et al.* 1984, 1988; Lock 1991).

A third principle states that the more intense the interaction between the student and its model(s), the more effective is the training. The intensity – i.e., the extent to which the tutor arouses a response in the student (Bandura 1977; see also Locke & Snow, Chapter 14) – can be measured from direct observations of the behavior of the interactants (e.g., emotional reactions) or from less direct measures (e.g., blood pressure or hormone levels). One implication, supported by some data reviewed by Pepperberg & Neapolitan (1988), is that, for both humans and birds, intense interaction requires one or more tutors per student (e.g., individualized instruction, a pair of interactive models (2 tutors : 1 student)). Limits may, of course, exist on the extent to which increasing the inten-

sity of an interaction increases learning: Overly nurturant or aggressive models may inhibit learning by, respectively, preventing the student from experimenting on his/her own or arousing fear or counter-aggression strong enough to block the processing of any input (for a possible example for white-crowned sparrows, see Casey & Baker 1993).

The fourth principle states that, if there exists some inhibition or resistance towards learning, the more important are the first three principles. The fourth principle is particularly relevant to *exceptional* avian communication (Pepperberg 1985, 1986a,b) because exceptional communication is usually characterized by vocal learning that, in the normal course of development, is thought unlikely to occur. For birds, exceptional learning is most often considered as (a) use of allospecific vocalizations by subjects generally expected to acquire functional use of only conspecific vocalizations (e.g., contextual use of song of unrelated species; for a description of exceptional song learning in a song sparrow in nature, see Baptista 1988 and for similar data on a Lincoln sparrow (*Melospiza lincolnii*), see Baptista *et al.* 1981), and (b) age-independent acquisition of vocalizations in species generally recognized as having a limited "sensitive phase" for vocal learning (for laboratory studies on white-crowned sparrows, see Baptista & Petrinovich 1984, 1986; Petrinovich 1988; cf. Marler 1970; for possible parallels in the wild, see Baptista 1985; Baptista & Morton 1988; for a review, see Baptista & Gaunt 1994). The term "exceptional" consequently implies that there exists some resistance toward acquiring the targeted behavior. Thus, for exceptional learning to occur, social modelling theory predicts that the tutor/model must be even more attuned to the behavior of the student, that the interactions must be even more intense, and that the demonstrations must be even more explicit as to the real world uses and consequences of the targeted behavior than in the case of normal learning. This principle has been particularly important for analyzing several instances of exceptional learning in passerine birds (Pepperberg 1985, 1986b, 1988b, 1991; Pepperberg & Neapolitan 1988; Pepperberg & Schinke-Llano 1991) and in understanding material that will be described in detail later – my research on a grey parrot's acquisition of referential, contextually applicable English vocalizations that are used to communicate with humans (e.g., Pepperberg 1990a, 1994b) and some of the studies on the acquisition of nonvocal, human-

---

[2] Although there is no evidence to suggest that adult birds intentionally teach song to juveniles or that adults take into account the level of juveniles' competence, one might easily imagine a scenario in which graded teaching occurs without intentionality or conscious awareness of another individual's state (see Caro & Hauser 1992). Imagine, for example, that something about the behavior, presence, or appearance of the juvenile unconsciously elicits singing and other behavior from the adult that differ slightly from what would be elicited by the presence of other adults, and that these behavior patterns alter in concert with the maturation of the juvenile. Post-"critical period" (>50 days) hatching-year white-crowned sparrows, for example, are brown-crowned and often "streaky" (L. Baptista, personal communication); such birds also act differently from considerably younger birds. Whether a brown-crown elicits slightly different singing and other behavior in adults and whether such behavior patterns would help the bird extend its learning period are unknown: the fine-grained analyses required for such a study have not been attempted to date.

based codes by nonhuman primates (e.g., Gardner & Gardner 1978; Savage-Rumbaugh *et al.* 1985).

SPECIFIC ASPECTS OF INPUT NECESSARY FOR EXCEPTIONAL LEARNING

In the above discussion, the relevant aspects of input that are necessary for learning via social interaction are embedded in the principles of the theory. The term "social interaction" as used above, moreover, refers not only to a social act but also to a constellation of behavioral features that influence the potency of a social act (Pepperberg 1993). Such features, although important, are not the main aspects of input that are necessary for learning. These secondary features may include the quality and quantity of input, the relative status of the interactants, the environment in which the interaction occurs, and the feedback received by the learners for their attempts at a targeted behavior.

To demonstrate the relative importance of various aspects of input for learning, one must first identify and separate these – and other – relevant aspects. From the principles of social modelling theory, one can extract three main factors (although more factors clearly exist) that subsume many of the (secondary) features I have mentioned and by which input can thus be characterized: (a) the degree to which input correlates with a specific aspect of an individual's environment (i.e., "referentiality", Smith 1991), (b) the extent to which input has functional meaning relevant to the individual's environment (known to psychologists as "contextual applicability"), and (c) the extent to which input is socially interactive. A detailed description of these factors follows immediately; a discussion of the consequences of omitting these factors from the learning situation comprises the final sections of this chapter.

Reference

Reference is, in part, what signals "are about" (Smith 1991). Reference concerns the direct relationship between a signal and an object or action. Reference is not always easily determined; e.g., when we say "match", we generally mean a specific object used to light a fire (what Smith (1991) labels an "external" referent), but we may also mean an action, as to "match" two socks. Similarly, a bird that emits an alarm call may refer to either the predator or the action it is about to take (or both). Thus, not all informa-

tion contained in a signal involves a single referent. The more explicit the referent of a signal, the more easily the signal appears to be learned.

Functionality

Functionality (contextual applicability) involves the pragmatics of signal use: when a signal is to be used and the effects of using information in the signal, i.e., when using a signal is advantageous and what advantage is accrued by using the signal. A male blue-winged warbler (*Vermivora pinus*), for example, must learn which of his songs is to be used for territorial defense and which is to be used for mate attraction (Kroodsma 1988); the proportion of different song types used by a territorial male and the context in which these songs are used appear to indicate his pairing status and would thus play a role in his reproductive success (Kroodsma 1981). Because use and effect of a signal may depend upon environmental context, functionality helps to define reference. In the above examples, context defines "match" as a noun or verb for humans. Contextual effects are also present for a bird such as the prairie warbler (*Dendroica discolor*): the so-called mate attraction song is occasionally used in territorial defense – but only in the presence of an unmated female (Nolan 1978). The more explicit a signal's functionality, the more readily the signal appears to be learned.

Social interaction

Social interaction has three major functions. First, social interaction can highlight which components of the environment should be noted ("The color is *blue*", Pepperberg 1983; neighbor versus stranger distinctions, Stoddard *et al.* 1990). Second, social interaction can emphasize common attributes – and thus possible underlying rules – of diverse actions (i.e., "*Throw* cork" versus "*Throw* corn", I. M. Pepperberg, unpublished data; the common need for evasive actions in response to different signals, Elowson & Hailman 1991). Third, social interaction can enable input to be continuously adjusted to match the level of the receiver ("*CH*ain" in response to "ain", I. M. Pepperberg, unpublished data; different song use as aggressive interactions escalate, Dabelsteen & Pedersen 1990). Interaction may also provide a contextual explanation of the reasons for an action and demonstrate its consequences (for details, see Pepperberg 1993). Interactive input thus facilitates learning.

In sum, reference and contextual applicability refer to real world use of input,[3] and social interaction highlights various components of the input. Researchers can specifically design input that varies with respect to these aspects and then evaluate the relative effects of such variation. To obtain information on the relative importance of these three aspects of input on learning in a mimetic species, I used several different conditions to train one adult and two juvenile grey parrots to produce English labels to identify various common objects. Studies to examine other aspects of input (e.g., how acquisition is affected by quantity of input, by the species identity of the model, and by input that reflects the level of competence of the student) are in progress.

THE EFFECTS OF DIFFERENT TYPES OF INPUT ON EXCEPTIONAL LEARNING IN GREY PARROTS

The juvenile grey parrots, Alo (female) and Kyaaro (male) were 10 and 6.5 months, respectively, at the beginning of the experiments; the adult bird, Alex, was about 12 years old. The study with Alex was completed before I obtained the juveniles, and the birds were isolated from one another during the experiments with the juveniles. The juveniles had had no formal training before these studies and had acquired no human vocalizations. Alex, however, had learned referential use of English labels for objects, shapes, colors, and numbers up to 6 to identify, request, refuse, and quantify objects, and had functional use of several phrases (e.g., "Come here," "You tickle," "What's that?", "Wanna go X" and "Want Y," where X and Y are location and object labels; Pepperberg 1987a, 1990a). He had been tested on abstract concepts such as the presence or absence of sameness and difference and could categorize objects with respect to color, shape, or matter (Pepperberg 1983, 1987b, 1988c). Other tests (Pepperberg 1990b) showed that he could comprehend as well as produce all his color, shape, material, and category labels.

Until the present experiments, Alex's training system

maximized reference, functionality, and social interaction. This system, called the model/rival procedure, was adapted from the work of Todt (1975). Model/rival training involves three-way interactions between two humans and the avian student. Model/rival training primarily introduces new labels and concepts but also aids in correcting pronunciation. Because the experiments described here are an in-depth comparison of training protocols, I provide details of the model/rival procedure, although the material is available elsewhere (Pepperberg 1981, 1983, 1990a,b,c, 1994a).

During model/rival training, humans demonstrate what a bird is to learn. In a session, a parrot observes two humans handling an object that the bird has previously freely selected (e.g., a preening implement, Pepperberg 1981). One human acts as a trainer, showing the item to the second human, who is both a model for the bird's responses and its rival for the trainer's attention. The trainer queries the model/rival about the item (e.g., "What matter?"), giving praise and the object to reward correct answers. A trainer shows disapproval for incorrect responses (e.g., deliberate errors) by scolding and temporarily removing the object from sight. A bird thus observes aversive consequences of errors. The model/rival is also asked to talk more clearly or try again when a response is incorrect or garbled, thereby allowing observation of "corrective feedback" (see Goldstein 1984; Vanayan et al. 1985). Because the bird's reward depends on the clarity of its response, and trainers slowly raise their criteria for reward, the protocol adjusts the level of training to the level of the bird.

By rewarding each correct identification with the item to which the label refers, the model/rival technique specifically demonstrates reference and functionality. In contrast, some programs designed to teach communication skills, for both humans and nonhumans, use unrelated rewards (see Pepperberg 1990a). In these programs, all correct responses to various commands are rewarded with a single item that neither directly relates to nor varies with the skill being taught (e.g., a subject receives an M&M (chocolate candy) whether it identifies a ball or picks up a block). Such unrelated rewards may delay learning by confounding the label or concept to be learned with some aspect of the reward item (Greenfield 1978; Pepperberg 1981; Miles 1983). My procedure, instead, provides the closest possible association of the label that is

---

[3] Note that reference and functionality cannot accurately be subsumed under the rubric "reinforcement." Reinforcement simply implies that some, not necessarily specific, positive outcome is associated with an action. Reference and functionality specify which particular positive outcome (i.e., one of several) is intimately associated with an object or action.

Table 9.1. *Components and results of tutoring for grey parrots*

	Reference	Functionality	Social interaction	Evidence for learning
Model/rival	Yes	Yes	Yes	Yes
Model/rival variant 1	No	No	Yes	Partial
Model/rival variant 2	Yes	Minimal	Minimal	No
Basic video	Yes	Partial	No	No
Video-variant 1	Yes	Partial	Minimal	No
Video-variant 2	Yes	Potential	No	No
Audiotape	No	No	No	No

being taught and the item to which it refers (Pepperberg 1981).

In addition, my protocol, unlike others (see e.g., Todt 1975; Goldstein 1984), requires repeating an interaction while reversing roles of the model/rival and human trainer and includes the parrot in the interactions. Thus, birds do not simply hear stepwise vocal duets, but observe and learn to engage in reciprocal communication that can be used by either party to request information or effect environmental change. Without role reversal, birds exhibit two types of behavior that are inconsistent with interactive, referential communication (Todt 1975): they do not transfer responses to anyone other than the human who poses the questions, and they do not learn both parts of the interaction.

Finally, the model/rival procedure also demonstrates comprehension, although to a limited extent. The bird observes one human say "Let's work on *key* today; please pick up the *key*" and the other human chooses a key from among a number of keys, corks, and pieces of paper. Thus, the bird sees that a particular type of object, and no other, is the item to which the label corresponds.

I designed the present studies to examine how various levels of reference, social interaction, and functionality might affect learning not only with respect to sound reproduction but also with respect to comprehension and appropriate use (i.e., actions that require cognitive processing). To provide input that varied with respect to these factors, I contrasted sessions of model/rival, videotape, and audiotape tutoring. Model/rival and videotape sessions could each have varying components of input (see Table 9.1).

In "model/rival-variant 1" sessions, I altered the usual model/rival procedure to eliminate as much reference and

functionality as possible (Pepperberg 1994a). Two humans enacted the same roles, but did not emphasize any connection between labels and specific objects or collections. One human asked a question and the other produced a string of labels in the absence of any objects. Correct replies garnered vocal praise and the opportunity to request anything desired (Pepperberg 1987a); errors elicited scolding and "time-outs".

In "model/rival-variant 2" sessions, the model/rival procedure was amended to eliminate some functionality and as much social interaction as possible. A single trainer sat with her back to the bird, who was seated on a perch within reach of an object (e.g., key) suspended from a pulley system. The trainer repeated various phrases about the object, e.g., "Look, a shiny *key*!", "Do you want the *key*?" etc. (see Pepperberg 1981), but did not make eye contact with the parrot nor did she ever present the object directly to the parrot. She would reward any attempt at the targeted label with vocal praise.

To provide training that closely followed the model/rival procedure but avoided social interaction and minimized contextual applicability, I used "basic video" sessions: I videotaped the adult trained parrot, Alex, during model/rival sessions and exposed the juvenile parrots to those tapes (Pepperberg 1994a). Although Alex already knew the targeted labels (Pepperberg 1990a,b), tapes did not present the targeted material as a review but followed the lines of actual training. Alex occasionally erred or interrupted with requests for other objects and changes of location (Pepperberg 1983, 1987a, 1994a) and, as in live model/rival presentations, trainers occasionally erred. The tapes also retained patterns of breaks for nonvocal exchanges (e.g., when trainers preened Alex) and trainers' departures by using, respectively, scenes of such nonvocal interactions or a blank screen. Juveniles watched the videos in isolation, so that no direct social interaction with trainers occurred. By watching a tape of a human or Alex produce a particular sound and either receive an object or be scolded, the juveniles saw but did not experience directly the effect of a vocalization. Videos, therefore, demonstrated reference but lacked clear functional meaning.

Interestingly, some studies have shown that children learn from television programs (e.g., *Sesame Street*) only when they view the programs with an *interactive* adult, who calls attention to objects ("See the flowers?"), responds to the child's questions ("[It's a] man with a balloon"), corrects the child ("No, that was a bunny"), requests information from the child ("What's that?" "What kind of boy is that?"), expands the child's utterances (Child: "Ball." Mother: "Ball. Three balls. One, two, three"), or repeats the child's utterances (Lemish & Rice 1986). Possibly the extent of interaction is critical, as not all children learned in the presence of adults (St Peters *et al.* 1989). I tested this premise in "video-variant 1." The juveniles' video sessions were repeated with "co-viewers" who merely ensured that the birds attended to the monitor. Trainers provided social approbation for viewing and pointed to the screen, making comments such as "Look what Alex has!", but did not repeat new labels, ask questions, or relate the content to other training sessions. Any attempt at the label would be rewarded with vocal praise and not the object. Thus, the amount of social interaction was limited and the amount of functional meaning was the same as in the basic videotape session.

In "video-variant 2," I ensured that lack of reward for an attempt at a targeted vocalization did not prevent learning from video. I used the basic videotape protocol, but included a reward system that enabled a parrot, in the absence of social interaction, to receive the item if it attempted to produce the label. The system was controlled by a student in another room who monitored the parrot's utterances through headphones. We audiotaped sessions to test for (inter)observer reliability.

Finally, to test the effects of total absence of reference, context, and social interaction, I exposed the juveniles to audiotapes. To ensure that these tapes were parallel to the model/rival and video procedures but eliminated all reference and context, audiotapes consisted of the audio portion of a basic videotape presentation. In these "basic audiotape" sessions, juveniles listened to the tapes in isolation, and no objects were associated in any way with the sounds presented over the speaker.

Experiments with Alo and Kyaaro

Each bird was trained under several conditions (Pepperberg 1994a). I chose labels that the aforementioned Alex could clearly produce (Pepperberg 1981, 1990a) to ensure that the vocalizations were within the capacity of the species. Each bird was trained on two labels in model/rival, basic video, video-variant 1, video-variant 2, and audiotape sessions. Neither juvenile received training on model/rival-variant 1. Kyaaro received two labels and Alo one label in model/rival-variant 2 sessions.

I counterbalanced labels, so that, with the exception of "paper" and "rock," labels used for one bird with one technique were used for the other bird with another technique. Both birds were exposed to "paper" via live tutors and to "rock" via audiotape in order to compare their speeds of learning. For each bird, I repeated one of the two labels from the basic video in the video-variant 1 condition to test for possible effects of co-viewers.

Alo's results

Alo never clearly produced, in the presence of trainers, any labels experienced via model/rival-variant 2 (key), any video (wood, nail, chalk, chain) or audio (key, rock) training. Tapes of her solitary sound productions also revealed that she did not "practice" (see Pepperberg *et al.* 1991) these labels. She failed to identify the objects or even produce an approximation of the correct labels on formal tests (I. M. Pepperberg, J. R. Naughton & P. A. Banta, unpublished data; Pepperberg & McLaughlin, 1996). In contrast, on tests on labels taught via the basic model/rival procedure, first trial scores were 34/40 for both cork and paper (details in Pepperberg 1994a).

Kyaaro's results

Kyaaro also did not produce, either in the presence of trainers or in private practice, labels he experienced via audio (wood, rock) or any form of video (cork, key, truck, bear) training. He attempted to produce labels taught via the model/rival technique (paper, nail), but, at the end of 11 months, ran them together ("ail-er") in a manner too difficult to distinguish by trainers for testing. He did, however, produce clearly differentiated versions of nail and paper during private practice. After additional months of training, his labels were at criterion for testing. On identification tests for items trained under non-(basic)model/rival procedures, Kyaaro scored zero on every trial. On tests given for labels taught via the basic model/rival procedure, his first trial scores were 34/40 for paper and 35/40 for nail (for details, see Pepperberg 1994a; Pepperberg & McLaughlin 1996).

Interestingly, in the first 11 months, Kyaaro did acquire a few extremely clear vocalizations from informal interactions with trainers: "Hi Kyo," "Want tickle," "Kiss." These utterances are often contextually appropriate; they are used, respectively, when we enter his room but ignore him (eg., during cleaning or a "time out"), while he bows his head and stretches toward our hands, and when he stretches his beak toward our faces. As Kyaaro always accepts tickles or beak rubs, such utterances cannot be tested and no claims can be made for their referentiality.

Experiments with Alex

As noted earlier, Alex had learned to produce and comprehend many labels and concepts under the basic model/rival condition. He was now taught a sequence of eight number labels that had no reference either to specific objects in the laboratory or to previously acquired English number labels. These labels were part of a study on ordinality, counting and serial learning (I. M. Pepperberg & J. Silverstone, unpublished data). The set, "il ee bam ba oo yuk chil gal," was derived from Korean count labels both to permit comparisons with children (Fuson 1988) and to be maximally different from English. "Bam" (pronounced /baem/) and "ba" were substituted for the Korean "sam" and "sa" because of Alex's occasional difficulty in producing "ss".

Alex was trained with model/rival-variant 1. We therefore did not initially model the connection between labels and the objects to which they referred. Alex did not attend to training (would request other items or to be moved to other locations), however, until we included a minimal point of reference: a piece of paper with the symbols 1–8 traced along the diagonal. The trainer held the paper in front of the model and stated, "Say number!" The model produced the altered Korean labels and was rewarded, or erred and was scolded. We routinely reversed roles of

model and trainer; Alex was also given a turn. Although we usually reward Alex either with the object that he has labeled *or* the opportunity to request a favored item ("I want X"; Pepperberg 1987a), here we used only the latter reward. Training, therefore, lacked the usual functional meaning and lacked all but minimal referentiality.

Alex learned the string of vocal labels that we modelled, but insisted on "nuk" in place of "yuk." Training took nine months, which was unusually long (Pepperberg 1981; I. M. Pepperberg & J. Silverstone, unpublished data). The most important finding, however, was that Alex could not subsequently learn to use these labels to refer to quantity: even after we modelled 1:1 correspondences between eight objects and the string of labels, he was unable to use elements in the string to refer to smaller quantities, e.g., to say "il ee bam" to three items. Alex thus learned to produce, but not comprehend these vocalizations (Pepperberg 1994a). Such behavior was all the more striking because, when taught by model/rival procedures, Alex had shown the capacity not only to answer questions by evaluating (processing) current information on the basis of some representation of prior experiences, but also to choose, from among various possible sets of rules that he has acquired, the set that appropriately governs the current processing of the data (Pepperberg 1990b). Thus, for example, without specific training, he was able to view a collection of multiple objects and answer any of a number of questions, taught under different circumstances, about these items ("How many?", "What object is color-X?", "What shape is object-Y?", "What object is bigger/smaller?," etc.). His current failure, therefore, could not be attributed to lack of cognitive skills.

Discussion and summary of the avian work

Social interaction, reference, and full functional meaning are all important factors in acquiring production and comprehension of an allospecific code, even for a mimic such as the grey parrot. Absence of some of these factors affects whether and how acquisition occurs. The effects of each condition can be described in some detail.

When all three factors are missing, as in audiotape presentations, mimetic learning does not occur. The juveniles whistled and squawked during sessions, but did not attempt to reproduce human sounds from the tape. Possibly such input was processed as background noise, corresponding to what would be environmental sounds in the wild. The birds were given no opportunity to deduce explicit meanings for the sounds and were not shown the purpose for which the sounds could be used. Their response to the sounds had no effect on what they subsequently heard or received, either vocally or physically.[4] Thus they had no reason to acquire the sounds. Even so, they might have learned the allospecific sounds from the tapes like some passerines (e.g., marsh wrens (*Cistothorus palustris*), Kroodsma & Pickert 1984a) and then either produced them at random or in connection with some irrelevant cue, or made some association between the novel sound and the novel object that was subsequently presented to them in testing. They did not, however, behave in any of these ways. Interestingly, no one has yet examined whether passerines that have learned allospecific songs from tape actually use these vocalizations in the contexts of mate attraction, mate guarding, or territorial defense; that is, whether contextual, referential use of an allospecific vocalization can occur if that vocalization is acquired without contextual, referential training. Contextual use with respect to territorial defense has, however, been documented for birds that learn allospecific songs from live tutors (e.g., song sparrow (*Melospiza melodia*), Baptista 1988).

The presence of reference and limited functional meaning in the absence of interaction (the basic video condition) is also not sufficient for psittacine vocal learning. The birds attended to the videos but did not acquire the sounds that were modelled. The juveniles could have failed to acquire vocalizations for at least three reasons: (a) they failed to realize that the interaction that they observed could be transferred to their own situation; (b) they could not determine exactly what aspect of Alex's behavior was actually causing transfer of the desired objects; or (c) they simply stopped responding to what they saw on tape because they received no encouragement for what could have been their first approximations to the targeted vocal-

ization. Alo's and Kyaaro's failure to acquire anything under variant 2 of the video condition, where attempts would be rewarded, shows that the third factor was irrelevant: no acquisition occurred that could be rewarded. Clearly, merely watching another individual receive objects for producing particular sounds provided insufficient input for acquisition. Even the presence of a trainer to maintain the birds' interest did not help.

The experiment with Alex showed that social interaction in conjunction with severely limited function and meaning engenders, at best, production but not comprehension of an allospecific code. Here Alex received positive feedback merely for making particular sounds in response to a specific cue; he was given no reason to work toward understanding either what he was saying or the appropriateness of his utterances. If acquired, such vocalizations are unlikely to be generalized to related situations. Such training represents most situations of mimetic birds that are pets, and explains why parrots were once thought incapable of doing more than randomly producing human speech sounds (see e.g., Lenneberg 1973).

Most likely, social interaction and functional use without reference will engender production without comprehension. Such is probably the case for pet birds that produce, for example, "Hello" or "Bye, bye" routines ("Good night dear," "Good-bye, and thank you"; Amsler 1947) appropriately but do not comprehend the use of the individual words in these routines. These birds may have a more general sense of the situations in which their vocalizations can be used than do birds taught without functionality, but do not fully comprehend that part of the allospecific code they have acquired.

Given the lack of learning under the model/rival-variant 2 conditions (reference, minimal functionality and minimal social interaction), I propose that grey parrots are also unlikely to acquire or comprehend elements of an allospecific code from input that is referential, fully functionally meaningful, but noninteractive. Thus, the presence or absence of an item that could be considered a reward is probably less important than the presence or absence of social interaction. Bandura (1977) stated that reinforcement cannot create novel behavior but will effectively shape behavior acquired in an approximate form or regulate behavior that has already been learned, and that, under natural conditions, people rarely learn exceptional behavior patterns they have never seen performed by

---

[4] Under some very specific conditions (birds were caged in auditory contact with conspecifics and could peck response keys for song exposure), zebra finches (for whom learning from tapes is exceptional; Price 1979) may learn vocalizations from audiotapes (Adret 1993). Not all birds learned, however, and slight changes in procedures (e.g., song duration, landing on perches rather than pecking, housing in vocal isolation) prevented learning from occurring (ten Cate 1991). The effects of these differences, particularly with respect to vocal interaction, remains unclear.

interactive models (e.g., videotapes can train a nonin-hibited behavior but not one that is inhibited; Bandura 1976). In basic model/rival training, for example, reward is not likely to be the critical factor for acquisition because a bird is rewarded only after it has attempted the targeted label; i.e., reward occurs only after some acquisition has taken place. The reward primarily reinforces referentiality.

Overall, I have sought to determine the conditions that will enable grey parrots to acquire a referential, allospecific communication code. Although mimetic birds are characterized by their extensive capacities to acquire allospecific vocalizations, grey parrots (at least) seem to learn to produce such a code most readily under certain environmental conditions and to acquire fully functional use of such a code under even more limited conditions. Although some combinations of conditions remain to be tested for these birds (e.g., reference and full functionality in the absence of social interaction; reference and limited functionality with full interaction; and the effectiveness of two- versus three-dimensional referents), input that is fully referential, functional, and socially interactive ensures that these parrots not only produce but also fully comprehend allospecific vocalizations (Pepperberg 1987a,b, 1990b, 1992, 1994a,b). Lack of some or all of these aspects will affect the course of acquisition and will probably prevent full allospecific acquisition.

All my experiments, however, involve teaching a refer-ential allospecific code to subjects in a laboratory. The question remains as to the external validity of my data: does my work relate to conditions in the real world, par-ticularly with respect to referentiality and constraints on acquisition?

No information exists on grey parrots' use of referen-tial vocalizations in the wild. Limited data on actions of other psittacine species (sentinel behavior (indigo macaws, *Anodorhynchus leari*, Yamashita 1987; Puerto Rican parrots, *Amazona vittata*, Snyder *et al.* 1987; maroon-fronted parrots *Rhynchopsittica terrisi*, Lawson & Lanning 1980; white-fronted Amazons, *Amazona albifrons*, Levinson 1980), individual recognition (Bahama Amazons, *Amazona leucocephala bahamensis*, Gnam 1988; glossy black cockatoos, *Calyptorhynchus lathami*, S. Pepper, personal communication, note Saunders 1983; galahs, *Cacatua roseicapillus*, Rowley 1980)), however, suggest that referentiality is a characteristic for which one might fruitfully search. Most likely, a grey parrot could

not acquire referential communication in the laboratory unless such behavior (see e.g., Pepperberg 1990a, 1992, 1994b) was based on a pre-existent cognitive architecture (Premack 1983, Rice 1980) involving perception, memory, and communicative intent.

If constraints found in the laboratory do indeed exist in the wild, an interesting scenario emerges (Pepperberg 1994c). Vocal learning, to be adaptive, must be both flex-ible and constrained: physically flexible to allow for the production of a range of sounds and constrained enough to be useful. Unconstrained, such learning would be mal-adaptive. What evolutionary pathway opted for wide-spread vocal learning in psittacids yet constrained it to appropriate situations? What mechanisms might be involved? An innate sound detector or filter could be too limiting and lack flexibility to change with varying environmental conditions. What if, instead, psittacine vocal learning occurred only in conjunction with cognitive choice or processing that relied upon meaning, function, and interaction (Pepperberg 1994c)?

A cognitive mechanism that works with respect to meaning and function would be simple and not very selec-tive; when social interaction is included, however, such a mechanism could be maximally adaptive. The meaning and use of most auditory signals are decoded by cognitive (information processing) mechanisms and relatively few signals would thereby be excluded as potential sources for learning. Social interaction, however, not only requires cognitive processing to assess the nature of the interaction, but also targets the input as worthy of processing. For example, a subject can, by assessing whether input is coming from a source that holds a dominant position, decide whether status or acceptance might be gained through imitation (Bandura 1977; see also Brown & Farabaugh, Chapter 7). In contrast, nonsocial background sounds (e.g., rustling of leaves) might be processed for their information content (the approach of a predator), and one might imagine a functional use for reproducing such sounds (deceiving a sympatric bird into taking cover and missing out on a food source), but the lack of social interaction would denote learning of such signals as mal-adaptive (i.e., rustling may have many implications) when compared to, for example, learning actual alarm calls (see Munn 1986).

In conclusion, determining how a particular species acquires and uses an allospecific communication code is not a simple matter. Might additional data weaken or strengthen the hypothesis that cognitive processing and social interaction are critical factors in such learning? Interestingly, data on nonvocal allospecific learning in nonhuman primates provide some support for this thesis.[5] Because the existence of advanced cognitive capacities have been demonstrated for nonhuman primates (e.g., Premack 1983), I concentrate on the effects of social interaction.

SOCIAL INTERACTION AND ACQUISITION OF NONVOCAL, HUMAN-BASED COMMUNICATION CODES BY NONHUMAN PRIMATES

Although the extent to which nonhuman primates can acquire a communication code isomorphic with that of human language is unresolved (see Terrace *et al.* 1979; Savage-Rumbaugh *et al.* 1980a,b; Greenfield & Savage-Rumbaugh 1991), the ability of such animals to acquire elements of a human-based code (e.g., the ability to label, request, refuse, and categorize objects and some actions and to comprehend some elements of the code) has been well documented (see Gardner & Gardner 1978; Savage-Rumbaugh *et al.* 1985). Not all programs to establish two-way communication with apes have, however, obtained comparable results. Animals in programs that relied most heavily on training related to operant conditioning (e.g., Premack 1976; Rumbaugh 1977; Terrace 1979a) demonstrated behavior that was far less flexible and less "language-like" (note Terrace 1979b) than apes trained within systems that more closely resembled the experiences of young children (Savage-Rumbaugh *et al.* 1985; Gardner & Gardner 1989). I analyze, below, some of these programs with respect to the aspects of input that I have shown to be important for establishing two-way vocal communication with birds; I then discuss the effect of social and environmental factors on establishing two-way communication with apes. In no sense is this discussion meant to be a detailed critique of the various chimpanzee studies; many other such critiques exist (e.g., Hill, 1995) and many other animals have been studied in addition to those I mention. The point of this section is merely to compare samples of the different types of input received

by some subjects and the level of communication that these subjects achieved.

Description of the nonhuman primate projects: Training with respect to reference, context, and social interaction

Programs designed to establish two-way communication with apes used different specific training techniques, even when the actual code being taught (e.g., American Sign Language, ASL) was identical. The conditions under which chimpanzees were taught human-based codes ranged from computer-driven systems with limited social interaction (Rumbaugh 1977) to cross-fostering programs in which the ape was raised in a situation much like that of a human child (Gardner & Gardner 1989; Savage-Rumbaugh 1991). Programs thus varied in the extent to which their techniques were referential, contextually applicable, and socially interactive. These differences ultimately affected some aspects of the animals' use of their acquired code, although all the animals acquired some or all of the capacities noted above.

Reference

The first goal of all the ape language projects was to establish some form of object–label association. In studies by Gardner & Gardner (1969, 1989), chimpanzees such as Washoe were presented with a particular exemplar (eg., one of several balls in their environment) and the ASL sign for BALL; she was then encouraged (often by molding her hand positions) to produce the sign appropriately. Her attention was also drawn to commonplace events and objects in a simple, succinct manner (THAT CHAIR). Similarly, in studies with the bonobo Kanzi, Savage-Rumbaugh (1991) has emphasized the use of particular symbols in particular contexts, although there appeared to be some initial conflation between the label for the fixed location where an object is to be found and the name of the object itself: the bonobo initially "may recognize and use the M&M symbol on the way to Flatrock [*the primary location where M&Ms were available*], but seems confused if the symbol is used during a game of tickle in the group room." (Savage-Rumbaugh, 1991, p. 221, bracketed material mine; note Savage-Rumbaugh 1987; Seidenberg & Petitto 1987). In contrast, Terrace's (1979b) regime for Nim often was to show him "how to make a well-formed

---

[5] For studies on cetaceans, see McCowan & Reiss, Chapter 10.

sign outside the context in which that sign was to be used (*ibid.*, p. 51)" so that "he would learn the mechanics of signing without being excited or distracted by the object referred to (*ibid.*, p. 52)." Nim's caregivers also signed about the objects and events in his environment, although possibly not in the most referential manner: Foods were identified before Nim could eat them, but while he was eating, his trainers signed EAT and not necessarily the labels of the foods he was in the process of eating. In the Rumbaugh (1977) study, Lana initially interacted with a computer that provided her with exactly the item or action she requested through a button press – i.e., hitting the sequence PLEASE MACHINE GIVE M&M provided her with a piece of chocolate candy, whereas PLEASE MACHINE GIVE RAISIN produced the dried fruit. However, in the initial stages of this study, no controls existed for determining the extent to which Lana actually intended to obtain the object she requested via computer. Thus, for example, the extent to which her symbol for M&M actually referred to the candy remained unknown. Premack's (1976) chimpanzees (Peony, Walnut, Elizabeth and Sarah) learned through a combination of errorless and choice trials. In the former, they were shown a piece of fruit and a magnetic-backed plastic geometric shape and obtained the fruit by placing the shape on a magnetic board. In the latter, they were shown two symbols and received the fruit for choosing the correct symbol. Sarah ultimately learned the correct associations via errorless trials, but learning was slow. In sum, reference was not clearly and consistently demonstrated to the same extent in all of the four projects.

Functionality

Functionality, as noted above, helps to define reference and for the animal subject defines the pragmatics of a symbol: how and when a signal should be used. In most cases, the chimpanzees learned that use of a specific symbol in a specific context provided a specific reward: the chance to eat the M&M or obtain a piece of apple. In the case of Nim, training did not always involve a precise explication of the functionality of the ASL signs he was to learn; thus CHAIR was signed before he was put in his highchair, but the sign could have meant any of a number of things associated with the entire procedure (e.g., being picked up, being put down in a seat). In the Premack and Rumbaugh studies, functionality was explicit, but generally only for a

very limited set of circumstances: little choice existed as to when a symbol should be used. Such was particularly the case in the Premack study, in which the choice of symbols was severely limited in each trial. Thus true pragmatics were part of training only in situations in which the chimpanzees were exposed to their human-based codes in situations resembling that of young human children; for example, when they saw the symbol MORE used both for more tickles and more of a specific food.

Social interaction

The extent of any form of social interaction differed dramatically among the projects, but even more dramatic was the extent to which the chimpanzees could observe the use of the code to be learned in a social setting; i.e., the extent to which they could base their behavior on models. Initially, in the Rumbaugh study, Lana was to have little interaction with the experimenters to avoid disrupting her strict training schedule. Technicians did, however, "'model' the correct behavior, taking her finger and pressing the correct key with it or pointing to the appropriate key or set of keys. They would also verbally admonish her for pressing the wrong key." (Gill & Rumbaugh, 1977, p. 159). Such interaction, however, was limited. Premack's Sarah received little modelling; she was, instead, predominantly given errorless trials during which her trainers rewarded her responses with appropriate objects (e.g., apple for correct placement of the appropriate symbol). Nim received different levels of interaction and modelling in different circumstances. While in the classroom, he was generally subjected to drills, often with trainers who were not fluent in sign; outside of the laboratory (and later in a restructured laboratory setting), he was able to observe interactions between more fluid and fluent signers. In contrast, the Gardners' cross-fostered chimpanzees constantly observed their human trainers signing to one another about their environment, and saw how humans answered their questions, complied with their requests, and expanded upon and corrected their dialogue (Gardner & Gardner 1989). Of particular interest is the case of Savage-Rumbaugh's bonobo, Kanzi, who spent approximately two years observing his mother's training (Savage-Rumbaugh *et al.* 1985). He watched as his mother used a keyboard to request different objects and actions; for example, he saw his mother press a key to obtain food that he would often steal; he saw the situations in which

Table 9.2. *Components and results of tutoring for primates*

	Reference	Functionality	Social modelling and interaction	Evidence for learning
Nim	Partial	Partial	Limited	Partial
Lana, Sarah	Yes	Partial	Limited	Partial
Washoe *et al.*, Kanzi	Yes	Yes	Yes	Yes

keyboard use was appropriate; he observed many inter-actions between his mother and her caregivers. Simply put, Kanzi experienced what may be the most effective form of modelling for the acquisition of interspecies communication: modelled interactions between the two relevant species (I. M. Pepperberg, unpublished data).

Analysis of the nonhuman primate projects

In no case did training in any of the nonhuman primate projects fail to incorporate some level of reference, functionality, and social interaction. However, the extent to which these aspects of input were included varied considerably. Moreover, the outcomes of the various studies are correlated with the extent to which these three aspects of training were emphasized (see Table 9.2). I discuss, below, the outcomes with respect to the simpler aspects of acquisition, that of production and comprehension of single labels; a discussion of combinatorial acquisition is beyond the scope of this paper.

Intermittent reference, intermittent functionality, and limited social modelling

This level of input describes that received by Nim, whose training often consisted of rote language drill. At the end of the project, Terrace (1979b) stated that Nim could produce 125 signs in appropriate contexts and that he comprehended 200 signs. However, the results of actual production and comprehension tests are not presented, and an analysis of a videotape by Savage-Rumbaugh & Sevcik (1984) suggests that Nim did not have receptive comprehension of some of these signs. Moreover, transcripts of Nim's laboratory productions (Sanders 1985) showed that approximately 44% of his utterances were simple repetitions of a previous utterance (see also Terrace *et al.* 1979). Nim's performance, however, may have been affected by the conditions of testing. When mothers attempt to direct the

attention of their children and elicit naming, the utterances produced are more simplistic than when the children are engaged in routine conversation (e.g., about their ongoing activities; Brown 1973; note Tomasello 1992 for such effects with respect to acquisition). Similar results have been seen for Nim: apparently, he would engage in more turn-taking discourse, make more spontaneous contributions, interrupt less, and give more novel (i.e., non-imitative) responses when placed in a social/play situation outside of the laboratory (O'Sullivan & Yeager 1989). No comparisons of data were made between the laboratory and nonlaboratory situation with respect to competence in comprehension or production, however, so that the actual extent of Nim's learning remains unclear.

Full reference, limited functionality, and limited social modelling

This level of input described that of Lana and Sarah. Lana was trained through a mostly noninteractive technique to produce symbols based on human language to answer specific questions or make requests (e.g., to use the symbol representing "M&M" to obtain the candy). She was, however, subsequently unable to comprehend the signal; that is, she could not choose M&M in the presence of the symbol (see Savage-Rumbaugh *et al.* 1980a,b). Although Sarah did appear to pass tests of comprehension as well as production (Premack 1976), all of the problems presented to her during a testing session were of the same type, generally only two alternatives were available for a response, and these alternatives were restricted to the appropriate category of answer (Premack 1976; Terrace 1979a). First-trial responses were rarely reported, and, given the 50% value of chance, the results are, on occasion, not statistically valid (P values greater than 0.05). Thus the actual abilities of these animals concerning their overall use of symbols may be more limited than was originally stated by the researchers.

Full references, full functionality, full social interaction

This level of input describes that experienced by the Gardners' cross-fostered chimpanzees (see e.g., Gardner & Gardner 1989) and Savage-Rumbaugh's bonobo (Savage-Rumbaugh et al. 1985; Greenfield & Savage-Rumbaugh 1993). Data presented on Kanzi demonstrate productive use and comprehension of labels (Savage-Rumbaugh et al. 1985; Savage-Rumbaugh 1987; but see Seidenberg & Petitto 1987). Positive evidence exists for label comprehension by several chimpanzees in the Gardners' study (Van Cantfort et al. 1989; Gardner et al. 1992; see also Fouts et al. 1976); the data are from an indirect form of testing similar to that used with a grey parrot (Pepperberg 1990b, 1992). Considerable data also exist for labelling capacity (Gardner & Gardner 1984). Moreover, rather than engaging in the word-for-word repetitions common in Nim's utterances (Sanders 1985), the Gardners' chimpanzees (Drumm et al. 1986) and Kanzi (Greenfield & Savage-Rumbaugh 1993) appeared to use partial repetitions of others' symbols, as do children (Casby 1986), to fulfil a variety of pragmatic discourse functions: express agreement, establish a shared topic, jointly focus attention.

PREDICTIONS FROM SOCIAL MODELLING THEORY

In sum, subjects exposed to input that is not fully referential, that does not provide adequate information on the functional use of the code that is being taught, and that fails to provide adequate social input (particularly modelled interactions) are not likely to acquire full competence in an allospecific communication code. A subject whose input lacks explicit, consistent reference will probably acquire some rote ability to emit the elements of the code but will be unable to use the code productively: the subject may lack both referential productive skills and true comprehension. A subject whose input lacks functionality will be unaware of the pragmatics of the code; that is, will lack knowledge of when and how to use a symbol, particularly in novel situations. A subject whose input lacks social interaction is not likely to acquire any aspects of the code: the subject will be unable to determine what is supposed to be acquired, have no motivation

to acquire the code, or, never having seen the code in use, not understand its function as a means of communication.[6] Input that combines only a few aspects or various limited aspects of referentiality, functionality, or social interaction will also lead to incomplete acquisition. For example, a subject that receives multimodal stimulation (e.g., an object (vision) and a sound (auditory)), attentional focusing (e.g., a point) and reinforcement (e.g., a favorite food for producing the sound) is not likely to acquire referential communication. The exchange provides only incomplete reference, partial interaction and confounded functionality; the subject may learn to produce the sound, but will associate it with the food reward rather than the targeted object. Such predictions are upheld by data from both avian and nonhuman primate subjects.

Clearly, this analysis has not examined the effects of all permutations of possible types of input, nor of the effects of limiting the various features of what constitutes ideal input. I have not, for example, contrasted in detail the effects of presenting input just slightly beyond the current level of the subject versus input that is far beyond its level (but see Pepperberg et al. 1991), nor have I examined the effects of varying other features of quality and quantity of input (e.g., clarity of input, number of exposures to the targeted behavior). My findings do suggest, however, that the acquisition of interspecies communication can greatly be facilitated by input that is well designed and that serves to maximize referentiality, functionality, and social interaction.

I have shown how social input affects exceptional learning in both birds and nonhuman primates; given that social modelling theory was first designed on the basis of human studies, much comparable data exist on parallels with human learning as well. The study of analogous behavior patterns in various animal species and humans has provided important insights into many cognitive and communicative processes (e.g., Premack 1976; Pepperberg

[6] Lenneberg purportedly recreated Premack's work with Sarah using college students, but I cannot locate the reference. Supposedly, the input was referential and contextually applicable but noninteractive, and although the students succeeded quite well on all their tasks, none realized afterwards that they had been taught a "language." How one might, in fact, go about teaching a subject that what is being taught is indeed a communication code in the total absence of interaction is not immediately obvious.

1983, 1990b,c, 1992, 1994b; Gardner & Gardner 1984, 1989; Griffin 1985; Greenfield & Savage-Rumbaugh 1993). Only recently, however, have researchers begun to examine parallels in the types of social constraints that may affect exceptional behavior in animals (e.g., Baptista & Petrinovich 1984, 1986; Pepperberg 1985, 1986a,b, 1988a,b; Neapolitan et al. 1988; Pepperberg & Neapolitan 1988; Zentall & Galef 1988; Pepperberg & Schinke-Llano 1991). Clearly, if social constraints are to be used to make predictions of behavior, further investigations are necessary into the adaptive significance of the effects of reference, functionality, and interaction on learning, the relative impact of various types of social interaction on learning, and the specific mechanisms whereby social input affects learning.

Moreover, because species-specific variations exist in the extent to which a behavior is exceptional, the degrees of social interaction, referentiality and contextual applicability necessary and sufficient for development may also vary across species and task; such differences must be examined in detail. For example, for an avian species with a sizeable repertoire, allospecific learning might be less exceptional than for a species with a repertoire as limited as that of the white-crowned sparrow: the extensive learning capacity required for a species to acquire a large conspecific repertoire might be correlated with an overall reduced selectivity toward what is learned, and allospecific learning might thus require less referentiality, contextual applicability, or interaction. In fact, when marsh wrens (birds with a large repertoire for whom allospecific learning in nature is unknown) were raised as a group with sedge wrens (*Cistothorus platensis*), tape-tutored with a mixture of marsh wren, sedge wren, Bewick's wren and swamp sparrow songs, and placed in isolation before they could hear the live sedge wren song, the marsh wrens acquired mostly marsh wren songs but did incorporate limited numbers of allospecific vocalizations (Kroodsma & Pickert 1984a). A correlation does, however, seem to exist between (a) the extent to which a behavior is indeed exceptional and (b) the degree of interaction, referentiality, and contextual applicability of the input needed to facilitate learning. When a behavior is *not* exceptional, interaction, referentiality and contextual applicability may simply facilitate or modify the course of development (e.g., the extent to which a male cowbird shifts to potent songs while singing to a female; West & King 1988); when a behavior *is*

exceptional, the type of input can determine whether learning will occur at all (e.g., for white-crowned sparrows, see Baptista & Petronovich 1984). Knowledge of these matters is likely to provide information applicable not only to the matters touched upon in this volume but also to the most general theories of learning.

ACKNOWLEDGEMENTS

The research reported here has been supported by NSF grants BNS 88–20098, 91–96066, IBN 92–21941, REU supplements, the University of Arizona Undergraduate Biology Research Program, and contributions to the Alex Foundation. Some of the research with Alex was presented as a senior thesis on serial learning by Jayme Silverstone at Northwestern University. I thank numerous undergraduates at Northwestern University and the University of Arizona for acting as trainers, Russ and Madonna LaPell of VIP Aviaries for donating Alo and Kyaaro, Gloria Dolan of Emerald Bird Caddy for donating equipment, and Pamela Banta, Martine Hausberger, Mary McLaughlin, and Chuck Snowdon for their constructive criticisms of earlier drafts of the manuscript.

REFERENCES

Adret, P. (1993). Vocal learning induced with operant techniques: An overview. *Netherlands Journal of Zoology* **43**: 125–42.

Amsler, M. (1947). An almost human grey parrot. *Aviculture Magazine* **53**: 68–9.

Bandura, A. (1971). Analysis of modeling processes. In *Psychological Modeling*, ed. A. Bandura, pp. 1–62. Aldine-Atherton, Chicago.

Bandura, A. (1976). Effecting change through participant modeling. In *Counseling Methods*, ed. J. D. Krumboltz & C. E. Thoresen, pp. 248–65. Holt, Rinehart & Winston, New York.

Bandura, A. (1977). *Social Modeling Theory*. Aldine-Atherton, Chicago.

Baptista, L. F. (1983). Song learning. In *Perspectives in Ornithology*, ed. A. H. Brush & G. A. Clark, Jr, pp. 500–6. Cambridge University Press, Cambridge.

Baptista, L. F. (1985). The functional significance of song sharing in the white-crowned sparrow. *Canadian Journal of Zoology* **63**: 1741–52.

Baptista, L. F. (1988). Imitations of white-crowned sparrow songs by a song sparrow. *Condor* **90**: 486–9.

Baptista, L. F. & Gaunt, S. L. L. (1994). Advances in studies of avian sound communication. *Condor* **96**: 817–30.

Baptista, L. F. & Morton, M. L. (1988). Song learning in montane white-crowned sparrows: From whom and when. *Animal Behaviour* **36**: 1753–64.

Baptista, L. F., Morton, M. L. & Pereyra, M. E. (1981). Interspecific song mimesis by a Lincoln sparrow. *Wilson Bulletin* **93**: 265–7.

Baptista, L. F. & Petrinovich, L. (1984). Social interaction, sensitive phases, and the song template hypothesis in the white-crowned sparrow. *Animal Behaviour* **32**: 172–81.

Baptista, L. F. & Petrinovich, L. (1986). Song development in the white-crowned sparrow: Social factors and sex differences. *Animal Behaviour* **34**: 1359–71.

Bedrosian, J. L., Wanska, S. K., Sykes, K. M., Smith, A. J. & Dalton, B. M. (1988). Conversational turn-taking violations in mother–child interaction. *Journal of Speech and Hearing Research* **31**: 81–6.

Brown, I. (1976). Role of referent concreteness in the acquisition of passive sentence comprehension through abstract modeling. *Journal of Experimental Child Psychology* **22**: 185–99.

Brown, R. (1973). *A First Language.* Harvard University Press, Cambridge, MA.

Caro, T. M. & Hauser, M. D. (1992). Is there teaching in nonhuman animals? *Quarterly Review of Biology* **67**: 151–74.

Casby, M. W. (1986). A pragmatic perspective of repetition in child language. *Journal of Psycholinguistic Research* **15**: 127–40.

Casey, R. M. & Baker, M. C. (1993). Aggression and song development in white-crowned sparrows. *Condor* **95**: 723–8.

Cazden, C. B. (1988). Environmental assistance revisited: Variation and functional equivalence. In *The Development of Language and Language Researchers: Essays in Honor of Roger Brown*, ed. F. S. Kessel, pp. 281–97. Erlbaum, Hillsdale, NJ.

Dabelsteen, T. & Pedersen, S. B. (1990). Song and information about aggressive responses of blackbirds, *Turdus merula*: Evidence from interactive playback experiments with territory owners. *Animal Behaviour* **40**: 1158–68.

Drumm, P., Gardner, B. T. & Gardner, R. A. (1986). Vocal and gestural responses of cross-fostered chimpanzees. *American Journal of Psychology* **99**: 1–29.

Dunham, P. J., Dunham, F. & Curwin, A. (1993). Joint-attentional states and lexical acquisition at 18 months. *Developmental Psychology* **29**: 827–31.

Elowson, A. M. & Hailman, J. P. (1991). Analysis of complex variation: Dichotomous sorting of predator-elicited calls of the Florida scrub jay. *Bioacoustics* **3**: 295–320.

Fey, M. E. (1986). *Language Intervention with Young Children.* College-Hill Press, San Diego, CA.

Fouts, R. S., Chown, G. & Goodin, L. (1976). Transfer of signed responses in American sign language from English stimuli to physical object stimuli by a chimpanzee. *Learning and Motivation* **7**: 458–75.

Furrow, D. & Nelson, K. (1986). A further look at motherese: A reply to Gleitman, Newport, and Gleitman. *Journal of Child Language* **13**: 163–76.

Fuson, K. C. (1988). *Children's Counting and Concepts of Numbers.* Springer-Verlag, New York.

Gardner, B. T. & Gardner, R. A. (1984). A vocabulary test for chimpanzees. *Journal of Comparative Psychology* **98**: 381–404.

Gardner, R. A. & Gardner, B. T. (1969). Teaching sign language to a chimpanzee. *Science* **165**: 664–72.

Gardner, R. A. & Gardner, B. T. (1978). Comparative psychology and language acquisition. In *Psychology: The State of the Art. Annals of the New York Academy of Sciences* **309**: 37–76.

Gardner, R. A. & Gardner, B. T. (1989). Early signs of language in cross-fostered chimpanzees. *Human Evolution* **4**: 337–65.

Gardner, R. A., Van Cantfort, T. E. & Gardner, B. T. (1992). Categorical replies to categorical questions. *American Journal of Psychology* **105**: 25–57.

Gill, T. V. & Rumbaugh, D. M. (1977). Training strategies and tactics. In *Language Learning by a Chimpanzee: The Lana Project*, ed. D. M. Rumbaugh, pp. 157–62. Academic Press, New York.

Gleitman, L. R., Newport, I. L. & Gleitman, H. (1984). The current status of the motherese hypothesis. *Journal of Child Language* **11**: 43–79.

Gnam, R. (1988). Preliminary results on the breeding biology of Bahama amazon. *Parrot Letter* **1**: 23–6.

Goldfield, B. A. (1993). Noun bias in maternal speech to one-year olds. *Journal of Child Language* **20**: 85–99.

Goldstein, H. (1984). The effects of modeling and corrected practice on generative language and learning of preschool children. *Journal of Speech and Hearing Disorders* **49**: 389–98.

Greenfield, P. M. (1978). Developmental processes in the language learning of child and chimp. *Behavioral and Brain Sciences* **4**: 573–4.

Greenfield, P. M. & Savage-Rumbaugh, E. S. (1991). Imitation, grammatical development, and the invention of protogrammar by an ape. In *Biological and Behavioral Determinants of Language Development*, ed. N. A. Krasnegor, D. M. Rumbaugh, R. L. Schiefelbusch & M. Studdart-Kennedy, pp. 235–58. Erlbaum, Hillsdale, NJ.

Greenfield, P. M. & Savage-Rumbaugh, E. S. (1993). Comparing communicative competence in child and chimp: The pragmatics of repetition. *Journal of Child Language* **20**: 1–26.

Griffin, D. R. (1985). The cognitive dimensions of animal communication. In *Experimental Behavioral Ecology and Sociobiology*, ed. B. Hölldobler & M. Lindauer, pp. 471–482. Fischer-Verlag, New York.

Harris, M. (1993). *Language Experience and Early Development: From Input to Uptake*. Erlbaum, Hillsdale, NJ.

Harris, M., Jones, D., Brookes, S. & Grant, J. (1986). Relations between non-verbal context of maternal speech and rate of language development. *British Journal of Developmental Psychology* **4**: 261–8.

Hill, J. H. (1995). Do apes have language? In *Research Frontiers in Anthropology*, ed. C. R. Ember, M. Ember & P. N. Peregrine, pp. 1–19. Prentice Hall, Englewood Cliffs, NJ.

King, A. P. & West, M. J. (1983). Epigenesis of cowbird song – A joint endeavor of males and females. *Nature* **305**: 704–6.

King, A. P. & West, M. J. (1989). Presence of female cowbirds (*Molothrus ater ater*) affects vocal imitation and improvisation in males. *Journal of Comparative Psychology* **103**: 39–44.

Krashen, S. D. (1980). The input hypothesis. In *Current Issues in Bilingual Education*, ed. J. E. Alatis, pp. 168–80. Georgetown University Press, Washington, DC.

Krashen, S. D. (1982). *Principles and Practices in Second Language Acquisition*. Pergamon Press, Oxford.

Kroodsma, D. E. (1974). Song learning, dialects, and dispersal in the Bewick's wren. *Zeitschrift für Tierpsychologie* **35**: 352–80.

Kroodsma, D. E. (1978). Aspects of learning in the ontogeny of bird song: where, from whom, when, how many, which, and how accurately? In *The Development of Behavior: Comparative and Evolutionary Aspects*, ed. G. M. Burghardt & M. Bekoff, pp. 215–30. Garland, New York.

Kroodsma, D. E. (1981). Geographical variation and function of song types in warblers (Parulidae). *Auk* **98**: 743–51.

Kroodsma, D. E. (1988). Song types and their use: Developmental flexibility of the male blue-winged warbler. *Ethology* **79**: 235–47.

Kroodsma, D. E. & Pickert, R. (1984a). Repertoire size, auditory templates, and selective vocal learning in songbirds. *Animal Behaviour* **32**: 395–99.

Kroodsma, D. E. & Pickert, R. (1984b). Sensitive phases for song learning: Effects of social interaction and individual variation. *Animal Behaviour* **32**: 389–94.

Kuczaj, S. A. II (1982a). On the nature of syntactic development. In *Language development: Syntax and Semantics*, ed. S. A. Kuczaj, pp. 37–71. Erlbaum, Hillsdale, NJ.

Kuczaj, S. A. II (1982b). Language play and language acquisition. In *Advances in Child Development and Behavior*, ed. H. Reese, pp. 197–233. Academic Press, New York.

Kuczaj, S. A. II (1983). *Crib Speech and Language Play*. Springer-Verlag, New York.

Kuczaj, S. A. II (1986). Discussion: On social interaction as a type of explanation of language development. *British Journal of Developmental Psychology* **4**: 289–99.

Laws, J. M. (1994). The effect of social interaction on song development in yearling lazuli buntings (*Passerina amoena*). Paper presented at the June 1994 North American Ornithological Conference, Missoula, MT.

Lawson, R. W. & Lanning, D. V. (1980). Nesting and status of the maroon-fronted parrot (*Rhynchopsitta terrisi*). In *Conservation of New World Parrots*, ed. R. F. Pasquier, pp. 385–92 ICBP Tech. Publ. no. 1.

Lemish, D. & Rice, M. L. (1986). Television as a talking picture book: A prop for language acquisition. *Journal of Child Language* **13**: 251–74.

Lenneberg, E. H. (1973). Biological aspects of language. In *Communication, Language, and Meaning*, ed. G. A. Miller, pp. 49–60. Basic Books, New York.

Levinson, S. T. (1980). The social behavior of the white-fronted Amazon (*Amazona albifrons*). In *Conservation of New World parrots*. ed. R. F. Pasquier, pp. 403–17. ICBP Tech. Publ. no. 1.

Lippitt, R., Polansky, N. & Rosen, S. (1952). The dynamics of power. *Human Relations* **5**: 37–64.

Lock, A. (1991). The role of social interaction in early language development. In *Biological and Behavioral Determinants of Language Development*, ed. N. S. Krasnegor, D. M. Rumbaugh, R. L. Schiefelbusch & M. Studdert-Kennedy, pp. 287–300. Erlbaum, Hillsdale, NJ.

Macphail, E. M. (1982). *Brain and Intelligence in Vertebrates*. Clarendon Press, Oxford.

Macphail, E. M. (1987). The comparative psychology of intelligence. *Behavioral and Brain Sciences* **10**: 645–95.

Mann, N. I. & Slater, P. J. B. (1994). What causes young male zebra finches, *Taeniopygia guttata*, to choose their father as song tutor? *Animal Behaviour* **47**: 671–7.

Mann, N. I. & Slater, P. J. B. (1995). Song tutor choice by zebra finches in aviaries. *Animal Behaviour* **49**: 811–20.

Marler, P. (1970). A comparative approach to vocal learning: Song development in white-crowned sparrows. *Journal of Comparative and Physiological Psychology* **71**: 1–25.

Marler, P. & Peters, S. (1988a). Sensitive periods for song acquisition from tape recordings and live tutors in the swamp sparrow, *Melospiza georgiana*. *Ethology* **77**: 76–84.

Marler, P. & Peters, S. (1988b). The role of song phonology and syntax in vocal learning preferences in the song sparrow, *Melospiza melodia. Ethology* **77**: 125–49.

Masur, E. F. (1988). Infants' imitation of novel and familiar behaviors. In *Social learning: Psychological and Biological Perspectives*, ed. T. R. Zentall & B. G. Galef, Jr, pp. 301–18. Erlbaum, Hillsdale, NJ.

Menyuk, P., Liebergott J. W. & Schultz, M. C. (1995). *Early Language Development in Full-term and Premature Infants*. Erlbaum, Hillsdale, NJ.

Miles, H. L. (1983). Apes and language. In *Language in Primates*, ed. J. de Luce & H. T. Wilder, pp. 43–61. Springer-Verlag, New York.

Mischel, W. & Liebert, R. M. (1967). The role of power in the adoption of self-reward patterns. *Child Development* **38**: 673–83.

Mowrer, O. H. (1954). A psychologist looks at language. *American Psychologist* **9**: 660–94.

Mowrer, O. H. (1958). Hearing and speaking: An analysis of language learning. *Journal of Speech and Hearing Disorders* **23**: 143–52.

Mundinger, P. (1979). Call learning in the Carduelinae: Ethological and systematic considerations. *Systematic Zoology* **28**: 270–83.

Munn, C. A. (1986). The deceptive use of alarm calls by sentinel species in mixed-species flocks of neotropical birds. In *Deception: Perspectives on Human and Nonhuman Deceit*, ed. R. W. Mitchell & N. S. Thompson, pp. 169–75. SUNY Press, Albany, NY.

Neapolitan, D. M., Pepperberg, I. M. & Schinke-Llano, L. (1988). Second language acquisition: Possible insights from how birds acquire song. *Studies in Second Language Acquisition* **10**: 1–11.

Nolan, V., Jr (1978). The ecology and behavior of the Prairie Warbler *Dendroica discolor. Ornithological Monographs* **26**.

O'Sullivan, C. & Yeager, C. P. (1989). Communicative context and linguistic competence: The effects of social setting on a chimpanzee's conversational skill. In *Teaching Sign Language to Chimpanzees*, ed. R. A. Gardner, B. T. Gardner & T. E. Van Cantfort, pp. 269–79. SUNY Press, Albany, NY.

Payne, R. B. (1978). Microgeographic variation in songs of splendid sunbirds *Nectarinia coccinigaster*: Population phenetics, habitats, and song dialects. *Behaviour* **65**: 282–308.

Payne, R. B. (1981). Song learning and social interaction in indigo buntings. *Animal Behaviour* **29**: 688–97.

Payne, R. B. (1982). Ecological consequences of song sharing: Breeding success and intraspecific song mimicking in indigo buntings. *Ecology* **63**: 401–11.

Payne, R. B. (1983). The social context of song mimicry: Song-matching dialects in indigo buntings (*Passerina cyanea*). *Animal Behaviour* **31**: 788–805.

Payne, R. B. & Groschupf, K. D. (1984). Sexual selection and interspecific competition: A field experiment on territorial behavior of nonparental finches (*Vidua* spp.). *Auk* **101**: 140–5.

Payne, R. B., Payne, L. L. & Doehlert, S. M. (1984). Interspecific song learning in a wild chestnut-sided warbler. *Wilson Bulletin* **96**: 292–4.

Payne, R. B., Payne, L. L. & Whitesell, S. (1988). Interspecific learning and cultural transmission of song in house finches. *Wilson Bulletin* **100**: 667–70.

Pepperberg, I. M. (1981). Functional vocalizations by an African grey parrot (*Psittacus erithacus*). *Zeitschrift für Tierpsychologie* **55**: 139–60.

Pepperberg, I. M. (1983). Cognition in the African grey parrot: Preliminary evidence for auditory/vocal comprehension of the class concept. *Animal Learning and Behavior* **11**: 179–85.

Pepperberg, I. M. (1985). Social modeling theory: A possible framework for understanding avian vocal learning. *Auk* **102**: 854–64.

Pepperberg, I. M. (1986a). Acquisition of anomalous communicatory systems: Implications for studies on interspecies communication. In *Dolphin Cognition and Behavior: A Comparative Approach*, ed. R. J. Schusterman, J. A. Thomas & F. G. Wood, pp. 289–302. Erlbaum, Hillsdale, NJ.

Pepperberg, I. M. (1986b). Sensitive periods, social interaction, and song acquisition: The dialectics of dialects? *Behavioral and Brain Sciences* **9**: 756–7.

Pepperberg, I. M. (1987a). Evidence for conceptual quantitative abilities in the African grey parrot: Labeling of cardinal sets. *Ethology* **75**: 37–61.

Pepperberg, I. M. (1987b). Acquisition of the same/different concept by an African grey parrot (*Psittacus erithacus*): Learning with respect to color, shape, and material. *Animal Learning and Behavior* **15**: 423–32.

Pepperberg, I. M. (1988a). An interactive modeling technique for acquisition of communication skills: Separation of "labeling" and "requesting" in a psittacine subject. *Applied Psycholinguistics* **9**: 59–76.

Pepperberg, I. M. (1988b). The importance of social interaction and observation in the acquisition of communicative competence: Possible parallels between avian and human learning. In *Social Learning: Pscyhological and Biological Perspectives*, ed. T. R. Zentall & B. G. Galef, Jr, pp. 279–99. Erlbaum, Hillsdale, NJ.

Pepperberg, I. M. (1988c). Comprehension of "absence" by an African grey parrot: Learning with respect to questions of same/different. *Journal of the Experimental Analysis of Behavior* 50: 553–64.

Pepperberg, I. M. (1990a). Some cognitive capacities of an African grey parrot (*Psittacus erithacus*). In *Advances in the Study of Behavior*, vol. 19, ed. P. J. B. Slater, J. S. Rosenblatt & C. Beer, pp. 357–409. Academic Press, New York.

Pepperberg, I. M. (1990b). Cognition in an African grey parrot (*Psittacus erithacus*): Further evidence for comprehension of categories and labels. *Journal of Comparative Psychology* 104: 41–52.

Pepperberg, I. M. (1990c). Referential mapping: A technique for attaching functional significance to the innovative utterances of an African grey parrot (*Psittacus erithacus*). *Applied Psycholinguistics* 11: 23–44.

Pepperberg, I. M. (1991). Learning to communicate: The effects of social interaction. In *Perspectives in Ethology*, Vol. 9, ed. P. P. G. Bateson & P. H. Klopfer, pp. 119–162. Plenum, New York.

Pepperberg, I. M. (1992). Proficient performance of a conjunctive, recursive task by an African grey parrot (*Psittacus erithacus*). *Journal of Comparative Psychology* 106: 295–305.

Pepperberg, I. M. (1993). A review of the effects of social interaction on vocal learning in African grey parrots (*Psittacus erithacus*). *Netherlands Journal of Zoology* 43: 104–24.

Pepperberg, I. M. (1994a). Vocal learning in grey parrots (*Psittacus erithacus*): Effects of social interaction, reference, and context. *Auk* 111: 300–13.

Pepperberg, I. M. (1994b). Evidence for numerical competence in an African grey parrot (*Psittacus erithacus*). *Journal of Comparative Psychology* 108: 36–44.

Pepperberg, I. M. (1994c). The African grey parrot: How cognitive processing might affect allospecific vocal learning. Paper presented at the Bielefeld Conference on Adaptive Behavior and Learning, Bielefeld, Germany.

Pepperberg, I. M., Brese, K. J. & Harris, B. J. (1991). Solitary sound play during acquisition of English vocalizations by an African grey parrot (*Psittacus erithacus*): Possible parallels with children's monologue speech. *Applied Psycholinguistics* 12: 151–78.

Pepperberg, I. M. & McLaughlin, M. A. (1996). Effect of avian–human joint attention on allospecific vocal learning by grey parrots (*Psittacus erithacus*). *Journal of Comparative Psychology* 110: in press.

Pepperberg, I. M. & Neapolitan, D. M. (1988). Second language acquisition: A framework for studying the importance of input and interaction in exceptional song acquisition. *Ethology* 77: 150–68.

Pepperberg, I. M. & Schinke-Llano, L. (1991). Language acquisition and use in a bilingual environment: A framework for studying birdsong in zones of sympatry. *Ethology* 89: 1–28.

Petrinovich, L. (1985). Factors influencing song development in white-crowned sparrows (*Zonotricia leucophrys*). *Journal of Comparative Psychology* 99: 15–29.

Petrinovich, L. (1988). The role of social factors in white-crowned sparrow song development. In *Social Learning: Psychological and Biological Perspectives*, ed. T. R. Zentall & B. G. Galef, Jr, pp. 255–78. Erlbaum, Hillsdale, NJ.

Piaget, J. (1954). Language and thought from the genetic point of view. *Acta Psychologica* 10: 51–60.

Premack, D. (1976). *Intelligence in Ape and Man*. Erlbaum, Hillsdale, NJ.

Premack, D. (1983). The codes of man and beast. *Behavioral and Brain Sciences* 6: 125–67.

Price, P. H. (1979). Developmental determinants of structure in zebra finch song. *Journal of Comparative and Physiological Psychology* 93: 260–77.

Rice, M. (1980). *Cognition to Language: Categories, Word Meanings, and Training*. University Park Press, Baltimore, MD.

Rowley, I. (1980). Parent–offspring recognition in a cockatoo, the galah, *Cacatua roseicapilla*. *Australian Journal of Zoology* 28: 445–56.

Rumbaugh, D. M. (ed.) (1977). *Language Learning by a Chimpanzee*. Academic Press, New York.

Ryan, J. (1973). Interpretation and imitation in early language development. In *Constraints on Learning*, ed. R. A. Hinde & J. Stephenson-Hinde, pp. 427–43. Academic Press, New York.

Sanders, R. J. (1985). Teaching apes to ape language: Explaining the imitative and non-imitative signing of a chimpanzee (*Pan troglodytes*). *Journal of Comparative Psychology* 99: 197–210.

Saunders, D. A. (1983). Vocal repertoire and individual vocal recognition in the short-billed white-tailed black cockatoo, *Calyptorhynchus funereus latirostris*. Carnaby. *Australian Wildlife Research* 10: 527–36.

Savage-Rumbaugh, S. (1987). Communication, symbolic communication, and language: Reply to Seidenberg and Petitto. *Journal of Experimental Psychology: General* 116: 288–92.

Savage-Rumbaugh, E. S. (1991). Language learning in the bonobo: How and why they learn. In *Biological and Behavioral Determinants of Language Development*, ed. N. A. Krasnegor, D. M. Rumbaugh, R. L. Schiefelbusch & M. Studdert-Kennedy, pp. 209–33. Erlbaum, Hillsdale, NJ.

Savage-Rumbaugh, E. S., Rumbaugh, D. M. & Boysen, S. (1980a). Do apes use language? *American Scientist* 68: 49–61.

Savage-Rumbaugh, S., Rumbaugh, D. M. & McDonald, K. (1985). Language-learning in two species of apes. *Neuroscience Biobehavioral Review* **9**: 653–65.

Savage-Rumbaugh, E. S., Rumbaugh, D. M., Smith, S. T. & Lawson, J. (1980b). Reference: The linguistic essential. *Science* **210**: 922–25.

Savage-Rumbaugh, E. S. & Sevcik, R. A. (1984) Levels of communicative competency in the chimpanzee: Pre-representational and representational. In *Behavioral Evolution and Integrative levels*, ed. G. Greenberg & E. Tobach, pp. 197–219. Erlbaum, Hillsdale, NJ.

Schwartz, R. G. & Terrell, B. Y. (1983). The role of input frequency in lexical acquisition. *Journal of Child Language* **10**: 571–88.

Scollon, R. (1976). *Conversations with a One-Year Old*. University Press of Hawaii, Honolulu.

Searcy, W. A., Marler, P. & Peters, S. S. (1985). Songs of isolation-reared sparrows function in communication, but are significantly less effective than learned songs. *Behavioral Ecology and Sociobiology* **17**: 223–29.

Seidenberg, M. S. & Petitto, L. A. (1987). Communication, symbolic communication, and language: Comment on Savage-Rumbaugh, McDonald, Sevcik, Hopkins, and Rupert (1986). *Journal of Experimental Psychology: General* **116**: 279–87.

Shiovitz, K. A. (1975). The process of species-specific song recognition by the indigo bunting, *Passerina cyanea*, and its relationship to the organization of avian acoustical behavior. *Behaviour* **55**: 128–79.

Smith, W. J. (1991). Animal communication and the study of cognition. In *Cognitive Ethology: The Minds of Other Animals*, ed. C. A. Ristau, pp. 209–30. Erlbaum, Hillsdale, NJ.

Snow, C. E. & Hoefnagel-Höhle, M. (1978). The critical period for language acquisition: Evidence from second language learning. *Child Development* **49**: 1114–28.

Snow, D. W. & Snow, B. K. (1983). Territorial song of the dunnock *Prunella modularis*. *Bird Study* **30**: 51–6.

Snyder, N. F., Wiley, J. W. & Kepler, C. B. (1987). *The Parrots of Luquillo: Natural History and Conservation of the Puerto Rican Parrot*. Western Foundation for Vertebrate Zoology, Los Angeles, CA.

St Peters, M., Huston, A. C. & Wright, J. C. (1989). Television and families: Parental coviewing and young children's language development, social behavior, and television processing. Paper presented at the Society for Research in Child Development, Kansas City, KS.

Stoddard, P. K., Beecher, M. D., Horning, C. L. & Willis, M. S. (1990). Strong neighbor–stranger discrimination in song sparrows. *Condor* **92**: 1051–6.

ten Cate, C. (1991). Behavior-contingent exposure to taped song and zebra finch song learning. *Animal Behaviour* **42**: 857–9.

Terrace, H. S. (1979a). Is problem-solving language? *Journal of the Experimental Analysis of Behavior* **31**: 161–75.

Terrace, H. S. (1979b) *Nim*. Knopf, New York.

Terrace, H. S., Petitto, L. A., Sanders, R. J. & Bever, T. G. (1979). Can an ape create a sentence? *Science* **206**: 891–902.

Thielcke, G. (1973). Uniformierung des Gesangs der Tannenmeise (*Parus ater*) durch Lernen. *Journal für Ornithologie* **114**: 443–54.

Todt, D. (1975). Social learning of vocal patterns and modes of their applications in grey parrots. *Zeitschrift für Tierpsychologie* **39**: 178–88.

Todt, D., Hultsch, H. & Heike, D. (1979). Conditions affecting song acquisition in nightingales. *Zeitschrift für Tierpsychologie* **51**: 23–35.

Tomasello, M. (1992). The social bases of language acquisition. *Social Development* **1**: 68–87.

Tomasello, M. & Farrar, J. (1986). Joint attention and early language. *Child Development* **57**: 1454–63.

Van Cantfort, T. E., Gardner, B. T. & Gardner, R. A. (1989). Developmental trends in replies to wh-questions by children and chimpanzees. In *Teaching Sign Language to Chimpanzees*, ed. R. A. Gardner, B. T. Gardner & T. E. Van Cantfort, pp. 198–239. SUNY Press, Albany, NY.

Vanayan, M. H., Robertson, A. & Biederman, G. B. (1985). Observational learning in pigeons: The effects of model proficiency on observer performance. *Journal of General Psychology* **112**: 349–57.

Veneziano, E. (1988). Vocal–verbal interaction and the construction of early lexical knowledge. In *The Emergent Lexicon: The Child's Development of a Linguistic Vocabulary*, ed. M. D. Smith & J. L. Locke, pp. 109–47. Academic Press, Orlando, FL.

Waser, M. S. & Marler, P. (1977). Song learning in canaries. *Journal of Comparative and Physiological Psychology* **91**: 1–7.

West, M. J. & King, A. P. (1988). Female visual displays affect the development of male song in the cowbird. *Nature* **334**: 244–6.

Williams, H. (1990). Models for song learning in the zebra finch: Fathers or others? *Animal Behaviour* **39**: 745–57.

Williams, H., Kilander, K. & Sotanski, M. L. (1993). Untutored song, reproductive success and song learning. *Animal Behaviour* **45**: 695–705.

Yamashita, C. (1987). Field observations and comments on the indigo macaw (*Anodorhynchus leari*), a highly endangered species from northeastern Brazil. *Wilson Bulletin* **99**: 280–2.

Zentall, T. R. & Galef, B. G. Jr (eds.) (1988). *Social learning: Psychological and Biological Perspectives*. Erlbaum, Hillsdale, NJ.

10 Vocal learning in captive bottlenose dolphins: A comparison with humans and nonhuman animals

BRENDA McCOWAN AND DIANA REISS

INTRODUCTION

Vocal learning involves the ability to modify and acquire new signals in an organism's vocal repertoire through the use of auditory information and feedback. Humans and many avian species, particularly songbirds, have demonstrated similarities and analogous patterns in the vocal acquisition of their respective repertoires. These similarities include the importance of auditory input, feedback, and social influences on vocal structure and acquisition, and stages of developmental overproduction, selective attrition, and vocal babbling/subsong (for reviews, see Kroodsma 1982; Pepperberg & Neapolitan 1988; Locke 1990, 1993a,b). Finding such parallels in phylogenetically distinct species is striking and suggests a convergence in strategies of vocal learning.

Evidence for vocal learning in other species is rare. Studies of vocal learning in nonhuman primates have suggested that learning plays a role in vocal development of contextual use and comprehension (Seyfarth et al. 1980; Cheney & Seyfarth 1982; Seyfarth 1986; Hauser 1988; Gouzoules & Gouzoules 1989) but clear evidence for the learning of vocal repertoires by nonhuman primates has been slow to emerge. However, recent results of studies of nonhuman primates (Elowson & Snowdon 1994; Snowdon et al., Chapter 12; Mitani & Brandt 1994) and birds (Brown & Farabaugh, Chapter 7) suggest greater vocal plasticity than was previously described and point to an importance of social factors on vocal structure and acoustic variability of calls. Therefore, to more clearly elucidate the phenomenon of vocal learning it is important to make a distinction between vocal learning (the ability to acquire new elements in one's vocal repertoire) and vocal plasticity (the ability to modify signal structure due to social or environmental conditions).

Cetaceans are the only mammals other than humans that demonstrate vocal learning (Payne et al. 1983; Tyack & Whitehead 1983; Richards et al. 1984; Bain 1986; Tyack 1986; Mobley et al. 1989; Caldwell et al. 1990; Reiss & McCowan 1993; McCowan & Reiss 1995a). Bottlenose dolphins (*Tursiops truncatus*) have demonstrated both vocal learning and a proclivity for vocal mimicry of both conspecific whistles and nonspecies-specific sounds (Richards et al. 1984; Tyack 1986; Caldwell & Caldwell 1972; Caldwell et al. 1990; Reiss & McCowan 1993). Thus, the dolphin may serve as a promising model for examining the processes of vocal learning in nonhuman animals in general, and in nonhuman mammals in particular.

BEHAVIORAL BIOLOGY OF THE BOTTLENOSE DOLPHIN

Bottlenose dolphins are highly social mammals that are widely distributed and reside throughout the world in temperate and tropical waters (Leatherwood & Reeves 1990). Longitudinal field studies have indicated a fission–fusion type social structure, showing social complexity rivalling that found in chimpanzee societies (Würsig 1978; Connor & Smolker 1985; Wells et al. 1987; Conner et al. 1992). Large groups are composed of smaller sub groups of age- or sex-related individuals or mother–young nursery groups. A strong mother–calf bond and unusually long lactation period of 1.5 to 4 years (Cockcroft & Ross 1990; D. Reiss, unpublished data) has been suggested as evidence for social learning (Brodie 1969; Smolker et al. 1993). Dolphins are physically precocial at birth; they must swim along side their mothers immediately following parturition (Cockcroft & Ross 1990). Physical development alone does not account for the prolonged lactation period in dolphins. The mother–infant bond is probably important for the infant's social development as well. Behavioral imitation by infants of their mothers' behaviors, even idiosyncratic and play behaviors, is widely observed in captive social groups (B. McCowan & D. Reiss, personal observation).

Observations of wild dolphins living in Sarasota Bay, Florida, have pointed to the presence of female subgroups composed of up to three generations of females (Wells *et al.* 1987). Females offspring tend to remain with their natal group longer than males, who tend to leave by the age of five years and form long-lasting male coalitions. Wells *et al.* (1987) reported higher survivorship rates in males that formed at least one strong long-lasting association with another male. Males in Shark Bay, Western Australia, also form strong long-lasting primary coalitions with one or two other males, and frequently join other males for shorter-term secondary alliances (Connor *et al.* 1992). Both primary and secondary coalitions are used during contexts of aggressive herding of individual females. Frequently, a primary coalition will follow a single cycling female for hours, days, or even weeks (Connor *et al.* 1992). Secondary alliances are often formed during competition between primary coalitions in herding a particular female (Connor *et al.* 1992). Variations in demographics, social structure, and behavioral ecology may vary in different populations; however, the formation and maintenance of social relationships and coalitions seems critical to the social lives of these mammals.

DOLPHIN VOCAL PRODUCTION

Bottlenose dolphins have a rich vocal repertoire and past studies have resulted in the broad categorization of vocalizations into three classes of signal: broad-band, short-duration clicks used in echolocation for navigation, orientation, and perception (Kellogg 1961; Norris 1969), other wide-band pulsed sounds such as squawks, yelps, squeals, and thunks used in social contexts (Caldwell & Caldwell 1977; McCowan & Reiss 1995b), and narrow-band, frequency-modulated whistles also used in social communication (Evans & Prescott 1962; Busnel & Dziedzic 1966; Caldwell & Caldwell 1968, 1972; Tyack 1976). Dolphin whistles are usually characterized by their relative changes in frequency over time, known as whistle "contour." Bottlenose dolphins are thought to develop an individually distinctive signature whistle contour, crystallized during their first year, which has been hypothesized to account for 70–95% of a dolphin's whistle repertoire, to function in individual recognition (Caldwell & Caldwell 1968; Tyack 1986; Caldwell *et al.* 1990; Sayigh *et al.* 1990; Smolker *et al.* 1993), and to facilitate reunion between separated individuals (Smolker *et al.* 1993). Tyack (1986) reported that dolphins spontaneously mimic the signature whistles of other dolphins. In two captive bottlenose dolphins in constant contact, Tyack reported that 20–30% of an individual's whistles were mimics of the other dolphin's signature whistle. He speculated that dolphins mimic each other's signature whistles to label conspecifics. In addition, Smolker (1993) reported that three adult males in close association showed convergence in the acoustic structure of their signature whistles over time (see also Tyack & Sayigh, Chapter 11).

Notably, most studies have focused solely on signature whistles and have termed all other whistles as "aberrant" or "variant" (for a review see Caldwell *et al.* 1990; Tyack 1986; Sayigh *et al.* 1990). The percentage of "aberrant" or "variant" whistles in a dolphin's whistle repertoire has been reported to range from approximately zero to 30% (for a review, see Caldwell *et al.* 1990; Tyack 1986; Sayigh *et al.* 1990). More recent research on signature whistles has indicated that some dolphins may possess "favored" whistles other than their signature whistle, termed "secondary" whistles (Tyack & Sayigh, Chapter 11). Several studies have indicated a predominant role for signature whistles in the dolphin's whistle repertoire (Caldwell & Caldwell 1968; Tyack 1986; Caldwell *et al.* 1990), but most studies on signature whistles have been confined to either small social groups ($n = 2$ individuals: Tyack 1986; Janik *et al.* 1994), isolated or restrained captive individuals (for a review, see Caldwell *et al.* 1990; Janik *et al.* 1994), and temporarily captured individuals (Sayigh *et al.* 1990; Tyack & Sayigh, Chapter 11). These contextual limitations were caused primarily by difficulty in identifying which one in a group of dolphins was vocalizing and have probably resulted in an underrepresentation of the types of whistle contours in the dolphin's repertoire.

Other past and current studies both in the field and in captivity observing dolphins in more varied contexts or interacting with conspecifics have suggested a larger whistle repertoire (Dreher & Evans 1964; Lang & Smith 1965; Dreher 1966; Bastian 1967; Burdin *et al.* 1975;

Kaznadzei *et al.* 1976; McCowan & Reiss 1995c; D. Wang & K. M. Dudzinski, unpublished data). Recent research by our study group on the whistle repertoires of captive adult dolphins, where positive identification of vocalizers was possible, demonstrated that individual dolphins produce a wide variety of whistle contours in socially interactive contexts and indicated that signature whistles may play a less predominant role in the adult repertoire than has been previously suspected (McCowan & Reiss 1995c). In addition, this research, with the use of a quantitative categorization technique, found that all adult dolphins shared a predominant whistle type across social groups and that some individuals shared other whistle types across social groups, shared whistle types exclusively within their social groups, and produced whistle types that were unique to individuals. Notably, the results from this study were consistent with studies published before the advent of the "signature whistle hypothesis" (Dreher & Evans 1964; Evans 1967; Burdin *et al.* 1975; Kaznadzei *et al.* 1976) in which investigators reported a large whistle repertoire within social groups, sharing of whistle types across social groups, and a predominant but not individualized whistle type. Thus, whether dolphins are using a wide repertoire of whistles in their vocal exchanges or primarily producing signature whistles remains unclear (for a full discussion of this issue, see McCowan & Reiss 1995c).

Additional quantitative analysis of whistle structure and function during normal socially interactive contexts is necessary to clarify the prevalence and role of signature whistles in the dolphin repertoire. Qualitative whistle categorization techniques used in these past studies on signature whistles may have led investigators to bias the categorization of whistle contours toward signature whistles. Past researchers have categorized both repetitions (with breaks) as well as continuous whistles (without breaks) as multilooped signature whistles (for a review, see Caldwell *et al.* 1990). Whistles that differed from signature whistles, but contained some apparent portion of signature whistle contours, were similarly categorized as signature whistles, termed "partials" or "deletions" of signature whistles (Tyack 1986). This type of categorization may have inflated the number of different whistles produced by one individual classified as the same signature whistle and the number of similar whistles produced by different individuals as different signature whistles. Until

appropriate perception and categorization experiments are conducted on dolphin whistles, it seems prudent to remain conservative and objective about categorizing whistle contours as independent productions and not group whistles based upon hypothesized "deletions" or "partials" of signature whistles.

Further, experimental research with computer-generated whistles has indicated that dolphins imitate and produce expansions and compressions of whistle contours, independent elements of these contours, and apparent combinations of two discrete contours (Reiss & McCowan 1993). Variations of imitations and productions of each computer-generated whistle were produced in the same behaviorally appropriate context (Reiss & McCowan 1993). Our results suggested that dolphin whistles may be composed of several discrete elements rather than deletions or partials of signature whistles and that the relative contour of whistles, not the absolute frequency and time measures, was an acoustic feature important to dolphins. Therefore, in developing a quantitative technique for analyzing the whistles of the dolphins' own natural communication system, we wanted to control for such structure and avoid forcing our results to agree with the signature whistle hypothesis (McCowan 1995). Repetitions of the same whistle were evaluated as separate whistles, whereas continuous whistle production was considered a multilooped whistle type, as would be the case for most studies on animal communication (e.g., syllable structure in songbirds). In fact, if signature whistles do play a predominant role in the dolphin whistle repertoire then our categorization scheme should have resulted in an elevated number of signature whistles represented in the repertoire, since repetitions of whistles were considered separately. Similarly, hypothesized "partials" or "deletions" were not categorized as signature whistles in our study but rather were considered as independent whistles, quantitatively classified into types by their similarity in contour. While our quantitative method may split whistle categories rather than lump them, categorization is based on objective mathematical and statistical rather than subjective qualitative measures of whistle similarity (McCowan 1995). Perception and categorization experiments on dolphin whistles are needed to validate or appropriately modify this technique for categorizing whistles based upon the dolphin's own perceptual system.

Previous studies into the ontogeny of dolphin

communication have focused primarily on whistles recorded after the first year of development. These studies have proposed that vocal learning during the first year probably contributes to the acquisition of the adult whistle repertoire (Caldwell *et al.* 1990; Sayigh *et al.* 1990; Reiss & McCowan 1993). In one study (Sayigh *et al.* 1990), male infant dolphins were found to develop a signature whistle similar in contour to that of their mothers, and female infants developed a signature whistle different in contour from their mothers', although more recently this gender difference has been found to be less absolute than was previously reported (L. S. Sayigh, personal communication; Tyack & Sayigh, Chapter 11). Because most developmental studies have focused solely on signature whistle development (Caldwell & Caldwell 1979; Sayigh *et al.* 1990; Tyack & Sayigh, Chapter 11), little attention has been directed to the ontogeny of the entire whistle repertoire within a socially interactive context. In addition, the extent to which this learning involves both structural and functional modifications of whistles during development has not been fully investigated.

In this chapter, we discuss methodological considerations for studies investigating vocal learning in dolphins. We present evidence for vocal learning in the dolphin from both observational ontogenetic studies and experimental studies and these findings are compared to previous and concurrent studies of vocal development in dolphins and other species. Finally, we discuss the evidence for convergent strategies in vocal learning in humans, birds, and dolphins.

METHODOLOGICAL CONSIDERATIONS FOR ONTOGENETIC STUDIES OF DOLPHIN VOCAL LEARNING

The study of vocal learning can be separated into three major aspects. Vocal learning can occur in the acquisition and modification of the acoustic structure of a signal, in the contextual use of signals, and in comprehension of signals. The occurrence of one aspect of vocal learning is not requisite for the emergence of others nor is it the case that all vocalizations within a repertoire be learned for evidence of learning of one call type (see also Seyfarth & Cheney, Chapter 13). In addition, the timing of learning in each of these aspects can be disparate; for example, comprehension of calls can precede production of vocal

signals as evidenced in human speech acquisition. Conversely, vocal production and mimicry can precede comprehension or contextual use (see also Snowdon *et al.*, Chapter 12; Seyfarth & Cheney, Chapter 13).

One important aspect of vocal acquisition involves the learning and modification of the acoustic structure of vocal signals. Vocal learning of acoustic structure can occur in different species, with differing degrees of vocal plasticity. Some species may exhibit a very "open" communication system in which new vocalizations are learned throughout life, whereas others may exhibit a more "closed" communication system in which only a few vocalizations are learned during specific periods in life or which are constrained by the acoustic features that can be learned. Traditional methods to assess the presence and degree of vocal learning in avian and nonhuman primate species have employed deafening (Talmage-Riggs *et al.* 1972; Steklis 1985), isolation rearing (Winter *et al.* 1973; Herzog & Hopf 1984; for a review, see Steklis *et al.* 1985), cross-fostering (Masataka & Fujita 1989; Owren *et al.* 1992), and conditioning (Hayes 1951; Sutton *et al.* 1981; Hopkins & Savage-Rumbaugh 1991). These experimental approaches can affect either the social environment or the biological state of the individuals under study and thus may hinder our ability to assess the features that are requisite for vocal learning in certain species. For instance, it has been well documented for many avian species that the presence of social companions or live interacting tutors can greatly influence the timing and type of vocal learning that occurs in birds (Baptista & Petrinovich 1984, 1986; Baptista 1988; Baptista & Morton 1988; Pepperberg & Neapolitan 1988; Pepperberg 1993, 1994; Chaiken *et al.* 1994; Baptista & Gaunt, Chapter 3; Pepperberg, Chapter 9). Therefore experimental paradigms with social species such as the dolphin need to address this issue squarely. How do we study the role of learning in species where isolation, cross-fostering, and social group manipulation can be potentially harmful to the infant's normal social development? New experimental paradigms must be developed that allow us to manipulate the acoustic environment of subjects sufficiently to investigate vocal learning without adversely altering the social environment.

Another important issue for ontogenetic studies involves differentiating between changes in vocalizations due to maturation of the vocal tract and those due to vocal learning. This is difficult, particularly in species where individuals share the same adult repertoire. Because the adult repertoire is not identically shared by different social groups of dolphins (McCowan & Reiss 1995c; Tyack & Sayigh, Chapter 13), the bottlenose dolphin provides a promising opportunity for the study of vocal learning in nonhuman mammals. The extent to which vocal learning modifies the developing repertoire can be determined by studying the ontogeny of whistles in infants from similar and different social and acoustic environments. In captive environments, infants are frequently exposed both acoustically and socially to individuals that are neither genetically related nor captured from a location similar to that of their genetic relatives. This artificial mix of individuals in captive social groups allows the documentation of noninheritable whistle acquisition and an investigation into the role of learning in whistle development.

VOCAL ONTOGENY IN DOLPHINS

Two studies will be presented here that address the role for learning in dolphin whistle structural and contextual development. The first section describes an experimental study that, with the use of an underwater computer keyboard, investigated the role and process of vocal imitation and spontaneous production of computer-generated whistles by two infant males (from 11 months of age) and two adult female captive bottlenose dolphins (Reiss & McCowan 1993). The second section presents data on an observational study of the whistle repertoires of ten captive adult bottlenose dolphins and the whistle ontogeny of eight captive-born infant bottlenose dolphins from three different captive social groups observed during normal social interactions from birth through their first year (McCowan & Reiss 1995a).

EXPERIMENTAL RESULTS: SPONTANEOUS VOCAL MIMICRY AND LEARNING THROUGH KEYBOARD USE

Introduction

Prior research has documented vocal mimicry in the dolphin but there has been little experimental work focusing on spontaneous (e.g., untrained) mimicry and its role in vocal ontogeny. In order to understand more clearly the

dolphins' proclivity for vocal mimicry and its relationship to vocal learning, we conducted a study that allowed us to systematically investigate vocal learning through an experimental paradigm (Reiss & McCowan 1993). An underwater keyboard system was designed in order to investigate the functional and developmental aspects of dolphin vocal learning and to provide dolphins with an environmental enrichment activity in which they could gain rudimentary control over certain aspects of their environment (for a detailed description of apparatus and methods, see Reiss & McCowan 1993). Through keyboard use, dolphins were exposed to a systematic chain of events: the dolphin's use of visual forms displayed on the face of the keyboard was followed by specific computer whistles generated underwater and in air, and corresponding objects (e.g., balls, rings, ring-floats, disks or an activity such as rubs) were given to the dolphin. The dolphins' use of the keyboard, general behavior, and vocal behavior were recorded during experimental sessions. Thus, we could record mimicry and productive use of novel signals and the environmental contexts in which this behavior would occur.

Computer-generated whistles were designed; they ranged from 2 to 16 kHz and had a duration of 0.4–0.8 s. Harmonic structure was also represented in these signals. The signals were designed to be similar to the frequency and temporal parameters of natural dolphin whistles (Herman & Tavolga 1980), yet different in the actual frequency modulation, the whistle contour, of their own signals as determined by our baseline recordings of the infants and adults. It was necessary that the computer-generated whistles were similar to, yet distinct from, whistles extant in the dolphins' own repertoires, for several reasons. First, although this system was not designed to elicit vocal mimicry, we wanted to use whistles that could be easily perceived and reproduced by the dolphins, thereby facilitating the dolphin's ability to process, produce, and remember these signals. Second, if the dolphins were to mimic and reproduce the model sounds, they had to be distinct from their own repertoire in order to demonstrate vocal mimicry, operationally defined as the copying of an otherwise improbable act or utterance (Thorpe 1963; Richards et al. 1984) rather than a mere elicitation of their own whistles (Andrew 1962).

During the experimental sessions, using a free-choice methodology, dolphins could choose to use or not use the

Table 10.1. *Composition and history of social groups 1, 2, and 3*

Animal name	Facility	Sex	Birth date	Collection date	Birthplace/acquisition location	Mother	Father
Group 1							
CIRCE	MWAUSA	F	–	7 Jun. '78	Compano Bay, TX	–	–
GORDO	MWAUSA	M	–	1969	Santa Barbara, CA	–	–
TERRY	MWAUSA	F	–	5 Jan. '80	Compano Bay, TX	–	–
DELPHI	MWAUSA	M	2 Aug. '83	–	Marine World Africa USA	CIRCE	GORDO
PANAMA	MWAUSA	M	31 Jul. '83	–	Marine World Africa USA	TERRY	GORDO
Group 2							
BAYOU	MWAUSA	M	–	7 Jun '78	Compano Bay, TX	–	–
CHELSEA	MWAUSA	F	–	23 Aug. '83	Gulfport Harbor, MS	–	–
SADIE	MWAUSA	F	–	23 Aug. '83	Gulfport Harbor, MS	–	–
SCHOONER	MWAUSA	M	–	7 Jun '78	Compano Bay, TX	–	–
STORMY	MWAUSA	F	–	7 Jun '78	Compano Bay, TX	–	–
DESI	MWAUSA	M	26 Apr. '89	–	Marine World Africa USA	STORMY	–
LIBERTY	MWAUSA	M	4 Jul. '90	–	Marine World Africa USA	CHELSEA	SCHOONER*
NORMAN	MWAUSA	M	11 May '91	–	Marine World Africa USA	STORMY	BAYOU*
SAM	MWAUSA	M	8 May '90	–	Marine World Africa USA	SADIE	–
Group 3							
ASTRA	SWO	F	3 Aug. '83	–	Sea World of California	CINDY	CLICKER
SCARLET	SWO	F	18 Oct. '79	–	Sea World of California	SCARBACK	DOC
NEPTUNE	SWO	M	21 Jan. '90	–	Sea World of Ohio	SCARLET	DOMINO
TASHA	SWO	F	15 Feb. '90	–	Sea World of Ohio	ASTRA	DOMINO

Notes:
MWAUSA, Marine World Africa USA; SWO, Sea World of Ohio; –, unknown or not applicable; *, preliminary sequence analysis (Cara Gubbins, unpublished data).

apparatus and could freely select any key. This system allowed dolphins to interact freely with a self-reinforcing system without any *explicit* training procedures. Here, we emphasize explicit because it has been well demonstrated that in any context with humans or animals, implicit demand characteristics may be operating for the individual (Orne 1981).

A social group of two female Atlantic bottlenose dolphins (*Tursiops truncatus*) – Terry, approximately 20 years old, and Circe, approximately 9 years old, and their two male offspring Panama and Delphi born in our research facility at Marine World Africa USA (Redwood City, California, in year 1 and Vallejo, California, in year 2) – were the subjects of the study (see Table 10.1). The research commenced when the young males were 11 months old and continued through their fifth year. Prior to the onset of this research, observations and recordings were conducted on vocal and behavioral ontogeny of the infants (Reiss 1988; McCowan & Reiss, 1995a,c) which served as a baseline for this study. The dolphins were research and exhibit animals and not participants in public demonstrations.

There were two stages of research: year 1 and year 2. Year 1 lasted from the time the males were 11 months through two years of age and year 2 lasted from four to five years of age. Research was interrupted for a one year period when the males were three to four years old due to a relocation of our research facility. Experimental sessions lasted for 30 minutes and were conducted at variable intervals. Outside of keyboard sessions the dolphins were not deprived of the items they could acquire through keyboard use.

The results of this research have been described in detail (Reiss & McCowan 1993). In this chapter we discuss the data as they are relevant to vocal learning in this species. Although all four dolphins were given the same opportunity to use the keyboard in year 1, only the two young males systematically used the system. The females used the keyboard only twice prior to the males' initial use. The females would occasionally position with the males as they used the keyboard and would often interact with the toys obtained through keyboard use by the males. Neither of the adult females were exposed to the keyboard during year 2. Analysis of audio and video recordings of experimental sessions demonstrated spontaneous vocal mimicry and productive use of facsimiles of the computer-generated model sounds, and evidence for vocal learning in this species. The term mimicry was designated to represent facsimile of the model sounds emitted by dolphins immediately following the model. Facsimiles emitted by dolphins that occurred in other contexts such as preceding key use, during object/activity interaction, or any time not immediately following the model sound were referred to as production.

Vocal mimicry and productive use of whistles

Early in the study, the dolphins began to mimic the model sounds. Analysis revealed an interesting process of initial mimicry. During the initial stage of vocal mimicry different aspects of the model sounds were produced. For example, the "ball" whistle was mimicked after 19 exposures. One of the young males used the ball key and after each model sound, reproduced the following elements of the model signal in four successive trials: (a) the end of the whistle, (b) the beginning of the whistle, (c) the entire whistle including fundamental and harmonic structure, and (d) the fundamental frequency contour without harmonic structure and showing a transposition to a slightly higher frequency than the model sound. Similar patterns were observed in the initial mimicry of other model sounds. After nine exposures the "rub" whistle was mimicked. In a single sequence following the model sound, the dolphin emitted successively three elements of the model in the following order: (a) the end component of the model, (b) the entire whistle contour compressed in both the time and frequency domain, and (c) the beginning component of the model. Most striking was the fact that these three successive instances of mimicry occurred within a 0.7-second duration, the same duration as the model sound. A third novel whistle paired with "rings" was subsequently mimicked after only two exposures to

the signal. In the first instance of mimicry the temporal domain was compressed; however, whistle contour was clearly reproduced. In the following instances of mimicry, after the third and fourth exposure to the model, the frequency modulation and temporal parameters were more closely matched. The fifth exposure was followed by a mimicked version in which both the duration and frequency were expanded.

Thus, from the onset of mimicry, the dolphins exhibited several patterns involved in their initial acquisition. First, they initially reproduced different segments of the whistle contour itself prior to mimicking the entire contour. Additionally, they tended to mimic the final segment of the model first and then mimic the latter segments, not unlike the recency/primacy effect in recall experiments in which the last element of a series is recalled and reproduced. They also showed the tendency to reproduce absolute frequency and temporal parameters of the model sounds, to transpose in the frequency domain, and to compress and expand both the frequency and temporal parameters in their whistled versions throughout this study, ending when the young males were five years of age.

The first evidence that dolphins were reproducing the model sounds was in the context of vocal mimicry as described above. However, the dolphins also began to *produce* facsimiles of the model sounds preceding use of the corresponding keyboard element or concurrent with interactions with objects obtained through keyboard use. The first instance of productive use occurred in the same session in which vocal mimicry was first evidenced. This session was marked by the multiplicity of contexts in which these facsimiles were first emitted. For example, during this early session the following sequence was recorded. One of the young males used the "ball" key, which was followed by mimicry of the model sound. The dolphin remained at the keyboard and interacting with the ball it had just obtained and remained in the general vicinity of the keyboard for several seconds. Two productions of the ball whistle were recorded while the dolphin pushed the ball about in the keyboard vicinity. The dolphin then reapproached the keyboard, at which time a facsimile of ball was produced, and immediately selected the ball key on the keyboard.

During year 1 the amount of vocal mimicry in relation to production followed a consistent relation across keyboard sessions (simple regression (values were natural log transformed to normalize data for simple regression), $r^2=0.539$, F test, $P=0.0118$). Across all year 1 sessions vocal mimicry ($n=165$) was about 19% higher than production ($n=139$). In contrast, during year 2 the occurrence of vocal mimicry in relation to production showed no consistent relation across sessions (simple regression $r^2=0.047$, F test, $P=0.8777$); however, vocal production ($n=210$) was approximately 10 times higher than vocal mimicry ($n=19$). There was a significant increase in the ratio of productions to vocal mimicry (χ^2 test, $P=0.0001$), and productions were 13.2 times more likely to occur in year 2 than in year 1 (odds ratio of 13.2, range 10.1–17.1, at 95% confidence intervals). These findings suggest that the prevalence of mimicry may be associated with earlier stages of vocal acquisition and that its relative frequency of occurrence diminishes during later stages of vocal production and use. The importance of imitation and auditory feedback of the individual's own vocal production in the process of acquisition of conspecific song has been well documented in birds (Marler & Waser 1977) and in the ontogeny of human vocal development (Bloom *et al.* 1974). Our findings suggest this process of mimicry is also important in the early stages of vocal learning in bottlenose dolphins.

Additionally, dolphin vocal mimicry and production showed a greater structural fidelity to the acoustic parameters of the model sounds in year 2 than in year 1. Instances of production and mimicry during year 2 matched the relative time and frequency parameters of the model sounds more closely than in year 1; however, the dolphins continued to expand and compress both of these acoustic parameters throughout the duration of the study (Fig. 10.1). Therefore, while these results indicate a general trend towards less variability between the model sounds and the dolphins' versions it is clear that these signals continue to be open to modification, showing plasticity in both temporal and frequency parameters. As reported in several other chapters (Brown & Farabaugh, Chapter 7; Pepperberg, Chapter 9; Snowdon *et al.*, Chapter 12), several species of bird, nonhuman primates, and dolphins show vocal plasticity and that this plasticity does not seem to be restricted to critical or sensitive periods.

Analysis of behavioral narratives recorded during keyboard sessions during year 2, under double-blind conditions (see Reiss & McCowan 1993), enabled us to determine the contextual use of the whistle production and mimicry. Our results revealed that the dolphins were using whistle facsimiles in behaviorally appropriate contexts. In a total 92 ball productions, 74 (80%) were emitted in contexts of ball play; of a total of 64 ring productions 47 (73%) were emitted during ring play, and out of a total of 5 rub productions all (100%) were produced during contexts of physical rubbing between the dolphin and investigators. Out of the 28 instances of productions of the novel ring–ball combination whistle, 23 (82%) were emitted during simultaneous ring and ball play. These findings suggested that the dolphins were not only learning novel signals but also forming associations between these signals and environmental conditions.

Combination whistles

We previously presented evidence for the learning of novel signals through exposure to model sounds. However, analysis of keyboard sessions during the second year revealed that a novel whistle type was being produced and appeared to be a combination of the ball and ring model sounds combined into one continuous whistle. There was always a minimum delay of 0.3 second between successive computer whistles being generated, preventing any two or more model sounds being produced in a continuous manner. Thus, the emergence of this apparent combination whistle could not have been acquired through initial stages of vocal mimicry alone. This type of whistle was first found during the twentieth session in year 2, a session in which five such combinations were produced. The production of ring–ball combination whistles persisted during year 2 and a total of 23 such whistle types were recorded primarily in the context of a new behavior pattern that had also emerged during year 2. The young males began a new type of object play in which they would interact with a ball and ring concurrently, holding both objects in their mouths or by interacting with both objects in a multitude of ways. The emergence of this novel combination whistle is important for several reasons. This type of whistle is apparently composed of two distinct segments that are also produced independently. To date, there have been no prior reports of segmental organization of dolphin whistle structure. Therefore, while this is only one novel whistle type, its specific contextual use and persistence across sessions suggests it may have some

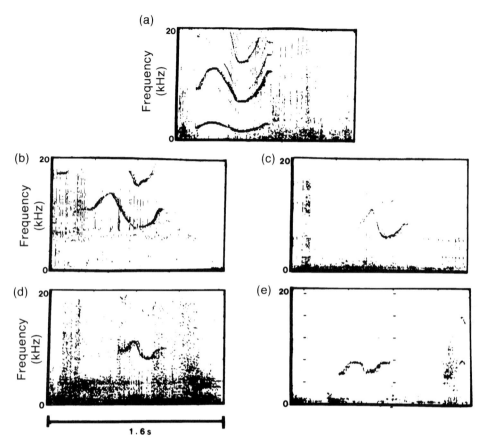

Fig. 10.1. Spectrograms of the computer-generated ball whistle and dolphin productions of ball facsimiles. (a) The model sound. (b) Spectrogram of ball facsimile production in which frequency modulation and duration closely approximated the model. (c) and (d): Spectrograms of ball facsimile production in which the duration and frequency range was compressed but the relative frequency modulation of the model was approximated. (e) Spectrogram of the ball facsimile produced by a dolphin at the beginning of year 2 after a single exposure to the model sound during a two-year hiatus. The relative frequency modulation of the signal was closely matched. (Copyright © 1993 by the American Psychological Association (from Reiss & McCowan 1993). Reprinted with permission.)

functional significance to the structure and organization of dolphin whistles.

Experimental evidence for vocal learning

The results of this study provide compelling evidence for vocal learning in this species. The contexts in which vocal mimicry and productive use occurred suggest that the dolphins were learning associations between the model sounds, their own vocal utterances, and the objects and activities resultant from keyboard use. The learning of novel signals and the acoustic variability in signal structure suggest that learning may be influenced both by environmental factors such as acoustic exposure to model sounds and by exposure to model sounds followed by specific contingencies and social factors such as interactions with objects, investigators and other dolphins in the environment. However, in this study we could not isolate the specific social factors from the more general environmental factors that might influence vocal learning.

The importance of social factors in vocal acquisition in birds has been well documented (for reviews, see Baptista

& Gaunt, Chapter 3; Pepperberg, Chapter 9). In contrast to the experimental paradigms used in studies of vocal learning in birds, previous experimental work investigating vocal learning in dolphins utilized a different approach. Due to the social and physiological requirements of young dolphins, social isolation is not possible and comparable studies using tutor tapes versus live tutors are not feasible. Even the use of cross-fostering is extremely rare in dolphins (Tyack & Sayigh, Chapter 11). Therefore, past experimental studies have generally focused instead on the extent to which dolphin vocal behavior could be modified using traditional shaping methodologies and food reinforcement (Richards *et al.* 1984; Sigurdson 1989). In contrast, our keyboard study was designed to encourage the dolphins to explore the contingencies of their behavior through keyboard use within a social environment and thus more closely approximating a social context in which learning might occur.

Notably, the results of our study using such a methodology were inconsistent with results reported for vocal learning in other past studies of Richards *et al.* (1984) and Sigurdson (1989). First, the rapidity with which the model sounds were spontaneously mimicked by the dolphins in our study greatly contrasts with results reported by both Richards *et al.* (1984) and Sigurdson (1989) in which over 1000 trials were required to train vocal mimicry. The disparity may be due to age- and sex-related factors or to the efficacy of different methodologies. In both the Richards *et al.* (1984) and Sigurdson (1989) studies, the subjects were older females (from five to seven years old). In our study only our two young males systematically used the keyboard while the older females showed minimal interest. However, little is known regarding gender and age differences in acquisition and use of species-specific vocalizations to support this view. Alternatively, the methodology used by Richards *et al.* (1984) and Sigurdson (1989) may not be appropriate for assessing vocal learning in this species. In these past studies, the investigators determined the vocal parameters that the dolphins would imitate and reproduce in response to a model sound, and selectively reinforced approximations of duration, fundamental frequency, and the modulation parameters of the model. In contrast, the dolphins in our study were fed prior to, and following, experimental sessions and could freely choose to use or not to use the keyboard and obtain or not to obtain model sounds and other contingencies. The dolphins

determined the type of input, the frequency of input, and the acoustic parameters they would imitate and reproduce. The efficacy of this methodology on vocal learning of zebra finches (*Taeniopygia guttata*) has also been reported (Adret 1993a,b). Zebra finches that could control the delivery of tape-recorded song demonstrated imitation in the absence of social interaction. Tape-recorded song alone did not elicit vocal mimicry in these species (Adret 1993a,b).

The importance of mimicry in vocal learning has been well demonstrated in the acquisition of human language (Valentine 1930; Bloom *et al.* 1974; Kuczaj 1987) and avian song (Marler & Peters 1982; Kroodsma & Pickert 1984; Baptista & Petrinovich 1986). Our results provide strong evidence for the use of vocal mimicry by dolphins in acquiring novel whistles. The contexts in which we observed the dolphins to interdigitate instances of mimicry and production, as evidenced by dolphin facsimiles preceding, following and overlapping the model sounds, could be interpreted as the model sounds serving as auditory input and feedback in comparison with their own renditions of the model sounds. Furthermore, results from this study and previous studies of acquisition of human language and avian song suggest that mimicry may be a more widespread strategy for vocal learning than previously suspected and that these divergent communication systems in three highly social species may share underlying mechanisms of vocal learning.

OBSERVATIONAL STUDY ON WHISTLE ONTOGENY IN DOLPHINS

Introduction

An observational study was conducted on the ontogeny of whistle repertoires, structure, and context in eight captive-born infant bottlenose dolphins from three different social groups (McCowan & Reiss 1995a). Because our computer keyboard study could address neither the individual changes nor the changes that occur normally across development, this study was conducted: (a) to compare species-typical whistle development in infants from similar and different captive social groups; and (b) to examine the role of learning in the acquisition of species-specific whistle repertoires, acoustic structure, and contextual use. Our captive social groups were composed of both related and unrelated individuals as well as individuals from different geographic locations (Table 10.1). Thus, we were able to examine the whistle ontogeny of infants exposed to the whistles of both related and unrelated individuals and compare the whistle repertoires of maternally related and unrelated infants raised in similar and different social and acoustic environments.

Seven infant males and one infant female from three different social groups of captive bottlenose dolphins were the subjects of this study. Social group 1 consisted of two male infants and three adults (one male, two females) housed in the research area at Marine World Africa USA, Redwood City, California. Social group 2 consisted of four male infants and five adults (two males, three females) housed in the show area at Marine World Africa USA, Vallejo, California. One adult female from social group 1 (Circe) was captured with three individuals from social group 2 (Bayou, Schooner, and Stormy). She was housed briefly with these individuals after acquisition. The second adult female from social group 1 (Terry) was acquired from another marine park. She was originally captured from a similar geographic location. She was never housed with individuals from social group 2. The adult male from social group 1 (Gordo), a Pacific bottlenose dolphin,[1] was captured from a second geographic location and never housed with individuals from social group 2. The other two adult females from social group 2 (Chelsea and Sadie) were captured from a third geographic location. The adult females from social group 3 were captive born and neither individual was housed with individuals from social groups 1 or 2.

Four infants (social group 1: $n=2$; social group 2: $n=2$) were observed from birth through their first year. Two infants from social group 2 were observed only through the first month. Two infants from Sea World of Ohio were observed from five to six months until the end of their first year. Ten adult dolphins (three males, seven females) housed with the infants, including their mothers, were simultaneously observed with the infants. Data were collected using a focal animal sampling regime (Altmann 1974) and a combination of continuous, event, and interval sampling through an underwater window or from an observation tower. Whistles were positively identified using simultaneous bubble stream emission during whistle

production in order to obtain reliable individual repertoires (McCowan 1995; McCowan & Reiss 1995a). Whistles were digitized, quantitatively measured, and categorized using the contour similarity technique (McCowan 1995; McCowan & Reiss 1995a,c) that statistically classifies whistles into types and compares whistle types based upon their similarity in contour (relative frequency changes over time).

Infant whistles were analyzed for each individual by month. Because no developmental landmarks have yet been identified for infant bottlenose dolphins, months were arbitrarily collapsed into three developmental periods for each infant: (a) months 1–4, (b) months 5–8, and (c) months 9–12. The collapsing of months was done to help illustrate trends over development and facilitate comparison across infants and between infants and adults. To examine the role of learning in infant whistle development, the whistle repertoires of adults in the infants' social groups were also analyzed for acoustic structure. Adult whistles were analyzed by individual, using the protocol described above and by McCowan (1995) and McCowan & Reiss (1995c). Infant whistle development was evaluated for evidence of vocal learning by comparing the whistle types and whistle structure within whistle types of infants from each social group and comparing them to the whistle types produced by infants and adults from similar and different social groups. Direct evidence for vocal learning would come from the existence of whistle types shared by infants and adults exclusively within social groups. Indirect evidence for vocal learning would come from ontogenetic changes in whistle structure by infants that could not be explained completely by maturation of the vocal tract or other physical development (e.g., respiratory control).

In addition, the contextual use of whistle types was analyzed for three major whistle categories in four infants and eight adults from social groups 1 (two infants, three adults) and 2 (two infants, five adults): whistle type 1, whistle type 2, and whistle types unique to individual (McCowan & Reiss 1995a). Only three super categories were used for this preliminary analysis due to small sample size: (a) contexts of separation (departure, approach, and swimming alone), (b) social interactions (aggressive, affiliative, and sexual), and (c) other (toy play, feeding, etc.). Detailed analyses of whistle types used in contexts of aggressive, affiliative, and sexual behavior will be con-

---

[1] All other adult dolphins are Atlantic bottlenose dolphins.

Table 10.2. *Percentage of total whistles for each whistle type for ten captive adult dolphins from three different social groups*

Whistle type	Social group 1			Social group 2					Social group 3		Percentage	No. of whistles
	Circe	Gordo	Terry	Bayou	Chelsea	Sadie	Schooner	Stormy	Astra	Scarlet		
Shared across social groups												
2	82.6%	25.0%	55.6%	23.8%	31.8%	80.0%	75.0%	40.0%	61.1%	28.6%	52.4	97
3			11.1%	9.5%	18.2%				5.6%		4.9	9
4	4.3%		5.6%								1.1	2
5				9.5%		10.0%			11.1%		3.2	6
6				9.5%				20.0%	5.6%		3.8	7
7	4.3%	12.5%	5.6%		22.7%			10.0%	16.7%	14.3%	8.1	15
103		6.3%									0.5	1
108		25.0%					15.0%				3.8	7
Shared with social groups												
67					4.8%						0.5	1
70			5.6%								0.5	1
109					4.5%						0.5	1
129		6.3%	5.6%								1.1	2
130		6.3%	5.6%								1.1	2
131		6.3%	5.6%								1.1	2
132				28.6%	4.5%		5.0%	25.0%			7.0	13
Unique to individuals	8.7%	12.5%	0.0%	14.3%	18.2%	10.0%	5.0%	5.0%	0.0%	57.1%	10.3	19
Total percentage	100	100	100	100	100	100	100	100	100	100	100	
Number of whistle types unique to individuals	2	2	0	2	1	1	1	1	0	3		
Total number of whistles	23	16	18	21	22	20	20	20	18	7	185	

Note:
Dark outlined boxes indicate that whistle type was a predominant whistle type.
Source: © Copyright 1995, *Ethology* (from McCowan & Reiss 1995c). Adapted with permission.

ducted once more data have been analyzed for adults (B. McCowan *et al.*, unpublished data). Infants were collapsed by infant and adults were collapsed by gender to facilitate comparison. It should be noted, however, that infants showed considerable variability in the contextual usage of these whistle types both across infants and developmental periods (B. McCowan *et al.*, unpublished data). Adults did not show significant differences and thus were collapsed to enhance sample size.

Whistle repertoires, whistle type use, and contextual usage in adult dolphins

The adult dolphins from three different social groups produced a wide variety of whistle types (total repertoire size for all adults=28 whistle types) that were either shared across social groups (29% of whistle types), exclusively shared within social groups (25%) or that were unique to individuals (46%) (Fig. 10.2, Table 10.2). Because categorization of whistle types was based upon contour alone and not the perceptual systems of dolphins, it is possible that some of these whistle types will be collapsed once perception and categorization experiments have been conducted. However, the most predominant whistle type for all adult dolphins was whistle type 2, which accounted for 52.4% of the combined adult repertoire (Table 10.2). Other whistle types accounted for only 0.5–10.3% of the total number of whistles in the combined repertoire. A spectrogram of whistle type 2 for each adult in our study group is provided in Fig. 10.3 for comparison. Clearly, the similarity of the whistle contour of this whistle type can be seen across individuals from similar and different social groups. Interestingly, we did not find a whistle type for each individual that matched the definition of a "signature" whistle

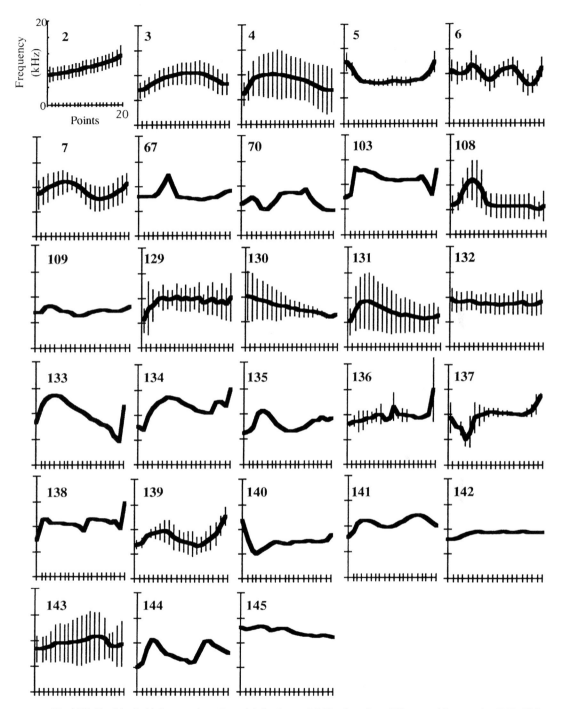

Fig. 10.2. Combined whistle repertoires of ten adult bottlenose dolphins from three different social groups (see Table 10.1). Whistle types are represented by the contour of the mean frequencies (line) and the standard deviations (bars) of the 20 frequency measurements (points). (Copyright © 1995, *Ethology* (from McCowan & Reiss 1995c). Reprinted with permission.)

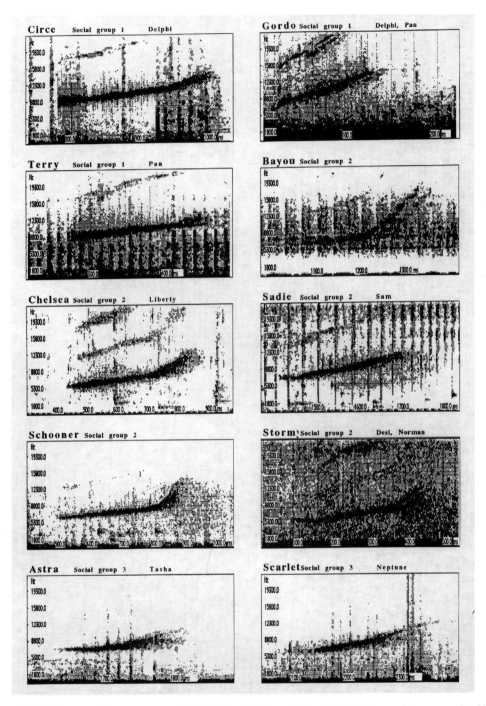

Fig. 10.3. Representative spectrograms of whistles from whistle type 2 for each adult. Name, social group membership, and names of infants, where applicable, are provided for each adult. (Copyright © 1995, *Ethology* (from McCowan & Reiss 1995c). Reprinted with permission.)

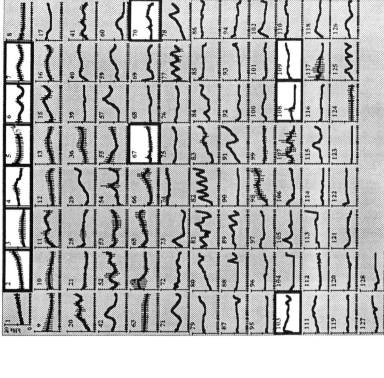

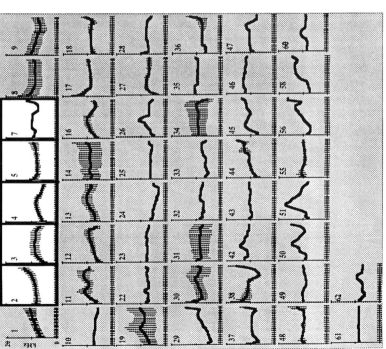

Fig. 10.4. Combined whistle repertoires of eight captive-born infant bottlenose dolphins from three different social groups. (a) Developmental period 1 (months 1–4); (b) Developmental periods 2 (months 5–8) and 3 (months 9–12). Whistle types are represented by the contour of the mean frequencies (line) and the standard deviations (bars) of the 20 frequency measurements (points). White boxes indicate whistle types that were also shared by adults. (Copyright © 1995 by the American Psychological Association (from McCowan & Reiss 1995a). Reprinted with permission.)

(Caldwell *et al.* 1990). Instead, despite our technique's bias for splitting whistle categories, whistle sharing occurred at both the species level and the social group level, and whistle types were also found that were unique to individuals. These data may suggest a functional or dialect difference for whistle types used at these various levels similar to that found for other cetaceans (Bain 1986) and some species of birds (Feekes 1982; Farabaugh *et al.* 1988; Adret-Hausberger *et al.* 1990; Brown & Farabaugh, Chapter 7).

For contextual usage, adult male and female dolphins differed in the use of whistle type 2, the most predominant whistle type in their repertoires. Whistle type 2 was produced primarily in contexts of separation by adult females (percentage of whistles: 93) and in social interactions by adult males (percentage of whistles: 61). This difference may suggest that whistle type 2 is used by females to contact individuals, i.e., infants or other social group members, and by males to maintain contact during association. This interpretation fits well with the respective association patterns of female and male dolphins. In the wild, adult females engage in more fluid association patterns and adult males in more long-term coalition patterns (Connor & Smolker 1985; Wells *et al.* 1987; Connor *et al.* 1992; Smolker *et al.* 1993). In contrast, whistle types unique to individuals were produced primarily during social interactions by both adult males (percentage of whistles: 50) and females (percentage of whistles: 67). Although difficult to interpret at this stage of analysis, these whistle types may be more functional in specific interactions between two or more individuals.

Whistle development in infant dolphins and evidence for vocal learning

Infant dolphins showed a complex pattern across development in the structure both of their whistle repertoires and of whistle types. The combined infant whistle repertoire across three social groups of dolphins consisted of 128 whistle types (Fig. 10.4, Table 10.3). Repertoires, like those of adults, were composed of whistle types shared across social groups (20%), whistle types shared exclusively within social groups (7%), and whistle types unique to individuals (73%). As for adults, it is possible that some of these whistle types will be collapsed once perception and categorization experiments have been conducted.

Some whistle types were present shortly after birth; others apparently emerged over development (Table 10.3). No evidence pointed to an influence on vocal learning by related individuals or close associates. Infants produced whistle types that were shared across social groups and within social groups, but no consistent relationship in whistle sharing was found between genetic relatives of infants and adults showing close association patterns (measured by the coefficient of association (Schaller 1972) within social groups (Gubbins 1993; McCowan & Reiss 1995a; see Table 10.1–10.3). This lack of association between social affiliation and vocal learning may reflect that in captive environments there is continuous visual and acoustic exposure to all individuals within a social group. Such continuous exposure may override the influence of particular social partners (see also Tyack & Sayigh, Chapter 11).

Nevertheless, several results from our analyses of the structure of whistles provided indirect and direct evidence for vocal learning in infant dolphins (Tables 10.2 and 10.3). First, infants showed great variability and plasticity in their whistling behavior, producing a variety of whistle types both early in and across development that were not identical with those found in the adult repertoires. Whistle contour production changed dramatically across development, showing a 70–86% turnover rate in the whistle contours representing infant repertoires across development. For example, infants produced several whistle types that were unique to individuals and changed across development or disappeared from the repertoire entirely, suggestive of a form of babbling. Second, infants across social groups shared whistle types that adults did not share, suggesting that such whistle types might be precursors to adult whistles. Third, infants produced whistle types shared with adults in later developmental stages that were not shared during earlier ones and this sharing did not appear to be entirely dependent upon maturation. The plasticity of infant whistle structure evident in these results provides indirect evidence for vocal learning, especially when one considers the experimental evidence for vocal mimicry in dolphins (Richards *et al.* 1984; Reiss & McCowan 1993). Direct evidence for vocal learning in our observational study also was found in the sharing of whistle types within social groups. Infants did not share whistle types with adults exclusive to their social groups until the second or third developmental period. This result

Table 10.3. *Developmental trends in whistle type frequency for eight captive-born bottlenose dolphins from three different social groups*

	Social group 1									Social group 2					Social group 3					
Shared across social groups	Delphi (Months)			Pan (Months)			Liberty (Months)			Norman (Months)			Desi (Month)	Sam (Month)	Neptune (Months)		Tasha (Months)		Percentage	No. of whistles
	1–4	5–8	9–12	1–4	5–8	9–12	1–4	5–8	9–12	1–4	5–8	9–12	1	1	5–8	9–12	5–8	9–12		
1	59%	10%		28%	3%	4%	52%	36%	47%	59%	11%	10%	88%	92%	34%		26%	67%	35.28	452
2	12%	19%	53%	26%	25%	36%	26%	43%	29%	17%	60%	55%	8%		28%	67%	34%	33%	30.68	393
3	9%	6%		1%						1%	1%								1.17	15
4	7%										2%								1.01	13
5	6%	16%	14%	6%		24%	6%	1%	2%	1%	3%	8%							4.61	59
6											3%								0.23	3
7	1%							5%	2%	1%	3%	3%					6%		1.48	19
103												1%							0.08	1
108									1%										0.16	2
8	6%	6%		3%		1%	1%		5%	1%	1%	1%			9%		3%		2.11	27
9		6%			3%	2%	3%	1%	2%	1%	2%	1%		2%			11%		1.87	24
10	3%					2%			2%		3%	2%			4%				0.86	11
11	6%	3%			6%	1%		1%			1%								0.47	6
12	3%					1%					1%						3%		1.09	14
13					13%	2%	2%	2%	1%	4%			5%		13%	25%	11%		2.81	36
16							1%	1%		2%					2%				0.94	12
28	1%			1%	3%				1%										0.23	3
36	4%			4%	3%							2%							0.47	6
52					3%						1%								0.16	2
55			6%				1%			1%		1%							0.39	5
60			3%											2%					0.16	2
63				2%		2%		1%											0.23	3
64				2%		2%			1%										0.23	3
65						4%						1%							0.31	4
66						1%						1%							0.16	2

Shared within social groups — whistle type repertoire table

Whistle type	17	31	36	74	32	85	143	153	161	138	104	105	40	48	47	12	35	3	Total %	N
Shared within social groups																				
14	6%																		0.16	2
15	13%	1%																	0.62	8
17	3%	1%		1%															0.39	5
18	6%				4%														0.55	7
20			9%			1%													0.39	5
53							1%	1%			1%	1%							0.16	2
67	6%						1%	1%			1%	2%							0.31	4
70										1%									0.16	2
109									1%										0.08	1
Unique to individuals																				
77			3%					1%	1%										0.16	2
41								1%	1%										0.16	2
98								1%	1%										0.16	2
Unique to development period	12%	13%	16%	28%	16%	16%	3%	6%	6%	11%	7%	15%		4%	11%	8%	6%		9.52	122
Total percentage	100	100	100	100	100	100	100	100	100	100	100	100	100	100	100	100	100	100	100.00	
Number of whistle types unique to individuals	1	4	4	13	14	4	2	14	8	14	7	11	0	0	4	1	2	0		3
Total number of whistles	17	31	36	74	32	85	143	153	161	138	104	105	40	48	47	12	35	3		1281

Notes:

Dark outlined boxes indicate that whistle type was predominant in repertoire.

Shaded whistle types indicate that whistle types were also shared with adults.

Source: © Copyright 1995 by the American Psychological Association (from McCowan & Reiss 1995a). Reprinted with permission.

suggests that infants learn specific whistle types from the adults of their social groups (Figs. 10.2, 10.4 and Tables 10.2, 10.3).

The evidence also suggests that infant dolphins do not show a critical or sensitive period for vocal learning during the first year of development (Sayigh *et al.* 1990; for a review, see Caldwell *et al.* 1990; Tyack & Sayigh, Chapter 11). Vocal learning probably continues well into, if not beyond, the second year (Reiss & McCowan 1993). Some whistle types showed adult patterns in structure during the first year but the entire whistle repertoire of infants was not similar to that of adults by the end of our observational period. For instance, some whistle types shared by adults of the same social group (e.g. whistle types 129, 130, 131, 132) were not represented in the infants' repertoires by the end of the first year (Tables 10.2 and 10.3). Yet, an analysis of the rate and amount of frequency modulation in infant whistles compared to adult whistles revealed that infants were capable, from a maturational standpoint, of producing the complexity (as measured by frequency modulation) of adult whistles by the second developmental period (McCowan & Reiss 1995a). This difference in repertoire composition between infants and adults is not surprising for several reasons. First, infant dolphins remain with their mothers from three to ten years after birth (Wells *et al.* 1987; Smolker *et al.* 1992) and have a strong mother–infant bond, with an unusually long lactation period. In addition, dolphins do not reach sexual maturity until approximately 8–11 years for males and 10–15 years for females (Mead & Potter 1990). Because dolphins live in complex and flexible social groups, infants probably need to learn how to integrate socially in their communities. The protracted mother–infant bond and the long prereproductive stage in dolphins may function, in part, to provide the infant with the necessary social tools. Vocal learning is possibly a requisite step in acquiring these tools because vocalizations play such a key role in mediating social interactions between dolphins. Second, experimental communication studies on vocal modification of repertoires have demonstrated that dolphins are clearly capable of mimicking both artificial and conspecific whistles well beyond the infant years (Richards *et al.* 1984; Tyack 1986; Reiss & McCowan 1993). Indeed, the dolphins' ability for vocal plasticity as juveniles, adolescents, and adults in what is probably an "open" communication system is related to their fission–fusion social structure and, specifically, to the flu-

idity of their short-term associations (Connor & Smolker 1985; Wells *et al.* 1987; Würsig 1978; Connor *et al.* 1992).

Modification of infant repertoires was also found in the frequency of use of whistle types and in the contextual use of whistle types. Marked changes in the frequency of use of whistle types occurred for all infants and showed a similar trend across development (Table 10.3). Whistle type 1, a predominant whistle type exclusive to and used by all infants (Table 10.2, Fig. 10.5), was displaced in developmental period 2 by whistle type 2, a predominant whistle type for all adults (Tables 10.1 and 10.2, Fig. 10.6). Like adults, infants produced other whistle types with relative infrequency (0.08–9.52%).

Infants also showed ontogenetic change in the contextual uses of whistle type 1, whistle type 2, and whistle types unique to individuals. However, a role for learning could not be implicated in the contextual usage of whistle types by infants at this early stage of analysis. In general, infants produced whistle type 1 in contexts of infant separation during developmental periods 1 and 2 (percentage of whistles: mean=60, SD=33), suggesting that this whistle type may act as an infant separation call during early development. Indeed, the structure and context of whistle type 1 (Fig. 10.5) are not unlike those of separation or isolation calls of other mammalian species (Walther 1984; Gelfand & McCracken 1986; Balcombe 1990; Balcombe & McCracken 1992; Scherrer & Wilkinson 1993). In contrast, whistle type 1 was produced in more variable contexts by infants during developmental period 3 (Separation: mean=44%, SD=22%. Social: mean=47%, SD=27%. Other: mean=9%, SD=17%.). The importance of this change to more variable contexts is unclear.

Whistle type 2 was generally produced in contexts of separation by infants (percentage of whistles: mean=74, SD=25%) during developmental periods 1 and 2, similar to adult females. By developmental period 3, infants produced whistle type 2 primarily during contexts of social interactions with conspecifics (percentage of whistles: mean=44, SD=20%), more similar to the adult males. However, considerable variability was found across infants and over development. In addition, differences found in the contextual usage of whistle type 2 between adult males and females in this preliminary analysis makes it difficult to assign a consistent social function for whistle type 2 in either adults or infants at this time. Yet, whistle type 2 is

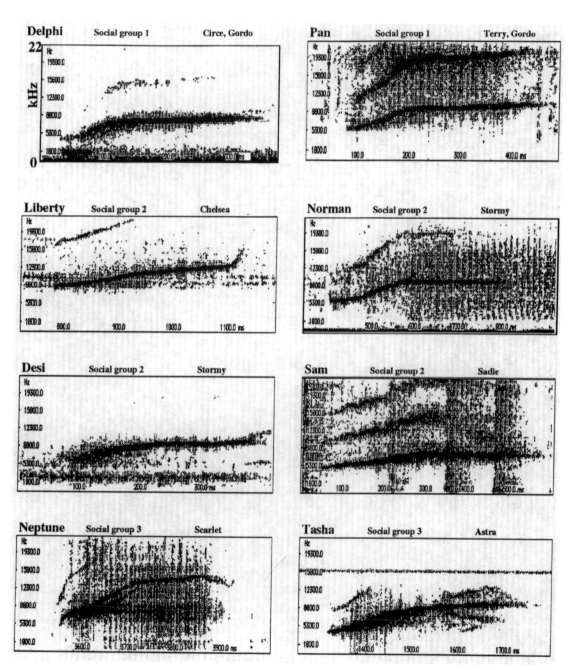

Fig. 10.5. Representative spectrograms of whistles from whistle type 1 for each infant. Name, social group membership, and names of mother and known father are provided for each infant. (Copyright © 1995 by the American Psychological Association (from McCowan & Reiss 1995a). Reprinted with permission.)

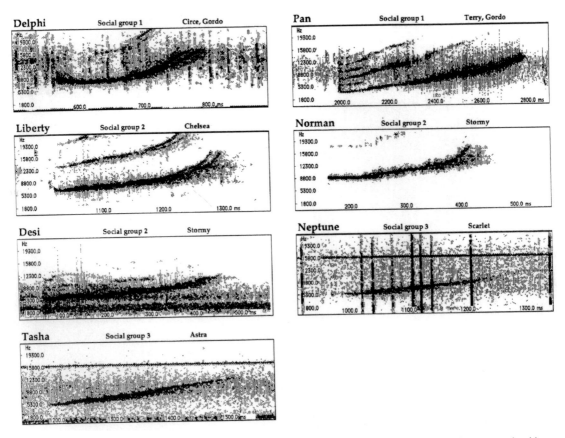

Fig. 10.6. Representative spectrograms of whistles from whistle type 2 for each infant. Name, social group membership, and names of mother and known father are provided for each infant. (Copyright © 1995 by the American Psychological Association (from McCowan & Reiss 1995a). Reprinted with permission.)

probably used to establish or maintain social contact between individuals.

Whistle types unique to individuals were produced generally during infant separation (percentage of whistles produced during separation: mean=66, SD=32%) during developmental periods 1 and 2. Again considerable variability was found across infants over development. By developmental period 3, however, all infants produced whistle types unique to individuals primarily during social interactions (percentage of whistles: mean=56, SD=10%), similar to both adult males and females. These results suggest that whistle types unique to individuals may function in specific types of social interactions between individuals for adult males and females, and for infants beginning in developmental period 3.

Because the frequency of use and contextual usage of whistles for some whistle types appeared to match the adult pattern during the first year, unlike the structure of repertoires, it may suggest that, like humans, the use and perhaps comprehension of whistle types are acquired within the first year and precede completed development of vocal production.

DISCUSSION

The dolphin is an exceptional model for the study of vocal learning for several reasons. Dolphins represent a highly social, large-brained, nonprimate species that has evolved to a totally marine existence. Their fission–fusion social structure coupled with their existence in an aquatic habitat makes it highly probable that acoustic communication plays an important role in the formation and maintenance of social relationships and behavior.

There is strong evidence for vocal learning in this species and experimental studies have demonstrated the dolphin's proclivity for vocal mimicry. However, the processes and stages of vocal development are not well understood. Many factors influence the vocal repertoires of both young and adult dolphins including acoustic, social, and environmental factors. Investigating the role of social influence on vocal development requires distinguishing ontogenetic changes solely due to maturation from those that are due to learning. Maturation and learning models need not be viewed as competing models but rather as complementary processes. Nevertheless, in order to determine the respective roles of maturation and learning and the processes by which such learning is achieved, we need first to determine what aspects of communication are more variable and "open" and what are more fixed or "closed."

Evidence for vocal modification and learning in dolphins

To date, a role for vocal learning and imitation has only been well documented in humans and select avian species (Marler 1977; Pepperberg & Neapolitan 1988; Locke 1990, 1993a,b). The relative roles of learning and maturation have not been well defined in the vocal development of most mammalian species. Most mammalian vocalizations are thought to be either fixed at birth (Talmage-Riggs *et al.* 1972; Winter *et al.* 1973; Newman & Symmes 1982; Herzog & Hopf 1984) or show modification of call features as a result of maturation (Seyfarth & Cheney 1986; Gouzoules & Gouzoules 1989; Elowson *et al.* 1992). However, some recent studies have pointed to a possible role for social influence on the vocal structure of specific call types in mammals (cats (Romand & Ehret; 1984); primates (Elowson *et al.* 1992; Elowson & Snowdon 1994; Snowdon *et al.*, Chapter 12; Seyfarth & Cheney, Chapter 13) and, in some cases, on entire vocal repertoires (cetaceans: Payne *et al.* 1983; Tyack & Whitehead 1983; Bain 1986; Mobley *et al.* 1989). For example, Seyfarth & Cheney (Chapter 13) have noted that subtle acoustic modifications in nonhuman primate vocalizations have been recently discovered, with the aid of new quantitative techniques, that are not inconsistent with a role for vocal learn-

ing. Similarly, Snowdon *et al.* (Chapter 12) have indicated that New World monkeys show evidence for vocal modification in some vocalization types that cannot be explained by maturation alone.

In contrast to other mammals, bottlenose dolphins have demonstrated vocal learning, especially under experimental conditions. The results of our experimental study (Reiss & McCowan 1993) and others (Richards *et al.* 1984; Sigurdson 1989) provide compelling evidence for vocal learning by this species. Although in our observational ontogenetic study (McCowan & Reiss 1995a) we were unable conclusively to show a role for imitation in dolphin whistle development, the pattern of whistle repertoire changes and specifically the exclusive sharing of certain whistle types between adults and infants of the same social group does suggest a role for vocal learning in dolphins. We are currently conducting an experimental study with two young captive-born infants (less than one year old) in which we acoustically expose infants to novel computer-generated whistles both with and without specific contingencies. This study should help us to show conclusively a role of learning and examine more precisely the roles of imitation and babbling in infant whistle development.

Babbling

Several avian species have been found to overproduce the sounds of their songs (with species' constraints imposed) prior to selective attrition of the appropriate song elements (Marler & Peters 1982; Marler & Sherman 1985; Marler 1990; Nelson 1992). This overproduction has been compared to the role of babbling in language acquisition by human infants. In fact, it has been suggested that such babbling in both birds and humans prepares the articulatory and perceptual systems for later imitation (Hauser & Marler 1992; Locke 1990). Babbling may also have similar importance to dolphins and, as discussed by Snowdon (1988) and Snowdon *et al.* (Chapter 12), to pygmy marmosets. One striking observation from our experimental study (Reiss & McCowan 1993) was that during the process of initial vocal mimicry of the computer model sounds, the dolphins appeared to selectively attend to and mimic different acoustic parameters in an interdigitated manner. This stage of initial mimicry and production could be interpreted as vocal play or babbling. The infants in our observational study (McCowan & Reiss 1995a) produced a wide variety of whistle contours, of which only a few were shared with the adults of similar and different social groups. These contours were not "infantile" renditions of whistle types extant in the adults' repertoire but were considerably different in structure. However, because we did not find an overall decrease in the number of whistle types that infants produced over the first year, it is difficult to say how this apparent overproduction functions in the dolphin vocal development. Further research on infant whistle development beyond the first year should help us to determine its role. Until we can experimentally determine the roles of auditory input and feedback, and selective attrition in the dolphins' repertoire, we cannot be certain that babbling plays an important role in bottlenose dolphins' development.

Thus, while babbling may serve to prime the production and perceptual systems of different species to learn fundamentally different sounds and the organization of such sounds, it appears to play an important and similar role for birds and humans and perhaps dolphins and pygmy marmosets in the acquisition of their species-typical vocalizations. Indeed, it may even be suggested that birds, humans, dolphins, and marmosets have an innate predisposition to overproduce the sounds of their respective communication systems in order to aid in learning their vocal repertoires. Functionally, the purpose of babbling may differ across species. Nevertheless, the process of babbling might be quite similar, including stages of overproduction, maintenance, and loss (Locke 1990; Marler 1990). In addition, there are several levels of function for which babbling may serve. One level might be more immediate or proximate, such as social reward mediated through "play," as has been proposed for human infants (Locke 1993b). Another level might be more removed or ultimate, such as the use of babbling to compare auditory input with auditory feedback and articulatory production (Marler 1990; Hauser & Marler 1992; Locke 1993a). These levels are not necessarily mutually dependent nor need they be identical for meaningful comparisons across different species.

Social imitation and mimicry

Past experimental research has indicated that older infants, juveniles, and adults are clearly capable of imitating whistles spontaneously (Reiss & McCowan 1993) and

as a result of training procedures (Richards *et al.* 1984; Sigurdson 1989). Interestingly, in our study on vocal mimicry and production of computer-generated whistles (Reiss & McCowan 1993), we found that our dolphins began mimicking the computer whistles relatively rapidly after exposure to the model sounds in contrast to the results of other experimental studies that required extensive training and thousands of trials to achieve good matching to the model sounds (Richards *et al.* 1984; Sigurdson 1989). We suggested that this difference may be due in part to significant differences in methodologies employed in each of these studies (Reiss & McCowan 1993).

The importance and efficacy of social interaction and the use of specific and functional contingencies in developing communicative behavior has been well demonstrated in studies of human language development (Locke 1990, 1993a,b; Locke & Pearson 1990; Meltzoff & Moore 1992) and in studies investigating the acquisition of artificial codes with animals (chimpanzees (Savage-Rumbaugh & Rumbaugh 1978; Savage-Rumbaugh 1986; Hopkins & Savage-Rumbaugh 1991); birds (Baptista & Petrinovich 1984, 1986; Pepperberg & Neapolitan 1988; Pepperberg 1994; Baptista & Gaunt, Chapter 3; Pepperberg, Chapter 9)). The methods we employed incorporated both of these features and permitted the dolphins to freely choose the acoustic features to reproduce. The discrepancy in results from different experimental studies suggests that dolphins may have a predisposition to mimic sounds that flourishes in an environment providing both social and environmental contingencies following vocal behavior. Conversely, by testing the efficacy of a methodology that simulated aspects of learning within a social environment, we gained support for the hypothesis that dolphins learn their whistle repertoires in socially and environmentally responsive contexts. This methodology may more closely simulate learning processes in a social environment and thus produce more rapid and effective learning for this species. Indeed, Baptista & Gaunt (Chapter 3) review avian literature in which a similar experimental paradigm yielded similar results in birds. Birds that were exposed to tape-recorded song under stimulus control of the birds or where birds were allowed the opportunity to observe the investigator operate a loud speaker acquired song from the tape. Tape-recorded song alone failed to elicit vocal learning in these

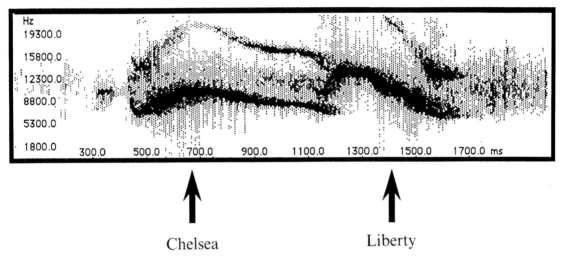

Chelsea   Liberty

Fig. 10.7. Whistle exchange between a mother (Chelsea) and her male infant (Liberty).

species (Todt *et al.* 1979; Adret 1993a,b).

In our observational study (McCowan & Reiss 1995a), however, the role of imitation was less clear. Using such a study design, we could not show conclusively that imitation played a role in vocal acquisition. However, because the two young dolphins in our experimental study were also two of the subjects of our observational study, some inferences can be drawn. Because these young males used imitation as evidenced in our experimental study (Reiss & McCowan 1993), they probably also employed it as a strategy in acquiring their species-specific whistles. This strategy may be used during earlier stages of development, perhaps in the second or third developmental period, as indicated by the sharing of whistle types exclusively between infants and unrelated adults of the same social group. In addition, further support for this view is derived from anecdotal evidence for imitation found in our recordings (see Fig. 10.7). A pattern of whistle exchange between mother and infant was found only occasionally. Yet, these immediate and overlapping whistle exchanges were similar to those found for the mimicry of computer whistle models by the one-year-old males. However, we cannot exclude the possibility that these were simply social whistle exchanges between mother and infant and not mimicry. We also cannot assume that because imitation occurs at later stages of development that it necessarily occurs at earlier ones, since, from a maturational standpoint, infants may be unable to imitate at earlier stages (Locke 1990;

Meltzoff & Moore 1992).

How do these data on dolphin mimicry relate to imitation in other species? Several experimental studies with birds and humans have indicated that imitation plays an important role in vocal development. Research has been conducted with several avian species in which the importance of self-hearing versus the hearing of others was experimentally manipulated and compared (Marler 1975; Marler & Waser 1977). Studies with human infants, either congenitally hearing impaired or tracheotomized, have also shed substantial light on the respective roles of auditory feedback from self-exposure or from hearing vocal models (Oller & Eilers 1988; Locke & Pearson 1990). Both sets of studies indicate that vocal models and thus imitation are important to appropriate acquisition of vocal repertoires in birds and humans. Delayed imitation of nonvocal behavior has been reported in newborn human infants (Meltzoff & Moore 1992). Imitation of some gestures may not emerge until later in development and others may improve over development toward more precise matching with adults. As in most cases concerning developmental processes, both maturational factors and learning probably contribute to the improvement of previously learned gestures and the emergence of new ones. These findings are important for three reasons. First, social imitation in infants was thought to not occur until 8 to 12 months of age. Meltzoff & Moore (1992) showed, using their new methodology, that social imitation is used by infants younger than one hour old. Second, the process of imitation observed gave insight into why infants imitate adults. These data suggested that infants use imitation to bond socially with, and identify, adults in their environments. Lastly, these results suggest that infants do not learn to imitate but rather have an innate predisposition to imitate the actions of socially important individuals. These data on human and animal imitation of vocal and nonvocal behavior suggest that imitation may not be learned by birds, dolphins, or humans, but rather that the predisposition to imitate is inherited.

Evolutionary parallels between dolphin and human vocal learning

If there is a convergence in the mechanisms of vocal learning in humans, birds, and dolphins, it may be a result of an underlying adaptive value of vocal learning and its evolution may have been based upon a similar set of social demands present in different species. It also is possible, given the phylogenetic distance between birds, dolphins, and humans, that vocal learning evolved under entirely different selection pressures and for different functions in each species, even if learning mechanisms are similar. However, a convergence in mammals, such as that between dolphins and humans, suggests some similar set of social pressures, namely a fluid social structure, possibly selected for a flexible communication system and, in turn, learning mechanisms to permit such flexibility. Dolphins, like humans, exhibit a fission–fusion social structure, where social groups show considerable fluidity in their composition and subgroups of individuals show long-term relationships (Tayler & Saayman 1972; Saayman & Tayler 1973; Würsig & Würsig 1977; Würsig, 1978; Wells et al. 1980, 1987; Ballance 1990; Smolker et al. 1992). Thus, the strategy of vocal learning may have evolved as a similar solution to a similar problem in these mammals. In the case of songbirds, such convergence may be the result of a similar solution to a very different problem. Vocal learning may have evolved under an entirely different set of selection pressures and for a different function, such as mate selection (Nottebohm 1972; Kroodsma 1983; West & King 1985; Brown & Farabaugh 1991). Indeed, humpback whale song and song learning by males in each season (Payne et al. 1983; Tyack & Whitehead 1983; Mobley et al. 1989) may be more functionally related to bird song than

to human language development, dolphin whistle development, or bird call development. However, we should be cautious in assigning selection pressures to the evolution of vocal learning in birds or mammals until a more comprehensive and comparative database is available. Using a comparative framework, we can test hypotheses on the presence of vocal learning across species. For instance, we would predict that vocal learning might be found in mammalian species where social group membership shows considerable fluidity (e.g., fission–fusion social systems – chimpanzees, spider monkeys) or where vocal communication is very important in mediating flexible social interactions, such as in social bonding or cooperative activities (e.g., squirrel monkeys, marmosets, wolves, coyotes). Species showing such characteristics warrant further investigation for evidence of vocal learning and modification.

Areas of future research for the study of vocal learning

Several areas of dolphin vocal ontogeny need to be further investigated. Vocal learning is clearly implicated in dolphin vocal ontogeny but the roles of selective reward or imitation and their biological constraints in shaping infant whistles in dolphins still remain unclear. Experimental evidence for the imitation of artificial signals by adults and juveniles is substantial (Richards et al. 1984; Reiss & McCowan 1993), but the use of such vocal imitation during early vocal development has still not been well documented. As indicated throughout this chapter, experimental studies similar to those conducted with avian species in which the infants' acoustic environment is manipulated are needed to examine the specific acoustic, social, and biological mechanisms of vocal learning in the developing whistle repertoire. As already mentioned, our study group is currently conducting an experiment with two captive-born infants using acoustic exposure of computer-generated whistles to examine the processes involved in dolphin whistle development. In addition, further research on the perception and categorization of dolphin whistles as well as the physiological and developmental mechanisms behind whistle production is necessary to determine the functional significance of whistle structure, with respect to both contour and other acoustic variables. Furthermore, analyses of whistles and whistle sequences need to be conducted in conjunction with playback experimentation in order to discern the social functions of whistles.

Only by combining experimental and observational approaches will we uncover the organization of dolphin whistle communication. A combined observational and experimental study on dolphin vocal ontogeny will help to determine how their whistles become integrated and organized across development, providing a more systematic approach to the study of whistle communication and perhaps a window into whistle structure and function in dolphins. In addition, further quantitative research on the vocal development of nonhuman primates and other mammalian species using a similar combination of paradigms is needed to compare the underlying mechanisms of vocal development across mammalian species and to examine the possible selection pressures that resulted in the evolution of vocal learning in humans and other animals.

ACKNOWLEDGEMENTS

This research was funded in part by Marine World Foundation, Harvard University, the Andersen Foundation, and Sigma Xi. We would like to thank Cara Gubbins, Kerri O'Brien, Carrie Lipe, Laura Edenborough, Spencer Lynn, the Project Circe volunteer staff, the Marine Mammal Department at Marine World Africa USA, Dave Miller, Sonny Allen, Jim Mullen, and the Animal Training Department at Sea World of Ohio.

REFERENCES

Adret, P. (1993a). Operant conditioning, song learning and imprinting to taped song in the zebra finch. *Animal Behaviour* 46: 149–59.

Adret, P. (1993b). Vocal learning induced with operant techniques: An overview. *Netherlands Journal of Zoology* 43: 125–42.

Adret-Hausberger, M., Güttinger, H. R. & Merkel, F. W. (1990). Individual life history and song repertoire changes in a colony of starlings (*Sturnus vulgaris*). *Ethology* 84: 265–80.

Altmann, J. (1974). Observational study of behavior: Sampling methods. *Behaviour* 49: 227–65.

Andrew, R. J. (1962). Evolution of intelligence and vocal mimicking. *Science* 137: 585–9.

Bain, D. E. (1986). Acoustic behavior of Orcinus: Sequences, periodicity, behavioral correlates, and an automated technique for call classification. In *Behavioral Biology of Killer Whales*, ed. B. C. Kirkevold & J. S. Lockard, pp. 335–71. Alan R. Liss, New York.

Balcombe, J. P. (1990). Vocal recognition of pups by mother Mexican free-tailed bats, *Tadarida brasiliensis mexicana*. *Animal Behaviour* 39: 960–6.

Balcombe, J. P. & McCracken, G. F. (1992). Vocal recognition in Mexican free-tailed bats: Do pups recognize mothers? *Animal Behaviour* 43: 79–88.

Ballance, L. T. (1990). Residence patterns, group organization and surfacing associations of bottlenose dolphins in Kino Bay, Gulf of Mexico. In *The Bottlenose Dolphin*, ed. S. Leatherwood & R. R. Reeves, pp. 267–83. Academic Press, San Diego, CA.

Baptista, L. F. (1988). Imitations of whistle-crowned sparrow songs by a song sparrow. *Condor* 90: 486–9.

Baptista, L. F. & Morton, M. L. (1988). Song learning in montane white-crowned sparrows: From whom and when. *Animal Behaviour* 36: 1753–64.

Baptista, L. F. & Petrinovich, L. (1984). Social interaction, sensitive phases, and the song template hypothesis in the white-crowned sparrow. *Animal Behaviour* 32: 172–81.

Baptista, L. F. & Petrinovich, L. (1986). Song development in the white-crowned sparrows: Social factors and sex differences. *Animal Behaviour* 34: 1359–71.

Bastian, J. (1967). The transmission of arbitrary environmental information between bottlenose dolphins. In *Animal Sonar Systems*, ed. R. G. Busnel, vol. 2, pp. 41–7. Laboratoire de Physiologie Acoustique, Jouy-en-Josas.

Bloom, L., Hood, L. & Lightbrown, P. (1974). Imitation in language development: If, when, and why. *Cognitive Psychology* 6: 380–420.

Brodie, P. F. (1969). Duration of lactation in Cetacea: An indicator of required learning. *American Midland Naturalist* 1: 312–13.

Brown, E. D. & Farabaugh, S. M. (1991). Song sharing in a group-living songbird, the Australian magpie, *Gymnorhina tibicen*. III. Sex specificity and individual specificity of vocal parts in communal chorus and duet songs. *Behaviour* 118: 224–74.

Burdin, V. I., Reznik, A. M., Skornyakov, V. M. & Chupakov, A. G. (1975). Communication signals in the Black Sea bottlenose dolphin. *Soviet Physics and Acoustics* 20: 314–18.

Busnel, R. G. & Dziedzic, A. (1966). Acoustic signals of the pilot whale *Globicephala melana* and of the propoises *Delphinus delphis* and *Phocena phocena*. In *Whales, Dolphins, and Porpoises*, ed. K. S. Norris, pp. 607–48. University of California Press, Berkeley, CA.

Caldwell, D. K. & Caldwell, M. C. (1977). Cetaceans. In *How Animals Communicate*, ed. T. A. Sebeok, pp. 794–808. Indiana University Press, Bloomington IN.

Caldwell, M. C. & Caldwell, D. K. (1968). Vocalizations of naive captive dolphins in small groups. *Science* **159**: 1121–3.

Caldwell, M. C. & Caldwell, D. K. (1972). Vocal mimicry in the whistle mode by an Atlantic bottlenose dolphin. *Cetology* **9**: 1–9.

Caldwell, M. C. & Caldwell, D. K. (1979). The whistle of the Atlantic bottlenosed dolphin (*Tursiops truncatus*) – Ontogeny. In *Behavior of Marine Animals*, ed. H. E. Winn & B. L. Olla, vol. 3, pp. 369–401. Plenum, New York.

Caldwell, M. C., Caldwell, D. K. & Tyack, P. L. (1990). Review of the signature-whistle hypothesis for the Atlantic bottlenose dolphin. In *The Bottlenose Dolphin*, ed. S. Leatherwood & R. R. Reeves, pp. 199–233. Academic Press, New York.

Chaiken, M., Böhner, J. & Marler, P. (1994). Repertoire turnover and the timing of song acquisition in European starlings. *Behaviour* **128**: 25–38.

Cheney, D. L. & Seyfarth, R. M. (1982). How monkeys perceive their grunts: Field playback experiments. *Animal Behaviour* **30**: 739–51.

Cockcroft, V. G. & Ross, G. J. B. (1990). Observations on the early development of a captive bottlenose dolphin calf. In *The Bottlenose Dolphin*, ed. S. Leatherwood & R. R. Reeves, pp. 461–79. Academic Press, New York.

Connor, R. C. & Smolker, R. A. (1985). Habituated dolphins (*Tursiops* spp.) in Western Australia. *Journal of Mammalogy* **66**: 398–400.

Connor, R. C., Smolker, R. A. & Richards, A. F. (1992). Dolphin alliances and coalitions. In *Coalitions and Alliances in Humans and Other Animals*, ed. A. H. Harcourt & F. B. M. de Waal, pp. 415–43. Oxford University Press, Oxford.

Dreher, J. J. (1966). Cetacean communication: Small-group experiment. In *Whales, Dolphins, and Porpoises*, ed. K. S. Norris, pp. 529–43. University of California Press, Berkeley, CA.

Dreher, J. J. & Evans, W. E. (1964). Cetacean communication. In *Marine Bioacoustics*, ed. W. N. Tavolga, vol. 1, pp. 373–99. Pergamon, Oxford.

Elowson, A. M. & Snowdon, C. T. (1994). Pygmy marmosets, *Cebuella pygmaea*, modify vocal structure in response to changed social environment. *Animal Behaviour* **47**: 1267–77.

Elowson, A. M., Snowdon, C. T. & Sweet, C. J. (1992). Ontogeny of trill and J-call vocalizations in the pygmy marmoset, *Cebuella pygmaea*. *Animal Behaviour* **43**: 703–15.

Evans, W. E. (1967). Vocalizations among marine mammals. In *Marine Bioacoustics*, ed. W. N. Tavolga, vol. 1, pp. 159–86. Pergamon, Oxford.

Evans, W. E. & Prescott, J. H. (1962). Observations of sound production capabilities of the bottlenose porpoise: A study of whistles and clicks. *Zoologica* **47**: 121–8.

Farabaugh, S. M., Brown, E. D. & Veltman, C. J. (1988). Song sharing in a group-living songbird, the Australian magpie. II. Vocal sharing between territorial neighbors, within and between geographic regions, and between sexes. *Behaviour* **104**: 105–25.

Feekes, F. (1982). Song mimesis within colonies of *Cacicus c. cela* (Leteridae, Aves). A colonial password? *Zeitschrift für Tierpsychologie* **52**: 119–52.

Gelfand, D. L. & McCracken, G. F. (1986). Individual variation in the isolation calls of Mexican free-tailed bat pups (*Tadarida brasiliensis mexicana*). *Animal Behaviour* **34**: 1078–86.

Gouzoules, H. & Gouzoules, S. (1989). Design features and developmental modification of pigtail macaque, *Macaca nemestrina*, agonistic screams. *Animal Behaviour* **37**: 383–401.

Gubbins, C. M. (1993). Ontogeny of behavior and social affiliations of captive bottlenose dolphins (*Tursiops truncatus*). Master's thesis. San Francisco State University.

Hauser, M. D. (1988). How vervet monkeys learn to recognize starling alarm calls: The role of experience. *Behaviour* **105**: 187–201.

Hauser, M. D. & Marler, P. (1992). How do and should studies of communication affect interpretations of child phonological development? In *Phonological Development: Models, Research, Implications*, ed. C. A. Ferguson, L. Menn & C. Stoel-Gammon, pp. 91–129. York Press, Timonium, MD.

Hayes, K. J. (1951). *The Ape in Our House*. Harper, New York.

Herman, L. M. & Tavolga, W. N. (1980). The communication system of cetaceans. In *Cetacean Behavior: Mechanisms and Functions*, ed. L. M. Herman, pp. 149–210. Wiley, New York.

Herzog, M. & Hopf, S. (1984). Behavioral responses to species-specific warning calls in infant squirrel monkeys reared in social isolation. *American Journal of Primatology* **7**: 99–106.

Hopkins, W. D. & Savage-Rumbaugh, E. S. (1991). Vocal communication as a function of differential rearing experiences in *Pan paniscus*: A preliminary report. *International Journal of Primatology* **12**: 559–83.

Janik, V. M., Dehnhardt, G. & Todt, D. (1994). Signature whistle variations in a bottlenosed dolphin, *Tursiops truncatus*. *Behavioral Ecology and Sociobiology* **35**: 243–8.

Kaznadzei, V. V., Krechi, S. A. & Khakhalkina, E. N. (1976).

Types of dolphin communication signals and their organisation. *Soviet Physics and Acoustics* **22**: 484–8.

Kellogg, W. N. (1961). *Porpoises and Sonar*. University of Chicago Press, Chicago.

Kroodsma, D. E. (1982). Learning and the ontogeny of sound signals in birds. In *Acoustic Communication in Birds: Song Learning and Its Consequences*, ed. D. E. Kroodsma & E. H. Miller, vol. 2, pp. 1–23. Academic Press, New York.

Kroodsma, D. E. (1983). The ecology of vocal learning. *Bioscience* **33**: 165–71.

Kroodsma, D. E. & Pickert, R. (1984). Sensitive phases for song learning: Effects of social interaction and individual variation. *Animal Behaviour* **32**: 389–94.

Kuczaj, S. A. II. (1987). Deferred imitation and the acquisition of novel lexical items. *First Language* **7**: 177–82.

Lang, T. G. & Smith, H. A. P. (1965). Communication between dolphins in separate tanks by way of an acoustic link. *Science* **150**: 1839–43.

Leatherwood, S. & Reeves, R. R. (eds.). (1990). *The Bottlenose Dolphin*. Academic Press, New York.

Locke, J. L. (1990). Structure and stimulation in the ontogeny of spoken language. *Developmental Psychobiology* **23**: 621–43.

Locke, J. L. (1993a). Learning to speak. *Journal of Phonetics* **21**: 141–6.

Locke, J. L. (1993b). *The Child's Path to Spoken Language*. Harvard University Press, Cambridge, MA.

Locke, J. L. & Pearson, D. M. (1990). Linguistic significance of babbling: Evidence from a tracheostomized infant. *Journal of Child Language* **17**: 1–16.

Marler, P. (1975). An ethological theory of the origin of vocal learning. *Annals of New York Academy of Sciences* **280**: 386–95.

Marler, P. (1977). The evolution of communication. In *How Animals Communicate*, ed. T. A. Sebeok, pp. 45–70. Indiana University Press, Bloomington, IN.

Marler, P. (1990). Innate learning preferences: Signals for communication. *Developmental Psychobiology* **23**: 557–68.

Marler, P. & Peters, S. (1982). Subsong and plastic song: Their role in the vocal learning process. In *Acoustic Communication in Birds: Song Learning and Its Consequences*, ed. D. E. Kroodsma & E. H. Miller, vol. 2, pp. 25–50. Academic Press, New York.

Marler, P. & Peters, S. (1988). The role of phonology and syntax in the vocal learning preferences in the song sparrow, *Melospiza melodia*. *Ethology* **77**: 76–84.

Marler, P. & Sherman, V. (1985). Innate differences in singing behaviour of sparrows reared in isolation from adult conspecific song. *Animal Behaviour* **33**: 57–71.

Marler, P. & Waser, M. S. (1977). Role of auditory feedback in canary song development. *Journal of Comparative and Physiological Psychology* **91**: 8–16.

Masataka, N. & Fujita, K. (1989). Vocal learning of Japanese and rhesus monkeys. *Behaviour* **109**: 191–9.

McCowan, B. (1995). A new quantitative technique for categorizing whistles using simulated signals and whistles from captive bottlenose dolphins (Delphindae, *Tursiops truncatus*). *Ethology* **100**: 177–93.

McCowan, B. & Reiss, D. (1995a). Whistle contour development in captive-born infant bottlenose dolphins: Role of learning. *Journal of Comparative Psychology* **109**: 242–60.

McCowan, B. & Reiss, D. (1995b). Maternal aggressive contact vocalizations in captive bottlenose dolphins (*Tursiops truncatus*): Wide-band low frequency signals during mother/aunt–infant interactions. *Zoo Biology* **14**: 293–310.

McCowan, B. & Reiss, D. (1995c). Quantitative comparison of whistle repertoires from captive adult bottlenose dolphins (Delphindae, *Tursiops truncatus*): A re-evaluation of the signature whistle hypothesis. *Ethology* **100**: 193–209.

Mead, J. G. & Potter, C. W. (1990). Natural history of bottlenose dolphins along the Central Atlantic coast of the United States. In *The Bottlenose Dolphin*, ed. S. Leatherwood & R. R. Reeves, pp. 165–95. Academic Press, New York.

Meltzoff, A. N. & Moore, M. K. (1992). Early imitation within a functional framework: The importance of person identity, movement, and development. *Infant Behavior and Development* **15**: 479–505.

Mitani, J. C. & Brandt, K. L. (1994). Social factors influence the acoustic variability in the long-distance calls of male chimpanzees. *Ethology* **96**: 233–52.

Mobley, J. R., Herman, L. M. & Frankel, A. S. (1989). Responses of wintering humpback *whales* (*Megaptera novaeangliae*) to playback of winter and summer vocalizations and synthetic sound. *Behavioral Ecology and Sociobiology* **23**: 211–23.

Nelson, D. A. (1992). Song overproduction and selective attrition lead to song sharing in the field sparrow. *Behavioral Ecology and Sociobiology* **30**: 415–24.

Newman, J. D. & Symmes, D. (1982). Inheritance and experience in the acquisition of primate acoustic behaviour. In *Primate Communication*, ed. C. T. Snowdon, C. H. Brown & M. R. Petersen, pp. 259–78. Cambridge University Press, New York.

Norris, K. S. (1969). The echolocation of marine mammals. In *The Biology of Marine Mammals*, ed. H. T. Anderson, pp. 391–423. Academic Press, New York.

Nottebohm, F. (1972). The origins of vocal learning. *American Naturalist* **106**: 116–40.

Oller, D. K. & Eilers, R. E. (1988). The role of audition in infant babbling. *Child Development* **59**: 441–9.

Orne, M. T. (1981). The significance of unwitting cues for experimental outcomes: Towards a pragmatic approach. In *Clever Hans Phenomenon: Communication with Horses, Whales, Apes, and People.* ed. T. A. Seboek & R. Roesenthal. Annals of the New York Academy of Sciences, vol. 364, pp. 152–9. New York Academy of Sciences, New York.

Owren, M. J., Dieter, J. A., Seyfarth, R. M. & Cheney, D. L. (1992). "Food" calls produced by adult female rhesus (*Macaca mulatta*) and Japanese (*M. fuscata*) macaques, their normally-raised offspring, and offspring cross-fostered between species. *Behaviour* **120**: 218–31.

Payne, K., Tyack, P. & Payne, R. (1983). Progressive changes in the songs of humpback whales (*Megaptera novaeangliae*): A detailed analysis of two seasons in Hawaii. In *Communication and Behavior of Whales* ed. R. Payne, pp. 9–57. AAAS Selected Symposium Series. Westview Press, Boulder, CO.

Pepperberg, I. M. (1993). A review of the effect of social interaction on vocal learning in African grey parrots (*Psittacus erithacus*). *Netherlands Journal of Zoology* **43**: 104–24.

Pepperberg, I. M. (1994). Vocal learning in grey parrots (*Psittacus erithacus*): Effects of social interaction, reference, and context. *Auk* **111**: 300–13.

Pepperberg, I. M. & Neapolitan, D. M. (1988). Second language acquisition: A framework for studying the importance of input and interaction in exceptional song acquisition. *Ethology* **77**: 150–68.

Reiss, D. (1988). Observations on the development of echolocation in young bottlenose dolphins. In *Animal Sonar: Processes and Performance*, ed. P. E. Nachtigall & P. W. B. Moore, pp. 121–7. Plenum, New York.

Reiss, D. & McCowan, B. (1993). Spontaneous vocal mimicry and production by bottlenose dolphins (*Tursiops truncatus*): Evidence for vocal learning. *Journal of Comparative Psychology* **107**: 301–12.

Richards, D. G,. Woltz, J. P. & Herman, L. M. (1984). Vocal mimicry of computer-generated sounds and labelling of objects by a bottlenose dolphin (*Tursiops truncatus*). *Journal of Comparative Psychology* **98**: 10–28.

Romand, R. & Ehret, G. (1984). Development of sound production in normal, isolated, and deafened kittens during the first postnatal months. *Developmental Psychology* **17**: 629–49.

Saayman, G. S. & Tayler, C. K. (1973). Social organization of inshore dolphins (*Tursiops aduncus* and *Sousa*) in the Indian Ocean. *Journal of Mammalogy* **54**: 993–6.

Savage-Rumbaugh, E. S. (1986). *Ape Language: From Conditioned Response to Symbol.* Columbia University Press, New York.

Savage-Rumbaugh, E. S. & Rumbaugh, D. M. (1978). Symbolization, language, and chimpanzees: A theoretical reevaluation based on initial language acquisition processes in four young *Pan troglodytes. Brain and Language* **6**: 265–300.

Sayigh, L. S., Tyack, P. L., Wells, R. S. & Scott, M. D. (1990). Signature whistles of free-ranging bottlenose dolphins *Tursiops truncatus*: Stability and mother-offspring comparisons. *Behavioral Ecology and Sociobiology* **26**: 247–60.

Schaller, G. B. (1972). *The Serengeti Lion: A Study of Predator–Prey Relations.* University of Chicago Press, Chicago.

Scherrer, J. A. & Wilkinson, G. S. (1993). Evening bat isolation calls provide evidence for heritable signatures. *Animal Behaviour* **46**: 847–60.

Seyfarth, R. M. (1986). Vocal communication and its relation to language. In *Primate Societies*, ed. B. B. Smuts, D. L. Cheney, R. M. Seyfarth, R. W. Wrangham & T. T. Struhsaker, pp. 440–51. Chicago University Press, Chicago.

Seyfarth, R. M. & Cheney, D. L. (1986). Vocal development in vervet monkeys. *Animal Behaviour* **34**: 1640–58.

Seyfarth, R. M., Cheney, D. L. & Marler, P. (1980). Vervet monkey alarm calls: Semantic communication in free-ranging primates. *Animal Behaviour* **28**: 1070–94.

Sigurdson, J. (1989). Frequency-modulated whistles as a medium for communication with the bottlenose dolphin (*Tursiops truncatus*). Paper presented at the Animal Language Workshop, Honolulu, Hawaii.

Smolker, R. A. (1993). Acoustic communication in bottlenose dolphins. Ph.D. dissertation, University of Michigan, Ann Arbor.

Smolker, R. A., Mann, J. & Smuts, B. B. (1993). Use of signature whistles during separations and reunions by wild bottlenose dolphin mothers and infants. *Behavioral Ecology and Sociobiology* **33**: 393–402.

Smolker, R. A., Richards, A. F., Connor, R. C. & Pepper, J. W. (1992). Sex differences in patterns of association among Indian Ocean bottlenose dolphins. *Behaviour* **123**: 38–69.

Snowdon, C. T. (1988). A comparative approach to vocal communication. In *Comparative Perspectives in Modern Psychology: The Nebraska Symposium on Motivation*, ed. D. L. Leger, pp. 145–99. University of Nebraska Press, Lincoln.

Steklis, H. D. (1985). Primate communication, comparative neurology, and the origin of language re-examined. *Journal of Human Evolution* **14**: 157–73.

Sutton, D., Trachy, R. E. & Lindeman, R. C. (1981). Primate phonation: Unilateral and bilateral lesion effects. *Behavioural Brain Research* **3**: 99–114.

Talmage-Riggs, G., Winter, P., Ploog, D. & Mayer, W. (1972). Effect of deafening on the vocal behavior of the squirrel monkey (*Samiri sciureus*). *Folia Primatologia* **17**: 404–20.

Tayler, C. K. & Saayman, G. S. (1972). The social organisation and behaviour of dolphins *Tursiops truncatus* and baboons *Papio ursinus*: Some comparisons and assessments. *Annals of the Cape Province Museum of Natural History* **92**: 11–49.

Thorpe, W. H. (1963). *Learning and Instinct in Animals*, 2nd ed. Harvard University Press, Cambridge.

Todt, D., Hultsch, H. & Heike, D. (1979). Conditions affecting song acquisition in nightingales (*Luscinia megarhynchos*). *Zeitschrift für Tierpsychologie* **51**: 23–35.

Tyack, P. L. (1976). Patterns of vocalizations in wild *Tursiops truncatus*. Senior thesis. Harvard University.

Tyack, P. L. (1986). Whistle repertoires of two bottlenose dolphins, *Tursiops truncatus*: Mimicry of signature whistles? *Behavioral Ecology and Sociobiology* **18**: 251–7.

Tyack, P. & Whitehead, H. (1983). Male competition in large groups of wintering humpback whales. *Behaviour* **83**: 132–54.

Valentine, C. W. (1930). The psychology of imitation with special reference to early childhood. *British Journal of Psychology* **21**: 105–32.

Walther, F. R. (1984). *Communication and Expression in Hoofed Mammals*. Indiana University Press, Bloomington.

Wells, R. S., Irvine, A. B. & Scott, M. D. (1980). The social ecology of inshore Odontocetes. In *Cetacean Behavior: Mechanisms and Functions*, ed. L. M. Herman, pp. 263–317. Wiley, New York.

Wells, R. S., Scott, M. D. & Irvine, A. B. (1987). The social structure of free-ranging bottlenose dolphins. In *Current Mammalogy*, ed. H. Genoways, pp. 247–305. Plenum Press, New York.

West, M. J. & King, A. P. (1985). Social guidance of vocal learning by female cowbirds: Validating its functional significance. *Ethology* **70**: 225–35.

Winter, P., Handley, P., Ploog, D. & Scott, D. (1973). Ontogeny of squirrel monkey calls under normal conditions and under acoustic isolation. *Behaviour* **47**: 230–9.

Würsig, B. (1978). Occurrence and group organization of Atlantic bottlenose porpoises (*Tursiops truncatus*) in an Argentine bay. *Biological Bulletin* **154**: 348–59.

Würsig, B. & Würsig, M. (1977). The photographic determination of group size, composition, and stability of coastal porpoises. *Science* **198**: 755–6.

11 Vocal learning in cetaceans

PETER L. TYACK AND LAELA S. SAYIGH

INTRODUCTION

Marine mammals stand out among nonhuman mammals in their abilities to modify their vocalizations on the basis of auditory experience. While there is good evidence that terrestrial mammals learn to comprehend and use their calls correctly, there is much less evidence for modification of vocal production (Seyfarth & Cheney, Chapter 13). In contrast, vocal learning has evolved independently in at least two marine mammal taxa, the seals and cetaceans, and is widespread among the whales and dolphins. We concentrate our focus in this chapter on vocal learning and development in the bottlenose dolphin (*Tursiops truncatus*) because it is the marine mammal species in which vocal learning and imitation has been best studied.

Dolphins produce a variety of sounds. The two predominant sound types are clicks, which can be used for echolocation, and frequency-modulated whistles, which are used for social communication. In addition to whistles, dolphins produce short frequency upsweeps that have been called chirps (Caldwell & Caldwell 1970). The dolphin vocal repertoire also includes a variety of burst pulsed sounds and combinations of pulses and whistles.

Captive bottlenose dolphins of both sexes are highly skilled at imitating synthetic pulsed sounds and whistles (Caldwell & Caldwell 1972; Herman 1980). Once a dolphin learns to copy a sound, the novel sound can be incorporated into its vocal repertoire, and the dolphin can produce the sound even when it does not hear the model. Bottlenose dolphins may imitate sounds spontaneously within a few seconds after the first exposure (Herman 1980), or after only a few exposures (Reiss & McCowan 1993). Dolphins can also be trained using food and social reinforcement to copy computer-generated whistle-like sounds (Evans 1967; Richards *et al.* 1984; Sigurdson

1993). After experience with imitation training, some dolphins imitate the model sound immediately after the initial presentation. Vocal responses are so easy to train in seals and dolphins that many commercial trainers prefer vocal responses instead of the response movements of limbs more typically trained in terrestrial mammals (Schusterman 1978; Defran & Pryor 1980). These results clearly demonstrate remarkable abilities of vocal imitation and of voluntary control over vocalization.

While there has been basic agreement about the echolocation functions of dolphin clicks, there has been less agreement about the social functions of whistles and of whistle imitation. Early work looked for context-specific whistle repertoires that were shared by all members of a species (Lilly 1963; Dreher & Evans 1964). In 1965, however, David and Melba Caldwell made a breakthrough in the study of whistles when they recorded whistles from five freshly caught wild dolphins in a variety of contexts. Caldwell & Caldwell (1965) reported that each individual dolphin tended to produce its own individually distinctive whistle, which they called a signature whistle. Caldwell *et al.* (1990) reviewed whistle repertoires from 126 captive bottlenose dolphins of both sexes and a wide range of ages. The primary method of identifying which dolphin produced a whistle was recording dolphins when they were isolated, and signature whistles made up about 94% of each individual's whistle repertoire in this context. These signature whistles were distinctive between individuals and stable over many years. However, Caldwell *et al.* (1990) also reported that most bottlenose dolphins produce an extremely variable array of whistles which differ from their signature whistles. These nonsignature whistles were called aberrant whistles by the Caldwells. We call them variant whistles in order not to imply that these whistles are aberrations of signature whistles. While variant whistles made up only about 6% of the whistles in the Caldwell *et al.* (1990) data set, which emphasized isolated animals, they are more common in other contexts. For example, Tyack (1986a) reported variant whistles made up 23% of the repertoire of two dolphins that were interacting socially, and Janik *et al.* (1994) reported that a dolphin produced 20% variant whistles when it was isolated, but that nearly 50% of the whistles it produced when it was being trained were variants. Variant whistles can be very diverse, but there can also be considerable overlap in the repertoires of variant whistles from different individuals.

Additional evidence for signature whistles comes from a study of free-ranging dolphins in inshore waters near Sarasota, Florida. This long-term field study of population biology and behavior tracks approximately 100 free-ranging dolphins, which can be identified through individually distinctive markings (Wells *et al.* 1987; Scott *et al.* 1990; Wells 1991). Once or twice yearly, dolphins are briefly held in a net corral to be measured, sampled, marked, and then released. During this time, we record vocalizations with suction cup hydrophones placed directly on the head of each animal. A library of 398 recording sessions, most containing hundreds of whistles from an identified individual, has been obtained from 134 known individuals. Many of these dolphins were first recorded at one to two years of age and have been recorded over spans of 10–20 years. All but a very few individuals have shown a stable and distinctive signature whistle (Sayigh *et al.* 1990), similar to that reported by Caldwell *et al.* (1990), over the entire span of recordings. Many of these dolphins have been recorded both when they are restrained by humans and when they are free-ranging, and their signature whistles are similar in both conditions (Sayigh 1992; Fig. 11.1). One of us (L.S.) followed groups of free-ranging Sarasota dolphins (average group size three to seven animals) and found that the signature whistles recorded from the individuals sighted in the group during these follows matched those recorded previously from the same individuals using suction cup hydrophones (Sayigh 1992). About half of the whistles from each group matched the signature whistles of group members; the other half were similar to the variant whistles described by the Caldwells. These results suggest a higher proportion of variant whistles during follows of

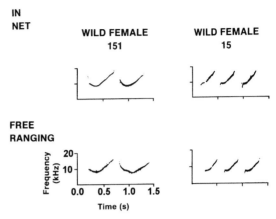

Fig. 11.1. Whistle contours recorded from two adult female bottlenose dolphins from a wild population near Sarasota, Florida. The left column shows whistles from female 151; the right column shows female 15. Note the similarity of contours recorded during temporary capture (top row) and under undisturbed free-ranging conditions (bottom row). Spectrograms were made with a sampling rate of 81 920 Hz, 256 point fast Fourier transform and Hamming window. (Modified from Sayigh 1992.)

wild dolphins than were typical either during the recordings of restrained wild dolphins or during the contexts in which the Caldwells recorded captive dolphins.

Caldwell & Caldwell (1979) conducted the first longitudinal study of whistle development in captive dolphins. They reported that dolphins whistle within a few days of birth, but these very young infants produced "tremulous, quavery" unstereotyped whistles that were more like variant whistles than like signature whistles. Most of the dolphin calves started to develop a stereotyped whistle within one to three months of age. This call then became a stable part of the dolphin's vocal repertoire. However, when Caldwell *et al.* (1990) compared whistle repertoires of infants around one year of age to those of juveniles, subadults, and adults, they found that production of a stable signature whistle was accompanied by an *increase* in the percentage of variant whistles with age.

Studies of vocal communication in terrestrial nonhuman mammals usually emphasize a stereotyped repertoire of species-specific vocalizations. If there is a systematic change in stereotypy over time, it usually decreases with age. Some calls may predominate in infancy and others in adulthood, but there is nothing comparable to the pattern reported for dolphin whistles. Dolphin calves start with a variable whistle repertoire. Within a few months to a year, most calves develop a stereotyped call out of this variable set. This stereotyped call is stable for decades, but is accompanied with age by steady and increasing production of highly variable calls. This communication system combines the capability to develop a highly stable stereotyped call with a lifelong ability to imitate sounds and an increasing proportion of variable sounds with age.

For the rest of this chapter, we first describe signature whistles in some detail, and then we focus on the following issues in the development of dolphin signature whistles:

1. Most dolphin calves develop signature whistles that differ from those of their parents.
2. Dolphin calves develop signature whistles that match acoustic models.
3. Dolphins vary widely in the timing of signature whistle development.
4. Both social and acoustic factors affect signature whistle development.
5. Dolphins maintain the ability to learn and produce new whistles throughout their lifetime.

In addressing the development of dolphin whistles, we include studies of captive dolphins, where the auditory and social environments can be manipulated and precisely specified, as well as studies of wild dolphins, observed in the context in which whistle function evolved. We then discuss the social functions of shared whistles and the evolution of vocal learning in marine mammals in general.

ACOUSTIC STRUCTURE OF SIGNATURE WHISTLES

There is a peculiar combination of stereotyped and variable features in signature whistles, and this complicates their definition. We discuss some examples in order to illustrate which features define the stereotypy of signature whistles. Figure 11.2 shows examples of signature whistles from two different individuals, a juvenile male born in captivity and an adult female from the wild Sarasota population. These are examples of whistles with repetitive elements, called loops by the Caldwells (Caldwell et al. 1990). The whistles shown here have one introductory loop, from zero to three repeated central loops, and one

terminal loop. For the whistles in the left column, the terminal loop is considered to be the last downsweep, but for the right column, the terminal loop is considered to be the upsweep–downsweep combination, because the terminal upsweep differs from the upsweep in the central loops. It should be noted that most of the sound spectrograms in Fig. 11.2 include several discrete whistle sounds separated by silence. We treat these whistle segments as one whistle if there is a regular repeated loop structure, just as discrete syllables of bird song are typically treated as one song if they occur in a regular and predictable series. Some dolphins, such as the one whose whistles are shown on the left, always separate loops by a silent interval. Others, such as the one on the right, do not always have gaps but may have fainter sections between loops. Depending upon the signal-to-noise ratio of the recording and the settings of the spectrographic analysis, these may appear continuous as on the top and middle right, or discontinuous as on the bottom right.

In a sample of signature whistles from 81 wild Sarasota dolphins, 59 (73%) had the repeated loop structure (Sayigh 1992). The remaining 22 (27%) produced signature whistles without repeated loops. Figure 11.3 shows four signature whistles without repeated loops, all of which were recorded from one wild dolphin. There is considerable variation in basic frequency and duration measures, presence of sidebands, and bandwidth of short segments of this signature whistle. For example, the central unmodulated segment of the whistle on the top right of Fig. 11.3 is much longer than that on the top left, while the earlier and later segments of the whistle are much less elongated. These kinds of variation in signature whistles are also thought to communicate factors such as the animal's motivational state (Caldwell et al., 1990; Janik et al. 1994). If animals know each other's signature whistles, then changes in whistle rate or in acoustic features of whistles may also broadcast information such as intention to join, affiliative or agonistic signals, level of arousal, or affective information.

Clearly there is considerable variability in what the Caldwells defined as one signature whistle, and there are other ways to approach whistle categorization. Some researchers, including Dreher & Evans (1964) and McCowan and Reiss (Chapter 10), treat each discrete tracing on the spectrogram as the unit of analysis for whistles. We have found problems with this approach,

LOOP STRUCTURE OF SIGNATURE WHISTLES

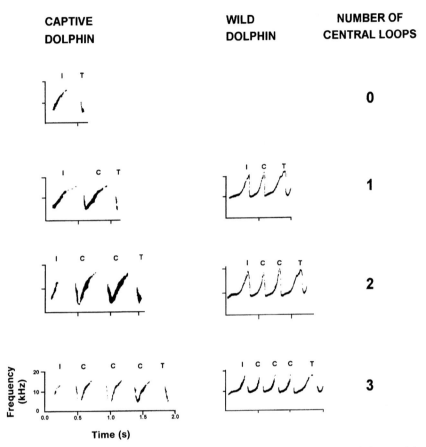

Fig. 11.2. Examples of signature whistles from two different individuals, a juvenile male from captivity (MLF 229 described in Caldwell *et al.* 1990) and an adult female from a wild population near Sarasota, Florida (Sarasota dolphin 35 described in Sayigh 1992). These are examples of what the Caldwells call repetitive whistles with one introductory (I) loop, 0–3 central (C) loops, and one terminal (T) loop. Spectrogram settings as in Fig. 11.1.

because it is very sensitive to recording and analysis conditions and would split what we consider one basic whistle pattern into a large number of subunits, whose cut-offs would vary depending upon where the whistle trace became too faint to be visible. For example, the whistle on the bottom right of Fig 11.2 might be considered as six different whistles, each one very different from the continuous whistle on the middle right. We tend to differentiate whistles based on the detailed structure of the loops, while allowing considerable variation in the number of loops. We even allow deletions of segments from a sig-

nature whistle (see e.g., Tyack 1986a). However, the continuous tracing approach may split what we call one whistle into subunits, and lump these subunits based on basic modulation patterns. For example, this approach might lump the introductory loop upsweeps or the downsweep–upsweep of the central loops of the two signature whistles shown in Fig. 11.2.

Many of the differences between McCowan & Reiss (Chapter 10) and this chapter probably stem from different approaches to categorizing whistles. Differences in the reported prevalence of upsweeps between our work and

VARIABILITY IN A NONREPETITIVE SIGNATURE WHISTLE

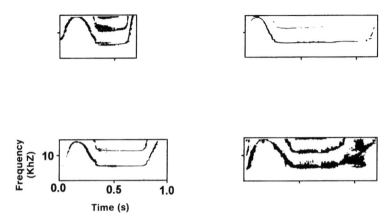

Fig. 11.3. Four examples of a signature whistle without repetitive loops. These spectrograms recorded from a wild dolphin (Sarasota 147) show the range of variability in duration and frequency of one young female's signature whistle. Spectrograms were made with a 90 Hz filter and a frequency range of 160–16 000 Hz. (From Sayigh *et al.* 1990.)

that of McCowan & Reiss (Chapter 10) may stem from two sources. We count many upswept introductory loops (e.g., Fig. 11.2) as parts of a signature whistle. We also might treat some of the short isolated upsweeps as chirps and not include them in our whistle tallies. Working with wild animals in a noisy environment, we may also fail to detect short highly modulated tonal sounds, but are more likely to detect longer less modulated sounds. Deciding which categorization scheme is most appropriate for a particular problem must ultimately depend upon how dolphins perceive and categorize whistles themselves.

SOCIAL CONTEXTS OF WHISTLE PRODUCTION AND INDIVIDUAL RECOGNITION

Enough is known about social behavior in wild *Tursiops* to place their acoustic communication in a social context. Bottlenose dolphins in coastal waters live in fission–fusion societies (Wells *et al.* 1987; Smolker *et al.* 1992). Groups in these societies are fluid; group composition often changes on a minute by minute basis. Bottlenose dolphins of either sex will have a shifting pattern of bonds with different individuals throughout their lifetimes. Within these fluid groups, however, certain individuals may have strong and stable relationships. For example, coalitions of two to three adult males may be sighted together more than 90% of the time for many years (Connor *et al.* 1992). One of the strongest bonds is the mother–calf bond. Bottlenose dolphins have extended maternal care, and a calf will typically stay with its mother until just before the next calf is born, for a period of approximately three to five years (full range from one to ten years: Wells *et al.* 1987).

Individual-specific social relationships thus appear to be a critical element of bottlenose dolphin societies. Maintaining these relationships in animals as mobile as dolphins requires a communication system of signature signals and individual recognition (Beecher 1989). Among terrestrial mammals, recognition can be achieved using visual, olfactory, or voice cues. Diving mammals have limited options for recognition. Dolphins lack an olfactory bulb and are thought to be anosmic. Vision is limited in most marine habitats down to several meters, presenting problems for visual communication between animals that frequently separate by tens of meters. Dolphin calves are precocial in locomotor skills, and even very young calves often separate tens of meters from their mothers, on whom they are completely dependent (Smolker *et al.* 1993). Wild dolphins are routinely out of range of companions for all sensory modalities except acoustic communication.

The signature whistle hypothesis proposes that dolphins use signature whistles in order to broadcast their identity and location to associates. Observations of captive dolphins suggest that whistles function to maintain contact, particularly between mothers and young (McBride & Kritzler 1951; Wood 1954). When mother and young are forcibly separated in the wild, both mother and calf whistle at high rates (Sayigh *et al.* 1990). However, when calves separate voluntarily from their mothers in Shark Bay, Western Australia, they seem to be able to keep track of their mothers without such a high whistle rate, for they often turn toward the mother before either produces a whistle (Smolker *et al.* 1993). During these voluntary separations, the calf often whistles only after turning toward the mother, as if to signal its intention to reunite with the mother.

The hypothesis that dolphins use signature whistles to broadcast individual identity rests on the assumption that dolphins can use acoustic features of whistles to recognize different individuals. Results from discrimination tasks show that captive dolphins can easily be trained to categorize signature whistles from the same individual as similar and signature whistles from different individuals as different, even if these are novel whistles from dolphins with which they have not interacted. A captive male bottlenose dolphin was able to discriminate signature whistles from two different males, in a design which used a large sample of whistles from both individuals (Caldwell *et al.* 1969). In tests with artificially truncated whistles, the duration of whistle required for discrimination was half a second, about the time required for one loop. Recall of this discrimination remained high for as long as the animal was retested, up to 22 days. Experimental playbacks with wild bottlenose dolphins demonstrated that mothers and offspring recognized each others' signature whistles even after calves became independent from their mothers (Sayigh 1992).

Quantitative signal processing also indicates that signature whistles provide sufficient information to distinguish individuals. The question of whether interindividual variability of signature whistles is greater than intraindividual variability was tested analytically by Buck & Tyack (1993), who developed a computer algorithm to compare similarity of the fundamental frequency of pairs of whistles. This algorithm measures differences in absolute frequency while allowing different segments of the whistle to vary their time axis to maximize fit in frequency. This was used to sort three randomly chosen whistles from each of ten dolphins. Five dolphins produced signature whistles without repetitive loops and the other five produced multiloop whistles. The algorithm correctly matched 15/15 of the whistles without repetitive loops and 14/15 central loops extracted from multiloop whistles.

STABILITY OF SIGNATURE WHISTLES

Signals used for individual recognition must remain stable over time. Stability in signature whistles, determined by visually comparing spectrograms of whistles recorded from the same individuals over time, has been documented for periods of more than a decade in both captive (Caldwell *et al.* 1990) and wild (Sayigh *et al.* 1990) dolphins. The combined data set of almost 300 individuals recorded either in captivity or in the wild indicates clearly that signature whistles, once developed, are stable throughout a dolphin's lifespan. An example of these data are presented in Fig. 11.4, which shows signature whistles recorded from an adult female over 11 years and a calf at one and three years of age.

There is also limited evidence that dolphins kept in unusual captive situations may stop producing their normal signature whistle. One male dolphin kept in isolation from one/two to eight/nine years of age stopped producing his signature whistle after prolonged exposure to recordings of whistles and artificial whistle-like sounds (Caldwell *et al.* 1990). Instead, he produced a large variety of whistles including imitations of the last stimulus to which he had been exposed. Tyack (1991a) reported a change in signature whistle production of a male bottlenose dolphin named Scotty after the death of his poolmate, named Spray. Two years after Spray's death, Scotty, produced only segments of his signature whistle, consisting of short upsweeps that had occurred at the end of the full whistle. Caldwell *et al.* (1990) suggested that some of the controversy surrounding signature whistles may stem from research with long-term captive animals either isolated or exposed to many artificial sounds. These abnormal conditions may lead to whistle repertoires that differ from those of dolphins raised in a more natural acoustic environment.

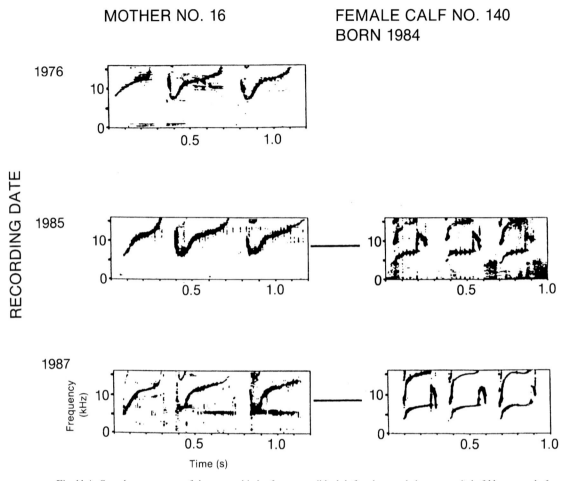

Fig. 11.4. Sound spectrograms of signature whistles from one wild adult female recorded over a period of 11 years and of one of her calves at one and three years of age. Note the stability of both signature whistles. Spectrogram settings as in Fig. 11.3. (From Sayigh *et al.* 1990.)

MOST DOLPHIN CALVES DEVELOP SIGNATURE WHISTLES THAT DIFFER FROM THEIR PARENTS

How can one evaluate the relative importance of genetic factors compared to auditory experience on the acoustic structure of vocalizations? Most mammals produce vocalizations very similar to those of their parents. Unfortunately, this similarity does not discriminate well between the two factors. However, if an animal produces vocalizations very different from those of its parents, this provides evidence against simple inheritance of call structure. We

discuss evidence that most dolphin calves develop signature whistles that are quite different from those of their parents.

Sayigh *et al.* (1995) compared the signature whistles of 42 wild dolphin calves to those of their mothers. Similarity was ranked by judges who rated pairs of spectrograms on a scale from 1 to 5, where 1 meant not similar and 5 meant very similar. Many judges ranked the same whistles to allow testing of the reliability of these similarity scores (mean effective reliability was 95%). Of the 42 calves, 31 (74%) produced signature whistle that were not judged to be similar to those of their mothers (score <3.66). However,

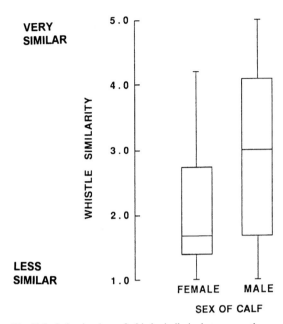

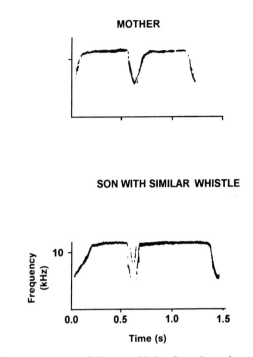

Fig. 11.5. Judges' ratings of whistle similarity between mothers and daughters versus mothers and sons in 42 mother–calf pairs from a wild population near Sarasota, Florida. A rating of 5 indicates high similarity; 1 indicates low similarity. Box plots show median, quartiles and extremes of ratings.

Fig. 11.6. Spectrograms of signature whistles of a mother and son from the wild population near Sarasota, Florida, rated as highly similar (average judges rating=5). Spectrograms were made with a sampling rate of 40 960 Hz, 256 point fast Fourier transform and Hamming window. (From Sayigh *et al.* 1995.)

there was a pronounced sex difference: males were more likely than females to produce whistles similar to those of their mothers, whereas females were more likely than males to produce whistles highly distinct from those of their mothers (Fig. 11.5). Figure 11.6 shows examples of whistles rated very similar (score=5) from a mother–son comparison and Fig. 11.7 shows examples of whistles rated very different (score=1) from a mother–daughter comparison.

In their longitudinal study of whistle development among newborn infant dolphins, Caldwell & Caldwell (1979) reported that one of the 14 infants showed an obvious match between its signature whistle and one of its parents. This one calf was raised alone with its mother (except for a brief period when an adult white-sided dolphin was also in the pool), and rapidly developed a whistle similar to its mother's. Sayigh (1992) analyzed whistles from nine dolphins born in a large community pool at the Miami Seaquarium. Of the five male and four female calves, only one calf, Dancer, developed a signature whistle somewhat similar to that of her mother Bebe (Fig. 11.8). One male, Papi, was known to be the father of all of these calves except

Dawn and Dancer. Only one male calf, Ivan, produced a whistle similar to that of Papi. Signature whistles of all of the calves and their known parents are shown in Figure 11.8. Thus, the body of evidence from wild and captive dolphins does not support the idea that dolphins inherit signature whistle contours similar to those of their parents.

DOLPHIN CALVES DEVELOP SIGNATURE WHISTLES THAT MATCH ACOUSTIC MODELS

Many studies of vocal development separate inheritance from experience through experiments in which the young are either isolated or cross-fostered with animals other than their relatives. Raising a highly social animal in isolation can lead to so many abnormalities that this paradigm has been questioned for the study of communicative behavior (for reviews, see West *et al.*, Chapter 4; Seyfarth & Cheney, Chapter 13). We doubt that raising a dolphin

MOTHER

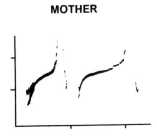

DAUGHTER WITH

DIFFERENT WHISTLE

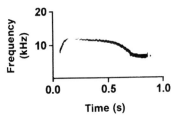

Fig. 11.7. Spectrograms of signature whistles of a mother and daughter from the wild population near Sarasota, Florida, rated as highly different (average judges rating=1 or least similar). Spectrogram settings as in Fig. 11.1. (From Sayigh *et al.* 1995.)

calf in isolation would provide a useful model for development in nature, even if such an experiment were judged to be practical and ethical. However, cross-fostering experiments have been productive for some social mammals. For example, primates raised with foster mothers of a different species still produce species-typical vocalizations (Owren *et al.* 1993; Seyfarth & Cheney, Chapter 13). Since signature whistles are individually distinctive, as opposed to species-specific, one can use cross-fostering with captive conspecific dolphins to evaluate the role of vocal learning in signature whistle development. We discuss data from one successful cross-fostering that took place during rehabilitation of a stranded one to two month old *Tursiops* calf. The top row of Fig. 11.9 shows the whistles of this calf, named April, as she arrived at the Gulfarium at Fort Walton Beach, Florida. April was bottle-fed by humans and initially kept in a pool with no other dolphins, but later she was put in a pool with one nulliparous adult female named Cindy. As was reported for two of the 120 dolphins studied by Caldwell *et al.* (1990), Cindy produced two

highly stereotyped whistle types. The primary whistle was the most common and was judged to be her signature whistle (third row of Fig. 11.9); the secondary stereotyped whistle was much rarer (fourth row of Fig. 11.9). By age six to seven months, April's whistle was quite different from when she was first recorded; it had become similar to the signature whistle of her foster mother (second row of Fig. 11.9). While we do not know the whistle of April's mother, it is highly likely that acoustic exposure to her foster mother influenced the course of April's whistle development.

There is a possible example of cross-species matching in the Caldwell & Caldwell (1979) study of vocal ontogeny in captive calves. One male *Tursiops* calf, raised in a pool with seven *Tursiops* and two Pacific white-sided dolphins (*Lagenorhynchus obliquidens*), had a whistle similar to the more vocal of the white-sided dolphins.

Some of the most convincing data for vocal learning in whistle ontogeny stems from comparisons of calf whistles to sounds present in their acoustic environments. Of the Miami Seaquarium calves discussed above, Tori and Sundance developed whistles very similar to the whistle used by trainers to signal a dolphin that it can approach for food after performing a requested task (Fig. 11.10). Another calf, named PJ, developed a whistle quite similar to the signature whistles of two subadults, Noel and Samantha, who themselves were raised together in the pool (Fig. 11.10).

These imitations are so closely matched that they provide compelling evidence for vocal learning in the development of signature whistles in bottlenose dolphins. The tendency of dolphin calves to match whistles other than those of their mothers as they develop their signature whistles presents an excellent alternative to separating the infant from its mother in isolation or cross-fostering experiments. Experiments that require separating the infant calf from its mother at birth not only may lead to abnormal development, but also do not isolate the calf from exposure to the mother's calls. Dolphin calves hear well at birth, so prenatal exposure to sounds *in utero* may affect vocal development.

DOLPHINS VARY WIDELY IN THE TIMING OF SIGNATURE WHISTLE DEVELOPMENT

There is extraordinary variability in the timing of when young dolphins acquire a signature whistle, from immedi-

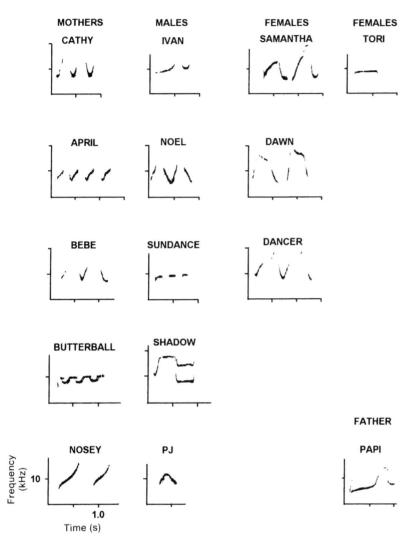

Fig. 11.8. Comparisons of the signature whistles of nine dolphin calves to those of their mothers. All of these calves were born in captivity at the Miami Seaquarium. Whistles from the calves of each mother are shown on the same row as the whistle of the mother. Also included in the lower right is a spectrogram of the signature whistle of the father, Papi, of seven of the nine calves. This male was not the father of Dawn and Dancer. Spectrogram settings as in Fig. 11.1.

ately after birth to over 17 months of age. Caldwell & Caldwell (1979) reported that most of the 14 captive infants in their study first produced signature whistles between 1.5 and 2.5 months of age, but that one dolphin had still not developed a stereotyped signature whistle at 17 months of age. The study of calves at the Miami Seaquarium suggests even earlier development of distinctive signature whistles. Two of the three calves were

recorded during the first week of life, at which time they already had relatively discriminable whistles (Fig. 11.11). A discriminant analysis of 33 acoustic features extracted from whistles produced during the first nine months of all three dolphins revealed low rates of misclassification of whistles and no tendency for misclassified whistles to come from early recordings (Fig. 11.12). The observation that two calves recorded in the first week of life already had

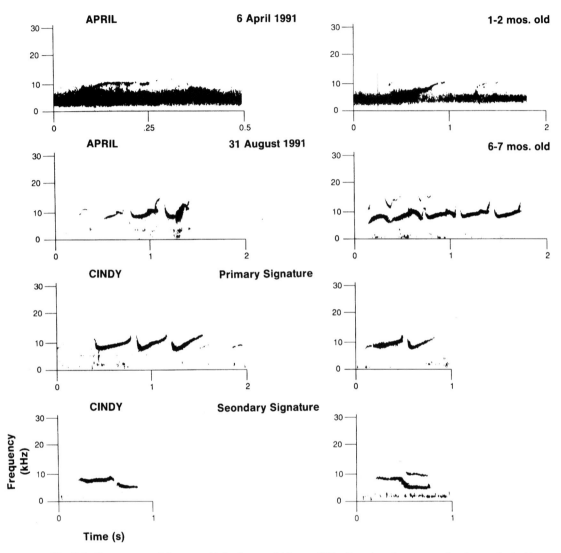

Fig. 11.9. Spectrograms of signature whistles from a wild-born calf (April) orphaned at two months of age and raised in captivity with a foster mother (Cindy). Over 90% of Cindy's signature whistles were of the primary signature contour, but she also regularly produced a secondary signature whistle. By six to seven months of age, the calf April had developed a signature whistle similar to Cindy's primary signature. Spectrogram settings as in Fig. 11.1.

whistles similar to their final signature whistle suggests that either prenatal exposure had an influence on whistle development and/or that they used a process of immediate imitation which must have been well developed at birth (such as the facial imitation produced by human neonates reported by Meltzoff & Moore 1977).

These results led us to design a study of the factors influencing the timing and outcome of signature whistle development in the wild. This study is conducted in inshore waters near Sarasota, and involves focal observations of wild mothers and calves during the period of whistle development, along with simultaneous acoustic recordings (Sayigh et al. 1993). Preliminary results from four calves indicate that there is considerable individual variability in the timing and outcome of signature whistle development (Table 11.1; Sayigh 1992). Two calves (one

Table 11.1. *Summary of whistle development data from a preliminary sample of four calves*

Calf ID no.	Sex	Time period of whistle crystallization	Contour similar to that of the mother?
203	M	1–2 months	Yes
212	?	1–2 months	Yes
209	?	3–4 months	No
210	F	13–24 months	No

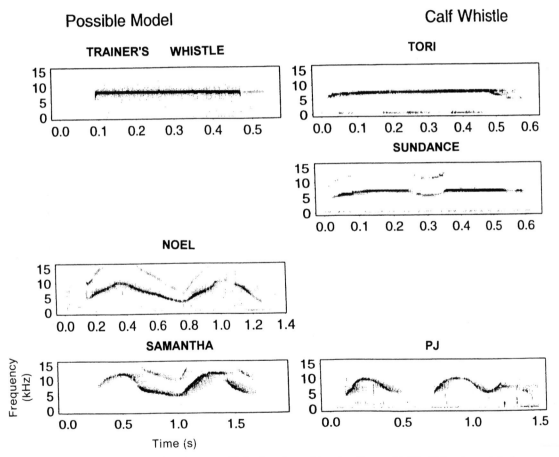

Fig. 11.10. Comparisons of the signature whistles from three of the calves illustrated in Fig. 11.8 to the most similar acoustic models present in the natal pool. These three calves were selected because they were born within a three-month period in the same pool. Spectrogram settings as in Fig. 11.1. (P. Tyack, J. McIntosh & K. Fristrup, unpublished data.)

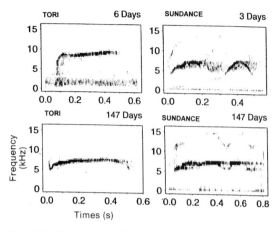

Fig. 11.11. Spectrograms of signature whistles from two captive-born calves recorded in the first week of life, with comparisons at 147 days of life. These were the only two of the nine calves illustrated in Fig. 11.8 that were recorded during the first week of life. Spectrogram settings as in Fig. 11.1. (P. Tyack, J. McIntosh & K. Fristrup, unpublished data.)

known to be male) exhibited relatively rapid (one to two months) whistle development and produced signature whistles that resembled those of their mothers, whereas two calves (one known to be female) exhibited more prolonged (3+ to 13+ months) whistle development and produced signature whistles that did not resemble those of their mothers.

BOTH SOCIAL AND ACOUSTIC FACTORS AFFECT SIGNATURE WHISTLE DEVELOPMENT

Preliminary data from our ongoing longitudinal study of wild dolphins in Sarasota indicate that some of the variability in timing and outcome of signature whistle development is correlated with the early social environment (Sayigh 1992). For example, the two calves just discussed above that exhibited rapid whistle development and produced signature whistles resembling those of their mothers (nos. 203, 212) may have had stronger bonds with their mothers than did the other two calves (nos. 209, 210). One indicator of the strength of a bond between individuals is the synchrony of their surfacings (Connor *et al.* 1992). Calves 203 and 212 showed higher levels of synchronous surfacing with their mothers during the period

of whistle development than did the other two calves (Sayigh 1992).

The number of different individuals that associated with the focal calves also may have affected whistle development. Calf 203, a known male, was consistently found in smaller groups than the other calves, and he spent a proportionately greater amount of time alone with his mother. Perhaps this contributed to his speedy whistle development and to his adoption of a signature whistle similar to that of his mother. However, calf 212 also exhibited early whistle development and produced a signature whistle similar to that of its mother, but had association patterns more similar to those of the other two calves.

The early acoustic environments of the calves, measured both in terms of whistle rates and the proportion of whistles heard by the calf that were the mother's signature whistle, also appeared to affect whistle development (Sayigh 1992). Calves 203 and 212 were exposed to the lowest overall whistle rates (0.64 and 0.73 whistles per minute) and the greatest percentages of their mother's signature whistle (20% and 18%; calculated by dividing the total number of signature whistles of the mother by the total number of whistles recorded, excluding calf whistles). At the other extreme, calf 210 was exposed to the highest average whistle rates (0.91 whistles per minute) and the lowest percentage of her mother's signature whistle (6%). Thus, the two calves that heard the fewest whistles and the greatest percentages of their mothers' signature whistle developed whistles resembling those of their mothers, and exhibited rapid whistle development. The calf that experienced the most diverse acoustic environment and the lowest percentage of her mother's signature whistle developed a whistle highly distinct from that of her mother, and exhibited prolonged (13+ months) and variable whistle development. The acoustic environment of calf 209 fell somewhere in the middle; it was second to calf 210 in overall whistle rates (0.79 whistles per minutes), and showed the second lowest percentage of signature whistles from the mother (14%). Calf 209's whistle development was intermediate in time course (three to four months) and its signature whistle was distinct from that of the mother.

What were the sources of the signature whistles adopted by calves 209 and 210? There was preliminary evidence for two possible sources (which were not mutually exclusive): "modelling" of a distinctive whistle by the mother, and learning from another individual present in

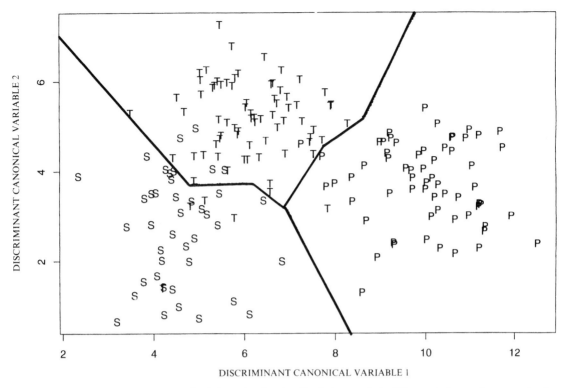

Fig. 11.12. Plot of the first two discriminant functions in a discriminant analysis of 33 acoustic features from a sample of whistles from the three calves illustrated in Fig. 11.10 during their first half-year of life. T represents whistles from Tori, S from Sundance, and P from PJ. Even the earliest whistles are well discriminated. (P. Tyack, J. McIntosh & K. Fristrup, unpublished data.)

the community. With respect to the first of these, the mother of calf 210 apparently produced a stereotyped whistle quite different from her own signature during her calf's first few months of life (Fig. 11.13). This whistle resembled the eventual signature whistle of the calf (Sayigh 1992). The mother may have acted as a social tutor, producing an acoustic model for her calf to imitate. Second, the signature whistles of both calves 209 and 210 showed strong similarities to whistles of young females present in the community. These could simply be chance similarities; there are bound to be accidental similarities among a community of 100 dolphins. The precision of whistle matching has not yet been quantified in a way that allows one to estimate how many dissimilar whistles can be produced by a large number of dolphins. Alternatively, the calves may have learned their signature whistles from these females. The signature whistle of calf 210 was similar to

that of female 175, who was present in five out of the 17 observation sessions. Calf 209, however, produced a whistle similar to that of a female (no. 140) with whom it was never observed in association (Fig. 11.14). Songbirds are able to learn songs to which they have received only very limited exposure (see e.g. Hultsch & Todt 1992; Peters et al. 1992; West & King 1990); perhaps a small number of particularly salient interactions with female 140 could have influenced whistle development. Alternatively, chance production of this whistle contour could have been reinforced by the mother, using a mechanism similar to that female cowbirds (Molothrus ater) use to modify male song (West & King 1988). Baptista and Gaunt (Chapter 3) and Snowdon et al. (Chapter 12) suggest that aggressive social interactions or the mere presence of an adult may inhibit the production of a call similar to that of the adult in some species. On the other hand, zebra finches (Taeniopygia

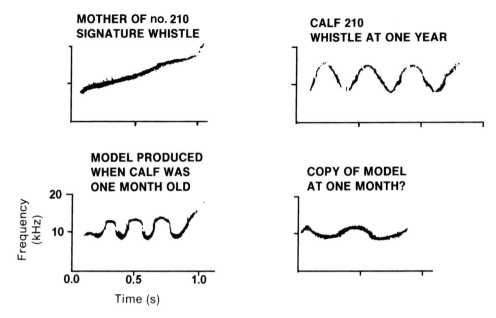

Fig. 11.13. Spectrograms of whistles produced by calf 210 and her mother, from the wild population near Sarasota, Florida. The top row shows the signature whistle of the mother and a whistle contour attributed to calf 210 as a yearling. The bottom row shows whistles suspected to be a model produced by the mother of calf 210 along with an apparent imitation by the calf of this possible model recorded when she was one month old. Spectrogram settings as in Fig. 11.1. (From Sayigh 1992.)

guttata) may selectively copy songs from an individual with which they interacted most aggressively (Clayton 1987). Payne (1983) also emphasizes a role for competitive imitation in indigo buntings (*Passerina cyanea*). In other species and contexts, presence of the adult coupled with affiliative interactions might encourage production of similar calls in the young. Further study of the early social and acoustic environments of developing dolphin calves is needed to elucidate which factor or combination of factors is responsible for the choice of contour by calves that do not produce whistles similar to those of their mothers.

DOLPHINS MAINTAIN THE ABILITY TO LEARN AND PRODUCE NEW WHISTLES THROUGHOUT THEIR LIFETIME

Bottlenose dolphins appear to develop a highly stable signature whistle within the first year or so of life, yet we have seen that captive dolphins are capable of imitating sounds throughout their lifespan (see e.g., Evans 1967; Richards *et al.* 1984; Sigurdson 1993). Why should dolphins maintain

such an active imitative ability into adulthood if they use primarily one stable whistle for individual identification? Observations of dolphins that are interacting socially and/or acoustically suggest that imitation may play an important role in the natural communication system of dolphins. In one study of two captive adult dolphins, Tyack (1986a) found that each imitated the signature whistle of the other at rates of about 25% (i.e., 25% of all occurrences of each signature whistle were imitations produced by the other dolphin). Other studies have reported rates of signature whistle imitation near 1% among captive dolphins that were in acoustic but not physical contact (Burdin *et al.* 1975; Gish 1979). These imitated signature whistles are not merely produced immediately after the partner makes its signature whistle, but they can become incorporated into a dolphin's whistle repertoire. For example, after a period of silence, dolphin A might produce a copy of dolphin B's signature whistle. The two animals in the Tyack (1986a) study were first housed together at about five years of age, well after signature whistles are developed. Thus, dolphins maintain the

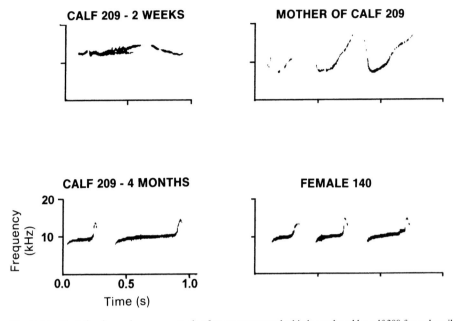

Fig. 11.14. The left column shows an example of an unstereotyped whistle produced by calf 209 from the wild population near Sarasota, Florida at an age of two weeks (top) and its signature whistle at an age of four months (bottom). The right column shows signature whistles from calf 209's mother (top), and subadult female 140 (bottom). The similarity in whistle contours between calf 209 and subadult 140 suggests that no. 209 may have learned its whistle from no. 140. However, no. 140 was not seen to associate with no. 209 during focal observations. Spectrogram settings as in Fig. 11.1. (From Sayigh 1992.)

ability well past infancy to add new whistles to their vocal repertoire through imitation of auditory models.

Since we already knew that captive dolphins imitate whistle-like sounds, it raises the question: is whistle imitation by adults an artefact of captivity or does it play a functional role in the wild? We have been able to address this question by studying wild dolphins from Sarasota. The whistles of known individuals can be identified when dolphins are recorded in a net corral. In these recordings, we have observed imitation of signature whistles between wild dolphins who share strong social bonds. For example, some pairs of adult males in Sarasota form a coalition and are usually sighted together (Wells *et al.* 1987). Signature whistles of each member of a pair of males (sighted together 75% of the time in Sarasota surveys) are shown in Fig. 11.15, along with imitations of the partner's whistle. Similar coalitions of two to three adult males are also reported from Western Australia, where males in a coalition often gain advantage in competition with other

males (Connor *et al.* 1992). One coalition regularly came in to a beach to be fed by humans, and the whistles of each of these three adult males could be recorded and identified in this setting. Smolker (1993) reported convergence of whistles among these three coalition members. A whistle initially produced by only one member of the coalition gradually became the most common whistle for all three males over two years. This provides another case of adult dolphins incorporating a new whistle into their repertoire.

Vocal development for many species has been characterized as a progressive narrowing from a large and variable (overproduced) repertoire to a more fixed and mature repertoire. While bottlenose dolphins do develop a stable signature whistle, variant whistles show an opposite pattern, with increasing variability in the vocal repertoire with age (Fig. 11.16). Caldwell *et al.* (1990) reviewed data on the percentage of variant whistles from 126 captive dolphins. The primary method used to identify which animal made a whistle was to record each dolphin when

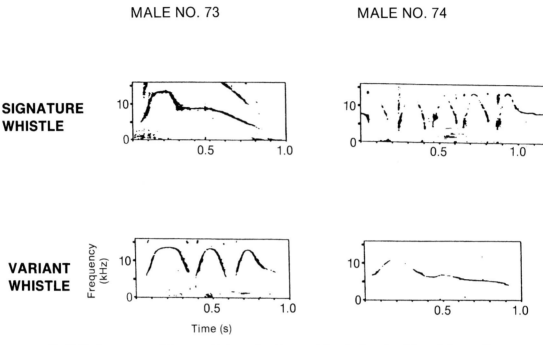

Fig. 11.15. Spectrograms of signature whistles produced by two adult males from the wild population near Sarasota, Florida, who were sighted together 75% of the time (Wells *et al*. 1987). Each male also produced whistles that were similar to the signature of the partner, and these variant whistles are interpreted as imitations of the partner's signature whistle. Spectrogram settings as in Fig. 11.3. (From Sayigh 1992.)

they happened to be briefly isolated, and they were recorded in a variety of behavioral contexts. This sample included 5 female infants and 8 males infants (up to one year of age), 12 female juveniles and 25 male juveniles (arbitrarily defined as animals 170–210 cm in length; or up to about 6 years of age), 25 female subadults (210–235 cm in length; between 6 and 12 years of age) and 22 male subadults (210–247 cm in length; between 6 and 13 years of age), and 18 female adults and 7 male adults (animals either known to have bred or larger than subadult size). Length is not an optimal criterion of age in dolphins, but this was required for animals of unknown age.

Nonsignature whistles were extremely rare among the female infants in this study, all of which had developed their signature whistle before they were first recorded. Only three variant whistles were recorded out of the 2968 whistles from female infants. Two of the smallest male infants appeared not yet to have developed signature whistles; the rest had between zero and 18% nonsignature whistles in this sample, which averaged over 600 whistles

per individual. There was a significant difference between the sexes, with males producing a larger proportion of variant whistles at each age. Males showed a steady increase in the percentage of variant whistles, while females showed a more complex pattern. While infant females had the fewest variant whistles and adult females had the highest percentage of variant whistles of all female age classes, juvenile females had slightly more variant whistles than did the subadults. The boxplots in Fig. 11.6 show that these differences in mean values stem from some outlying subadult females with many variant whistles and a relatively high percentage of variant whistles for the upper quartile of adult females.

Sayigh *et al*. (1990) found a similar pattern in wild dolphins, with males producing more variant whistles than females did. Maturing male calves showed a particular increase in the number of variant whistles after separation from their mother when they presumably formed a broader network of social relationships (Fig. 11.17). Some of these new whistles appeared to match those of other

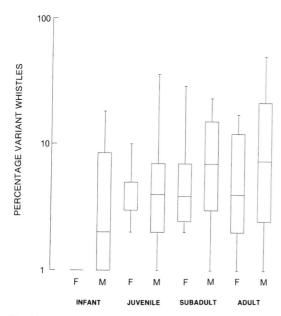

Fig. 11.16. Variation in percentage of variant whistles as a function of age and sex class in captive dolphins (data from Caldwell *et al.* 1990). At each age, males have more variant whistles. Both sexes show an increase in variant whistles with age. Box plots show medians, quartiles and extremes of ratings.

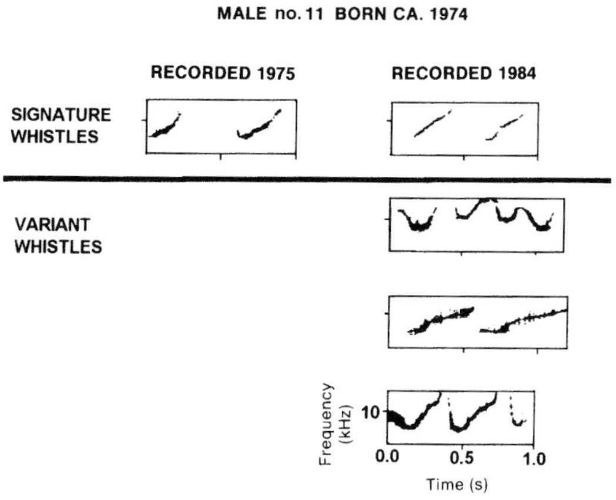

Fig. 11.17. Spectrograms of the signature whistle of male dolphin 11 from the wild population near Sarasota, Florida, recorded in 1975 and 1984, along with variant whistles recorded in 1984. No variant whistles were sampled when this calf was one year old. They were more common when he was ten years old and independent from his mother. Spectrogram settings as in Fig. 11.3. (Modified from Sayigh *et al.* 1990.)

animals caught at the same time, and could be categorized as imitations of signature whistles, as in Fig. 11.15.

We currently know little about the social functions of variant whistles. Some variant whistles are highly variable and seldom repeated, whereas others may be repeated and relatively stereotyped. Tyack (1986a) and Janik *et al.* (1994) defined several simple categories such as upsweeps (rise whistles), downsweeps, and sinusoidal patterns of frequency modulation that differed from the signature whistle but were repeated. McCowan & Reiss (Chapter 10) report repeated upsweep whistles, some of which are similar to what we have categorized as rise whistles. They also appear to include chirps and what we would categorize as elements of signature whistles into these more generic categories. More research needs to be conducted on these variants whistles that are repeated and are shared by many individuals.

SOCIAL FUNCTIONS OF SHARED WHISTLES AMONG DOLPHINS

The results on imitation of signature whistles between dolphins sharing strong social bonds suggest a link between vocal imitation and social interaction similar to that seen in a variety of bird species. Convergence of calls among birds within a flock has been reported for a large variety of species (Mundinger 1970; Mammen & Nowicki 1981; Nowicki 1989). Within such groups, birds that tend to associate closely also are more likely to share calls (Mundinger 1970; Hausberger 1993; Farabaugh *et al.* 1994; West *et al.*, Chapter 4; Brown & Farabaugh, Chapter 7). Both females and males may produce these calls, which often appear to function in recognition and maintenance of contact with particular companions.

Thorpe & North (1965) suggested that "the extreme developments of the imitative ability occur where the main function of the song is to provide for social recognition and cohesion . . ." In some bird species, a male and female pair may learn to sing together in a highly synchronized duet, with each partner having a different part in the duet. Both Gwinner & Kneutgen (1962) and Thorpe & North (1966) described interactions where one bird imitates the song of the absent partner, after which the absent partner "would return as quickly as possible as if called by name" (Thorpe & North 1966). Tyack (1993) presented a similar interaction, in which one adult female wild dolphin showed highly specific vocal responses to apparent imitation of her signature whistle by another adult female.

Richards *et al.* (1984) demonstrated that dolphins have the cognitive abilities required to perform the vocal labelling or naming suggested by Thorpe & North (1966). Richards and his colleagues trained a dolphin to imitate artificial whistle-like sounds. When they gave a dolphin the command to imitate a sound, they would simultaneously play the model sound and hold up an arbitrary object. For example, they might show a frisbee when they played an unmodulated tone, and a pipe when they played a warble. After the dolphin performed well on this task, Richards *et al.* started occasionally to show the object but not play the model sound. The dolphin had to remember how to reproduce the model sound that had been associated with the object in order to make the correct vocal response. After sufficient training, the dolphin succeeded in learning to label each arbitrary object with an arbitrary whistle sound. More detailed study in birds and dolphins is required to determine whether individuals imitate each other's distinctive sounds to achieve a naming function of this sort.

Imitation of signature whistles may also function as an affiliative or empathic signal (Bavelas *et al.* 1987). In human mother–infant interaction, imitation is thought to play a critical role in matching or regulating affective states (Stern 1985). In our work with wild dolphins held in net corrals, mothers and calves exchange whistles every few seconds when involuntarily separated. This rate seems to maintain contact and may reflect a similar rule for affective communication. Vocal convergence of speech sounds is also reported among adult humans, and this has been interpreted as an affiliative signal (Locke 1993; Locke & Snow, Chapter 14). Brown & Farabaugh (Chapter 7) argue that affiliative interactions may also be particularly important for vocal learning in birds.

Detailed behavioral observations need to be conducted on the contexts and consequences of vocal mimicry in order to test the predicted links with functional reference or naming and with affiliative behavior and affect. These referential or affiliative roles of imitation are not mutually exclusive, and may reinforce each other (Owings 1994).

THE EVOLUTION OF VOCAL LEARNING IN MARINE MAMMALS

When evaluating the remarkable imitative abilities of bottlenose dolphins, it is important to recognize that most mammals solve their communication problems with limited abilities to modify vocal production. Even such drastic treatments as deafening at birth, which would derail vocal development in humans and many bird species, do not prevent terrestrial nonhuman mammals from developing normal vocalizations. Auditory deprivation may produce only minor modifications, such as slightly longer mews in deafened kittens (Buchwald & Shipley 1985). The importance of genetic factors compared to auditory experience is suggested by studies of primate hybrids, whose calls blend features from both parents but match those of neither (Newman & Symmes 1982; Brockelman & Schilling 1984; Seyfarth & Cheney, Chapter 13). Primates raised in isolation or with foster mothers of a different species still produce species-typical vocalizations (Winter *et al.* 1973; Owren *et al.* 1993). The hybrid and cross-fostering results are particularly striking, since these animals are constantly exposed, and must learn to respond to vocalizations that differ from the ones they produce themselves.

Much more striking effects of auditory input are reported for vocalizations from a variety of marine mammal species. As we have noted, many repeated experiments with artificial acoustic models have demonstrated that adult dolphins can acquire novel vocalizations by imitation. A captive harbor seal (*Phoca vitulina*), was reported to imitate human speech with a regional accent (Ralls *et al.* 1985). Captive beluga whales (*Delphinapterus leucas*) are also reported to imitate human speech well enough for caregivers to "perceive these sounds as emphatic human conversation" (Ridgway *et al.* 1985) or even for words to be recognized (Eaton 1979). The only other animals with such highly developed skills of vocal imitation are the most accomplished avian mimics such as parrots and mynahs.

Many songbirds have geographic dialects in their songs, and vocal learning has been shown to play a role in the formation of dialects in many species (see e.g., Mundinger 1982). Similar geographic dialects have been found among marine mammals. Reproductive advertisement displays of Weddell seals (*Leptonychotes weddelli*) show regional dialects (Thomas & Stirling 1984). Recordings of bearded seals (*Erignathus barbatus*) separated by 150–350 km showed clear differences (Cleator *et al.* 1989). The songs of humpback whales (*Megaptera novaeangliae*) also show regional dialects; within a population, they change progressively over time (Payne *et al.*

1983). Songs from different oceans or decades share so few sounds, that even the most cursory listening or scanning of spectrograms would reveal significant differences. These observations of obvious geographic dialects in the vocalizations of marine mammals contrast with the more subtle differences described for nonhuman primates (Mitani *et al.* 1992; Snowdon *et al.*, Chapter 12). However, while vocal dialects have often been suggested as evidence for vocal learning in birds (see e.g., Kroodsma & Baylis 1982), they do not in themselves provide strong evidence, for they could reflect genetic differences between adjacent populations. The vocal convergence at any one time within a population of singing humpback whales coupled with rapid changes in the song over time provide stronger evidence for vocal learning in these animals.

These data suggest that a diverse group of mammals, all inhabiting the marine environment, have evolved highly developed vocal learning abilities more similar to those of humans and some birds than of their terrestrial ancestors. What selection pressures have driven such diverse groups as birds, dolphins, and humans to rely on vocal learning in development of their vocalizations? There are both costs and advantages to a communication system which can rapidly be modified through learning rather than requiring generations of selection to modify fixed innate systems. This is an example of a general developmental issue. Developmental pathways can vary from highly canalized and not modifiable by experience, to highly open and modifiable. Canalized pathways are more appropriate where there is one solution to a stable problem and where it is critical to produce the correct response upon the first encounter with the problem. For example, the first time a solid object lodges in your trachea, you must dislodge it within seconds or you are likely to asphyxiate. This problem is a good candidate for a reflex. More open pathways show their benefits where the optimal response varies depending upon conditions, where errors are not lethal, and where there are repeated opportunities to hone ever more successful responses.

If we now narrow our focus to communication, several factors have been suggested as favoring more open communication systems. We will focus on two factors: mobility and individual recognition. Both of these factors emphasize the unpredictability of the optimal response. Mobility has been suggested as a factor affecting differences in flexibility of learning ability for communicative signals among birds (Marler 1987) and primates (Snowdon and Elowson 1992), with more mobile populations needing to adapt vocalizations to local conditions. The prediction that more mobile animals would have more open communication is supported by contrasting marine vs terrestrial mammals. Marine mammals as a group tend to be much more mobile than terrestrial mammals; relatively few terrestrial mammals are migratory, yet the majority of marine mammals are. Marine mammals live in a fluid moving medium. Even marine mammals without seasonal migrations may cover tens to hundreds of kilometers per day.

The mobility hypothesis would also predict the differences in whistle repertoires observed between male and female bottlenose dolphins. For example, in the resident Sarasota dolphin community described by Wells et al. (1987), adult females associate in groups termed female "bands", in which several matrilines are represented. Although subadult females associate with other subadult animals, they tend to return to their matrilineal band at maturity. Adult males, however, frequently spend time alone or with one or two other males, and tend to have much larger home ranges than do females. Both in Sarasota and in Western Australia, males appear to have more variable whistle repertoires (Sayigh et al. 1990; Smolker 1993).

Communication is fundamentally a social interaction, so one might expect social functions for vocal learning to be particularly important. Individual recognition is one such function that is of particular relevance for signature whistles. While species recognition can be achieved using fixed or canalized signals and recognition mechanisms, individual recognition requires learning. In order to recognize an individual, an animal must be able to learn to discriminate the signature calls of that individual from those of others. Species that require individual recognition may evolve more distinctive signature signals (i.e., higher variability between individuals and lower variability within an individual's own repeated signals), as well as heightened perceptual abilities to discriminate signature signals (Loesche *et al.* 1991). While vocal learning might help an animal to produce a particularly distinctive or stable signature vocalization, many species rely upon involuntary voice cues to recognize one another. These voice cues derive from slight differences in the vocal tracts of individuals. Diving mammals may not be able to rely upon these involuntary voice cues, because the slight differences in resonances of gas-filled vocal tracts may be trivial compared to modifications produced by the compression of these structures during diving (Tyack 1991b). Instead, if diving mammals are to rely upon individually distinctive vocalizations while underwater, they may need to use vocal learning to produce distinctive signals under voluntary control. Vocal learning may thus play an unusually important role in the maintenance of individually distinctive signature vocalizations for marine mammals as they dive at sea.

Open communication systems may be favored by other elements of individual-specific social relationships in which animals recognize one another and modify how they respond to an individual based upon a remembered history of interaction. Animals that form social relationships face a tricky balancing act between confluences and conflicts of interest. They must rely upon their groups to survive and reproduce, but success in competing with conspecifics is also a critical component of fitness. Byrne & Whiten (1988) reviewed arguments that these social relationships create a strong selection pressure for flexible and strategic behavior. This logic also applies to communication systems. Successful communication requires shared understanding between signaller and recipient. However, an animal with a closed innate link between signal and response may be vulnerable to a competitor who could manipulate these inflexible responses. If this creates a selection pressure for more strategic use of a more flexible communication system, then each individual may develop slightly different communication patterns, which may put a premium not only on identifying each individual but also on being able to predict and interpret that particular individual's signal or response.

Animals that maintain relationships have an opportunity to learn how a specific partner uses or reacts to a vocalization. This allows the development of flexible and subtle communication. Snowdon *et al.* (Chapter 12) and Seyfarth & Cheney (Chapter 13) emphasize how much developmental flexibility primates appear to retain by learning how to comprehend and how to use their relatively fixed patterns of vocal production. However, the ability to modify one's vocal repertoire introduces significant additional flexibility that may be particularly important for animals that can try new signals and can learn over time how others react to their own vocalizations in particular contexts (West & King 1990).

DIVERSITY IN FUNCTIONS OF VOCAL LEARNING AMONG CETACEANS

There is a great diversity of social systems, life history, and mobility among those marine mammal species that are skilled at vocal learning (Tyack 1986b). Data on acoustic communication and social behavior also suggest a broad range of learned signals, from highly individually distinctive signals, to group-specific vocal repertoires, to signals shared among groups over broad geographic areas, and to vocal dialects in different areas. Socioecological comparisons between these species suggest diverse functions for vocal learning among marine mammals. We review four of the best known species, from the relatively fluid groups of humpback whales to bottlenose dolphins with strong relationships within fluid groups, to groups that are stable for days or more such as those of the sperm whale (*Physeter macrocephalus*), and finally to the extremely stable groups of the killer whale (*Orcinus orca*).

Most baleen whales migrate thousands of kilometers between summer feeding areas and winter breeding areas. Except for the mother–calf bond, which lasts less than one year, most groups of baleen whales last only hours, with little evidence for individual-specific associations. Sounds on the feedings grounds are poorly understood by researchers, but several different species of baleen whales appear to produce reproductive advertisement displays. The behavioral context and acoustic structure of the songs of humpback whales are the best understood advertisement display in baleen whales. During the breeding season, lone adult male humpbacks repeat long complex songs. They usually stop when they join with other whales. Aggressive behavior is often seen after a singer joins with another male. After the rare times a female joins with the singer, behavior associated with sexual activity has been observed (Tyack 1981). While breeding males are not territorial, song clearly mediates both intrasexual competitive interactions and some intersexual associations on the breeding ground. As in many songbirds, vocal learning appears to function to produce more complex displays through sexual selection. However, rather than producing larger song repertoires, humpbacks continually change their songs. There appears to be a very strong force for vocal convergence at any one time, coupled with progressive change in all aspects of the song over time. Similar changes have been reported for the simpler songs of bowhead whales (*Balaena mysticetus*; Ljungblad *et al.* 1982). Finback whales (*Balaenoptera physalus*) produce long repeated series of 20 Hz pulses during their breeding season. The pulse intervals of these finback series also show significant changes from year to year (Watkins *et al.* 1987).

Bottlenose dolphins have a fission–fusion society in which groups change composition on time scales of minutes to hours (Wells *et al.* 1987; Smolker *et al.* 1992). Within these changing groups, there are very stable social bonds, particularly between a mother and her infant or between members of a coalition of adult males. Adult females form a relatively stable network of associations with one another, and social protection of calves appears to influence female grouping patterns. Individual-specific social relationships are important to bottlenose dolphins of both sexes from infancy to adulthood. The signature whistles described in this chapter can help to maintain these relationships within fluid groupings.

Sperm whales are the largest toothed whale. They inhabit deep water ocean, with females and young favoring water over 1000 m deep and latitudes of less than 40 degrees. After dispersing from their natal groups, male sperm whales feed in higher latitudes, especially in summer, and sexually mature males migrate to tropical breeding grounds in the winter. Female sperm whales are typically sighted in groups of 20 or so animals. Whitehead (1989) suggested that there might be advantages to coordinating foraging with nearby animals, and that foraging strategy might determine the size and structure of these groups. Groups off the Galapagos are typically formed of two matrilineal units that stay together for several days. The matrilineal units themselves are very stable, with pairs of individual females being sighted within the same unit for many years. Gordon (1987) argued that a primary function for these kin-based matrilineal units is vigilance for predators and social defense of calves. While sperm whales have relatively stable matrilineal groups, most individuals will have a variety of associates throughout their lifespan, with varying degrees of association. Sperm whales produce distinctive rhythmic patterns of clicks, called codas, often in exchanges between individual whales (Watkins & Schevill 1977). While Watkins & Schevill (1977) described codas as individually distinctive, they also described an exchange in which each whale matched the coda of the other whale. Moore *et al.* (1993) described two

coda patterns that comprised more than 50% of the codas from many individual whales within many different groups recorded over a large part of the southeast Caribbean. Weilgart & Whitehead (1993) described different shared coda patterns for the Galapagos whales, and Weilgart *et al.* (1993) described geographic variation in the proportional usage of different codas. More work is needed to track coda usage of individual sperm whales, within matrilines, and over large areas, but the current evidence suggests possible variety in usage for individual and regional identification that is consistent with the variety of problems posed by sperm whale societies.

The most stable groups documented among mammals occur among fish-eating killer whales in the inshore waters of the Pacific Northwest. Neither sex disperses from its natal group; the only way a group changes is by birth, death, or rare fissions of very large groups (Bigg *et al.* 1990). The potential for cooperative breeding has been suggested by the frequent association of young with pod members other than the mother (Haenel 1986), and by anecdotal reports of apparent teaching (Lopez & Lopez 1985). Killer whales produce distinctive discrete calls, which are very stereotyped. Each pod of killer whales has a group-specific repertoire of discrete calls which is stable for many years (Ford 1990). These group-specific repertoires are thought to indicate pod affiliation, maintain pod cohesion, and to coordinate activities of pod members. When killer whales from two different pods are housed together in captivity, they may imitate each other's calls, and preliminary observations of call development also suggest that these calls are learned (Bowles *et al.* 1988).

There is a clear correlation between the types of social bond and communication signal seen in different cetacean groups (Tyack 1986b). Individual-specific signals have been reported for species with strong individual social bonds; group-specific signals have been reported for species with stable groups, and population-specific advertisement displays have been reported among species with fluid bonds. The diversity of functions of vocal learning in present-day cetacean species complicate study of the evolutionary origins of this skill. It is possible that vocal learning evolved *de novo* in these different groups as independent solutions to the different problems posed by differing social organization. However, once a more flexible system of vocal development evolved to solve one problem, it may have allowed more flexibility to solve different problems. As the descendants of animals that evolved the trait branched into other niches, they may have used vocal learning to other ends. Even within a species, the optimal social structure may vary with ecological conditions over time scales that are short compared to evolutionary change. Vocal learning may provide a mechanism whereby the vocal repertoire can develop to match the particular social system experienced by an individual.

SUMMARY

Dolphins use whistles in an unusually open system of vocal communication. Most dolphin calves develop individually distinctive signature whistles that differ from those of their parents but that match acoustic models present in their early auditory environment. The lack of correlation in signature whistles of parents and offspring and the strong correlation with auditory models provide strong evidence for vocal learning in the development of signature whistles. Dolphins vary widely in the timing of signature whistle development. Both acoustic and social factors appear to modify the timing and outcome of whistle development. Dolphins also maintain the ability to acquire new whistles into their vocal repertoire throughout their lifetime. After a dolphin develops its signature whistle, the diversity of whistles in its repertoire increases with age. Dolphins imitate the whistles of animals with which they share strong bonds, and the diversity of whistles in a dolphin's repertoire appears to be related to the size of its social network. A diving marine mammal may not be able to rely upon involuntary voice cues for individual recognition, but may rather require vocal learning to maintain a stable signature as its vocal tract changes shape with increasing pressure. Many dolphins live in societies in which they may form new individual-specific social relationships throughout their lifetime. This may provide an important social factor for the development and maintenance of this open communication system.

ACKNOWLEDGEMENTS

The research discussed here which was conducted by P. Tyack and L. Sayigh was supported by ONR grant N00014–87–K–0236, NIH grant 1R29NS25290, NSF Doctoral Dissertation Grant BNS–9014545, and the Education Office of the Woods Hole Oceanographic Institution. The longitudinal study of whistles of dolphin calves from the Miami Seaquarium was a collaboration of Tyack, Sayigh, Kurt Fristrup, and Janet McIntosh. Thanks to Joan Caron and the staff of Seaquarium for help during the long period when data were collected. The Cindy/April study was conducted at the Gulfarium in Fort Walton Beach, Florida. Thanks to Forrest Townsend and the staff of Gulfarium for all of their help with this study. Thanks also to Randy Wells and all of his colleagues at the Dolphin Biology Research Institute for the unparalleled research opportunities with their long term study of identified bottlenose dolphins near Sarasota, Florida. Sarasota follows were conducted under Marine Mammal Permit 638 from the US National Marine Fisheries Service to Randall Wells. Thanks to Charles Snowdon, Martine Hausberger, and Vincent Janik for reviewing an earlier version of this chapter. This is contribution number 9247 from the Woods Hole Oceanographic Institution.

REFERENCES

Bavelas, J. B., Black, A., Lemery, C. R. & Mullett, J. (1987). Motor mimicry as primitive empathy. In *Empathy and its Development*, ed. N. Eisenberg & J. Strayer, pp. 317–38. Cambridge University Press, Cambridge.

Beecher, M. D. (1989). Signalling systems for individual recognition: An information theory approach. *Animal Behaviour* 38: 248–61.

Bigg, M. A., Olesiuk, P. F., Ellis G. M., Ford, J. K. B. & Balcomb, K. C. III (1990). Social organization and genealogy of resident killer whales (*Orcinus orca*) in the coastal waters of British Columbia and Washington state. *Reports of the International Whaling Commission Special Issue* 12: 383–405.

Bowles, A., Young, W. G. & Asper, E. D. (1988). Ontogeny of stereotyped calling of a killer whale calf, *Orcinus orca*, during her first year. *Rit Fiskideildar* 11: 251–75.

Brockelman, W. Y. & Schilling, D. (1984). Inheritance of stereotyped gibbon calls. *Nature* 312: 634–6.

Buchwald, J. S. & Shipley, C. (1985). A comparative model of infant cry. In *Infant Crying*, ed. B. M. Lester & C. F. Z. Boukydis, pp. 279–305. Plenum, New York.

Buck, J. R. & Tyack, P. L. (1993). A quantitative measure of similarity for *Tursiops truncatus* signature whistles. *Journal of the Acoustical Society of America* 94: 2497–506.

Burdin, V. I., Reznik, A. M., Shornyakov, V. M. & Chupakov, A. G. (1975). Communication signals of the Black Sea bottlenose dolphin. *Soviet Physics and Acoustics* **20**: 314–18.

Byrne, R. & Whiten, A. (1988). *Machiavellian Intelligence.* Clarendon Press, Oxford.

Caldwell, M. C. & Caldwell, D. K. (1965). Individualized whistle contours in bottlenosed dolphins (*Tursiops truncatus*). *Science* **207**: 434–5.

Caldwell, M. C. & Caldwell, D. K. (1970). Etiology of the chirp sounds emitted by the Atlantic bottlenosed dolphin: a controversial issue. *Underwater Naturalist* **6**: 6–9.

Caldwell, M. C. & Caldwell, D. K. (1972). Vocal mimicry in the whistle mode by an Atlantic bottlenosed dolphin. *Cetology* **9**: 1–8.

Caldwell, M. C. & Caldwell, D. K. (1979). The whistle of the Atlantic bottlenosed dolphin (*Tursiops truncatus*) – ontogeny. In *Behavior of Marine Animals*, vol. 3 Cetaceans, ed. H. E. Winn & B. L. Olla, pp. 369–401. Plenum, New York.

Caldwell, M. C., Caldwell, D. K. & Hall, N. R. (1969). An experimental demonstration of the ability of an Atlantic bottlenosed dolphin to discriminate between whistles of other individuals of the same species. *Los Angeles County Museum of Natural History Foundation Technical Report* **6**.

Caldwell, M. C., Caldwell, D. K. & Tyack, P. L. (1990). A review of the signature whistle hypothesis for the Atlantic bottlenose dolphin, *Tursiops truncatus*. In *The Bottlenose Dolphin: Recent Progress in Research*, ed. S. Leatherwood & R. Reeves, pp. 199–234. Academic Press, San Diego.

Clayton, N. S. (1987). Song tutor choice in zebra finches. *Animal Behaviour* **35**: 714–21.

Cleator, H. J., Stirling, I. & Smith, T. G. (1989). Underwater vocalizations of the bearded seal (*Erignathus barbatus*). *Canadian Journal of Zoology* **67**: 1900–10.

Connor, R. C., Smolker, R. A. & Richards, A. F. (1992). Aggressive herding of females by coalitions of male bottlenose dolphins (*Tursiops* sp.). In *Coalitions and Alliances in Humans and Other Animals*, ed. A. H. Harcourt & F. B. M. de Waal, pp. 415–43. Oxford University Press, Oxford.

Defran, R. H. & Pryor, K. (1980). The behavior and training of cetaceans in captivity. In *Cetacean Behavior: Mechanisms and Functions*, ed. L. M. Herman, pp. 319–62. Wiley-Interscience, New York.

Dreher, J. & Evans, W. E. (1964). Cetacean communication. In *Marine Bioacoustics*, vol. 1, ed. W. N. Tavolga, pp. 373–93. Pergamon Press, Oxford.

Eaton, R. L. (1979). A beluga whale imitates human speech. *Carnivore* **2**: 22–3.

Evans, W. E. (1967). Vocalization among marine mammals. In *Marine Bioacoustics*, vol. 2, ed. W. N. Tavolga, pp. 159–86. Pergamon Press, Oxford.

Farabaugh, S. M., Linzenbold, A. & Dooling, R. J. (1994). Vocal plasticity in budgerigars (*Melopsittacus undulatus*): Evidence for social factors in the learning of contact calls. *Journal of Comparative Psychology* **108**: 81–92.

Ford, J. K. B. (1990). Vocal traditions among resident killer whales (*Orcinus orca*) in coastal waters of British Columbia. *Canadian Journal of Zoology* **69**: 1454–83.

Gish, S. L. (1979). A quantitative description of two-way acoustic communication between captive Atlantic bottlenosed dolphins (*Tursiops truncatus* Montagu). Ph.D. dissertation, University of California at Santa Cruz.

Gordon, J. C. D. (1987). Behaviour and ecology of sperm whales off Sri Lanka. Ph.D. dissertation. University of Cambridge.

Gwinner, E. & Kneutgen, J. (1962). Über die biologische Bedeutung der "zweckdienlichen" Anwendung erlernter Laute bei Vögeln. *Zeitschrift für Tierpsychologie* **19**: 692–6.

Haenel, N. J. (1986). General notes on the behavioral ontogeny of Puget Sound killer whales and the occurrence of allomaternal behavior. In *Behavioral Biology of Killer Whales*, ed. B. C. Kirkevold & J. S. Lockard, pp. 285–300, Zoo Biology Monographs, vol. 1. Alan R. Liss, New York.

Hausberger, M. (1993). How studies on vocal communication in birds contribute to a comparative approach of cognition. *Etología* **3**: 171–85.

Herman, L. M. (1980). Cognitive characteristics of dolphins. In *Cetacean Behavior: Mechanisms and Functions*, ed. L. M. Herman, pp. 363–429. Wiley-Interscience, New York.

Hultsch, H. & Todt, D. (1992). The serial order effect in the song acquisition of birds: relevance of exposure frequency to models. *Animal Behaviour*, **44**: 590–2.

Janik, V. M., Denhardt, G. & Todt, D. (1994). Signature whistle variations in a bottlenosed dolphin, *Tursiops truncatus*. *Behavioral Ecology and Sociobiology* **35**: 243–8.

Kroodsma, D. E. & Baylis, J. R. (1982). Appendix: a world survey of evidence for vocal learning in birds. In *Acoustic Communication in Birds*, vol. 2, *Song Learning and its Consequences*, ed. D. E. Kroodsma & E. H. Miller, pp. 311–37. Academic Press, New York.

Lilly, J. C. (1963). Distress call of the bottlenose dolphin: stimuli and evoked behavioral responses. *Science* **139**: 116–18.

Ljungblad, D. K., Thompson, P. O. & Moore, S. E. (1982). Underwater sounds recorded from migrating bowhead whales, *Balaena mysticetus*, in 1979. *Journal of the Acoustical Society of America* **71**: 477–82.

Locke, J. L. (1993). *The Child's Path to Spoken Language.* Harvard University Press, Cambridge, MA.

Loesche, P., Stoddard, P. K., Higgins, B. J. & Beecher, M. D. (1991). Signature versus perceptual adaptations for individual vocal recognition in swallows. *Behaviour* **118**: 15–25.

Lopez, J. C. & Lopez, D. (1985). Killer whales (*Orcinus orca*) of Patagonia and their behavior of intentional stranding while hunting nearshore. *Journal of Mammalogy* **66**: 181–3.

Mammen, D. L. & Nowicki, S. (1981). Individual differences and within-flock convergence in chickadee calls. *Behavioral Ecology and Sociobiology* **9**: 179–86.

Marler, P. (1987). Sensitive periods and the roles of specific and general sensory stimulation in bird-song learning. In *Imprinting and Cortical Plasticity*, ed. J. P. Rauschecker & P. Marler, pp. 99–135. Wiley, New York.

McBride, A. F. & Kritzler, H. (1951). Observations on the pregnancy, parturition, and postnatal behavior in the bottlenose dolphin. *Journal of Mammalogy* **32**: 251–66.

Meltzoff, A. N. & Moore, M. K. (1977). Imitation of facial and manual gestures by human neonates. *Science* **198**: 75–8.

Mitani, J. C., Hasegawa, T., Gros-Louis, J., Marler, P. & Byrne, R. (1992). Dialects in wild chimpanzees? *American Journal of Primatology* **27**: 233–43.

Moore, K. E., Watkins, W. A. & Tyack, P. L. (1993). Pattern similarity in shared codas from sperm whales (*Physeter catodon*). *Marine Mammal Science* **9**: 1–9.

Mundinger, P. C. (1970). Vocal imitation and individual recognition of finch calls. *Science* **168**: 480–2.

Mundinger, P. C. (1982). Microgeographic and macrogeographic variation in acquired vocalizations of birds. In *Acoustic Communication in Birds*, vol. 2, *Song Learning and its Consequences*, ed. D. E. Kroodsma & E. H. Miller, pp. 147–208. Academic Press, New York.

Newman, J. D. & Symmes, D. (1982). Inheritance and experience in the acquisition of primate acoustic behavior. In *Primate Communication*, ed. C. T. Snowdon, C. H. Brown & M. R. Peterson, pp. 259–78. Cambridge University Press, Cambridge.

Nowicki, S. (1989). Vocal plasticity in captive black-capped chickadees: The acoustic basis and rate of call convergence. *Animal Behaviour* **37**: 64–73.

Owings, D. H. (1994). How monkeys feel about the world: A review of *How monkeys see the world*. *Language and Communication* **14**: 15–30.

Owren, M. J., Dieter, J. A., Seyfarth, R. M. & Cheney, D. L. (1993). Vocalizations of rhesus (*Macaca mulatta*) and Japanese (*Macaca fuscata*) macaques cross-fostered between species show evidence of only limited modification. *Developmental Psychobiology* **26**: 389–406.

Payne, K., Tyack, P. & Payne, R. (1983). Progressive changes in the songs of humpback whales (*Megaptera novaeangliae*): a detailed analysis of two seasons in Hawaii. In *Communication and Behavior of Whales*, ed. R. Payne, pp.

9–57. AAAS selected symposium series, Westview Press, Boulder CO.

Payne, R. B. (1983). The social context of song mimicry: song matching dialects in indigo buntings (*Passerina cyanea*). *Animal Behaviour* **31**: 788–805.

Peters, S., Marler, P. & Nowicki, S. (1992). Song sparrows learn from limited exposure to song models. *Condor* **94**: 1016–19.

Ralls, K., Fiorelli, P. & Gish, S. (1985). Vocalizations and vocal mimicry in captive harbor seals, *Phoca vitulina*. *Canadian Journal of Zoology* **63**: 1050–6.

Reiss, D. & McCowan, B. (1993). Spontaneous vocal mimicry and production by bottlenose dolphins (*Tursiops truncatus*): Evidence for vocal learning. *Journal of Comparative Psychology* **107**: 301–12.

Richards, D. G., Wolz, J. P. & Herman, L. M. (1984). Vocal mimicry of computer-generated sounds and vocal labelling of objects by a bottlenosed dolphin, *Tursiops truncatus*. *Journal of Comparative Psychology* **98**: 10–28.

Ridgway, S. H., Carder, D. A. & Jeffries, M. M. (1985). Another "talking" male white whale. *Abstracts: Sixth Biennial Conference on the Biology of Marine Mammals*, p. 67.

Sayigh, L. S. (1992). Development and functions of signature whistles of free-ranging bottlenose dolphins, *Tursiops truncatus*. Ph.D. dissertation, MIT/WHOI Joint Program, WHOI 92–37.

Sayigh, L. S., Tyack, P. L. & Wells, R. S. (1993). Recording underwater sounds of free-ranging bottlenose dolphins while underway in a small boat. *Marine Mammal Science* **9**: 209–13.

Sayigh, L. S., Tyack, P. L., Wells, R. S. & Scott, M. D. (1990). Signature whistles of free-ranging bottlenose dolphins, *Tursiops truncatus*: Stability and mother–offspring comparisons. *Behavioral Ecology and Sociobiology* **26**: 247–60.

Sayigh, L. S., Tyack, P. L., Wells, R. S., Scott, M. D. & Irvine, A. B. (1995). Sex difference in whistle production in free-ranging bottlenose dolphins, *Tursiops truncatus*. *Behavioral Ecology and Sociobiology* **36**: 171–7.

Schusterman, R. J. (1978). Vocal communication in pinnipeds. In *Behavior of Captive Wild Animals*, ed. H. Markowitz & V. J. Stevens, pp. 247–308. Nelson-Hall, Chicago.

Scott, M. D., Wells, R. S. & Irvine, A. B. (1990). A long-term study of bottlenose dolphins on the west coast of Florida. In *The Bottlenose Dolphin*, ed. S. Leatherwood & R. Reeves, pp. 235–44. Academic Press, New York.

Sigurdson, J. (1993). Whistles as a communication medium. In *Language and Communication: Comparative Perspectives*, ed. H. L. Roitblat, L. M. Herman & P. Nachtigall, pp. 153–73. Erlbaum, Hillsdale, NJ.

Smolker, R. A. (1993). Acoustic communication in bottlenose dolphins. Ph.D. dissertation, University of Michigan.

Smolker, R. A., Mann, J. & Smuts, B. B. (1993). Use of signature whistles during separation and reunions by wild bottlenose dolphin mothers and infants. *Behavioral Ecology and Sociobiology* **33**: 393–402.

Smolker, R. A., Richards, A. F., Connor, R. C. & Pepper, J. P. (1992). Sex differences in patterns of association among Indian Ocean bottlenose dolphins. *Behaviour* **123**: 38–69.

Snowdon, C. T. & Elowson, A. M. (1992). Ontogeny of primate vocal communication. In *Topics in Primatology, vol. 1, Human Origins*, ed. T. Nishida, W. C. McGrew, P. Marler, M. Pickford & F. B. M. De Waal, pp. 279–90. University of Tokyo Press, Tokyo.

Stern, D. N. (1985). *The Interpersonal World of the Infant*. Basic Books, New York.

Thomas, J. A. & Stirling, I. (1984). Geographic variation in the underwater vocalizations of Weddell seals (*Leptonychotes weddelli*) from Palmer Peninsula and McMurdo Sound, Antarctica. *Canadian Journal of Zoology* **61**: 2203–14.

Thorpe, W. H. & North, M. E. W. (1965). Origin and significance of the power of vocal imitation: With special reference to the antiphonal singing of birds. *Nature* **208**: 219–22.

Thorpe, W. H. & North, M. E. W. (1966). Vocal imitation in the tropical boubou shrike, *Laniarius aethiopicus, major* as a means of establishing and maintaining social bonds. *Ibis* **108**: 432–5.

Tyack, P. (1981). Interactions between singing Hawaiian humpback whales and conspecifics nearby. *Behavioral Ecology and Sociobiology* **8**: 105–16.

Tyack, P. (1986a). Whistle repertoires of two bottlenosed dolphins, *Tursiops truncatus*: Mimicry of signature whistles? *Behavioral Ecology and Sociobiology* **18**: 251–7.

Tyack, P. (1986b). Population biology, social behavior, and communication in whales and dolphins. *Trends in Ecology and Evolution* **1**: 144–50.

Tyack, P. (1991a). Use of a telemetric device to identify which dolphin produced a sound. In *Dolphin Societies: Discoveries and Puzzles*, ed. K. Peyor & K. S. Norris, pp. 319–44. University of California Press, Berkeley, CA.

Tyack, P. (1991b). If you need me, whistle. *Natural History*, August, pp. 60–1.

Tyack, P. (1993). Why ethology is necessary for the comparative study of language and communication. In *Language and Communication: Comparative Perspectives*, ed. H. L. Roitblat, L. M. Herman & P. Nachtigall, pp. 115–52. Erlbaum, Hillsdale, NJ.

Watkins, W. A. & Schevill, W. E. (1977). Sperm whale codas. *Journal of the Acoustical Society of America* **62**: 1485–90.

Watkins, W. A., Tyack, P., Moore, K. E. & Bird, J. E. (1987). The 20-Hz signals of finback whales (*Balaenoptera physalus*). *Journal of the Acoustical Society of America* **82**: 1901–12.

Weilgart, L. & Whitehead, H. (1993). Coda vocalizations in sperm whales (*Physeter macrocephalus*) off the Galapagos Islands. *Canadian Journal of Zoology* **71**: 744–52.

Weilgart, L. S., Whitehead, H., Carler, S. & Clark, C. W. (1993). Variations in the vocal repertoires of sperm whales (*Physeter macrocephalus*) with geographic area and year. *Abstract. Tenth Biennial Conference on the Biology of Marine Mammals*, p. 112.

Wells, R. S. (1991). The role of long-term study in understanding the social structure of a bottlenose dolphin community. In *Dolphin Societies*, ed. K. Pryor & K. S. Norris, pp. 199–235. University of California Press, Berkeley, CA.

Wells, R. S., Scott, M. D. & Irvine, A. B. (1987). The social structure of free-ranging bottlenose dolphins. *Current Mammalogy* **1**: 247–305.

West, M. J. & King, A. P. (1988). Female visual displays affect the development of song in the cowbird. *Nature* **334**: 244–6.

West, M. J. & King, A. P. (1990). Mozart's starling. *American Scientist* **78**: 106–14.

Whitehead, H. (1989). Formations of foraging sperm whales, *Physeter macrocephalus*, off the Galapagos Islands. *Canadian Journal of Zoology* **67**: 2131–9.

Winter, P., Handley, P., Ploog, D. & Schott, D. (1973). Ontogeny of squirrel monkey calls under normal conditions and under acoustic isolation. *Behaviour* **47**: 230–9.

Wood, F. G., Jr (1954). Underwater sound production and concurrent behavior of captive porpoises, *Tursiops truncatus* and *Stenella plagiodon*. *Bulletin of Marine Science, Gulf and Carribean* **3**: 120–33.

12 Social influences on vocal development in New World primates

CHARLES T. SNOWDON, A. MARGARET ELOWSON, AND REBECCA S. ROUSH

INTRODUCTION

The study of vocal development in nonhuman animals has focused primarily on vocal production. However, vocal development involves not only production, but development of appropriate usage and appropriate responses to the calls of others. The relative neglect of usage and responses has led to a distorted view of vocal development as emphasized by Seyfarth & Cheney (Chapter 13) for monkeys and West *et al.* (Chapter 4) for birds. The argument is often made that nonhuman mammals differ fundamentally in vocal development from both birds and human beings but, as we have argued previously (Snowdon & Elowson 1992) and Seyfarth & Cheney (Chapter 13) argue, primates are similar to humans and birds if all three components of vocal development – production, usage and response – are evaluated.

In this chapter we argue that when functionally similar vocalizations are chosen from species that have similar social organization, then similar developmental processes will be found regardless of the taxon studied. We examine the idea that nonhuman primate vocal structures are fixed and review evidence that vocal plasticity may be quite common. We then provide three examples of phenomena from our research with marmosets and tamarins that illustrate how plasticity in pygmy marmoset trill vocalizations can be influenced by changes in social companions, how infant babbling in pygmy marmosets is a social interaction, and how the structure and usage of food-associated calls is acquired by cotton-top tamarins, and how social environments can inhibit the expression of these calls.

PLASTICITY OF VOCAL PRODUCTION

Most of the avian developmental research has focused on song in passerine birds. While song is a highly audible and important part of the vocal repertoire, song has very specialized functions in territory defense and maintenance and in mate attraction. Song in most temperate zone species is frequently limited to males (but see Hausberger, Chapter 8, on European starlings for a notable exception), and appears to be under the control of testosterone. In nonhuman primates there are several examples of vocalizations that might be functionally equivalent to song: howls of howler monkeys, songs of gibbons, duets of titi monkeys, and the long calls or loud calls of many primates. In cercopithecine monkeys, Gautier & Gautier-Hion (1977) have described loud calls given only by males. These calls are highly stereotyped. However, except for gibbon song (where there is strong evidence for a genetic influence on vocal structure (Brockelman & Schilling 1984; Giessmann 1984; Tenaza 1985)), we know little about the ontogeny of long or loud calls in nonhuman primates. Often these calls are shown to be genetically stable (e.g., type 1 Loud Calls of male cercopithecine monkeys (Gautier & Gautier-Hion 1977) when appropriate developmental studies are done.

However, there are many other vocalizations that are given by birds of both sexes throughout the year in a wide variety of contexts (e.g., food calls of chickens, Marler *et al.* (1986a,b); calls of goldfinches (*Spinus tristis*), Mundinger (1970); and "chick-a-dee" calls of black-capped chickadees (*Parus atricapillus*) (Nowicki 1989) that are used to promote social interactions within pairs or flocks. We have relatively little information concerning the ontogeny of these other vocalizations in comparison to what we know about song development, but we do know that some of these calls can be modified by adult birds in response to changed social environments. Thus, Mundinger (1970) reported that goldfinches acquire some call types of their mate during pair formation, and Nowicki (1989) has shown that chickadees change subtle features of the "dee" note of the "chick-a-dee" call when they form winter flocks.

Many of the primate calls that have been studied to date are functionally more similar to the nonsong vocalizations of birds. Thus, those studying bird song ontogeny and those studying primate vocal development are studying very different phenomena.

Many of the passerine birds that have been studied are territorial and seasonally monogamous, rearing one clutch at a time, with the young of the year dispersing before the next breeding season. Thus, most passerine birds spend relatively little time in close contact with their families after the first year. However, most primate groups are stable multigenerational groups and young animals might spend their entire lives within a group, or, if they do disperse, it is generally not until after several younger siblings have been born. Most primates spend a longer proportion of time in contact with parents and with younger siblings than do most of the passerine birds studied. These differences of social organization and group stability may lead to differences in the ontogeny of communication as well. Birds that disperse in the first year of life to form new territories may display a more time limited vocal development, with a high degree of stereotypy subsequent to plastic song (see Nelson, Chapter 2) compared both to more sedentary species of birds and mammals.

Group-living avian species (see Baptista & Gaunt, Chapter 3; West et al., Chapter 4; Zann, Chapter 6; Brown & Farabaugh, Chapter 7; Hausberger, Chapter 8) appear to be where we find exceptions to the "typical" pattern of ontogeny in birds. A direct comparison of the ontogeny of functionally similar calls in group-living and cooperatively breeding birds with similar mammalian species shows that similarities outweigh differences in vocal development.

Seyfarth & Cheney (Chapter 13) note that 79% of the studies supporting the idea that call structure of nonhuman primates is fixed at birth were published prior to 1987, and that 71% of the studies published after 1987

documented some sort of developmental modification in vocal production. Thus, the apparent lack of vocal variability often perceived to be the case with nonhuman primates and other mammals may have been an artefact of the availability of appropriate quantitative analytic methods to document the variability, or it may be the result of a changing perspective of scientists so that we now seek vocal variation instead of assuming stereotypy. Many of the studies in this volume reporting changes in vocal production independent of critical or sensitive periods have used more precise quantitative analytic methods and have displayed an explicit interest in understanding how vocal development occurs within a social context.

In nonhuman primates and other mammals, there have been relatively few demonstrations of population or group variation or of extensive within-individual variation that approach the complexity of song repertoires in birds (but see McCowan & Reiss, Chapter 10; Tyack & Sayigh, Chapter 11, for examples of individual variation in whistle structure in dolphins (*Tursiops truncatus*); and see Seyfarth & Cheney, Chapter 13, for evidence of matrilineal signatures in macaques). Green (1975) described population differences in the vocalizations given by Japanese macaques (*Macaca fuscata*) during food provisioning, suggesting that learning was the most plausible mechanisms for these differences. Masataka (1992) has described how human caregivers of Japanese macaques have conditioned individual monkeys to alter vocalizations in a feeding context. Hodun et al. (1981) described subspecies variation in the long calls of wild saddle-back tamarins (*Saguinus fuscicollis*), but reported that a few individuals had long call structures intermediate between two adjacent subspecies. Subsequent recordings of captive-born tamarins found that hybrids of these same two subspecies had long call structures that were more similar to those of the most predominant subspecies in the colony, and not to the call structure of either parental subspecies, suggesting social learning as a plausible mechanism. Mitani et al. (1992) have described population differences in pant hoot vocalizations of Gombe Stream chimpanzees (*Pan troglodytes*) compared to those at Kasoje in the Mahale mountains a few hundred kilometers away. Further search for examples of population differences in vocalizations of primate species and studies of the development of this variation will be important to see whether there is some continuity between avian and primate development.

Finally, many of the calls that have been investigated developmentally in nonhuman primates have been infant separation calls in squirrel monkeys, macaques and common marmosets (see e.g., Newman 1995) or alarm calls (Herzog & Hopf 1984; Seyfarth & Cheney, Chapter 13). These types of vocalization might be expected to be highly conservative compared with affiliative vocalizations and, thus, less likely to be modified through development. As several of the chapters in this volume show (West *et al.*, Chapter 4; Zann, Chapter 6; Brown & Farabaugh, Chapter 7; Hausberger, Chapter 8) vocal structures are often modified through affiliative interactions with social companions (see also Mundinger 1970; Nowicki 1989). The search for developmental parallels in nonhuman primates is likely to be more successful with examination of vocalizations used in affiliative as well as agonistic contexts.

WHY STUDY MARMOSETS AND TAMARINS?

Marmosets and tamarins are small primates found only in Central and South America. The monkeys range from 100 to 750 g adult weight and are exclusively arboreal. Each species that has been studied has a highly complex system of vocal communication, with a large "vocabulary" of individual calls and several examples of calls produced in fixed sequences. All of the species are cooperative breeders. Generally there is a single reproductive pair (though several exceptions have been described) and other group members defer reproduction and assist in caregiving to infants. In captivity, females appear to be reproductively inhibited, rarely ovulating while in a subordinate role within a group (Ziegler *et al.* 1990; Abbott *et al.* 1993). Fathers and other males in a group do the majority of infant caregiving, and it is necessary for young animals to have direct experience taking care of someone else's infants if they are to become successful parents themselves (Epple 1978; Snowdon 1990). The reproductive female typically gives birth to twins that weigh 20–25% of her body weight at birth, and the female can be pregnant with her next litter while she is still nursing her current litter, making necessary the extensive infant caregiving by other group members.

Migrations between groups generally occur at rela-

tively low rates (one animal per group per year) except in times of extreme weather conditions (such as a drought) when group composition may change very rapidly (Savage *et al.* 1996). This life history poses several interesting developmental points. Infant monkeys grow up with several different caregivers and do not develop an exclusive relationship with the mother. After reaching independence, a monkey is actively involved in taking care of younger siblings within its natal group. At the same time these juveniles and subadults must subordinate their own reproduction to that of the breeding adults. After puberty monkeys may disperse to other groups where they must enter as subordinate animals and assist with infant care and defense in the new groups if they are to obtain a breeding position, while others remain in natal groups until a breeding vacancy occurs. The social roles played by individuals as subordinate caregivers who might rapidly acquire a breeding role create interesting problems for vocal development. Which caregivers have an influence on behavioral development? How do animals acquire an adult repertoire without appearing to challenge the breeding adults? How do they adjust their communication and behavior in order to be accepted into other groups?

We have developed three model systems to study the ontogeny of primate vocal communication in marmosets and tamarins: (a) the development of trill vocalizations in pygmy marmosets, (b) "babbling" behavior in pygmy marmosets, and (c) ontogeny of food-associated calls in cotton-top tamarins. Each of these is a form of communication that is affected by social interactions with other group members. There is ample opportunity for transgenerational influences on vocal development. Each of these model systems has provided a demonstration of social influences on vocal development, and each has clear parallels to results found in birds (both song and calls), as well as to other mammals, including humans.

DEVELOPMENT OF TRILLS IN PYGMY MARMOSETS

The trills of pygmy marmosets (*Cebuella pygmaea*) have been well studied in adults. Pola & Snowdon (1975) described four types of trill-like vocalizations and subsequently showed (Snowdon & Pola 1978) that two of these trill forms: Open Mouth and Closed Mouth trills were labelled categorically by the monkeys. The trills have individual specific differences, and playback studies demonstrated that monkeys could perceive and respond to individual differences (Snowdon & Cleveland 1981; Snowdon 1987). We have observed trills given by adults in an antiphonal form with clear orders of turn-taking so that among a group of three adults, it was rare to find an animal giving a second trill before each of the other two group members had trilled. Furthermore, one sequence of turn-taking was much more common than other sequences (Snowdon & Cleveland 1984). In a field study, the trills were shown to be very important in maintaining cohesion among group members and for signalling safely. Furthermore, monkeys altered the structure of the trills to provide more cues for sound localization, the further they were from other members of their group (Snowdon & Hodun 1981). The use of these calls in affiliative contexts, the modifiability in different environmental conditions, and the individual identity of trills all suggested that the trills would be very interesting to study developmentally.

We studied nine infants from five litters in our captive colony, from birth through sexual maturity at two years (Elowson *et al.* 1992). Recordings were made from monkeys for 30–60 minutes per week in undisturbed conditions. Only calls that could be precisely identified as coming from an infant were analyzed. We suggested four developmental models that could be evaluated from our data.

1. Call structures are fixed at birth with little or no subsequent modification.
2. Call structure changes with physical maturation. This model predicts that all infants would show similar changes at similar times, and that with increasing body size there would be a decrease in pitch and an increase in duration (due to greater lung capacity).
3. A closed vocal development system similar to that typically reported for bird song with evidence of a sensitive period followed by stereotypy or crystallization of call structure.
4. An open vocal development system in which call structure remains variable and flexible without becoming stereotyped.

We made spectrograms and measured vocal parameters of 880 trills and found that there were significant changes in call structure over time. However, these changes were not consistent across litters. While several litters did show decreases in pitch and increased duration as predicted by the maturational model, some litters showed increased pitch or decreased duration over time. The first two models were therefore rejected.

With the naturalistic paradigm we used we could not evaluate whether or not there was a sensitive period, but we used measures of coefficient of variation to evaluate whether calls become increasingly stereotyped over development. Only one of the eight parameters we measured displayed a decreased coefficient of variation over age (indicating greater stereotypy of call structure). All other parameters showed no change in variability over development, indicating a lack of stereotypy. We concluded that development of trill vocalizations in pygmy marmosets was most consistent with an open vocal development model, suggesting that changes in call structure were possible at any time in development.

SOCIAL INFLUENCES ON TRILL PLASTICITY IN ADULTS

As noted in several other chapters (West *et al.*, Chapter 4; Payne & Payne, Chapter 5; Brown & Farabaugh, Chapter 7; Hausberger, Chapter 8; McCowan & Reiss, Chapter 10; Tyack & Sayigh, Chapter 11) birds and dolphins appear to be sensitive to changes in the social environment. Songs, other calls and whistles can all change when an individual has a different social companion. We have found two phenomena that provide evidence of similar changes in vocal structure in adult pygmy marmosets.

We used the gift of two groups of pygmy marmosets from Dr John Newman of the National Institutes of Health, Bethesda, Maryland, to look at the influence that new conspecific neighbors might have on trill structure (Elowson & Snowdon 1994). While the NIH animals were in quarantine in an environment very similar to those of our regular colony room, we recorded trill vocalizations from as many animals as possible in both our original colony and among the new arrivals. We collected recordings from individual monkeys in our colony over a two to five month period without any change in their environment, to provide data in the stability of trills and thus serve as a control. After baseline data were collected over a nine week period of quarantine, the NIH animals were moved into the same colony room as our original monkeys. The groups were housed in identical cages and visually separated from each other by partitions, but now the NIH

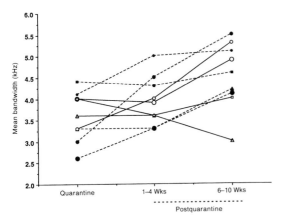

Fig. 12.1. Mean bandwidth of trills given by nine pygmy marmosets is each of three time blocks. Quarantine: prior to housing both colonies in the same room. Post quarantine: after both colonies were housed in the same room. Open symbols represent University of Wisconsin animals; closed symbols represent NIH animals. Large circles, breeding females; triangles, subadult males; small circles, juveniles; small squares, infants. (From Elowson & Snowdon 1994.)

groups and our original groups were in vocal contact for the first time. We recorded again from the same individuals for ten weeks afterwards.

We analyzed only trills that we were absolutely sure were given by a known individual and compared only the nine individual monkeys for whom we had a large enough sample of trills (at least ten trills for each phase of the study). We found a significant shift in all monkeys, regardless of age or group of origin, toward higher-pitched trills and toward trills with greater bandwidth (see Fig. 12.1) during the first ten weeks after the groups were housed in the same colony room. Recordings from monkeys in stable social environments over the equivalent time periods showed much smaller changes in both of these parameters, and the changes were in the opposite direction. Thus, the changes we observed in pitch and bandwidth appear to be only in response to a change in social environment and do not represent normal drifting in vocal structure. The fact that age was not a significant variable indicates that infant, juvenile, subadult and adult pygmy marmosets all changed their vocalizations in the same direction and to the same degree. There was therefore little evidence that vocal plasticity was limited to a sensitive period.

We have completed the first part of a second study to examine the effects of pairing individuals with a new mate on trill structure (C. T. Snowdon *et al.*, unpublished data). Mundinger (1970) has reported that American gold-finches, when paired, incorporated parts of each others' call notes. We were interested in whether pairing would influence trill structure in pygmy marmosets in the same way. We recorded trill vocalizations from four male and four female pygmy marmosets before and after pairing and looked for changes in trill structure. None of the animals that were paired had had social experience with the other until the time of pairing. We measured four parameters for each of the trills recorded for each of the eight monkeys before and after pairing, and we found significant changes in trill structure in all four pairs.

We defined three possible types of change: convergence, where the two animals differed significantly on a parameter prior to pairing but converged on the same value after pairing; divergence, where the two animals had similar structure before pairing, but shifted to different structures after pairing; tracking, where, similar to the Elowson & Snowdon (1994) study described above, animals showed parallel shifts in the value of a vocal parameter (e.g., both increasing bandwidth). Of 16 possible parameters (4 parameters×4 pairs), 13 showed significant change (convergence, divergence or tracking) within six weeks after pairing. As with the previous study, the magnitude of changes shown after pairing was much greater than changes observed over a comparable or longer time period without any change in social environment. If a change in social environment is the major influence on changes in vocal structure, then the trills, once altered at the time of pairing should remain consistent for a relatively long period afterwards, provided no further change in social environment has occurred. Some monkeys showed stable trill structure three years later.

Each pair produced a unique pattern of change, so the results are not as clearly interpretable as those of Mundinger (1970), where each mate acquired the call notes of the other or Nowicki (1989), where all chickadees in a flock converged toward the same fine structure of the "dee" note. In human speakers there is a phenomenon called "optimal convergence" (Giles & Smith 1979). Speech style provides cues about group membership, and there is some evidence that we are attracted to others who share our speech style. One response to joining a new group is therefore to imitate some aspects of their speech, but a newcomer who imitated all dimensions of the speech of the group is often viewed negatively (Giles & Smith 1979). A speaker who makes some attempt to accommodate to a group is viewed positively, but one who shows too much accommodation is viewed as insincere or patronizing. For monkeys, dolphins and birds, we also hypothesize the value of changing some aspects of vocal production to accommodate to a new group while maintaining individuality, but we cannot always predict which features would change and which would not. The process of vocal accommodation between new social companions might be viewed as a negotiation process that leads to a companion-specific pattern of accommodation (see also examples in Hausberger, Chapter 8; Locke & Snow, Chapter 14).

OTHER EXAMPLES IN MAMMALS

The phenomenon of adjusting call structures to changes in social environments is not unique to pygmy marmosets. Randall (1989) showed that banner-tail kangaroo rats (*Dipodomys spectabilis*) use foot-drumming as a form of seismic communication, and that there are consistent individual differences in the drumming signatures of kangaroo rats. She has also shown that other kangaroo rats can discriminate between playbacks of individual foot-drumming signatures (Randall 1994). Recently she found that these individual signatures can be modified in response to two important social changes (Randall 1995). First, juvenile kangaroo rats have more variable foot-drumming signatures than adults, and individual juveniles developed more stereotyped forms of foot-drumming as they become adults. Second, when juveniles and adults changed territories, they also changed the structure of foot-drumming to minimize overlap and to maximize differentiation with the foot-drumming signatures of their new neighbors. Territorial changes occurred more frequently during times of high population density, suggesting that population density is another, possibly indirect, factor that can lead to change in the structure of signals.

A second example is in the pitch of the advertisement calls of fat-tailed dwarf lemurs (*Cheirogaleus medius*, Stanger 1993). A group of 11 male dwarf lemurs was housed in individual cages in the same room, and the pitch of their whistles varied between individuals. We have cal-culated the coefficients of variation on Stanger's data and these averaged 1.53($\pm$0.51)% over the 11 males. The males were then subdivided with the three males with highest pitch and the three with lowest pitch housed in individual cages in one new room, while the five animals intermediate in pitch were housed individually in a second room. Seven of the males continued to vocalize; four ceased to vocalize. We have calculated a variety of statistics from Stanger's data after the change in social housing. All seven lemurs showed significant changes in pitch (mean change in pitch 1.33($\pm$1.1) kHz, z scores; mean 5.99, range 2.09 to 17.78). The coefficients of variation for the calls after the environmental change were still quite low (2.66($\pm$1.44)%) indicating stability within individual call structure. The animals in each of the two groups tended to converge toward a common frequency, with the animals originally representing the extremes of pitch demonstrating the greatest change. Furthermore Stanger reported that during call exchanges between pairs of animals, there was a tendency for the pitch of calls to converge during an exchange, with exchanges often being terminated when convergence was reached.

Summary

The results from the pygmy marmosets taken together with those from kangaroo rats and dwarf lemurs reported here and the results on dolphins and whales described in the chapters by Tyack & Sayigh (Chapter 11) and McCowan & Reiss (Chapter 10) suggest that vocal plasticity can be found in mammals and that this plasticity may be most evident during periods of change in the social environment. Furthermore, the results suggest that vocal plasticity is possible at any age or any stage in development and is not restricted to a critical or sensitive period. However, in each of these examples from mammals, the basic call structure is broadly adult-like in very young animals (see Seyfarth & Cheney, Chapter 13). There is no evidence at present that new calls are learned *de novo*. Rather the plasticity is reflected in subtle changes in vocal structure similar to those found in bird calls (e.g., Nowicki 1989). In the past, qualitative methods have been used to evaluate the degree of change or of stereotypy in vocalizations. It may be necessary to use quantitative measurement methods and to focus on more subtle changes in structure in order to document the social effects and lifelong potential for vocal change that appears in all of the avian and mammalian examples.

"BABBLING" IN PYGMY MARMOSETS

The definition of babbling in humans appears to be based on several factors. According to Locke (1993), babbling begins precipitiously and is rhythmical. It is independent of the child's ambient language, yet canonical babbling is described as the production of well-formed syllables that contain segments resembling exemplars of adult phonemes. Babbling is differentiated from cooing and crying by showing resemblance to adult phonemes. Canonical babbling begins at about five to seven months of age and progresses from production of a restricted set of sounds, stops, nasals and glides through the formation of recognizable words.

Babbling is thought to play an important role in the development of human language (Oller *et al.* 1976). While we typically think of babbling as a vocal activity, manual "babbling" has also been described in deaf children (Petitto & Martentette 1991). According to Locke (1993), babbling also serves to maintain parental proximity to the infant, promotes social exchanges and assists in the development of social and emotional ties between the infant and its caregiver. There is evidence that attention given by caregivers toward infants provides a reinforcement of babbling (Todd & Palmer 1968; Oller *et al.* 1976). Locke (1993) viewed a vocal dialogue between infant and caregiver as a necessary context for language acquisition. Vocal turn-taking first appears in human infants between 12 and 18 weeks of age. Infants initiate babbling more when they are alone than when others speak, but the initiation leads to social interactions.

To date, the major nonhuman animal model of babbling has been the subsong and plastic song of birds (Marler 1970; Marler & Peters 1982), but there are several reasons why a nonhuman primate model of babbling would be a useful supplement to the traditional bird song model. First, subsong and plastic song appear in birds at the time of sexual maturity far later than the onset of babbling in humans. Second, in most temperate zone birds, subsong and plastic song are produced only by males, yet human speech occurs in both sexes. Third, babbling covers the entire range of sounds that will eventually become a part of the child's language (although not other

nonlinguistic vocalizations) whereas the bird song model is restricted to a single part of the bird's vocal repertoire, song. Fourth, because nonhuman primates develop much faster than humans and can be studied under controlled conditions, a primate model can be useful in understanding functional and structural aspects of babbling much more quickly than would be possible in human infants.

Several years ago we reported that pygmy marmosets showed babbling behavior and we presented some qualitative data (Snowdon 1988). Pygmy marmosets produce a long series of vocalizations that seem to juxtapose a wide variety of adult-like vocalizations with a variety of recognizable but not quite adult calls and some totally idiosyncratic vocalizations (see Fig. 12.2). The babbling bouts begin before weaning and continue for several weeks after weaning when infants begin to move independently.

We are completing a systematic study of babbling in infant pygmy marmosets (A. M. Elowson *et al.*, unpublished data in preparation). We tape-recorded eight focal infants for two 30-minute sessions each week during the first 30 weeks of life. We examined all calling bouts of at least three different call types with at least 30 seconds of separation between calls serving to define a bout. After both acoustic and contextual analyses (described in more detail below), we labelled these bouts "babbling" since 90% of the individual calls produced were recognizable from the adult repertoire, and since these calls were clearly produced in a social context and elicited social responses from older group members. Although not exactly equivalent to the babbling of human infants (where individual phonemes rather than words may be repeated extensively), we consider that this vocal behavior of infants of a species that has a nonphonetic communication system provides a closer analogy to human babbling than does the subsong or plastic song of birds.

While we were recording infant vocalizations, we took scan samples every 30 seconds to record what the infant was doing and the reactions of other family members to it. These behavioral categories were analyzed according to the vocal activity of the infant (babbling or not babbling) and summarized in two-week blocks from week 1 to 20. We also selected one-minute samples from four babbling bouts within each two-week block to analyze the acoustic structure of the babbling in relation to the description of calls in the adult repertoire (Pola & Snowdon 1975). Calls were

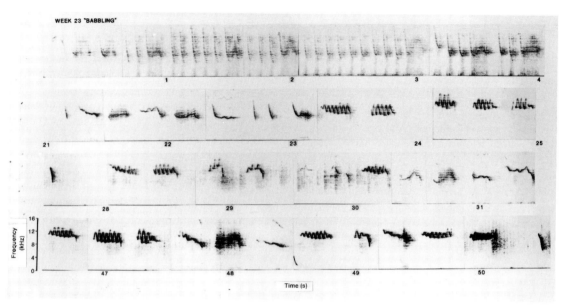

Fig. 12.2. A sequence of babbling behavior from one bout of an infant pygmy marmoset. Note the high rate of vocalization and the great diversity of call structures. (From Snowdon 1988; reprinted from *Comparative Perspectives in Modern Psychology* (Nebraska Symposium on Motivation, 1987, vol. 35), by permission of the University of Nebraska Press. Copyright 1988 by the University of Nebraska Press.)

identified according to whether they were similar enough to adult calls to be recognizable, or other calls, not clearly identifiable from the adult repertoire. Since weaning occurs at about eight weeks of age and infants are generally independent of caregivers after eight weeks, we have analyzed results in terms of weeks 1–8 versus weeks 9–20 to see whether the behavioral transition to independence had any effect on the context or structure of babbling.

Babbling was observed in all but one infant within the first two weeks of life. There was an average of 1.2 bouts per hour in the first two weeks increasing to 4–4.5 bouts per hour in weeks 3–6 and peaking at 6 bouts an hour in weeks 7–8. Babbling remained constant at 5–6 bouts per hour through week 20. Mean bout duration was 42 seconds in the first two weeks and ranged between 75–110 seconds per bout for the rest of the sampling. There was great individual variability both in frequency and duration of babbling bouts. The average babbling bout consisted of a mean of 9.8 different call types per bout and there was no difference in number of different call types or in the Shannon Weaver diversity index between dependent and independent conditions. Dependent infants produced calls at the rate of 3.28 calls per second, significantly

greater than the rate of 2.69 calls per second produced by independent infants.

Babbling occurred in a variety of contexts. Dependent infants were left alone by caregivers for several minutes at a time, and they immediately started to vocalize. Caregivers appeared to ignore these vocalizations but when there was any disturbance in the colony room, the infants were retrieved quickly. This behavior suggested that some of what we were labelling as "babbling" might really be distress vocalizations or crying. But when we compared the distribution of call types given during these bouts with bouts given by infants that were not left alone or with bouts observed after infants were independent and not displaying distress behavior, we found no differences in structure. The proportion of calls clearly different from calls in the adult repertoire was small and consistent at 10%. The finding that more than 90% of all calls analyzed in babbling bouts were similar to or identical with adult forms suggests that the definition of canonical babbling in humans can be applied to the babbling of pygmy marmosets at both dependent and independent stages.

In our behavioral observations, we found that babbling

was more likely to occur when an infant was alone and locomoting independently, but the result of babbling was an increased probability of the infant approaching or being approached by an older animal, and of increased huddling and grooming. Affiliative social responses were common. Within a babbling bout, an infant did not display any functional separation of affiliative and agonistic calls (as evaluated by reference to adult usage of these call types and the high diversity of call types given within a bout). During dependent weeks there was an increased likelihood of a babbling infant being picked up and carried by an adult. Thus, babbling in pygmy marmosets appears to occur in the same social contexts and with the same social consequences as Locke (1993) described for human babbling behavior.

Summary

The vocal bouts of pygmy marmosets appear to be analogous to babbling in humans. Most of the vocalizations are identifiable as similar to the sounds in the adult repertoire, suggesting a similarity to canonical babbling. The babbling bouts are given most often when infants are alone and locomoting, and the babbling leads to increased contact with caregivers, as demonstrated by calling infants being picked up and carried by adults or approaching, huddling and grooming with other family members. There are still many more questions we need to ask. Is there evidence of adults reinforcing infants more when they produce an adult form of a call? Is there evidence of progressive development toward more precise adult forms of calls with increased babbling experience? What are the effects on adult communication skills of the great individual variability in babbling that we have observed? Is there a correlation between caregiver response to infants and amount of babbling shown? Babbling begins at an early developmental stage in pygmy marmosets, it involves both sexes, and the infant babbling demonstrates the production of a large range of vocalizations from the adult repertoire. There is a clear social component to babbling, but whether this involves social reinforcement of babbling or not, requires more study. Babbling appears to be a real phenomenon in pygmy marmosets, and the pygmy marmoset can provide us with a research model that is more analogous to human babbling than subsong and plastic song in birds.

ONTOGENY OF FOOD-ASSOCIATED CALLS IN COTTON-TOP TAMARINS

Food-associated calls have been described and studied in several species (chickens, *Gallus domesticus* Marler *et al.* 1986a,b; toque macaques, *Macaca sinica* (Dittus 1984); rhesus macaques, *Macaca mulatta* (Hauser & Marler 1993a,b; spider monkeys, *Ateles geoffroyi* (Chapman & Lefebvre 1990); chimpanzees (Hauser & Wrangham 1987; Hauser *et al.* 1993); and golden lion tamarins (*Leontopithecus rosalia*) Benz 1993; Benz *et al.* 1992)). In most of these studies animals have calls that are generally specific to food, and the rate of calling is generally related to either the quantity or preference ranking of food suggesting that food-associated calls are honest signals about food quality and quantity.

Food-associated calls in adult cotton-top tamarins

Cleveland & Snowdon (1982) first described food-associated vocalizations in cotton-top tamarins, noting that there were two types of chirps, C-chirps and D-chirps, given in association with feeding. Subsequently Elowson *et al.* (1991) completed an experimental analysis of food-associated vocalizations. Adult cotton-top tamarins (*Saguinus oedipus*) were presented with six different types of food (standard tamarin chow, peanuts, peaches, egg, etc.) as well as some nonfood items that were the same size and as easy to manipulate as food pieces (paper clips, nut shells). These items were presented to the tamarins one at a time for five minutes following a five minute baseline condition and the rate of C- and D-chirps as well as the context of both call types was recorded. The data from the five-minute period prior to presentation of foods or objects provided a no-food, no-object baseline condition. Subsequent to the recording of responses to different foods and objects, the monkeys were tested with every possible pairing of food types to determine a preference hierarchy.

The adult cotton-top tamarins used the C- and D-chirps almost exclusively with food. We rarely recorded these chirps during times when food was not available to the monkeys and we found a very low rate of these calls given when nonfood manipulable objects were presented. More than 97% of all of the C- and D-chirps recorded

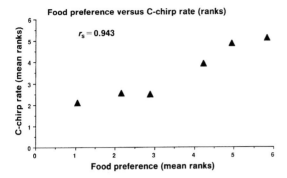

Fig. 12.3. Correlation of mean food preference rank with mean C-chirp rate rank for adult cotton-top tamarins. (From Elowson *et al.* 1991.)

were given in the context of feeding. These do appear to be food-associated calls. There were clear context differences in the use of each type of chirp. C-chirps were given mainly when a monkey approached a food bowl or sorted through the bowl to find a piece of food. The D-chirps were given usually after the monkeys had left the food bowl and were eating the food.

There were clear individual differences between adult cotton-top tamarins in their preference for different foods, but there was a positive correlation for each animal between the rate of C-chirps given and its preference ranking for a food. Averaged over all of the subjects, there was a significant positive correlation between food preference rank and the rate of chirps observed (see Fig. 12.3). Thus, it appeared that for adult cotton-top tamarins the C- and D-chirps function as representational signals that indicate the preference of food, with the rate of vocalization providing information about the motivation of the caller for the food. Food-associated calls in tamarins thus provide an honest indicator of food preference.

Food-associated calls in juvenile and subadult tamarins

We have become interested in the ontogeny of this calling system. How does the structure, function and usage of this call vary with development and is there any indication that social interactions with other monkeys affect the development of this call system? Cotton-top tamarins nurse from their mothers for the first eight weeks and they are gradually weaned as their fathers and older siblings offer them

solid food starting in the fifth or sixth week. By three months of age the monkeys are usually feeding on their own. The monkeys reach puberty at about 18 months of age, but females remain reproductively inhibited and do not ovulate while there is a reproductive female present (Ziegler *et al.* 1990). Females can ovulate and become pregnant within eight days of being removed from living in a group with a reproductive female and being placed with a novel male. This transition is independent of age and puberty (Ziegler *et al.* 1990).

As a first step to understanding the ontogeny of food-associated calls Roush & Snowdon (1994) studied 14 juvenile (prepubertal) and subadult (postpubertal) monkeys ranging in age from 4 months to 29 months of age. The same procedures were used as in the adult study (Elowson *et al.* 1991). Animals were presented with six different foods and with nonfood manipulable objects. Subsequent to the recording of vocalizations in conjunction with food presentations, each animal was tested for food preferences by presenting all possible pairs of foods in a choice apparatus.

Similar to adults, the juvenile and subadult tamarins gave most of their C- and D-chirps to food, and rarely gave these calls in control conditions. However, in contrast to the adult, they did give a significant number of C- and D-chirps to nonfood, manipulable objects, though at a significantly lower rate than they called to the food items. There were two major differences between young tamarins and adults in the production of calls. When the acoustic parameters (duration, start frequency, peak frequency) of juvenile C- and D-chirps were evaluated using a criterion of the mean ± 1 SD of adult values, a large number of both juvenile and subadult calls fell outside the range of the adult calls. It would be expected that about 32% of calls would be outside of this distribution by chance for adults. The percentage of calls outside of the adult distribution ranged from 28% to 75% for young tamarins, with only four young tamarins being within the range of adult variability in call structure. Interestingly, there was no correlation between age and the percentage of adult-like chirps produced (see Fig. 12.4). Of the four most adult-like infants, two were 8 months old, one 20 and one 29 months old. Of the four least adult-like callers, one was 4 months, one 20 months and two 26 months old. This lack of relationship between stereotypy of call production and age was surprising, since we had hypothe-

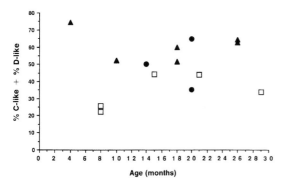

Fig. 12.4. Proportion of C-like and D-like-chirps (>±1 SD from adult parameters) in response to food for individual infant and juvenile in three families of cotton-top tamarins as a function of age (From Roush & Snowdon 1994.)

sized a linear progression of competence in production with increasing age.

Immature tamarins gave a much higher proportion of other calls (not C-chirps or D-chirps) in feeding contexts. Whereas only about 13% of the calls recorded from adults in feeding situations were neither C- nor D-chirps, 32% of the calls of juveniles and subadults were neither C- nor D-chirps, and there was no distinction between older and younger immatures. Furthermore, we found that food preferences measured in a two-choice test apparatus were much less stable in immature tamarins with greater numbers of ties and violations of transitivity than we had found in adults, and there was no correlation between food preference and the rate of vocalization.

These results suggested that a social inhibitory process was active in the immature tamarins. Immature animals at all ages had the capacity to produce clearly identifiable food-associated calls. This is not an inability to produce the food-associated calls, but rather an inconsistency in production. The fact that many other nonfood calls were given in feeding contexts and that many of these nonfood-associated calls were frequently submissive vocalizations (R. S. Roush, unpublished data) suggests that young monkeys are inhibited when feeding in the presence of adults. Furthermore, the lack of a linear age-related development in both the structure of C- and D-chirps and the usage of these calls suggests that a function of this immature form of calling may be to signal subordinate reproductive status to the adults. As noted above, reproductive inhibition is a common finding in cooperatively

breeding mammals, and the use of immature forms of vocal communication may either be a consequence of this reproductive inhibition or serve as a social mechanism to avoid appearing to challenge the reproductive adults.

We have found similar results suggesting a social inhibitory process in two other signal systems. Snowdon (1988) reported that there was an immature form of the long call used in territory defense. Adult forms of long calls were rarely observed in reproductively inhibited cotton-top tamarins, but the frequency of observation of each of the forms of territorial long calls increased after animals were reproductively active. As part of the study on trill development in pygmy marmosets (Elowson *et al.* 1992) we found that trills which are used in a turn-taking pattern by adult pygmy marmosets were typically given by juvenile and subadult pygmy marmosets in a series of vocalizations, including many other vocal forms. This "immature" form of usage continued until the animals were removed from a subordinate role and paired as breeders.

The results from the food-associated call study, the long call study and the trill study all suggest that social environments can provide an inhibition to vocal development as well as a facilitation of vocal development. The hypothesis that the presence of adults inhibits adult structure and usage of food-associated calls in young tamarins can be tested in two ways: by removing adult monkeys temporarily from family groups and testing to see whether the oldest offspring use food-associated calls in an adult-like way, or by testing monkeys in the weeks immediately before and immediately after they are removed from a subordinate role in a group and allowed to become breeders.

We have some preliminary results from the second test (R. S. Roush, unpublished data). Nine monkeys (five female and four male) were tested weekly for food-associated vocalizations during the three weeks in their natal families immediately before and eight weeks after separation from their families and pairing with a new mate. Within two weeks after pairing there was a significant decrease in the number of nonfood-associated calls from levels observed when monkey were subordinate within a family. The animals reached adult levels a mean of 3.2 weeks after removal from the family. In contrast the proportion of C-chirps and D-chirps that were more than ±1 SD of adult values remained at high levels through six to nine weeks. Measures of affiliative interactions between monkeys and dates of the female's first ovulation suggested that the calls reached adult form more quickly in those pairs that formed close affiliative relationships but not where females ovulated soonest. Two separate developmental processes appear: an immediate reduction of non C- and D-chirp vocalizations in feeding contexts when the social inhibition from the family was removed, but a more gradual development of precise adult forms of C- and D-chirps that appeared to be related to the quality of affiliative interations within a pair.

First appearance of food-associated calling in infant tamarins

As noted above, infant tamarins begin the transition from nursing to eating solid food at about five weeks of age, and an important component of the weaning process is food-sharing behavior initiated by adult animals (Feistner & Chamove 1986). We have studied ten infant monkeys during the transition from nursing through first feeding on solid food, recording monkeys both when food was offered to infants by adult caregivers and when the family was presented with high or low preference foods (R. S. Roush, unpublished data).

Although nursing infants gave a variety of vocalizations, we did not record any C- or D-chirps until the infants began to feed independently. As with the older juveniles and subadults we have studied, the infants gave a variety of other vocalizations during feeding. Approximately 40% of the vocalizations during independent feeding were C-chirps and D-chirps, and the remaining 60% were other types of vocalization. And like juveniles, subadults and adults, infants were only observed to give C- and D-chirps while feeding. We have not observed these calls from infants in any other contexts.

The limitation of infants giving C- and D-chirps only upon their first independent feeding on solid food might be interpreted as a preprogrammed response to solid food, but, prior to feeding independently, the infants have received considerable experience with solid food and food-associated vocalizations during food transfer. Food transfer attempts can be initiated by either the infant or an adult with food. We found that adults produced C- and D-chirps on only 30% of infant-initiated food transfer attempts, but adult C- and D-chirp vocalizations accompanied more than 60% of all successful food transfers.

Thus, adult food vocalizations are associated with successful food transfer attempts and may serve to reinforce the infant's behavior. In several of the successful food transfers the adult offering the food gave a long series of C- and D-chirps in a chatter-like sequence, and on many of these trials, several other group members were observed to orient toward the adult offering the food and to give chirps as well. The food transfer bouts are therefore generally associated with adults giving food-associated vocalizations and, in many cases, the adult is giving long strings of food-associated calls accompanied by visual orientation and calling from other group members. From the infant's perspective this behavior provides a rich context for learning several things: which items are appropriate to eat, when it will be possible to obtain food from an adult, and what vocalizations are associated with feeding. The coincidence of adult food-associated vocalizations and transfer of food provides an attentional focus and reinforcement for infant learning.

Summary

The study of the ontogeny of food-associated calls in cotton-top tamarins provides a model system for studying the development of production, comprehension, and usage of two call types. The C- and D-chirps (used by adult tamarins to communicate about the presence of food and the individual's preference for different foods) do not appear in the infant repertoire until the infant starts eating solid food. Independent feeding, in turn, occurs only after a period when other group members have shared food with the infant. The probability of food sharing is greater after an adult gives a food-associated call, and frequently several group members vocalize together with series of food-associated calls when food is being offered to an infant, thus creating a context for learning both the structure and usage of the calls. However, reproductively inhibited monkeys regardless of age produce many food-associated calls that do not match the precision of structure shown by adults and they do not use the calls in precisely the same way as adults (in a way analogous to plastic song in birds). The fact that they are often able to produce calls with adult structure and can use the calls in appropriate contexts suggests that consistent adult structure and usage are being inhibited by the social environment rather than suggesting that younger animals are incapable of producing and using the calls appropriately. It will be important to find ways of distinguishing between competence and performance in other studies of vocal development where younger organisms appear not to have adult-like skills.

CONCLUSIONS

Many of the differences in vocal development between birds and mammals previously reported are due to a restricted focus on studying vocal production rather than studying the ontogeny of usage and comprehension as well. In addition the use of modern acoustic analysis techniques coupled with increased attention to the social context in which vocal development occurs has led to increasing evidence that vocalizations can be modified subtly in response to social influences. As noted by several authors in this volume, a consensus is emerging that mammalian and avian species living in complex social environments have at least some vocalizations (generally used in affiliative contexts) that are responsive to changes in social environment. There is evidence of vocal accommodation to social companions to (presumably) indicate some degree of affiliation with others while still maintaining an individual signature. This vocal accommodation is often subtle and often revealed only by quantitative analysis, and these changes are not age specific but possibly occur throughout an individual's life. When the ontogeny of usage and of response to vocalizations is examined, the parallels between birds, nonhuman mammals and humans become even stronger (West *et al.*, Chapter 4; Seyfarth & Cheney, Chapter 13).

We have reviewed our current work with pygmy marmosets that demonstrates in the trill vocalizations an open developmental process that does not lead to stereotyped or crystallized calls, but rather a flexible call that can be modified to environmental or social changes. We have described the babbling behavior of infant pygmy marmosets that seems to be analogous to the babbling of human infants, in that canonical syllables (calls) are produced and that babbling occurs in a rich social context which leads older group members to interact socially with infants. Future work on babbling should focus on the specific relationship between call types given by the infant and social responses. Is there evidence that adults reinforce infants for vocalizing and is there evidence of improved vocal performance by infants as a result of adult response

to babbling? Finally, we have discussed our work on the development of food-associated calls in cotton-top tamarins, where the food-sharing behavior of adults with infants provides a rich context for learning both the structure and usage of these vocalizations, but where social factors relating to the subordinate status of juveniles and subadults appear to prevent the consistent expression of adult structure and adult usage, until animals reach breeding status. These examples taken together suggest that social influences can be important for creating environments for young animals to learn the structure and usage of calls, for altering vocal structures at all stages of development, as well as for inhibiting the adult production and usage of calls in subordinate animals. The role of social companions is critical for vocal development in a wide variety of species, and there is no longer any need to consider nonhuman primates and other mammals as fundamentally different from birds and human beings in the process of vocal development.

ACKNOWLEDGEMENTS

We thank Cristina Lazaro-Perea, Andrew Walls, Jennifer Greene, Margaret George, Carol Sweet, Pamela Tannenbaum, and Kay Stopfer for their assistance with the studies reported here. Supported by US Public Health Service Grant MH 29,775, Research Scientist Award MH 00,177, and funds from the University of Wisconsin Graduate School Research Committee.

REFERENCES

Abbot, D. H., Barrett, J. & George, L. M. (1993). Comparative aspects of the social suppression of reproduction in female marmosets and tamarins. In *Marmosets and Tamarins: Systematics, Behaviour and Ecology*, ed. A. B. Rylands, pp. 152–63. Oxford University Press, Oxford.

Benz, J. J. (1993). Food-elicited vocalization in golden lion tamarins: Design features for representational communication. *Animal Behaviour* **45**: 443–55.

Benz, J. J., Leger, D. W. & French, J. S. (1992). Relation between food preference and food-elicited vocalization in golden lion tamarins (*Leontopithecus rosalia*). *Journal of Comparative Psychology* **106**: 142–9.

Brockelman, W. Y. & Schilling, D. (1984). Inheritance of stereotyped gibbon calls. *Nature* **312**: 634–6.

Chapman, C. A. & Lefebvre, L. (1990). Manipulating foraging

group size: Spider monkey food calls at fruiting trees. *Animal Behaviour* **39**: 891–6.

Cleveland, J. & Snowdon, C. T. (1982). The complex vocal repertoire of the adult cotton-top tamarin (*Saguinus oedipus*). *Zeitschrift für Tierpsychologie* **58**: 231–70.

Dittus, W. P. J. (1984). Toque macaque food calls: Semantic communication concerning food distribution in the environment. *Animal Behaviour* **32**: 470–7.

Elowson, A. M. & Snowdon, C. T. (1994). Pgymy marmosets, *Cebuella pygmaea*, modify vocal structure in response to changed social environment. *Animal Behaviour* **47**: 1267–77.

Elowson, A. M., Sweet, C. S. & Snowdon, C. T. (1992). Ontogeny of trill and J-call vocalizations in the pygmy marmoset (*Cebuella pygmaea*). *Animal Behaviour* **42**: 703–15.

Elowson, A. M., Tannenbaum, P. & Snowdon, C. T. (1991). Food associated calls correlate with food preferences in cotton-top tamarins. *Animal Behaviour* **42**: 931–7.

Epple, G. (1978). Reproductive and social behavior of captive tamarins with special reference to captive breeding. *Primates in Medicine* **10**: 50–62.

Feistner, A. T. C. & Chamove, A. S. (1986). High motivation toward food increases food-sharing in cotton-top tamarins. *Developmental Psychobiology* **19**: 439–52.

Gautier, J.-P. & Gautier-Hion, A. (1977). Vocal communication in Old World primates. In *How Animals Communicate*, ed. T. A. Sebeok, pp. 890–964. Indiana University Press, Bloomington.

Giessmann, T. (1984). Inheritance of song parameters in gibbon song analyzed in 2 hybrid gibbons (*Hylobates pileatus*×*Hylobates lar*). *Folia Primatologica* **24**: 216–35.

Giles, H. & Smith, P. (1979). Accommodation theory: Optimal levels of convergence. In *Language and Social Psychology*, ed. H. Giles & R. N. St Clair, pp. 45–65. Basil Blackwell, Oxford.

Green, S. (1975). Dialects in Japanese monkeys: Vocal learning and cultural transmission of locale-specific behavior? *Zeitschrift für Tierpsychologie* **38**: 304–14.

Hauser, M. D. & Marler, P. (1993a). Food-associated calls in rhesus macaques. I. Socioecological factors. *Behavioral Ecology* **4**: 194–205.

Hauser, M. D. & Marler, P. (1993b). Food-associated calls in rhesus macaques. II. Costs and benefits of call production and suppression. *Behavioral Ecology* **4**: 206–12.

Hauser, M. D., Teixidor, P., Fields, L. & Flaherty, M. (1993). Food elicited calls in chimpanzees: Effects of food quantity and divisibility. *Animal Behaviour* **45**: 817–19.

Hauser, M. D. & Wrangham, R. W. (1987). Manipulation of

food calls in captive chimpanzees. *Folia Primatologica* **48**: 207–10.

Herzog, M. & Hopf, S. (1984). Behavioral responses to species-specific warning calls in infant squirrel monkeys reared in social isolation. *American Journal of Primatology* **7**: 99–106.

Hodun, A., Snowdon, C. T. & Soini, P. (1981). Subspecific variation in the long calls of the tamarin, *Saguinus fuscicollis*. *Zeitschrift für Tierpsychologie* **57**: 97–110.

Locke, J. (1993). *The Child's Path to Spoken Language*. Harvard University Press, Cambridge, MA.

Marler, P. (1970). A comparative approach to vocal development: Song development in the white-crowned sparrow. *Journal of Comparative and Physiological Psychology*, Supplement **71**: 1–25.

Marler, P., Dufty, A. & Pickert, R. (1986a). Vocal communication in the domestic chicken: I. Does a sender communicate information about the quality of food to a receiver? *Animal Behaviour* **34**: 188–93.

Marler, P., Dufty, A. & Pickert, R. (1986b). Vocal communication in the domestic chicken: II. Is a sender sensitive to the presence and nature of receiver? *Animal Behaviour* **34**: 194–8.

Marler, P. & Peters, S. (1982). Subsong and plastic song: Their role in vocal processes. In *Acoustic Communication in Birds, Vol. 1, Song Learning and its Development*, ed. D. E. Kroodsma & E. H. Miller, pp. 25–50. Academic Press, New York.

Masataka, N. (1992). Attempts by animal caretakers to condition Japanese macaque vocalizations result inadvertently in individual-specific calls. In *Topics in Primatology*, Vol. 1, *Human Origins*, ed. T. Nishida, W. C. McGrew, P. Marler, M. Pickford & F. B. M. de Waal, p. 271–8. University of Tokyo Press, Tokyo.

Mitani, J. C., Hasegawa, T., Gros-Louis, J., Marler, P. & Byrne, R. (1992). Dialects in wild chimpanzees? *American Journal of Primatology* **27**: 233–43.

Mundinger, P. (1970). Vocal imitation and recognition of finch calls. *Science* **168**: 480–2.

Newman, J. D. (1995). Vocal ontogeny in macaques and marmosets: Convergent and divergent lines of development. In *Current Topics in Primate Vocal Communication*, ed. E. Zimmermann, J. D. Newman & U. Jürgens, pp. 73–97. Plenum, New York.

Nowicki, S. (1989). Vocal plasticity in captive black-capped chickadees: The acoustics basis of call convergence. *Animal Behaviour* **37**: 64–73.

Oller, D. K., Wieman, L. A., Doyle, W. J. & Ross, C. (1976). Infant babbling and speech. *Journal of Child Language* **3**: 1–11.

Petitto, L. A. & Martentette, P. F. (1991). Babbling in the

manual mode: Evidence for the ontogeny of language. *Science* 251: 1493–6.

Pola, Y. V. & Snowdon, C. T. (1975). The vocalizations of pygmy marmosets, *Cebuella pygmaea. Animal Behaviour* 23: 826–46.

Randall, J. A. (1989). Individual foot-drumming signatures in banner-tail kangaroo rats, *Dipodomys spectabilis. Animal Behaviour* 38: 620–30.

Randall, J. A. (1994). Discrimination of foot-drumming signatures by kangaroo rats, *Dipodomys spectabilis. Animal Behaviour* 47: 45–54.

Randall, J. A. (1995). Modification of foot-drumming signatures by kangaroo rats: Changing territories and gaining new neighbours. *Animal Behaviour* 49: 1227–37.

Roush, R. S. & Snowdon, C. T. (1994). Ontogeny of food-associated calls in cotton-top tamarins. *Animal Behaviour* 47: 263–73.

Savage, A., Giraldo, L. H., Soto, L. H. & Snowdon, C. T. (1996). Demography, group composition and dispersal in wild cotton-top tamarin (*Saguinus oedipus*) groups. *American Journal of Primatology* 38: 85–100.

Snowdon, C. T. (1987). A naturalistic view of categorical perception. In *Categorical Perception*, ed. S. Harnad, pp. 332–54. Cambridge University Press, New York.

Snowdon, C. T. (1988). A comparative approach to vocal communication. In *Comparative Perspectives in Modern Psychology: The Nebraska Symposium on Motivation*, ed. D. L. Leger, pp. 145–99. University of Nebraska Press, Lincoln.

Snowdon, C. T. (1990). Mechanisms maintaining monogamy in monkeys. In *Contemporary Issues in Comparative Psychology*, ed. D. A. Dewsbury, pp. 225–51. Sinauer Associates, Sunderland, MA.

Snowdon, C. T. & Cleveland, J. (1981). Individual recognition of contact calls in pygmy marmosets. *Animal Behaviour* 28: 717–27.

Snowdon, C. T. & Cleveland, J. (1984). "Conversations" among pygmy marmosets. *American Journal of Primatology* 7: 15–20.

Snowdon, C. T. & Elowson, A. M. (1992). Ontogeny of primate vocal communication. In *Topics in Primatology*, Vol. 1, *Human Origins*, ed. T. Nishida, W. C. McGrew, P. Marler, M. Pickford & F. B. M. de Waal, pp. 279–90. University of Tokyo Press, Tokyo.

Snowdon, C. T. & Hodun, A. (1981). Acoustic adaptations in pygmy marmoset contact calls: Locational cues vary with distance between conspecifics. *Behavioral Ecology and Sociobiology* 9: 295–300.

Snowdon, C. T. & Pola, Y. V. (1978). Interspecific and intraspecific responses to synthesized pygmy marmoset vocalizations. *Animal Behaviour* 26: 192–206.

Stanger, K. F. (1993). Structure and function of the vocalizations of nocturnal prosimians (Cheirogaleidae). Ph.D. dissertation, Eberhard-Karls-Universität, Tübingen, Germany.

Tenaza, R. (1985). Songs of hybrid gibbons (*Hylobates lar×H. muelleri*). *American Journal of Primatology* 8: 249–53.

Todd, G. A. & Palmer, B. (1968). Social reinforcement of infant babbling. *Child Development* 39: 591–6.

Ziegler, T. E., Snowdon, C. T. & Uno, H. (1990). Social interactions and determinants of ovulation in tamarins (*Saguinus*). In *Primate Socioendocrinology*, ed. T. E. Ziegler & F. B. Bercovitch, pp. 113–33. Alan R. Liss, New York.

13 Some general features of vocal development in nonhuman primates

ROBERT M. SEYFARTH AND DOROTHY L. CHENEY

INTRODUCTION

Striking parallels exist between the development of speech in human infants and the development of song in birds. Many sparrows, for example, learn their songs more readily during a sensitive period than at other times during development, require practice, and must hear themselves sing for normal song to develop (Baptista & Petrinovich 1986). These same features characterize both the earliest speech of human infants (see e.g., Ferguson *et al.* 1992) and second language learning, whether spoken or signed, among older individuals (Johnson & Newport 1989). Song production in zebra finches and canaries, like speech production in humans, is under lateralized neural control (Arnold & Bottjer 1985; Nottebohm 1991). Damage to any one of these areas, like damage to Broca's or Wernicke's area in (usually) the left temporal cortex of the human brain (for reviews, see Caplan 1987, 1992), produces highly specific deficits in the production or processing of communicative sounds.

As a result of these parallels in both behavior and neurobiology, studies of avian song development currently provide the best animal model for research on the mechanisms underlying speech development (Marler 1987). In contrast, while nonhuman primates are our closest living relatives and have often been used as animal models for the study of human social development (see e.g., Hinde 1984), their vocal communication is generally thought to provide no useful parallels with the development of human speech. This is because, unlike the songs of many birds, the vocalizations of most infant monkeys and apes typically appear fully formed, with the same acoustic features as adults' calls, and seem to undergo relatively little modification in the months and years thereafter (for reviews, see Newman & Symmes 1982; Seyfarth 1987; Snowdon 1987, 1988,

1990). It is not that monkeys and apes fail to develop *language* – songbirds don't, either – but that their vocal development seems much more innate and hard-wired than the development of either human speech or avian song.

These generalizations are derived from studies of vocal *production*, which we define here as the delivery of calls with a particular set of acoustic features. By contrast, relatively few studies of either nonhuman primates or birds have focused on two other features of vocal communication: vocal *usage*, defined as the use of particular calls in specific social or ecological circumstances, such as interacting with a dominant individual as opposed to a subordinate, or when confronted with a leopard as opposed to an eagle; and the *responses* that animals show to the vocalizations of others. In this chapter we argue that vocal production, vocal usage, and responses to vocalizations develop at different rates in nonhuman primates and are affected by different social and physiological mechanisms. To illustrate this point, we review studies of vocal development in vervet monkeys, rhesus macaques, Japanese macaques, and infants cross-fostered between these two macaque species. In all cases, vocal production seems most innate (although it is by no means always fixed at birth) and shows the least modification over time. By contrast, vocal usage is only partially innate and more clearly affected by experience, while infants' responses to the calls of others are almost entirely determined by experience. A review of 34 studies conducted on 14 different primate species suggests that these conclusions are broadly applicable across the entire Order Primates.

We suggest that, by focusing exclusively on vocal production, comparative studies have highlighted the one area in which human and nonhuman private vocal development differ most conspicuously and ignored the many ways in which they are alike. In their overall organization and in many features of their development, vocal communication in human and nonhuman primates resemble each other considerably more than they differ.

VOCAL DEVELOPMENT IN VERVET MONKEYS

Free-ranging adult vervet monkeys (*Cercopithecus aethiops*) living in Amboseli National Park, Kenya, give acoustically different vocalizations to at least five different types of predator: large terrestrial carnivores, eagles, snakes, baboons, and unfamiliar humans. Within the first three classes, adults restrict their alarm calls to one or a small number of species. Terrestrial predator alarm calls are given primarily to leopards (*Panthera pardus*), eagle alarm calls to martial and crowned eagles (*Polemaetus bellicosus* and *Stephanoetus coronatus*), and snake alarm calls to pythons (*Python sebae*). Each alarm call type elicits a different adaptive response from those nearby. When monkeys on the ground hear a leopard alarm call, they run into trees. Eagle alarm calls cause them to look up into the air or run into a bush. If the monkeys are already in a tree, eagle alarm calls cause them to run out of it, onto the ground, and into a bush. Snake alarm calls, in contrast, cause monkeys to search the area around them, often as they stand on their hind legs (Struhsaker 1967; Seyfarth *et al.* 1980).

In addition to their alarm calls, adult vervets grunt to one another in a variety of different circumstances. Individuals may grunt when approaching a dominant member of their group, when approaching a subordinate, when beginning to move onto an open plain or watching another animal do so, and upon sighting another group. Even to an experienced observer, all of these grunts sound very similar, across both different individuals and different contexts (Struhsaker 1967). Field experiments and acoustic analysis, however, have shown that vervets actually make use of four different grunt types, each of which elicits a subtly different response from animals nearby (Cheney & Seyfarth 1982; Seyfarth & Cheney 1984).

Finally, adult vervet monkeys use a distinctive, trilled vocalization called a "wrr" (Cheney & Seyfarth 1988; Hauser 1989) or an intergroup wrr (Struhsaker 1967) to signal the presence of another vervet group. Alarm calls, grunts, and wrrs are not the only calls in the vervets' vocal repertoire but they are the vocalizations whose development has been studied in the greatest detail. They are, as a result, the focus of subsequent sections.

The development of vocal production

To test whether infant vervets (defined as animals aged under 12 months) need "practice" before they can pronounce alarm calls correctly, one would ideally tape-record alarm calls from infants at various ages beginning shortly after birth and examine whether the acoustic properties of these calls became more adult-like over time in ways that could not be explained simply by the maturation of the vocal tract. For two reasons, however, such data are difficult to acquire. First, alarm calls occur unpredictably and are difficult to tape-record systematically. Second, infant vervet monkeys give alarm calls far less often than adults do, and rarely give them when they are very young. Infants younger than one month of age have never been observed to give an alarm call upon sighting a predator or to give a call in any context that sounds like an alarm, and only a few alarm calls from infants under six months of age have ever been tape-recorded.

Those alarm calls that have been recorded from vervet infants are acoustically similar to the alarm calls given by adults (Fig. 13.1), suggesting that the production of vervet alarm calls may be largely innate. Moreover, alarm calls given by infants are almost as effective as those of adults in eliciting the appropriate escape response from others nearby (Seyfarth & Cheney 1986). Upon hearing an alarm call from an infant, the initial response from adults is to do what they would do if they heard an adult's alarm: look up, for example, or run toward a tree. The adults' response is rarely complete, however, suggesting that adults attend both to information encoding predator identity and to information encoding the identity and reliability of the caller.

Grunts provide a more detailed picture of the development of vocal production because infant vervets begin grunting at high rates from the day they are born. In one respect the production of grunts seems at least partially innate, because the grunts of day-old infants are easily recognizable to human listeners as grunts and not one of the other calls in the vervets' repertoire. In the fine details of their acoustic features, however, infant grunts differ in

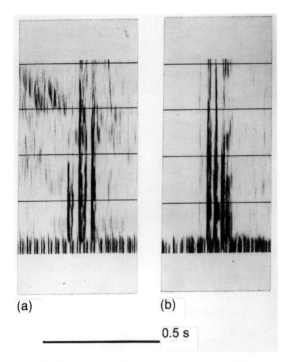

(a)   (b)

0.5 s

Fig. 13.1. Comparison of (a) an alarm call given by a 5.5-month-old male infant vervet monkey to a marabou stork with (b) an alarm call given by an adult male vervet to a martial eagle. In each wide-band (300 Hz filter) spectrogram the *x*-axis represents time and the *y*-axis represents frequency in units of 1 kHz. (Reprinted with permission from Seyfarth & Cheney 1986).

many respects from those of adults. Compared with adult grunts, for example, infant grunts are longer. They also have more acoustic units per call and longer interunit intervals. As infants grow older, the acoustic features of their grunts gradually come to resemble those of adults. Some features become adult-like by 12 weeks, others by one year, and still others when the infants are between two and three years old. These changes presumably reflect a combination of the maturation of the infant's vocal tract, the young monkey's growing coordination in the use of its vocal apparatus, social experience, and perhaps learning from older group members (Seyfarth & Cheney 1986).

The development of intergroup wrrs is similar to the development of grunts, but with an intriguing twist (Hauser 1989). Infants from birth to three months of age give trilled, wrr-like calls in two different contexts, both associated with distress. "Lost wrrs" are given by infants who are separated from their mothers, while "contact wrrs" are given by infants (usually 1.5 months of age or younger) who are in contact with their mother (Hauser 1986, 1989). Acoustically, contact wrrs differ from adult intergroup wrrs because they have a higher, less stable fundamental frequency, higher stressed harmonics, and greater frequency modulation (Fig. 13.2). Lost wrrs, in contrast, are much more adult like; in particular, they exhibit the distinct pulses alternating with periods of silence that characterize adult intergroup wrrs (Fig. 13.2). Almost from birth, then, infant vervets produce both a trilled call that differs from, and a trilled call that is strikingly similar to, the adult wrr.

Between roughly 10 and 18 months of age, young vervets rarely give any sort of trilled vocalization, and the adult-like, lost wrr seems to disappear from their repertoire. When young juveniles resume the production of trilled vocalizations at around 18 months of age, they do so largely in the context of intergroup interactions and produce a trill which, like the juveniles' grunts, requires many months before it gradually develops the acoustic features of the adult intergroup wrr (Hauser 1989).

The development of vocal usage

Vervet monkeys in Amboseli regularly come into contact with over 150 species of birds and mammals, only some of which pose any danger to them. Vervets therefore face an intriguing problem in classification: they must learn which species are predators and which are not and, within the former class, which species fall into the categories designated by each of their five different alarm calls. When infant vervets first begin giving alarm calls, they often give alarm calls to harmless species such as warthogs, small hawks, or pigeons. The range of species eliciting alarm calls from infants, however, is not entirely random.

Figure 13.3 compares the species that elicited eagle alarm calls from infant, juvenile, and adult vervets over two nine-month periods in 1980 and 1983. Confronted with the same array of real and potential predators, adults were most selective. They gave alarm calls almost exclusively to raptors (family Falconidae), a group whose members are distinguished from other birds by their relatively large size, curved beaks, and talons. Within this class, adults gave alarm calls most often to the vervets' two confirmed predators, martial and crowned eagles. Juveniles were less selective, but more likely to give alarm calls to

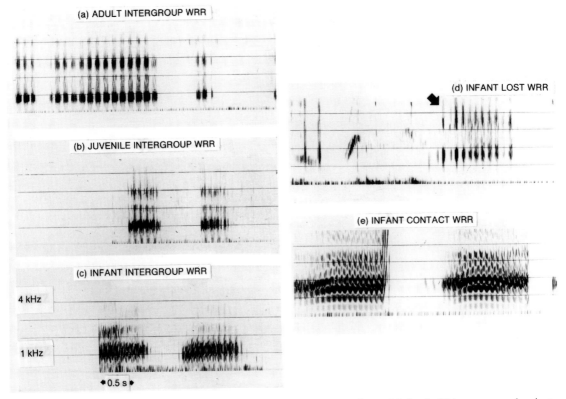

Fig. 13.2. Sound spectrograms of vervet monkey wrrs: (a) intergroup wrr by an adult female; (b) intergroup wrr by a juvenile; (c) intergroup wrr by an infant 10–18 months old; (d) lost wrr by an infant <3 months old; (e) contact wrr by an infant <1.5 months old. (Reprinted with permission from Hauser 1989.)

raptors than to nonraptors. Infants were the least selective and did not distinguish between these two broad classes of bird. Nevertheless, infants gave eagle alarm calls only to birds and objects in the air (e.g., a falling leaf), and never to animals on the ground (Seyfarth & Cheney 1980, 1986).

Similarly, although infants gave leopard alarm calls to a variety of species that posed no danger to them (warthogs, for example) they restricted their leopard alarm calls primarily to terrestrial mammals. Finally, they gave snake alarm calls exclusively to long snake-like objects. In other words, infants behaved as if they were predisposed from birth to divide other species into different classes: predators versus nonpredators and, within the former class, terrestrial carnivores, eagles, and snakes.

This mixture of relatively innate mechanisms and experience in the development of vocal usage is not unique to alarm calls. When infant vervets first begin to grunt,

they use many of the acoustically different grunt subtypes found in adult communication. For each of these calls the relation between grunt type and social situation is imprecise, but still not entirely random. Although adults give the "grunt to an animal moving into an open area" only when they themselves or another individual are moving into an open area, infants between one and four months of age use this call as they move into a new area, follow their mothers, or follow a juvenile playmate (Seyfarth & Cheney 1986). Similarly, infant vervets give another type of grunt in a context that can broadly be defined as "distress or the proximity of an unfamiliar conspecific." Over time, this category is further divided to incorporate subtly different grunts for "proximity of a dominant," "proximity of a subordinate," and "proximity of another group." Hauser (1989) described a similar development progression in the vervets' use of intergroup wrrs. Once again, in their use of

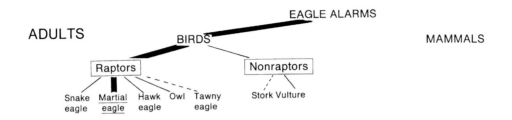

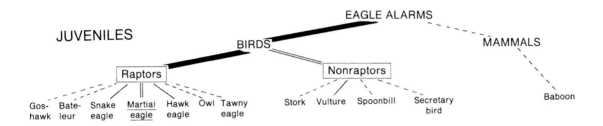

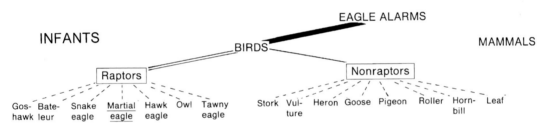

Fig. 13.3. Avian species that elicited eagle alarm calls from infants, juveniles, and adults over two nine-month periods in 1980 and 1983. Thickness of lines indicates number of alarm calls given to each species. Broken line, less than 5; single line, 6–10; double line, 11–15; thick solid line, >15. (Reprinted with permission from Seyfarth & Cheney 1986.)

grunts and wrrs infants behave as if they are predisposed to divide social situations into broad categories.

The development of grunts and wrrs illustrates the complex learning that must occur if a young monkey is to begin using vocalizations appropriately. Correct usage requires that an animal distinguish group members from nonmembers, and, within the former category, dominant individuals from subordinates. Unlike ground and aerial predators, however, animals in these classes are not grossly different morphologically, and there is considerable evidence that young primates need social experience before they know, for example, which members of their group rank above and below them (see e.g., Cheney 1977; Berman 1982; Datta 1983).

How, then, do infant vervets develop correct call usage? The most likely answer seems to be observation and experience. Just as infant monkeys learn their dominance ranks by interacting with others and observing interactions in which they are not themselves involved (Berman 1982; Datta 1983; Chapais 1988a,b), so do infants learn which individuals should receive a grunt to a dominant as opposed to a grunt to a subordinate and which should receive an intergroup wrr. In Hauser's (1989) study, infants exposed to higher rates of intergroup encounters (and thus more intergroup wrrs) produced wrrs in the appropriate context at earlier ages than infants who experienced such encounters at lower rates, suggesting that infants do indeed profit from their exposure to adult vocalizations.

Data on the development of alarm calls raise the further possibility that infants learn correct call usage as a result of active intervention or "teaching" by adults. As noted earlier, alarm calls by infants elicit many of the same responses from others as do alarm calls by adults. If an infant is the first member of its group to give an eagle alarm call, for example, nearby adults respond by looking up. If the infant has made a mistake and alarm-called at a harmless raptor, adults quickly return to what they were doing. If, however, the infant has spotted a martial eagle, adults are very likely to give an alarm call themselves (Seyfarth & Cheney 1986). These second alarm calls might serve as reinforcers that guide the infant's developing recognition of the relation between different alarm calls and their referents. An infant may find it reinforcing, for example, to see its alarm call prompt a wave of alarm-calling throughout the group, just as a human toddler often perversely seems to be reinforced when it upends a jug of water and nearby adults begin shouting.

While the responses of adults may facilitate the development of call usage, however, we have found no evidence that adults explicitly teach infants. Adults are as likely to give second alarm calls following a correct alarm call by another adult as they are following a correct alarm call by an infant or juvenile (Seyfarth & Cheney 1986). The tendency of adult animals to give alarm calls selectively thus creates a situation in which social learning by observation can easily occur, but such learning is, from the adults' perspective, inadvertent. There is no indication that adult vervets, like adult human teachers, can attribute ignorance to infants or recognize that infants are particularly in need of encouragement, correction, or instruction (Cheney & Seyfarth 1990; for a definition of teaching based on its function rather the underlying causal mechanisms – a definition by which the adult vervets' behavior would qualify as "teaching" – see Caro & Hauser 1992).

The development of responses to the calls of others

Adult vervet monkeys respond differently to different alarm calls. To test whether these responses appear fully developed in infants or are modified during ontogeny, we carried out a series of playback experiments using infants and their mothers as subjects. At monthly intervals, we played to infants aged three to seven months a leopard alarm call, an eagle alarm call, and a snake alarm call originally given by an adult member of their own group. All infants were within 5 m of their mothers, though not in contact with them, when the alarm calls were played. We divided infants' responses into three categories: (a) run to mothers; (b) adult-like responses, defined as the typical responses of juveniles and adults to playbacks of each alarm call type; and (c) "wrong" responses, defined as responses likely to increase the risk of being killed, given the hunting strategy of each predator. For example, since leopards hunt vervets by hiding in bushes and eagles are skilled at taking vervets in trees, running into a bush at the sound of a leopard alarm call or running into a tree at the sound of an eagle alarm call may actually increase an infant's risk of being taken.

Figure 13.4 summarizes infant responses. At three to four months of age most infants ran to their mothers. Few showed any adult-like responses. Between four and six months of age, running to the mother decreased and a higher proportion of subjects responded as adults would have under similar circumstances. Many infants, however, also responded in potentially dangerous ways. Among infants over six months old, running to the mother was rare, wrong responses declined in frequency, and most infants behaved like adults (Seyfarth & Cheney 1986).

As with call usage, there were hints that learning from adults may have played a role in the infants' development. When we examined films of infant behavior in the seconds immediately after an alarm call had been played, we found that infants who responded only after first looking at an adult were much more likely to respond correctly than were infants who responded independently, without first looking at another animal (Seyfarth & Cheney 1986). This observation does not prove that exposure to other animals is necessary for the development of correct responses, since infants might develop normal behavior even in the absence of cues received from others. It does, however, indicate that helpful cues are available from those nearby and that infants may take advantage of them.

Once again, however, we found no evidence of active pedagogy on the part of adults. When we compared a mother's behavior toward an infant who had responded correctly with the same mother's behavior toward an infant who made an inappropriate response, we found no indication that mothers paid particular attention to infants who behaved incorrectly (Seyfarth & Cheney 1986).

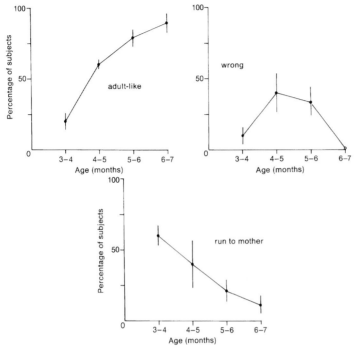

Fig. 13.4. The proportion of infant subjects showing different classes of response to playback of alarm calls at different ages. Responses are defined in the text. Values shown at each age represent proportion of subjects (means and standard errors). (Reprinted with permission from Seyfarth & Cheney 1986.)

Just as vervets need experience before they can respond appropriately to each others' vocalizations, they require experience before they respond to the calls of other species. For example, vervets in Amboseli are sympatric with a brightly colored songbird, the superb starling (*Spreo superbus*). Like vervets, starlings give acoustically different alarm calls to terrestrial and avian predators. If adult vervets hear the starlings' terrestrial predator alarm call, they often run toward trees; if they hear the starling's avian predator alarm call, they look up in the air (Cheney & Seyfarth 1985). Vervets hear starling alarm calls at different rates in different habitats (Hauser 1988). Groups living exclusively in dry woodlands hear starling terrestrial predator alarm calls roughly half as often as groups whose territories are closer to swamps. To test whether such differences in exposure affect the age at which infants begin to respond appropriately to starlings' alarm calls, Hauser (1988) played starling alarm calls to infants at successively older ages and filmed their responses. Infants living near swamps responded appropriately at signifi-

cantly earlier ages than did infants inhabiting dry woodland. Auditory experience seemed to be the crucial variable, since there was no evidence that infants in swamp groups were developmentally more advanced than those living in the woodland.

Summary

In vervet monkeys, the developmental course of vocal production, usage, and response are strikingly different. Each component of vocal communication develops at a different rate and is affected by different causal mechanisms. Compared with usage and response, the production of calls appears to be most innate: much of the adult vocal repertoire is recognizable in the calls of even very young infants. In contrast, call usage appears to develop as the result of a mixture of innate mechanisms and experience. Young monkeys seem predisposed from a very early age to use each of their calls in certain broadly defined social or ecological settings, but require many months before they display the precise relation between call type and social context seen in adult vocalizations. Finally, vervets show no predisposition to respond in particular ways to the different vocalizations of other individuals; instead, their distinctive responses to different call types seem to develop entirely as the result of experience.

VOCAL DEVELOPMENT IN CROSS-FOSTERED MACAQUES

While field studies of vervet monkeys suggest that the development of vocal production is largely innate, they cannot rule out the possibility that the earliest sounds a monkey makes are influenced by auditory experience, either in the first few weeks after birth or *in utero*. Moreover, while the development of correct call usage and response seem to be influenced by cues that infant vervets receive from nearby adults, it is also possible that such social learning is entirely unnecessary; normal development might occur even if adults were absent. Purely observational data do not allow us to specify which cues are necessary and which are sufficient for the emergence of adult-like communication.

Faced with similar problems, scientists working on birds have turned to experiments in which individuals are raised in auditory isolation, or reared with tape-recorders or live conspecifics as song "tutors." These manipulations have helped enormously to clarify which components of song production are innate and which depend more strongly upon social and/or auditory experience (see e.g., Marler & Sherman 1985). Unfortunately, such tests are difficult to carry out on nonhuman primates. As the pioneering work of Harlow (1958) was the first to demonstrate, monkeys reared in isolation exhibit many gross social and behavioral abnormalities, including some abnormalities in vocal production (Newman & Symmes 1974). Under such conditions, it is difficult to tell whether abnormal vocal communication results exclusively from an impoverished auditory environment (as seems to be true for birds) or occurs as part of the more general consequences of extreme social deprivation.

In an attempt to overcome the problems created when monkeys are reared in isolation, in 1986 we joined with Michael Owren to begin a study in which infant rhesus (*Macaca mulatta*) and Japanese (*M. fuscata*) macaques were cross-fostered shortly after birth and raised in groups of another species. The goal was to provide infants with a social environment that was almost identical with that of their own species and an acoustic environment that was substantially different from the one they would normally have experienced.

Both Japanese and rhesus macaques use calls that can be divided into the same general acoustic categories (excluding alarm calls): coos, gruffs, geckers, screams, barks, and creaks (rhesus macaques (Rowell & Hinde 1962; Peters 1983; Gouzoules *et al.* 1984; Hauser 1991, 1992); Japanese macaques (Takeda 1965; Green 1975a,b; Kawabe 1973; Owren & Casale 1994)). Cross-fostered infants would, therefore, be physically capable of producing the same calls as their foster species did. However, the frequency with which Japanese and rhesus macaques use each of these six call types differs substantially. Rhesus macaques, for example, tend to use gruffs in the same social contexts that Japanese macaques use coos. As a result, cross-fostered infants would be confronted with social situations in which their own calls would differ markedly from those used by their peers. If these infants adopted their foster species' call types, this would provide evidence of far greater vocal learning than had previously been demonstrated in any nonhuman primate.

The study population consisted of two social groups of each species, housed at the California Regional Primate Research Center. Each of the four groups contained a single adult male, four to five adult females, two to five immature offspring, and up to four unrelated immature males. These individuals provided baseline data on the vocal repertoires of adults and on the species-typical course of vocal development in infants and juveniles.

Four infants, two from each species, were fostered from one species to the other during the first week of life. They received normal maternal care, gained weight normally, and were, by many behavioral measures, fully integrated into their adopted groups. Cross-fostered animals played, groomed, and were groomed at rates that were indistinguishable from their normally raised peers, and like their peers they were supported by their mothers in aggressive alliances. Perhaps as a result, they acquired dominance ranks immediately below those of their mothers (Owren & Dieter 1989; M. J. Owren *et al.*, unpublished data). With the rare exception of an alarm bark, cross-fostered animals could not hear calls given by members of their own species. To study the development of vocal production and vocal usage, we tape-recorded the calls of all cross-fostered subjects and their normally raised peers during their first two years of life. To study the development of responses to the calls of others, we carried out playback experiments in which cross-fostered and normally raised individuals heard the calls of their mothers and other adult females.

The development of vocal production

We tape-recorded over 3800 calls from 16 normally raised infants in each of the two species during their first year and found that over 80% could be classfied into one of the six categories described above. The ease with which infants' calls could be sorted into the same categories as those of adults suggests that call production has a substantial innate component. We were, however, particularly interested in the acoustic features of calls produced by cross-fostered infants, since these animals were unable to hear vocalizations from any member of their own species.

Analysis was complicated by the fact that many of the calls given by Japanese and rhesus macaques are acoustically very similar. For example, although the "food coos" produced by rhesus and Japanese macaque infants differed statistically between species and among individuals, they nonetheless showed substantial overlap on every acoustic feature measured (Owren *et al.* 1992). And while the food coos of adult female rhesus and Japanese macaques were individually distinctive, they showed no species differences.

This extensive overlap between species in the acoustic features of food coos had two unfortunate consequences. First, it meant that cross-fostered subjects had no model for vocal learning that differed from the one they would normally have experienced. Second, similarities in rhesus and Japanese macaque food coos made it difficult for us to determine whether cross-fostered infants adhered to their own species' food coo or modified them to resemble the food coo of their foster species. On the one hand, the two cross-fostered rhesus subjects continued to produce calls that were typical of infants in their own species, supporting the view that coo production is innate and unmodified by auditory experience. On the other hand, the two cross-fostered Japanese macaques produced calls that were, by some measures, closer to the mean values of their foster species, supporting the view that coos can be modified.

Even these measurements, however, were still within the observed range of variation shown by normally raised Japanese macaques (Owren *et al.* 1992). In sum, while the results are complicated by acoustic similarities between the two species, they are most consistent with the view that the development of vocal production is largely innate.

In marked contrast, Masataka (1992; Masataka & Fujita 1989) has claimed that cross-fostered rhesus and Japanese macaque infants in his study learned to produce a food coo with a fundamental frequency (F0) that was different from that of their own species' and similar to that of their foster species. If these results could be supported, they would provide striking evidence for the modification of vocal production as a result of auditory experience. There are, however, flaws in Masataka's study. First, his data suggest a degree of call stereotypy in rhesus and Japanese macaque food coos that has not been found in any other study (e.g., Green 1975a,b; Peters 1983; Inoue 1988; Hauser 1991; Owren *et al.* 1992). Perhaps more important, Masataka provided no evidence of a species difference in adult females' food coos. Our own study revealed no species differences. If the food coos given by adults in Masataka's study showed a similar overlap between species, it would have been impossible for cross-fostered infants to modify their calls as a result of auditory experience. Finally, Masataka failed to explain how an adult female's F0 which is typically one-half that of a juvenile's could usefully serve as a model to be copied during infant call development, particularly as the infant's own F0 is known to change with age (Inoue 1988; Owren *et al.* 1992; for comments on Masataka's statistical analysis, see also Snowdon 1990).

More convincing evidence that vocal production may be modified by auditory experience is provided by Hauser's (1992) study of food coos in the rhesus macaques of Cayo Santiago, Puerto Rico. In a large group of over 300 individuals, the female members of one matriline produced a "nasalized" food coo unlike the food coos given by other individuals. Because such calls were produced only by animals who interacted with one another at high rates, Hauser speculates that auditory experience has a major effect on the development of this particular acoustic cue (see also Green 1975b). This hypothesis, however, has not yet been tested experimentally, nor has the possibility of heritable morphological determinants that might lead to nasalization been ruled out.

Table 13.1. *Results of statistical tests comparing the proportion of coos used by cross-fostered (C-F) and normally raised (NR) individuals during their first and second year of life*

Year	Context	C-F rhesus macaques tested against:		C-F Japanese macaques tested against:	
		NR rhesus macaques	NR Japanese macaques	NR rhesus macaques	NR Japanese macaques
1	Calm		*	*	
	Cagemate		*	*	
	Play		+	+	
2	Calm	*	+	*	
	Cagemate		*	*	
	Alpha male			+	
	Play	*		*	

Note:
*, $P < 0.05$; +, $P < 0.06$. A blank indicates results were not significant.
Source: Reprinted with permission from Owren *et al.* (1993).

The development of vocal usage

Normally raised rhesus and Japanese macaques between the ages of one and two years displayed a striking species difference in call usage. In four well-defined social contexts, Japanese macaques gave coos almost exclusively, while rhesus macaques gave a mixture of coos and gruffs (Owren *et al.* 1993; Owren & Casale 1994). These species differences provided an excellent opportunity to determine whether cross-fostered infants' call usage was modified as a result of experience. Table 13.1 summarizes the results of a comparison between cross-fostered subjects and (a) normally raised juveniles of their own species, and (b) normally raised juveniles of their foster species.

Cross-fostered subjects generally adhered to their own rather than their adopted species' call usage. Cross-fostered Japanese macaques rarely used gruffs, even in contexts in which gruffs were commonly given by their foster mothers and other cagemates. It was not unusual, for example, to see a cross-fostered Japanese macaque giving coos as it played with its rhesus macaque peers, even though the rhesus playmates were themselves giving gruffs. Cross-fostered rhesus macaques also showed little modification in call usage, though in a few contexts they used coos and gruffs at amounts that were intermediate between those of their own and of their foster species (Table 13.1; Owren *et al.* 1993).

The lack of modification in the vocal usage of cross-fostered subjects cannot be explained on the basis of species differences in the physiology of sound production because animals in both species were capable of producing both coos and gruffs. Indeed, the demands placed on cross-fostered subjects to modify their vocal usage were relatively slight: an individual could have acquired its foster species' pattern of vocal usage simply by modifying the rate at which it used calls that were already in its repertoire. On the whole, however, such modification did not occur.

The cross-fostered animals' failure to modify their vocal usage can also not be explained as the result of abnormalities in social experience, since by virtually all behavioral measures cross-fostered individuals were fully integrated into their adopted groups. Instead, the evidence from rhesus and Japanese macaques, like the evidence from vervet monkeys, suggests that there is a strong innate component to the development of vocal usage, and that the development of vocal usage is only partially affected by auditory or social experience.

The development of responses to the calls of others

Despite their strikingly different use of vocalizations, cross-fostered animals were fully integrated into their adopted social groups. This level of social integration suggests that foster mothers had learned to recognize the calls of their offspring and that cross-fostered infants had learned to identify the calls not only of their mothers but also of groupmates other than their mothers. To test these hypotheses, we conducted playback experiments using food coos, coos given in play (called "play coos"), and threat gruffs as stimuli. Subjects were cross-fostered juveniles, normally raised juveniles, foster mothers, and other adult females. The playback experiments had two aims: to test whether cross-fostered subjects were predisposed to attend preferentially to calls of their own as opposed to their foster species; and to determine whether cross-fostered subjects and their mothers had learned to recognize and respond appropriately to each others' vocalizations.

There was considerable variation in the responses elicited by call playbacks. Many subjects showed no obvious response, while in a few cases subjects approached the side of the cage from which the call had been played. Some playbacks elicited counter-calling by others in the group, always in the form of coos. The most consistent response to playback was looking in the direction of the speaker. We defined the strength of response as the length of time that subjects looked in the direction of the speaker and/or approached the speaker in the 10 seconds following call playback minus the duration of these same behaviors in the 10 seconds before playback. By defining a response in this way, we controlled for the subject's behavior at the time a playback was conducted.

Responses to own and adopted species' calls

Cross-fostered animals' call production and usage were relatively unaffected by auditory experience. If their responses to calls were equally unaffected, they should have reacted more strongly than their foster peers to unfamiliar calls of the cross-fostered animals' own species. To test this hypothesis, each of the four cross-fostered juveniles was played two different food coos given by an unfamiliar juvenile member of its own species. As a control, a normally raised cagemate was played the same calls. In separate trials, therefore, a cross-fostered rhesus macaque and her Japanese macaque cagemate were played an unfamiliar rhesus food coo, while a cross-fostered Japanese macaque and his rhesus cagemate were played an unfamiliar Japanese macaque food coo. Recall that earlier analysis

had shown that, despite considerable variation among individuals, the food coos of infant rhesus and Japanese macaques showed a clear species difference (Owren *et al.* 1992).

There was no difference in the strength with which cross-fostered animals and their normally raised cagemates responded to these playbacks. Cross-fostered animals responded for a mean of 0.8 seconds, normally raised animals responded for a mean of 1.3 seconds (Mann–Whitney U test, $n_1=8$, $n_2=8$, $U=28$, $P>0.10$). There was, as a result, no indication that cross-fostered juveniles had retained a predisposition to attend preferentially to their own, as opposed to their foster, species' vocalizations. By contrast, other experiments demonstrated clearly that cross-fostered subjects and their foster mothers had learned to recognize each others' vocalizations. Indeed, if anything, cross-fostered juveniles and their adopted mothers responded more strongly to each others' calls than did normally raised juveniles and their mothers.

Responses by juveniles to calls of their mothers

For example, cross-fostered juveniles responded more strongly to their mother's food coos than to the food coos of other adult females (Fig. 13.5: $n_1=8$, $n_2=8$, $U=20.5$, $0.05<P<0.10$) and they also responded more strongly to their mother's threat gruff (Fig. 13.6: $n_1=8$, $n_2=8$, $U=13$, $P<0.05$). Normally raised juveniles showed no such bias in response. The ability of cross-fostered juveniles to distinguish among the threat gruffs of adult females is striking, because the threat gruffs of adult female rhesus and Japanese macaques exhibited a clear species difference (M. J. Owen, unpublished data). Cross-fostered juveniles therefore learned to recognize individual idiosyncrasies in a call whose acoustic properties differed from the one they would normally have heard.

Responses by mothers to calls of their offspring

Even more striking was the ability of foster mothers to recognize the calls of their adopted offspring. Unlike other adult females, foster mothers responded more strongly to their adopted offspring's food coos than to the food coos of other juveniles (Fig. 13.7: $n_1=8$, $n_2=8$, $U=20$, $0.05<P<0.10$). And like the mothers of normally raised immatures (Fig. 13.8: $n_1=12$, $n_2=22$, $U=63$, $P<0.01$), foster mothers responded more strongly to their offspring's

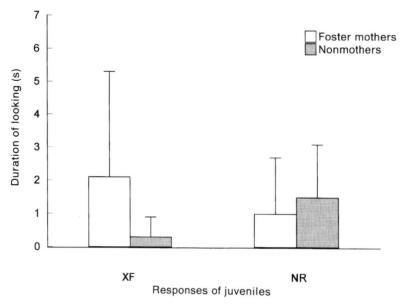

Fig. 13.5. Duration of responses shown by cross-fostered (XF) and normally raised (NR) juveniles to playback of food coos given by their foster mothers and by nonmothers. Histograms show means and standard deviations.

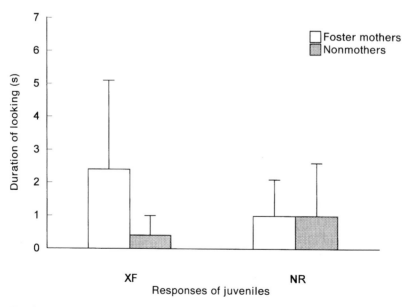

Fig. 13.6. Duration of responses shown by cross-fostered (XF) and normally raised (NR) juveniles to playback of threat gruffs given by their foster mothers and by nonmothers. Histograms as for Fig. 13.5.

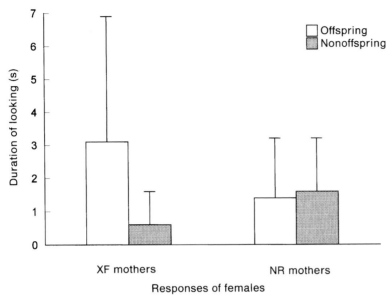

Fig. 13.7. Duration of responses shown by the mothers of cross-fostered juveniles (XF mothers) and the mothers of normally raised juveniles (NR mothers) to playback of food coos given by their offspring and by nonoffspring. Histograms as for Fig. 13.5.

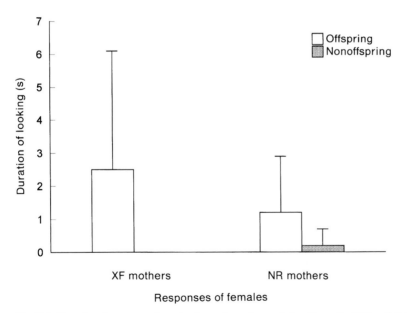

Fig. 13.8. Duration of responses shown by the mothers of cross-fostered juveniles (XF mothers) and the mothers of normally raised juveniles (NR mothers) to playback of play calls given by their offspring and by nonoffspring. Histograms as for Fig. 13.5.

play calls than to the play calls of other juveniles (Fig. 13.8: $n_1=6$, $n_2=6$, $U=9$, $P<0.05$). In doing so, however, foster mothers had to learn the individually distinctive features of a call that they would normally not encounter in these circumstances. Whereas normal rhesus mothers discriminated their offspring's play gruff from the play gruffs of others, foster mothers discriminated their offspring's play *coo* from the play gruffs of others. Similarly, whereas normal Japanese macaque mothers responded more strongly to their offspring's play coo than to the play coos of others, foster mothers responded more strongly to their offspring's play *gruff* than to the play coos of others. Foster mothers, therefore, learned to recognize their offspring's calls even though they were acoustically different from species-typical calls.

In sum, these experiments provide two sorts of evidence in support of the view that monkeys' responses to the calls of others depend strongly upon experience. First, cross-fostered juvenile macaques show no tendency to respond preferentially to calls of their own, as opposed to their adopted, species. Second, both juvenile and adult macaques can learn to recognize and respond selectively to individual idiosyncrasies in the calls of animals from another species. At least partly as a result, cross-fostered infants are able to develop bonds with their foster mothers that are as close as the bonds formed between normally raised infants and their mothers.

Summary

Compared with field research on vervet monkeys, cross-fostering experiments use an entirely different technique to study vocal development. Nonetheless, they yield similar results. Just as field observations suggest that vocal production in vervets is largely innate, cross-fostering experiments indicate that the acoustic features of macaque food coos are only slightly altered, if they are altered at all, when animals are raised in an auditory environment different from the one they would normally experience. Moreover, just as field observations reveal a predisposition among vervets to use particular vocalizations in specific social or ecological circumstances, cross-fostering experiments show that macaques are strongly predisposed to use particular calls in specific social situations. This predisposition is modified only slightly, if at all, during development. Finally, just as field observations demonstrate that vervets' responses to calls, including the calls of other species, are modified as a result of experience, cross-fostering experiments indicate that macaques, too, can learn to recognize and respond selectively to a wide variety of unfamiliar auditory stimuli.

COMPARISON WITH OTHER PRIMATE SPECIES

The studies reviewed above suggest that at least three general features characterize the development of vocal communication in nonhuman primates. First, vocal development is the result of three separate processes: the development of production, usage, and response to the calls of others. Second, different causal mechanisms underlie each of these processes. Third, because they are affected by different causal mechanisms, the three components of vocal development emerge in young animals at different rates.

To test the generality of these conclusions, we examined data from 34 studies of 11 other primate species and divided them according to whether they dealt with production, usage, or response. In each case, we considered whether the component(s) of vocal communication under study: (a) appeared fully formed at a very early age and subsequently showed little or no change as animals grew older; (b) appeared initially in a form that was different from that of adults but shared some adult-like features, and subsequently showed gradual modification toward the adult pattern as animals grew older; or (c) appeared initially in a form that exhibited few if any adult-like features, and subsequently showed substantial modification toward the adult pattern as individuals grew older. Table 13.2 presents the results of arranging studies in this way. For the sake of completeness, we include in Table 13.2 both the vervet and macaque studies reviewed above (used to formulate our hypothesis) and data from other studies of many different species (used to test it). To illustrate further the logic underlying our approach, we briefly review some of the typical results in each category.

Vocal production: Evidence

Among studies of vocal production, Winter *et al.*'s (1966, 1973) research on squirrel monkeys (*Saimiri sciureus*) offers perhaps the classic example of calls that appear with

Table 13.2. *A classification of 34 studies of vocal develoment in 14 nonhuman primate species, divided according to whether they dealt with vocal production, usage, or response, and according to whether they found that call development was adult-like at first emergence, with virtually no modification during development, broadly similar to adults at first emergence with some limited modification during development, or exhibited no resemblance to adults at first emergence and showed substantial modification during development*

	Adult-like, no modification during development	Broadly adult-like, some modification during development	No resemblance to adults, substantial modification during development
Vocal production	Squirrel monkeys: all calls (1–5) Spider monkeys: all calls (6) Cotton-top tamarins: long calls (7) Pygmy marmosets: J-calls (7) Talapoin monkeys: all calls (8) Rhesus macaques: all calls (10, 11) Japanese macaques: all calls (10, 11) Stumptailed macaques: chirps, geckers, screams (26) Vervets: alarm calls (13) Gibbons: great calls (25, 29)	Marmosets and tamarins: chirps and long calls (14) Pygmy marmosets: trills (7, 15, 31), chirps (7), J-calls (15) Tamarins: long calls (16) Japanese macaques: food coos (17) Rhesus macaques: "contact calls" (9), food coos (18) Pigtailed macaques: screams (17, 18, 36) Vervets: grunts (13), wrrs (30) Chimpanzees: pant-hoots (34) Bonobos: various calls (32)	
Vocal usage		Cotton-top tamarins: long calls (14), alarm calls (33) Cotton-top tamarins: C- and D-chirps (28) Japanese macaques: "uh" call (27), coos (11) Rhesus macaques: coos, gruffs (11) Pigtailed macaques: screams (19, 20) Vervets: alarm calls (12, 13, 21), grunts (13), wrrs (30) Chimpanzees: pant hoots (35)	
Responses to the calls of others			Goeldi's monkeys: alarm calls (22) Japanese macaques: food coos, gruffs (10, 23) Japanese macaques: alarm calls (24) Rhesus macaques: food coos, gruffs (10) Vervets: alarm calls (13)

Notes:
Key to references: (1) Winter *et al.* 1973; (2) Talmage-Riggs *et al.* 1972; (3) Lieblich *et al.* 1980; (4) Symmes *et al.* 1979; (5) Hertzog & Hopf 1984; (6) Eisenberg 1978; (7) Snowdon 1987; (8) Gautier 1974; (9) Newman & Symmes 1974; (10) Owren *et al.* 1992; (11) Owren *et al.* 1993; (12) Seyfarth & Cheney 1980; (13) Seyfarth & Cheney 1986; (14) Snowdon *et al.* 1985; (15) Elowson *et al.* 1991; (16) Hodun *et al.* 1981; (17) Green 1975b; (18) Hauser 1992; (19) Gouzoules & Gouzoules 1989a; (20) Gouzoules & Gouzoules 1989b; (21) Brown *et al.* 1992; (22) Masataka 1983a; (23) Masataka 1985; (24) Masataka 1983b; (25) Brockelman & Schilling 1984; (26) Chevalier-Skolnikoff 1974; (27) Green 1981; (28) Roush & Snowdon 1994; (29) Tenaza 1985; (30) Hauser 1989; (31) Elowson & Snowdon 1994; (32) Hopkins & Savage-Rumbaugh 1991; (33) Heymann 1990; (34) Mitani & Brandt 1994; (35) Pusey 1990; (36) Gouzoules & Gouzoules 1996.

adult-like acoustic properties shortly after birth and subsequently show little developmental modification. Winter and colleagues found that of the eight call types that made up the squirrel monkey's reportoire, six appeared on the first day of life, with acoustic features that were indistinguishable from those of adult calls. One call was recorded on the third day of life but took up to three months before it assumed its final, adult form. A final call type, the twitter/trill, first appeared at three months of age and stabilized after nine months (Winter *et al.* 1966, 1973). These results recall field data on the predator alarm calls of vervet monkeys, which, at their first appearance, were indistinguishable in their acoustic properties from the calls of adults. The vervets' alarm calls, like most of the squirrel monkeys' calls, subsequently showed no developmental change over time.

Data of a different sort from studies of "hybrid" squirrel monkeys and gibbons provide further evidence in support of innate vocal production with little if any developmental modification. Winter (1969) found that two subspecies of squirrel monkeys that differed in facial features – the "Roman arch" and "Gothic arch" forms – also differed in the acoustic features of their isolation peeps. These differences emerged on the first day of life (Symmes *et al.* 1979; Lieblich *et al.* 1980). Among hybrid offspring of Roman arch×Gothic arch matings, those with Roman arch facial features almost invariably gave isolation peeps of the Roman arch variety, while those with Gothic arch facial features tended to give isolation peeps of the Gothic arch variety (Newman & Symmes 1982). Similarly, in hybrid zones where two species of gibbon, *Hylobates lar* and *Hylobates pileatus*, overlap, the acoustic features of hybrid individuals' great calls could accurately be predicted by their pelage (Brockelman & Schilling 1984). Being careful to distinguish between the development of call production and call usage, Brockelman & Schilling (1984, p. 636) concluded: "although singing with parents may be important . . . for learning [the great call's] proper timing and context, the acoustic pattern of the final result seems to be resistant to alteration by learning . . . and under strong genetic control".

These observations can be contrasted with those obtained by Snowdon and colleagues (Snowdon 1987; Elowson *et al.* 1991; Elowson & Snowdon 1994; Snowdon *et al.*, Chapter 12) working on the "trills" of pygmy mar-

mosets (*Cebuella pygmaea*). When infant pygmy marmosets aged 0–14 weeks first began giving trills, their calls were broadly similar to those given by adults. Infants' trills, for example, were easily recognizable to a human listener as trills and not some other vocalization. Infants' trills nevertheless differed from those of adults according to a number of acoustic features, and not simply because of their relatively higher F0. Over time, these features gradually changed. Infants' and juveniles' trills became more variable, not less, with age and did not stabilize and become fully adult-like until roughly two years of age, when individuals were sexually mature. Even among adults, moreover, trills could be modified by social experience (Elowson & Snowdon 1994). Like the production of grunts and wrrs among infant vervet monkeys, the development of trill production among infant pygmy marmosets was characterized by a mixture of relatively innate features and gradual modification (Snowdon *et al.*, Chapter 12). From a very early age and with little or no experience, infants were predisposed to utter a call that had the same general features as the comparable call given by adults, but that was nevertheless not yet fully adult-like in its acoustic properties. Over time, through a process that is still unknown, these acoustic features changed to become increasingly like those of adults.

Such modification within constraints appears in many studies of primate vocal production (Table 13.2). By contrast, we could find no study of nonhuman primates in which the earliest vocalizations of infants consistently bore little or no resemblance to any calls in the adult vocal repertoire. It remains to be determined whether this represents a genuine feature of primate vocal production or is simply a consequence of the limited number of species that have been sampled.

Vocal production: Discussion

Completely innate vocal production is certainly possible in principle. It occurs, for example, in many insects (see e.g., Huber *et al.* 1994), frogs (see e.g., Ryan 1988), and some birds (see e.g., Kroodsma 1982). In practice, however, completely innate vocal production is difficult to demonstrate. Although the earliest calls of infant monkeys may sound adult like, detailed analysis can nonetheless reveal subtle acoustic cues that distinguish infant calls from those of adults. Winter's early analysis of infant squirrel monkeys' calls, for example (Winter *et al.* 1966, 1973; Winter 1969), made no use of the modern techniques of digital signal processing, nor did he have any reliable information on the range of vocal variability in adulthood. He may therefore have overstated the extent to which infants' calls were completely adult like in their acoustic structure. A similar problem arises in species where infants do not use certain call types before they have reached a particular age (see e.g., Green 1981). When the predator alarm calls of infant vervet monkeys first appear at two to three months of age, they may sound like those of adults and look like them when displayed on a sound spectrograph, but we cannot rule out the possibility that weeks or months of auditory experience have been necessary for normal development to occur.

Whatever their faults, however, the data from many primate species provide an extremely strong case for vocal production that is either wholly or in large part genetically fixed (Table 13.2). The evidence is particularly persuasive in studies of hybrid and cross-fostered individuals.

At the other extreme, we should not expect to find many species in which the acoustic structure of infant vocalizations is completely different from those of adults. Infants and adults of the same species presumably share, for the most part, a similar vocal apparatus, and similar physiology will inevitably lead to infant and adult vocalizations that resemble each other to some degree.

This said, it is nonetheless apparent that vocal production in humans and many songbirds is so labile that the earliest communicative sounds by infants can, and often do, include noises that are seldom given by adults. Avian subsong, for example, often "bears little or no resemblance to natural adult song" and unlike mature song "is often similar across species" (Marler 1990, p. 13). Similarly, while the babbling sounds of human infants "are highly structured and a large proportion are clearly phonetically related to human speech," human infants also produce many communicative sounds that are not obviously early versions of what will later become speech (Oller 1981, p. 85; see also Kent 1981). Some authors (e.g., Takeda 1965, 1966; Kawabe 1973) described similarly unconstrained babbling in nonhuman primate infants, but at present there are few quantitative spectrographic data to support such assertions. No monkey or ape has thus far demonstrated as much flexibility in the ontogeny of vocal production as is evident in humans or many songbirds.

In sum, the development of vocal production in nonhuman primates is clearly more fixed, genetically, than that in humans, but perhaps not as rigidly fixed as Winter's early papers implied. Moreover, increasing subtlety in acoustic techniques may be allowing investigators to uncover new evidence of more modifiable vocal production in monkeys and apes. Of the 14 studies cited in support of the view that call production is adult like at birth and shows no developmental modification, 11 are dated before 1987; of the 14 studies that provide evidence of more substantial modification during development, only four are dated before 1987. The great majority have been carried out since then, and unlike most earlier studies have involved the use of modern, computer-based digital signal processing.

Vocal usage: Evidence

Many adult monkeys use particular vocalizations only in specific, narrowly defined contexts. Adult vervet monkeys give snake alarm calls only to pythons; adult rhesus and pigtailed (*Macaca nemestrina*) macaques give certain types of screams only during physical fights with higher-ranking opponents (Gouzoules & Gouzoules 1989a); adult cottontop tamarins (*Saguinus oedipus*) give C- and D-chirps only when feeding or upon the discovery of food (Roush & Snowdon 1994). By contrast, we found no studies in which immature primates with no prior experience restricted calls of a specific type to one narrowly defined stimulus, like a particular species of predator, food, or a particular social circumstance; that is, we found no studies in which vocal usage fell into the category "adult-like, no modification during development" (Table 13.2).

Instead, young monkeys tend to use different vocalizations in contexts that are broadly similar, though not exactly the same, as the contexts in which the same calls are used by adults. The alarm calls of infant vervet monkeys provide perhaps the best-known example (reviewed above; see also Brown *et al.* 1992, who elicited terrestrial and avian alarm calls from captive vervets by showing them cardboard silhouettes of different predators). An equally clear case, however, comes from Gouzoules & Gouzoules' (1989a,b; 1996) studies of pigtailed macaque screams. Observations indicated that adults gave acoustically different screams in four distinct contexts, depending on whether they were interacting with an opponent of higher or lower dominance rank and whether the interaction did or did not involve physical contact. Immature animals were generally correct in their use of the different scream types; like adults, they gave screams almost exclusively in agonistic interactions. However, they were much "sloppier" in their use of a particular acoustic variant in a specific social situation. Immature animals acted as if they were predisposed to use a general class of calls in a broadly defined set of circumstances, but they required age and experience before they were able to limit the use of a particular subtype of scream to a certain narrowly defined category of social interaction.

Studies in which infants show some early predisposition to give certain calls in particular broad social contexts may be contrasted with a few studies in which the earliest relation between call type and social situation was entirely random. In a study of pygmy marmosets, Snowdon (1987; see also Snowdon et al., Chapter 12) found that infants gave trills, J-calls, and chirps "utterly indiscriminately." Over time, the contexts in which trills and J-calls were given gradually changed, so that by the end of their first year the infants' calls had begun to exhibit the same relation between acoustic subtypes and social context as found in adult vocalizations. Chirps, however, took much longer to develop, and exhibited no differentiation of call subtype by context even after 12 months.

Vocal usage: Discussion

Comparative research suggests that completely innate vocal usage is unlikely to be found in any animal species. The history of ethology includes many patterns of behavior, such as the pecking response of a herring gull chick to the red spot on its mother's beak (Hailman 1969) or the feeding and begging behaviors of ring doves (Klinghammer & Hess 1972; Lehrman 1972), that were originally thought to involve an innate response to one specific, narrowly defined stimulus but which were eventually shown to require at least some learning or experience on the part of young animals (for a review, see Gould 1982). Given the preponderance of such results from so many other species, it is not surprising that we were unable to find any examples of rigidly fixed, innate vocal usage among nonhuman primates.

There are other, more functional, reasons why we should not expect vocal usage to be fixed and adult-like at first emergence. To begin with, environments change, and animals that use calls to signal about features of their environment must be able to adjust accordingly. Vervet monkeys in Amboseli, Kenya, are preyed upon by only two species of eagle, the crowned eagle and the martial eagle, whose features are largely white when seen from below (Seyfarth & Cheney 1980, 1986). In contrast, vervets in Zimbabwe, Botswana, and South Africa are also preyed upon by Vereaux's eagle (*Aquila verreauxi*), whose features are largely black (Gargett 1971). The distribution of these raptors, however, is not clear cut. Martial and crowned eagles do occur in southern Africa, and Verreaux's eagles occur in the more mountainous areas of East Africa. Under these conditions, vervets clearly benefit by using alarm calls that, though closely tied to a particular class of referents (large raptors that are primarily white when seen from below), can nonetheless be modified to include new and physically different exemplars within each class (Marler (1977) makes a similar point with reference to birds).

Many other primate vocalizations are given only to particular individuals, like a frequent grooming partner (Smith *et al.* 1982), or only to particular individuals in specific circumstances (e.g., lower-ranking animals during aggression without physical contact; Gouzoules *et al.* 1984). In these cases, it would obviously be impossible for selection to favor innate vocal usage, since infants require time and experience before they develop close associations and learn who is higher or lower ranking then they are.

Given the impossibility of innate vocal usage and the obvious advantages of a flexible relation between call and referent (or call and social context), the tendency of young, inexperienced monkeys to use specific calls in particular social contexts is striking. Studies by Snowdon and colleagues on pygmy marmosets (Snowdon 1987; Snowdon *et al.*, Chapter 12) currently provide the only case in which calls by infants were used in a way that was entirely random with respect to social context and thus showed no relation whatever to adult vocal usage. In contrast, most other primates, when confronted with a novel or unexpected situation, use a call that maintains the existing relation in their repertoire between call type and social context. Captive vervet monkeys living in Southern California leave the closed, indoor part of their cages each morning as soon as a technician has opened the door to their outside arena. As they move through this door they give the same vocalization that East African vervets give when moving out into an open savanna (personal observation).

The predisposition of young animals to give certain sorts of calls in certain broad contexts may arise in part because of what Morton (1977) has termed "motivation–structural rules." Hauser (1993), for example, found that in many primate species calls with a high F0 tended to be produced by fearful individuals, whereas calls with a low F0 tended to be produced by aggressive individuals. This sort of link between acoustic structure and motivational state could easily lead to a general tendency to use certain call types in particular social situations – in other words, to a pattern of vocal usage that is similar to that of adults at first emergence.

Though broadly predictive, Morton's suggested relation between motivation and acoustic structure breaks down in many cases where acoustic properties and call usage have been closely studied (Hauser 1993). Vervet monkeys for example, use acoustically similar grunts in at least four very different social contexts; their eagle alarm also resembles these grunts in bandwidth and spectral features (Seyfarth & Cheney 1984b; Owren & Bernacki 1988). Calls in each of these circumstances are acoustically similar, but the motivation underlying their production probably differs markedly from one context to the next. Similarly, vervets use acoustically similar chutters when threatening a member of their own group, a member of another group, an unfamiliar human, or snakes (Cheney 1984). Once again, there is no general rule based on motivation that either links the contexts in which chutters are given or distinguishes the usage of chutters from the usage of grunts. Finally, both rhesus and pigtailed macaques give acoustically different screams in aggressive encounters with different classes of opponents, but the acoustic features of screams used by the two species in identical agonistic contexts are very different (Gouzoules & Gouzoules 1989a). In each of these cases the relation between acoustic structure and call usage may be weak because an animal's motivation is more closely associated with acoustic features other than fundamental frequency, or because the acoustic structure of calls is more closely associated with particular external referents than with the caller's motivation (Hauser 1993).

In sum, nonhuman primate infants seem predisposed from birth to use different vocalizations in particular broadly defined social or ecological circumstances. With some exceptions, however, no general "phonological" rule allows us to predict the type of call that will be given in any situation. This partially innate development of vocal usage undoubtedly allows young animals to communicate more effectively with adults than they would if the relation between call type and social context were entirely random. Over time, through a process that is not well understood, young animals sharpen the relation between cell type and social context. Such gradual development is essential in species where animals give different calls to different individuals, and where infants need time before they can recognize others and learn about their own social position. The development of vocal usage in primates may reflect a compromise between the evolutionary advantages of completely innate and entirely learned communication.

Response to the calls of others: Evidence

Adult squirrel monkeys give two acoustically distinct alarm calls to different classes of predator. Alarm peeps are given to flying birds or fast moving objects, while yaps are given to mammals, snakes, or strange objects moving on the ground (Winter et al. 1966; Jürgens 1979). Monkeys that hear alarm peeps typically run to cover or shelter and then become immobile; monkeys that hear yaps climb upward if possible, look toward the predator, and begin yapping themselves. Herzog & Hopf (1984) played alarm peeps and yaps to six infant squirrel monkeys that had been reared in social isolation but had occasional access to a cloth "mother surrogate." Infants who heard alarm peeps responded by immediately running to the mother surrogate, where they became immobile. Playback of two other auditory stimuli, screams and a "control tone," failed to elicit this response. Infants who heard yaps "briefly startled or interrupted their activities, then looked around in a comparatively relaxed way" (Herzog & Hopf 1984, p. 101). Responses to yap alarms were stronger (and more like those of adults) when playback of yaps was combined with presentation of a stuffed model cat or snake.

Because subjects in these experiments had been reared in isolation, Hertzog & Hopf (1984, p. 99) concluded that "the perception of both alarms calls is innate." Their conclusion seems unwarranted for at least two reasons. First, responses to yap alarms clearly were not specific to this call or identical with the responses of adults. Second, the authors noted that running to the mother and becoming immobile were not elicited by screams or control tones and thus "appeared exclusively as a response to alarm peeps." This is a curious result, because running to the mother is a general response to distress in most infant primates. If normally raised infant squirrel monkeys are like most primate infants in running to their mothers at the sound of many calls, the infants' responses to alarm peeps in this experiment can probably not be taken as proof of an "innate reaction." Because of difficulties in interpreting these data, we have omitted Hertzog's & Hopf's study from Table 13.2.

By contrast, all other studies that have examined infants' responses to vocalizations have found that infants' earliest reactions bear little relation to the responses shown by adults. Vervet infants, for example, respond to playback of leopard, eagle, and snake alarm calls by running to their mothers or showing some other, often inappropriate, reaction (see above). Infant vervets are four to five months of age before they respond appropriately to the different alarm calls given by superb starlings (Hauser 1988). Similarly, infant Goeldi's monkeys (*Callimoco goeldi*; Masataka 1983a) and Japanese macaques (Masataka 1983b) show no response at all to their own species' different alarm calls. Finally, infant Japanese macaques require experience before they are able to recognize their mothers' calls (Masataka 1985). In none of these studies did infants seem predisposed to respond in particular ways to particular types of call.

Response to the calls of others: Discussion

By the time they have reached one year of age, monkeys respond in predictably different ways to the calls of others. Some of these different responses depend on the call's acoustic properties: alarm calls elicit flight, screams and grunts elicit vigilance, intergroup calls draw animals to the source of the vocalization. Responses also depend crucially, however, on the identity of the caller. Territorial gibbons (*Hylobates* spp.) and titi monkeys (*Callicebus moloch*) respond more strongly to the calls of a strange male from a border area than to the calls of a neighbor (Robinson 1981; Mitani 1990). Female vervets and macaques respond more strongly to their offspring's scream than to the scream of an unrelated juveniles (Cheney & Seyfarth 1980; Gouzoules *et al.* 1984). Among vervets, the threat vocalization of an unrelated animal is generally ignored unless that animal has recently groomed with the listener, in which case the listener responds more strongly (Seyfarth & Cheney 1984a). Female baboons (*Papio cynocephalus*) typically ignore grunts from an unrelated animal, but if that individual has recently behaved aggressively toward the listener the grunt acts to reconcile the two animals, restoring their relationship to its earlier condition (Cheney *et al.* 1995).

Nonhuman primates thus live in groups where the most adaptive responses to vocalizations can emerge only after considerable social experience. Young animals must learn who are members of their group and who are not, which individuals are their close associates (generally matrilineal kin) and which are not, where they themselves stand in a dominance hierarchy, and who has interacted with them in a friendly or aggressive way in the recent past. Under these circumstances it is not surprising that natural selection has not favored a fixed, unmodifiable link between hearing a vocalization and responding to it.

The relatively greater developmental flexibility in response to calls as opposed to call production and call usage is perhaps most clearly demonstrated in the study of cross-fostered rhesus and Japanese macaques (see above). These individuals continued to produce their own species' calls and use them in species-typical social contexts, but they showed much greater flexibility in their ability to learn the vocalizations of another species and to recognize the calls of their foster mothers.

GENERAL DISCUSSION

The conclusions that emerge from work on vervet monkeys, rhesus macaques, and Japanese macaques are broadly applicable across a wide variety of nonhuman primate species.

Different causal mechanisms underlie call production, usage, and response; as a result, these components of vocal ontogeny develop at different rates. Vocal production is the component of vocal development under the strongest genetic control, and as a result becomes fully adult like at an earlier age than does vocal usage or response. Most nonhuman primates are able to make recognizably adult-like calls from a very early age, long before they can use these sounds in the appropriate circumstances or respond in an adult-like manner to the calls of others. Granted, the fine-tuning of call production often takes much longer, and it may be many months or even years before an immature animal's calls are exactly like those of adults; nevertheless, infant monkeys begin life with a much greater head start in call production than they exhibit in either call usage or response.

Compared with call production, usage and response develop much later. Usage appears to be only partially under genetic control, while response seems to be affected almost entirely by environmental cues, including social cues received from other group members. In two studies where usage and response were compared directly, correct call usage developed later than correct responses to the calls of others (Seyfarth & Cheney 1986; Hauser 1989; see also Marler *et al.* 1981).

From this we may conclude further that *nonhuman primate vocal development is broadly similar to human language development.* Comparative studies of vocal development in human and nonhuman primates have concentrated almost exclusively on vocal production, and as a result have concluded that monkeys and apes are strikingly different from humans. Typically, scientists have identified a characteristic of human speech development – say, the "babbling" stage that appears to be a necessary precursor of normal speech, or the disastrous effects of early deafness, which illustrate the need for auditory feedback – then searched for parallels in the vocal development of one or more nonhuman species. Their results have produced a list of characteristics in which nonhuman primates (or birds) do or do not mirror the human condition.

Suppose, in contrast, that we step back from these particular features, take a broader view, and note first, that in both human and nonhuman primates the development of vocal production, usage, and response are affected by different causal mechanisms.

Second, in both human and nonhuman primates the relative roles of innate and learned mechanisms in production, usage, and response are similar. For example, when compared with language usage and the responses of humans to the utterances of others, human language production is, relatively speaking, the most innate. Babbling begins at roughly the same age across cultures, and in its earliest stages involves many of the same sounds, regardless of the infant's linguistic environment (for a review, see Locke 1993). Deaf infants and those with neonatal brain damage begin to babble at a later age than normal infants, but "their syllabic and segmental patterning broadly

resemble [those] of hearing infants and children in many ways . . ." (Locke 1993, p. 180). Once infants have begun to babble extensively, trained phoneticians can distinguish Swedish babbling, for example, from Japanese babbling (de Boysson-Bardies & Vihman 1991), but the very earliest sounds of infants seem relatively unaffected by their auditory environment. In contrast, human language usage and human responses to the remarks of others are, obviously, entirely determined by experience.

Third, because production, usage, and response are, in both human and nonhuman primates, controlled by different causal mechanisms, they become adult like at different ages and follow the same order of emergence. In both human and nonhuman primates, infant begin by making sounds that share many features of the adult "language" they will eventually use. Learning to respond appropriately to the calls of others, however, takes more time, and depends not only on the development of comprehension but also the development of motor skills. Finally, correct call usage depends on both adult-like pronunciation and the young organism's ability to identify and classify features of its environment. In both human and nonhuman primates, the development of adult-like comprehension (or response) precedes the development of correct call usage (Wanner & Gleitman 1982; Seyfarth & Cheney 1986; Hauser 1989).

To make the same point, suppose we adopted an evolutionary perspective and imagined that vocal development in contemporary monkeys and apes approximates the ancestral hominoid condition before language appeared. How easy would it be for evolution to effect a transition from this simian form to the modern human state? The answer is: quite easy indeed. Selection favoring a larger vocabulary of words, for example, would necessarily have to favor both a less rigidly fixed process of vocal production and a more open, arbitrary relation between calls and their referents. In their general characteristics, these are the only features that separate the vocal development of monkeys from the development of speech in human infants, at least until the emergence of syntax.

With just a slight difference in perspective, then, the most striking differences in vocal development between human and nonhuman primates diminish in importance when they are compared with the many overall similarities. Human and nonhuman primates differ, it is true, in the degree to which vocal production is innate. And it is no doubt also true that the greater motor plasticity in human infants vastly increases the potential complexity of their vocal usage and response to vocalizations. At the same time, it is increasingly apparent that human and nonhuman primates are similar in the overall organization of vocal development; the separate causal mechanisms affecting production, usage, and response; their order of emergence; and the relative importance of innate and learned factors in the development of usage and response. Concentrating exclusively on the details of vocal production misses the forest for the trees: in fact, the processes of vocal development in human and nonhuman primates share some striking similarities along with their differences.

ACKNOWLEDGEMENTS

Our research on vervet monkeys was supported by NSF, the Wenner-Gren Foundation, the Harry Frank Guggenheim Foundation, and the Research Foundation of the University of Pennsylvania; research on rhesus and Japanese macaques was supported by NIH and the University of Pennsylvania. Earlier drafts of this chapter profited from the comments of the editors and of Harold and Sally Gouzoules, Marc Hauser, Peter Marler, John Mitani, and Michael Owren.

REFERENCES

Arnold, A. P. & Bottjer, S. (1985). Cerebral lateralization in birds. In *Cerebral Lateralization in Nonhuman Species*, ed. S. D. Glick, pp. 11–39. Academic Press, New York.

Baptista, L. F. & Petrinovich, L. (1986). Song development in the white-crowned sparrow: Social factors and sex differences. *Animal Behaviour* **34**: 1359–71.

Berman, C. M. (1982). The ontogeny of social relationships with group companions among free-ranging infant rhesus monkeys. I. Social networks and differentiation. *Animal Behaviour* **30**: 149–62.

Brockelman, W. Y. & Schilling, D. (1984). Inheritance of stereotyped gibbon calls. *Nature* **312**: 634–6.

Brown, M. M., Kreiter, N. A., Maple, J. T. & Sinnott, J. M. (1992). Silhouettes elicit alarm calls from captive vervet monkeys (*Cercopithecus aethiops*). *Journal of Comparative Psychology* **106**: 350–9.

Caplan, D. (1987). *Neurolinguistics and Linguistic Aphasiology*. Cambridge University Press.

Caplan, D. (1992). *Language: Structure, Processing, and Disorders*. MIT Press, Cambridge, MA.

Caro, T. M. & Hauser, M. D. (1992). Is there teaching in nonhuman animals? *Quarterly Review of Biology* **67**: 151–74.

Chapais, B. (1988a). Experimental matrilineal inheritance of rank in female Japanese macaques. *Animal Behaviour* **36**: 1025–37.

Chapais, B. (1988b). Rank maintenance in female Japanese macaques: Experimental evidence for social dependency. *Behaviour* **104**: 41–59.

Cheney, D. L. (1977). The acquisition of rank and the development of reciprocal alliances among free-ranging immature baboons. *Behavioral Ecology and Sociobiology* **2**: 303–18.

Cheney, D. L. (1984). Category formation in vervet monkeys. In *The Meaning of Primate Signals*, ed. V. Reynolds & R. Harre, pp. 58–72. Cambridge University Press, Cambridge.

Cheney, D. L. & Seyfarth, R. M. (1980). Vocal recognition in free ranging vervet monkeys. *Animal Behaviour* **28**: 362–7.

Cheney, D. L. & Seyfarth, R. M. (1982). How vervet monkeys perceive their grunts: Field playback experiments. *Animal Behaviour* **30**: 739–51.

Cheney, D. L. & Seyfarth, R. M. (1985). Social and non-social knowledge in vervet monkeys. *Philosophical Transactions of the Royal Society, Series B* **308**: 187–201.

Cheney, D. L. & Seyfarth, R. M. (1988). Assessment of meaning and the detection of unreliable signals by vervet monkeys. *Animal Behaviour* **36**: 477–86.

Cheney, D. L. & Seyfarth, R. M. (1990). *How Monkeys See the World*. University of Chicago Press, Chicago.

Cheney, D. L., Seyfarth, R. M. & Silk, J. B. (1995). The role of grunts in reconciling opponents and facilitating interactions among adult female baboons. *Animal Behaviour* **50**: 249–57.

Chevalier-Skolnikoff, S. (1974). The ontogeny of communication in the stumptailed macaque (*Macaca arctoides*). *Contributions to Primatology* **2**: 1–174.

Datta, S. (1983). Relative power and the acquisition of rank. In *Primate Social Relationships*, ed. R. A. Hinde, pp. 93–102. Blackwell Scientific, Oxford.

de Boysson-Bardies, B. & Vihman, M. M. (1991). Adaptation to language: Evidence from babbling and first words in four languages. *Language* **67**: 297–319.

Eisenberg, J. F. (1978). Communication mechanisms in New World primates with special reference to vocalizations in the black spider monkey (*Ateles fusciceps robustus*). In *Aggression, Dominance, and Individual Spacing*, ed. L. Krames, P. Pliner & T. Alloway, pp. 31–48. Plenum, New York.

Elowson, A. M. & Snowdon, C. T. (1994). Pygmy marmosets, *Cebuella pygmaea*, modify vocal structure in response to changed social environment. *Animal Behaviour* **47**: 1267–77.

Elowson, A. M., Tannenbaum, P. L. & Snowdon, C. T. (1991). Food associated calls correlate with food preferences in cotton-top tamarins. *Animal Behaviour* **42**: 931–7.

Ferguson, C., Menn, L. & Stoel-Gammon, C. (1992). *Phonological Development*. York Press, Toronto.

Gargett, V. (1971). Some observations on black eagles in the Matopos, Rhodesia. *Ostrich*, Suppl. **9**: 91–124.

Gautier, J. P. (1974). Field and laboratory studies of the vocalizations of talapoin monkeys (*Miopithecus talapoin*). *Behaviour* **51**: 209–73.

Gould, J. L. (1982). *Ethology: The Mechanisms and Evolution of Behavior*. W. W. Norton, New York.

Gouzoules, H. & Gouzoules, S. (1989a). Design features and developmental modification of pigtail macaque, *Macaca nemestrina*, agonistic screams. *Animal Behaviour* **37**: 383–401.

Gouzoules, H. & Gouzoules, S. (1989b). Sex differences in the acquisition of communicative competence by pigtail macaques (*Macaca nemestrina*). *American Journal of Primatology* **19**: 163–74.

Gouzoules, H. & Gouzoules, S. (1996) Recruitment screams of pigtail monkeys (*Macaca nemestrina*): Ontogenetic perspectives. *Behaviour*, in press.

Gouzoules, S., Gouzoules, H. & Marler, P. (1984). Rhesus monkey (*Macata mulatta*) screams: Representational signalling in the recruitment of agonistic aid. *Animal Behaviour* **32**: 182–93.

Green, S. (1975a). Variation of vocal pattern with social situation in the Japanese monkey (*Macaca fuscata*): A field study. In *Primate Behavior*, vol. 4, ed. L. A. Rosenblum, pp. 1–104. Academic Press, New York.

Green, S. (1975b). Dialects in Japanese monkeys: Vocal learning and cultural transmission of locale-specific vocal behavior? *Zeitschrift für Tierpsychologie* **38**: 304–14.

Green, S. (1981). Sex differences and age gradations in vocalizations of Japanese and lion-tailed monkeys (*Macaca fuscata* and *Macaca silenus*). *American Zoologist* **21**: 165–83.

Hailman, J. (1969). The ontogeny of an instinct. *Behaviour Supplement* **15**: 1–159.

Harlow, H. F. (1958). The nature of love. *American Psychologist* **13**: 673–85.

Hauser, M. D. (1986). Male responsiveness to infant distress calls in free-ranging vervet monkeys. *Behavioral Ecology and Sociobiology* **19**: 65–71.

Hauser, M. D. (1988). How infant vervet monkeys learn to recognize starling alarm calls: The role of experience. *Behaviour* **105**: 187–201.

Hauser, M. D. (1989). Ontogenetic changes in the comprehension and production of vervet monkey (*Cercopithecus aethiops*) vocalizations. *Journal of Comparative Psychology* **103**: 149–58.

Hauser, M. D. (1991). Sources of acoustic variation in rhesus macaque vocalizations. *Ethology* **89**: 29–46.

Hauser, M. D. (1992) Articulatory and social factors influence the acoustic structure of rhesus monkey vocalizations: A learned mode of production? *Journal of the Acoustical Society of America* **91**: 2175–9.

Hauser, M. D. (1993). The evolution of nonhuman primate vocalizations: Effects of phylogeny, body weight, and motivational state. *American Naturalist* **142**: 528–42.

Herzog, M. & Hopf, S. (1984). Behavioral responses to species-specific warning calls in infant squirrel monkeys reared in social isolation. *American Journal of Primatology* **7**: 99–106.

Heymann, E. W. (1990). Reactions of wild tamarins, *Saguinas mystax* and *Saguinas fuscicollis*, to avian predators. *International Journal of Primatology* **11**: 327–37.

Hinde, R. A. (1984). Biological bases of the mother–child relationship. In *Frontiers of Infant Psychiatry*, vol. 2, ed. J. Call, E. Galensen & R. L. Tyson, pp. 21–34. Basic Books, New York.

Hodun, A., Snowdon, C. T. & Soini, P. (1981). Subspecific variation in the long calls of the tamarin, *Saguinus fuscicollis*. *Zeitschrift für Tierpsychologie* **57**: 97–110.

Hopkins, W. D. & Savage-Rumbaugh, E. S. (1991). Vocal communication as a function of differential rearing experiences in *Pan paniscus*: A preliminary report. *International Journal of Primatology* **12**: 559–84.

Huber, F., Moore, T. E. & Loher, W. (1993). *Cricket Behavior and Neurobiology*. Cornell University Press, Ithaca, NY.

Inoue, M. (1988). Age gradations in vocalization and body weight in Japanese monkeys. *Folia Primatologica* **51**: 76–86.

Johnson, J. & Newport, E. L. (1989). Critical period effects in second language learning: The influence of maturational state on the acquisition of English as a second language. *Cognitive Psychology* **21**: 61–99.

Jürgens, U. (1979). Neural control of vocalization in non-human primates. In *Neurobiology of Social Communication in Primates*, ed. H. D. Steklis & M. Raleigh, pp. 11–44. Academic Press, New York.

Kawabe, S. (1973). Development of vocalization and behavior of Japanese macaques. In *Behavioral Regulators of Behavior in Primates*, ed. C. R. Carpenter pp. 164–84. Associated Universities Press, Lewisburg, PA.

Kent, R. D. (1981). Articulatory–acoustic perspectives on speech development. In *Language Behavior in Infancy and Early Childhood*, ed. R. E. Stark, pp. 105–21. Elsevier, New York.

Klinghammer, E. & Hess, E. (1972). Parental feeding in ring

doves (*Streptopelia roseogrisea*): Innate or learned? In *Control and Development of Behavior: An Historical Sample from the Pens of Ethologists*, ed. P. H. Klopfer & J. P. Hailman, pp. 35–61. Addison-Wesley, Reading, MA.

Kroodsma, D. E. (1982). Learning and the ontogeny of sound signals in birds. In *Acoustic Communication in Birds*, vol. 2, ed. D. E. Kroodsma & E. H. Miller, pp. 1–23. Academic Press, New York.

Lehrman, D. (1972). The physiological basis of parental feeding in the ring dove (*Streptopelia risoria*). In *Control and Development of Behavior: An Historical Sample from the Pens of Ethologists*, ed. P. H. Klopfer & J. P. Hailman, pp. 62–76. Addison-Wesley, Reading, MA.

Lieblich, A. K., Symmes, D., Newman, J. D. & Shapiro, M. (1980). Development of the isolation peep in laboratory-bred squirrel monkeys. *Animal Behaviour* 28: 1–9.

Locke, J. L. (1993). *The Child's Path to Spoken Language*. Harvard University Press, Cambridge, MA.

Marler, P. (1977). Development and learning of recognition systems. In *Recognition of Complex Acoustic Signals*, ed. T. H. Bullock, pp. 77–96. Dahlem Konferenzen, Berlin.

Marler, P. (1987). Sensitive periods and the role of specific and general sensory stimulation in birdsong learning. In *Imprinting and Cortical Plasticity*, ed. J. P. Rauschecker & P. Marler, pp. 235–47. Wiley, New York.

Marler, P. (1990). Innate learning preferences: Signals for communication. *Developmental Psychobiology* 23: 557–68.

Marler, P. (1991a). The instinct for vocal learning: Songbirds. In *Plasticity of Development*, ed. W. S. Hall & R. J. Dooling, pp. 36–73. MIT Press, Cambridge, MA.

Marler, P. (1991b). Song-learning behavior: The interface with neuroethology. *Trends in Neuroscience* 14: 199–206.

Marler, P. & Sherman, V. (1985). Innate differences in singing behavior of sparrows reared in isolation from adult conspecific song. *Animal Behaviour* 33: 57–71.

Marler, P., Zoloth, S. & Dooling, R. (1981). Innate programs for perceptual development: An ethological view. In *Developmental Plasticity: Behavioral and Biological Aspects of Variations in Development*, ed. G. Gollin, pp. 135–72. Academic Press, New York.

Masataka, N. (1983a). Psycholinguistic analyses of alarm calls of Japanese monkeys (*Macaca fuscata*). *American Journal of Primatology* 5: 111–25.

Masataka, N. (1983b). Categorical responses to natural and synthesized alarm calls in Goeldi's monkeys (*Callimico goeldi*). *Primates* 24: 40–51.

Masataka, N. (1985). Development of vocal recognition of mothers in infant Japanese monkeys. *Developmental Psychobiology* 18: 107–14.

Masataka, N. (1992). Attempts by animal caretakers to condition Japanese macaque vocalizations result inadvertently in individual-specific calls. In *Topics in Primatology*, vol. 1, ed. T. Nishida, W. C. McGrew, P. Marler, M. Pickford & F. de Waal, pp. 271–8. University of Tokyo Press, Tokyo.

Masataka, N. & Fujita, K. (1989). Vocal learning of Japanese and rhesus monkeys. *Behaviour* 109: 191–9.

Mitani, J. C. (1990). Experimental field studies of Asian ape social systems. *International Journal of Primatology* 11: 103–26.

Mitani, J. C. & Brandt, K. L. (1994). Social factors influence the acoustic variability in the long-distance calls of male chimpanzees. *Ethology* 96: 233–52.

Morton, E. S. (1977). On the occurrence and significance of motivational-structural rules in some birds and mammal sounds. *American Naturalist* 111: 855–69.

Newman, J. & Symmes, D. (1974). Vocal pathology in socially deprived monkeys. *Developmental Psychobiology* 7: 351–8.

Newman, J. & Symmes, D. (1982). Inheritance and experience in the acquisition of primate acoustic behavior. In *Primate Communication*, ed. C. T. Snowdon, C. H. Brown & M. R. Petersen, pp. 259–78. Cambridge University Press, New York.

Nottebohm, F. (1991). Reassessing the mechanisms and origins of vocal learning in birds. *Trends in Neuroscience* 14: 206–11.

Oller, D. K. (1981). Infant vocalizations: Exploration and reflexivity. In *Language Behavior in Infancy and Early Childhood*, ed. R. E. Stark, pp. 85–104. Elsevier, New York.

Owren, M. J. & Bernacki, R. (1988). The acoustic features of vervet monkey alarm calls. *Journal of the Acoustical Society of America* 83: 1927–35.

Owren, M. J. & Casale, T. M. (1994). Variation in fundamental frequency peak position in Japanese macaque (*Macaca fuscata*) coo calls. *Journal of Comparative Psychology* 108: 291–7.

Owren, M. J. & Dieter, J. A. (1989). Infant cross-fostering between Japanese (*Macaca fuscata*) and rhesus macaques (*Macaca mulatta*). *American Journal of Primatology* 18: 245–50.

Owren, M. J., Dieter, J. A., Seyfarth, R. M. & Cheney, D. L. (1992). "Food" calls produced by adult female rhesus (*Macaca mulatta*) and Japanese (*M. fuscata*) macaques, their normally raised offspring, and offspring cross-fostered between species. *Behaviour* 120: 218–31.

Owren, M. J., Dieter, J. A., Seyfarth, R. M. & Cheney, D. L. (1993). Vocalizations of rhesus (*Macaca mulatta*) and Japanese (*M. fuscata*) macaques cross-fostered between species show evidence of only limited modification. *Developmental Psychobiology* 26: 389–406.

Peters, E. H. (1983). Vocal communication in an introduced

colony of feral rhesus monkeys (*Macaca mulatta*). Doctoral dissertation, University of Florida.

Pusey, A. E. (1990). Behavioral changes at adolescence in chimpanzees. *Behaviour* 115: 203–46.

Robinson, J. G. (1981). Vocal regulation of inter- and intragroup spacing during boundary encounters in the titi monkey, *Callicebus moloch*. *Primates* 22: 161–72.

Roush, R. S. & Snowdon, C. T. (1994). Ontogeny of food associated calls in cotton-top tamarins. *Animal Behaviour* 47: 263–73.

Rowell, T. E. & Hinde, R. A. (1962). Vocal communication by the rhesus monkey (*Macaca mulatta*). *Proceedings of the Zoological Society of London* 138: 279–94.

Ryan, M. J. (1988). Constraints and patterns in the evolution of anuran acoustic communication. In *The Evolution of the Amphibian Auditory System*, ed. B. Fritzsch, M. J. Ryan, W. Wilczynski, T. E. Hetherington & W. Walkowiak, pp. 121–46. Wiley, New York.

Seyfarth, R. M. (1987). Vocal communication and its relation to language. In *Primate Societies*, ed. B. B. Smuts, D. L. Cheney, R. M. Seyfarth, R. W. Wrangham & T. T. Struhsaker, pp. 440–51. University of Chicago Press, Chicago.

Seyfarth, R. M. & Cheney, D. L. (1980). The ontogeny of vervet monkey alarm calling behavior: A preliminary report. *Zeitschrift für Tierpsychologie* 54: 37–56.

Seyfarth, R. M. & Cheney, D. L. (1984a). Grooming, alliances, and reciprocal altrusim in vervet monkeys. *Nature* 308: 541–3.

Seyfarth, R. M. & Cheney, D. L. (1984b). The acoustic features of vervet monkey grunts. *Journal of the Acoustical Society of America* 75: 1623–8.

Seyfarth, R. M. & Cheney, D. L. (1986). Vocal development in vervet monkeys. *Animal Behaviour* 34: 1640–58.

Seyfarth, R. M., Cheney, D. L. & Marler, P. (1980). Vervet monkey alarm calls: Semantic communication in a free-ranging primate. *Animal Behaviour* 28: 1070–94.

Smith, H. J., Newman, J. D. & Symmes, D. (1982). Vocal concomitants of affiliative behavior in squirrel monkeys. In *Primate Communication*, ed. C. T. Snowdon, C. H. Brown & M. R. Petersen, pp. 30–49. Cambridge University Press, New York.

Snowdon, C. T. (1987). A comparative approach to vocal communication. *Nebraska Symposium on Motivation* 35: 145–200.

Snowdon, C. T. (1988). Communication as social interaction: Its importance in ontogeny and adult behavior. In *Primate Vocal Communication*, ed. D. Todt, P. Goedeking & D. Symmes, pp. 108–22. Springer-Verlag, Berlin.

Snowdon, C. T. (1990). Language capacities of nonhuman animals. *Yearbook of Physical Anthropology* 33: 215–43.

Snowdon, C. T., French, J. A. & Cleveland, J. (1985). Ontogeny of primate vocalizations: Models from bird song and human speech. In *Current Perspectives in Primate Social Dynamics*, ed. D. Taub & F. King, pp. 389–402. Van Nostrand Rheinhold, New York.

Struhsaker, T. T. (1967). Auditory communication in vervet monkeys (*Cercopithecus aethiops*). In *Social Communication Among Primates*, ed. S. A. Altmann, pp. 281–324. University of Chicago Press, Chicago.

Symmes, D., Newman, J. D., Talmadge-Riggs, G. & Leiblich, A. K. (1979). Individuality and stability of isolation peeps in squirrel monkeys. *Animal Behaviour* 27: 1142–52.

Takeda, R. (1965). Developmental of vocal communication in man-raised Japanese monkeys. I. From birth to 6 weeks. *Primates* 6: 337–80.

Takeda, R. (1966). Development of vocal communication in man-raised Japanese monkeys. II. From seventh to thirteenth week. *Primates* 7: 73–116.

Talmage-Riggs, G., Winter, P., Ploog, D. & Mayer, W. (1972). Effects of deafening on the vocal behavior of the squirrel monkey (*Saimiri sciureus*). *Folia Primatologica* 17: 404–20.

Tenaza, R. (1985). Songs of hybrid gibbons (*Hylobates lar×H. muelleri*). *American Journal of Primatology* 8: 249–53.

Wanner, E. & Gleitman, L. (1982). *Language Acquisition: The State of the Art*. Cambridge University Press, New York.

Winter, P. (1969). Dialects in squirrel monkeys: Vocalization of the Roman arch type. *Folia Primatologica* 10: 216–29.

Winter, P., Handley, P., Ploog, D. & Schott, D. (1973). Ontogeny of squirrel monkey calls under normal conditions and under acoustic isolation. *Behaviour* 47: 230–9.

Winter, P., Ploog, D. & Latta, J. (1966). Vocal repertoire of the squirrel monkey (*Saimiri sciureus*): Its analysis and significance. *Experimental Brain Research* 1: 359–84.

14 Social influences on vocal learning in human and nonhuman primates

JOHN L. LOCKE AND CATHERINE SNOW

INTRODUCTION

In this chapter, we review work on the nature of vocal learning in human primates, comparing them *en passant* to nonhuman primates who share many of their capacities but are both less eager and less successful vocal learners. The basic question underlying this review is whether the precocious and prolific vocal learning of human primates can be explained by biological mechanisms that are specific to the language system or whether it relates to more general social capacities and to the particular social context of vocal learning in humans.

We know that young human primates are particularly good at vocal learning. One bit of evidence in support of this contention is that all national languages are spoken, even though extremely subtle articulatory and auditory discriminations are relied on to carry meaning in spoken languages. In addition, babbling and vocal play are early developmental activities universally observed in normally developing children (Locke 1992; Locke & Pearson 1992). Imitative vocal behavior is also universal in young children and common even in more mature language users. Furthermore, language learning, particularly word learning, by young children, is quite rapid and efficient.

Although there is much emphasis on children's preparedness for *language* learning, in fact children everywhere seem to enter the language system of conventional words use through the use of vocal forms that are more like adult forms in sound than in semantic or syntactic function. These early forms could be argued, though, to foreshadow a major function of oral language even in adulthood, namely to effect participation in social interaction rather than transmission of information. We argue, then, that vocal and social capacities are prior to, potential sources of,

and better candidates for, biological buffering than are grammatical capacities.

At the same time as young human primates show a universal interest in and talent for vocal learning, their caregivers show considerable interest in stimulating the vocal channel as a form of social interaction. Talking to infants is universal and has universal characteristics when it occurs, although its frequency varies enormously across cultures. Interpreting infants' word-like vocalizations as intentional and meaningful is a normal, almost unavoidable adult activity. Adults also universally provide infants with simplified forms that typically serve as models for the first conventional words, forms that often serve little communicative function other than allowing children to participate in social interactions. Adults also are very likely to respond to child vocalizations with imitation or with "expansions" into more conventional adult-like forms.

Given the remarkable ease of, and preference for, vocal learning in human primates, we ask on what specific skills and capabilities this ease and preference rest. To what extent is vocal learning dependent, for example, on social influences – exposure to adult talk, involvement in oral language-based interactions during infancy and early childhood, adult tendencies to model oral language, to teach words, and to correct child vocal productions? In other words, is vocal development merely an early stage in language acquisition, or is it an independent accomplishment that may have some role in generating language as a side-effect? The latter interpretation opens up the possibility of a language-free conception of vocal behavior for consideration in a comparative perspective. We might also ask whether the difference between humans and nonhuman primates in their reliance on and development of vocal channels of communication to the point where language becomes inevitable is a simple function of biological differences or whether it is mediated by differences in social interaction and exposure to vocal language.

VOCAL LEARNING IN HUMAN AND NONHUMAN PRIMATES

Early listening in humans

The first vocal accomplishment of human infants is in fact perceptual – learning to distinguish language- and speaker-specific acoustic stimuli. Vocal learning begins prenatally, when the fetus learns enough about its mother's voice to recognize it after birth. The three-day-old neonate's preference for maternal voice (DeCasper & Fifer 1980; DeCasper & Spence 1986; DeCasper et al. 1994) is presumed to occur because intonation and rhythmic features of the mother's airborne voice resembles her voice as transmitted prenatally through the soft tissues and fluids of her body. At two to four days of age, there also is a documented preference for the voice of an adult female that is speaking the language spoken by the mother prenatally (Mehler et al. 1988; Moon et al. 1993).

Mature vervet (Cercopithecus aethiops) and squirrel monkeys (Saimiri sciureus) apparently can identify juvenile and infantile offspring from their vocalizations (Waser 1977; Cheney & Seyfarth 1980, 1990; Snowdon et al. 1985; Symmes & Biben 1985), and there is some indication that cross-fostered juvenile macaques respond preferentially to the sound of their mother's voice (cf. Seyfarth & Cheney, Chapter 13). However, there seems to be no evidence relating to the effects of maternal voice on the nonhuman primate fetus, although investigators have found cardiac accelerations to conspecific threat calls in neonatal chimpanzees (Pan troglodytes) that were not found for other conspecific calls and noises and not obtained in orangutans (Pongo pygmaeus; Berntson & Boysen 1989). At three and four months, isolation-reared chimpanzees showed different cardiac responses to conspecific screams and laughter, pointing to the possible presence of specialized perceptual processing mechanisms in these species (Berntson et al. 1990).

That human neonates prefer the language heard in utero at the same age as they prefer the voice or speech heard prenatally suggests that, early on, maternal and linguistic preference may be one and the same – a single set

of cues may distinguish individuals and languages. Given the filtering of soft tissues, these cues would be of the relatively low frequency type that reveal little about specific consonants and vowels but preserve linguistic prosody, i.e., the frequency and intensity of the voice as well as temporal information relating to the rhythm and rate of speech. This is shown by recordings of intrauterine maternal voice transmission (Querleu et al. 1985) and by electronic (low-pass) filtering of speech, which preserves differences between vocal emotions (Ross et al. 1973; Scherer et al. 1972) and languages (Cohen & Starkweather 1961). Prosody thus conveys to the fetus information about the identity and superficial linguistic behaviors of its host (cf. Fernald 1992a,b).

In the present context, it is important to consider the indexical significance of language. When people speak, they provide clues as to their age, social class, education, region of birth or residence, politics, and national affiliation (see below). That is to say, upon hearing individuals speak we learn something about their group. The mother speaks a particular language and also has her own unique vocal signature. The neonate's listening preference thus rests, potentially, on a single set of vocal characteristics that distinguish its mother from (a) all other speakers of her language and (b) speakers of all other languages. The infant's initial preference for the mother's "voice" is, thus, a preference for the sound pattern of her language as well as any personally identifying information. According to this view, the infant's preference for a language – an identifying characteristic of his or her social group – follows from familiarization with a specific individual, the infant's mother or primary caregiver. Moreover, this individual-to-group sequence conforms to a general pattern that applies to avians as well as primates (Bolhuis 1991). For example, ducklings appear to respond preferentially to conspecific calls because of their similarity to the duckling's own vocalizations (Gottlieb 1991a,b).

The three-day-old human also looks preferentially at its mother, relative to other mothers, but facial learning can only occur postnatally, and therefore must commence immediately following birth (Field et al. 1984; Bushnell et al. 1989). One asumes that the face has been of such importance phylogenetically that newborns are pre-adapted to look at it, and thus to extract information from it. It is an ability that humans share with other primates (Sackett et al. 1970, cited by Sackett & Ruppenthal 1974), who prefer the faces of conspecifics. They also share the human infant's interest in the eyes, a preference that is undiminished by the existence of mouth movements (Haith *et al.* 1977; Keating & Keating 1982, 1993; Mendelson *et al.* 1982). Whether nonhuman primate infants prefer the face of their mother presently seems to be unknown.

In the second six months of life, when brain developments enable human infants to inhibit attention to irrelevant stimulation (Chugani *et al.* 1987) – and thus, presumably, to listen selectively – they again demonstrate a preference for the sound of the ambient language over a foreign one (Jusczyk *et al.* 1993a,b). There are suggestions in the literature that infants may be aware of the relationship between prosodic cues and the linguistic units (clauses, grammatical phrases) with which they co-occur (Hirsh-Pasek *et al.* 1987; Goodsit *et al.* 1993), although the research reporting this result has been questioned (Fernald & McRoberts 1994). During this time, listening experience strengthens some adult-like perceptual categories that are evident in the first few weeks of life and weakens others. In effect, native language listening experience has the effect of shifting the perceptual boundaries of vocalic sounds at about six months (Kuhl *et al.* 1992), and modifies the boundaries of consonantal sounds at about 10 to 12 months of age (Werker & Polka 1993).

This might suggest that the year-old human infant is already deeply involved in *linguistic learning*, but this is not strictly true. In a model of phonetic perceptual development, Werker & Pegg (1992) proposed that infants progress through four types of perceptual learning on their way to full linguistic capacity. In these types, the infant is influenced by ambient stimulation, but this effect is primarily *auditory* and relatively nonspecific. Since, in the case of consonant-like sounds, these phases occur by or before the age of ten months – an age in which comprehension of at least a few dozen words is usually demonstrable – the learning of words antedates the kinds of perceptual learning that are most straightforwardly linguistic.

As for production, infants may occasionally reproduce maternal intonation contours in the first six months of life (Masataka 1992), and reveal in spontaneous utterances their assimilation of ambient sound patterns in the second six months (de Boysson-Bardies *et al.* 1992). During the vocal learning phase, infants thus become acquainted with vocal cues that distinguish people, transmit emotion, regulate social behavior, and superficially characterize their native language.

It should be recognized that the vocal learning that leads to spoken language occurs long before infants possess the cognitive capability to fully appreciate its existence. The behavioral precursors to language must therefore be motivated by "nonlinguistic" factors. For example, the frightened infant may be reassured by maternal vocalization; the risk-taking infant is cautioned by it (Locke 1996).

Early learning in nonhuman primates

There have been many studies of the vocal learning of nonhuman primates, presumably because they are so obviously proficient in, and in their daily lives reliant on, vocal communication. Studies of their expressive vocal learning – largely negative overall – have been reviewed in detail elsewhere (Snowdon *et al.*, Chapter 12; Seyfarth & Cheney, Chapter 13). Thus, we concentrate here on important questions left unanswered about the possibility of receptive vocal learning. For example, it makes sense to ask whether neonatal monkeys and apes react selectively to maternal voice to the same degree as human newborns. No answer seems to be available at the moment, but it is important to find out, for if young primates are less reactive to maternal voice, one might not expect that with increasing fearfulness they would be selectively reassured by her voice. And if her voice failed to offer the infant any needed information, there would be less reason for infant apes and monkeys to monitor maternal vocal activity, and thus to learn from and about it.

Infants prefer the company of their mother over that of a stranger, and thus we may assume that they prefer to hear, and therefore actively listen to, her voice as well as her speech (Bowlby 1969). For this reason alone, one might expect infants to store more maternal speech than nonmaternal speech. But they also may have more opportunity to learn from their mother, since her presence would encourage quiet alertness – which other research indicates may be necessary for efficient vocal learning (Kugiumutzakis 1993) – rather than an agitated or fretful state.

The storage phase

We saw earlier that vocal learning begins as early as the prenatal period, when experience begins to tune vocal learning mechanisms to the prosody of the mother and, incidentally, to her language. This is followed by a phase in the development of linguistic capacity that is massively dedicated to storage of utterances (Locke 1994a,b). Typically, this phase is revealed in the form of lexical comprehension in infants as young as 4.5 months, when there may be selective responding to their own name (Mandel et al. 1995). At eight months, the normally developing infant responds appropriately to an average of 36 words and at ten months to about 67 words (Bates et al. 1994).

The true number of words that has been stored may be greater than this, for to be credited with lexical storage infants must respond appropriately to words. Appropriate responding cannot occur unless supporting contextual and referential information also was stored, along with the words, and the infant is sufficiently responsive (e.g., by looking at or pointing to named objects) that its comprehensions can be inferred by adult observers. Speaking is, thus, preceded by a long interval of lexical storage. A somewhat analogous period has been observed in the vocal development of songbirds (Marler & Peters 1982).

It is appropriate to note that one nonhuman primate – the bonobo, Kanzi – has demonstrated storage and comprehension of a fairly large vocabulary of spoken words that were not specifically taught, but used in the context of his everyday social interactions with language researchers (Savage-Rumbaugh et al. 1993). This behavior may have been facilitated by a human-like enculturation that encouraged joint attention to objects (Tomasello et al. 1993).

Vocal productions in infancy: Turn-taking

Although language was once thought to begin with the production of the child's first word (Jakobson 1968), research conducted over the last 30 years suggests that spoken language in children is continuous with earlier vocal and gestural behaviors in form and/or function. In fact, one of the great unresolved mysteries of research in language acquisition is the issue of identifying the first word – first words turn out to be socially constructed rather than objective phenomena.

During the first months of life, infant vocal productions are likely to occur during face to face interaction with an adult. Several investigators have found a marked increase in vocal turn-taking between 12 and 18 weeks of age (Ginsburg & Kilbourne 1988). Studies suggest that, at this age, infants begin to forestall vocalization if their mother is speaking and to fall silent if their mother starts to speak (Moran et al. 1987), a practice that intensifies in subsequent months (Papoušek & Papoušek 1989).

The development of prefrontal cortex may contribute to these ontogenetic processes and events. It is now understood that this area figures importantly in the development of social and emotional responding both in nonhuman primates and humans (Kalin & Shelton 1989; Kalin et al. 1991). It also makes possible inhibition of attention to irrelevant stimulation and therefore contributes to the infant's capacity for selective attention (Goldman-Rakic 1987; Diamond 1990). Since, as seen earlier, activity of the prefrontal cortex increases greatly during the 7 and 12 months interval, this area may play a major role in selective attention to speech, thus enabling utterance storage and the attentional tuning of perceptual mechanisms (Werker & Pegg 1992).

Turn-taking also occurs in nonhuman primates, at least in mature animals. In the vocal exchanges between squirrel monkeys, there is evidence for a conversation-like temporal structure (Biben et al. 1986), and what may be a turn-taking cue was reported recently by Hauser (1992a). In vervets and rhesus macaques (Macaca mulatta), he found that a decline in the fundamental frequency of certain call types was associated with vocal interruption by another animal.

Vocal productions: Control

Much attention has been devoted to the role of vocal tract anatomy and peripheral effector systems in relation to the hominid's emerging capacity for spoken language. However, the importance of the pharynx may have been overstated (cf. Lieberman 1984). There are documented cases of children who have learned to speak without a tongue (MacKain 1984), and it is possible to learn language without a left cerebral hemisphere (Dennis & Kohn 1975). But it is essential that the vocal tract be hooked up to a reasonably competent control center somewhere in the brain.

Children increasingly develop vocal control over the first half-dozen years of life. The process seems to take a step forward when babbling begins, usually at about six to seven months (Oller & Eilers 1988). Reduplicated babbling refers to the alternate depression and elevation of the mandible, with points of oral contact, while phonating. To the ear, this activity produces syllabic elements, e.g., dadada, that resemble speech. Although a behavior referred to as "lip-smacking" or "tongue-smacking" has been described in the nonhuman primate literature (Anthoney 1968; Redican, 1975; Marler & Tenaza, 1977; Kenney *et al.* 1979), it is not clear that this activity, which seemingly lacks phonatory support, is quite the same thing as babbling. However, certain of the vocal behaviors of young pygmy marmosets seem to affect their parents in human-like ways by promoting proximity and care (cf. Snowdon *et al.*, Chapter 12).

Infant vocalizations prior to about six months, when canonical babbling (the first vocalizations that have a syllabic, hence speech-like character) emerges, can easily be distinguished as "distress" versus "nondistress" vocalizations. Mean incidence of nondistress vocalizations increases between two and five months, although with considerable individual variation at both ages (Bornstein & Tamis-LeMonda 1990). This finding was replicated among infants in the USA, France, Israel and Japan, with no intercountry differences in frequency or length of vocalization bouts (Bornstein *et al.* 1992). Bornstein & Tamis-LeMonda (1990) found no correlation between distress and nondistress vocalizations, suggesting that these are subserved by different systems, nor did they identify health or demographic indicators of vocalization frequency. The inexorable but variably timed emergence of sophisticated, consonantal, syllabic, and prosodically organized units among infant vocalizations during the first year of life is also attested by Papoušek's careful study of 18 German infants between 2 and 15 months (Papoušek 1994).

Conditioning studies suggest that nonhuman primates are able to learn when and with what duration to vocalize, as participants in carefully controlled experiments (Sutton *et al.* 1973; Sutton *et al.* 1978), but there is little evidence that they are able to make fine adjustments in their vocal quality in response to feedback or reinforcement. When efforts have been made to condition vocal behavior, and then to bilaterally lesion areas of the neocortex that are homologous to Broca's, no change has occurred either in the number or duration of learned (or spontaneous) vocalizations. Such changes as these occurred only when lesions

were made in the area of the anterior cingulate gyrus (Sutton *et al.* 1981; Trachy *et al.* 1981), a structure associated with volitional speech in humans. If neocortex plays little or no role in conditional vocalization, as the work of Sutton and his colleagues suggests, it seems unlikely that nonhuman primates would be capable of the rampant vocal learning that occurs in language development.

Because the left sylvian fissure may be longer than the right in great apes (LeMay & Geschwind 1975) and Old World monkeys (Falk 1978), it is unclear how to interpret similar asymmetries in the human brain. What, then, about the neurophysiology of phonation in nonhuman primates? Fortunately, we are beginning to obtain some preliminary information on which parts of the primate brain control laryngeal and respiratory activity associated with phonation. Research by Larson (1988; Larson & Kistler 1986) and by Jürgens and his associates (cf. Jürgens 1995) suggests that much of phonatory activity is controlled by the midbrain periaqueductal gray, which receives projections from the limbic system. This is not surprising, considering the close relationship between emotionality and vocal behavior, and Sutton's work suggesting that spontaneous vocalization survives bilateral lesions of the cortex. It would be interesting to see what parts of monkey brain are activated by conspecific and heterospecific vocalizations of various types.

Vocal productions: Accommodation

Stored material begins to become expressible through a process that has been termed vocal accommodation (Locke 1993), an infantile variant of what social psychologists have called "speech accommodation." Vocal accommodation refers to minor adjustments, usually unintentional, in one's vocal output that cause it to be more like that of an influential individual with whom one is interacting. This process requires very little articulatory control, since it affects a range of mainly phonatory qualities, including the pitch, loudness, and length of utterances.

There is evidence that vocal accommodation occurs initially in the human infant before talking begins (Kuhl & Meltzoff, 1982, 1988; de Boysson-Bardies *et al.* 1989; Masataka, 1992). Exposure to stereotyped utterances in contextually limited circumstances typically produces an initial stock of word-like vocal behaviors that antedate development of phonology and morphology by many months. Vocal accommodation is therefore linguistically significant in the developing human, even though it is not an example of linguistic behavior per se.

There is evidence to suggest that vocal accommodation also occurs in nonhuman primates. For example, when a squirrel monkey engages in "dialogue" with a conspecific cagemate, the frequency and morphology of its vocalizations may be influenced by immediately preceding vocal behaviors of the other animal (Maurus *et al.* 1987). Some of the vocal parameters on which there is evidence of accommodation, e.g., fundamental frequency modulations (Maurus *et al.* 1988), also are used contrastively in human languages.

In a study of coo calls in free-living rhesus monkeys, Hauser (1992b) observed some behaviors that seem to qualify as vocal accommodations. By noting which animals vocalized, and their genetic relationship to other animals, Hauser was able to discover something quite remarkable. His sound spectrographic analyses revealed that coo calls from the members of one matriline differed systematically from the coos of other matrilines. The acoustic basis for differences lay in the spectral prominences above the second harmonic, proportion of missing harmonics, intensity of the fundamental frequency and second harmonic, and energy between harmonics. The result of these variations was that, to the human listener, one matriline sounded "nasal" and the others did not. Although Hauser could not rule out an organic explanation, these effects may well have been the result of infants' accommodations to their mother's nasal quality.

As we will see below, it is ontogenetically significant that vocal accommodation, as an *accommodation*, belongs to a category of behaviors that are absorbed by the *dependent or subordinate* member of a dyadic relationship.

SOCIAL SUPPORT FOR VOCAL LEARNING: SKILLS AND CAPABILITIES

Humans have a rather special social setting for learning to vocalize

Infant vocalizations emerge and develop with a high level of reliability and considerable universality across populations and settings. Nonetheless, it is possible to identify massive social influences on vocal development in humans. These are of two sorts: social stimuli that are universally available to normally developing infants, and social stimuli that are related to culture, social class, or other factors that vary across infants.

In their study of two- and five-month-olds, Bornstein & Tamis-LeMonda (1990) found that maternal production of infant-directed speech as well as maternal behaviors encouraging attention to the environment at both two and five months predicted frequency of nondistress vocalization at five months. Unsurprisingly, frequency of infant nondistress vocalization also related to maternal infant-directed speech, with imitation of infant noises by far the most frequent maternal response to nondistress vocalizations; distress vocalizations were more likely to elicit nurturing or caregiving behavior (Bornstein *et al.* 1992).

Infants seem to understand the nature of social and communicative interactions from an early age (cf. Trevarthen 1979). There is some evidence that infant vocalizations are more speech like (e.g., more likely to have consonantal closures) when they occur in interactive settings, and after adult imitative responses to previous vocalizations (Dodd 1972; Bloom *et al.* 1987; Veneziano 1988; for a review see Koopmans-van Beinum 1993). Between two and five months, infant vocalizations become increasingly similar to preceding adult vocalizations in intonational features (Masataka 1992), suggesting imitation or incorporation by infants of adult prosody. It is also now well attested that deaf infants' prelexical vocalizations fail to differentiate in the same way as those of their hearing peers (Stoel-Gammon & Otomo 1986; Stoel-Gammon 1988). However, in most cases it does not seem that individual differences in children's early phonetic and phonological output are related in any specific way to individual differences (as opposed to language differences) in the phonetic and phonological characteristics of input to them.

The special place of infant vocalizations as potential social, communicative responses is perhaps clarified by considering the emergence of vocalizations as related to gestures in early communication. It might seem obvious that children in the sensorimotor stage, and children who furthermore spend much of their time and energy operating on the world manually, would consider gestures as likely media for communication. Young children do often adopt gestures, both "iconic" and somewhat conventionalized ones, for early communicative purposes, but gestures used communicatively in fact emerge after words, and are used intensively during the one word period mostly *in combination with* words or vocalizations (Carpenter *et al.* 1983), not on their own. Thus, it seems that the vocal channel, despite the articulatory challenges it presents, is somewhat privileged for communication, although of course children growing up in a family or society where gestural communication systems are used will learn those instead of or in addition to vocal systems and children growing up without access to spoken language will rely on gesture (Goldin-Meadow, Chapter 15).

As a result of their precocious commitment to social, typically vocal, interaction, infants collaborate with their adult caregivers to create recurrent opportunities for practicing vocal and verbal interactions. Of children's early acquired words, the vast majority are somewhat marginal as conventional linguistic phenomena (Ninio 1993; Ninio & Snow, 1996). Examples include particles and interjections (down! or oops!) or words learned and used exclusively in recurrent interaction routines ("Woof-woof" as a response to "What do doggies say?"). Standard procedures for young children's vocal productions include borrowing a word from the preceding adult utterance ("Do you want some juice?" "Juice!"), and using a very limited verbal repertoire quite promiscuously (e.g., the use of "Mommy" for all requests, even if not addressed to mother).

There is considerable bias now to describe the circumstances under which children's lexical learning (which we think of as a more sophisticated version of vocal learning, although it obviously has elements beyond the exclusively vocal) is promoted interactionally. Young children who learn words faster are those who hear more words in total (Huttenlocher *et al.* 1991), who hear concrete nouns while they and their mothers are attending to the object named (Tomasello & Farrar 1986), and who hear action verbs just before the action named is witnessed or undertaken (Tomasello & Barton 1994). Children's earliest words are indeed very likely to be the words they themselves have heard most often (Hart 1991). In fact, children's one-word utterances consist of precisely the same words as mothers use in their one-word utterances to express the same communicative intent (Ninio 1992).

All these findings suggest that early lexical learning is deeply embedded in shared social activities, and that children's early words are acquired precisely to enable them to participate more effectively in the social interations they would have been engaging in anyway using babbling,

gesture, or action on objects. Of course, later lexical acquisition is a process that might be described as more "cognitive" and less social, precisely because words acquired after the first few thousand are unlearnable through joint attentional procedures. This is because they are much more likely to have abstract rather than concrete referents, e.g., to be superordinates rather than basic object level terms, or mental rather than action verbs. Acquisition of these more sophisticated lexical items also requires exposure and relates to the frequency with which they are heard (Beals & Tabors 1995), but information about the meanings of these words is evidently available from their subcategorization structures as well as their mappings to the word events (Gleitman 1990) and fast mapping (acquisition after a single exposure) of at least some word meanings becomes possible after the lexical system is sufficiently structured that gaps are evident.

Attachment, emotionality, and fear in vocal learning

At present, no research shows a positive relationship between strength of maternal attachment and rate of language development, and at least one study is negative on this point (Bretherton *et al.* 1979). And yet it is logical that an infant's relationship with its mother would affect its vocal learning, and vice versa, at least through a link with vocal turn-taking. First, Bowlby (1969) suggested that attachment begins in earnest when infants' preference for their mother's company is satisfied either facially or vocally, i.e., not just facially. Since the phase in which this happens usually begins at about 12 weeks, attachment may commence at about the same time as vocal turn-taking. Second, both vocal learning and turn-taking involve emotional interpretation and inhibition, functions made possible by development of prefrontal cortex. Studies suggest that prefrontal cortex becomes unusually active as early as six to seven months (Chugani *et al.* 1987; Bell & Fox 1992), perhaps beginning sometime after three months (Chugani & Phelps 1986).

During the second six months of life, there are increases in vocal monitoring and vocal and lexical learning. Maternal facial monitoring and, presumably, attention to affective and indexical properties of the mother's voice reach a peak between seven months and a year. Both perceptual learning and vocal production are first conspicuously reorganized by listening experience, and the rate of lexical storage probably increases during this period. One might ask why all these important preparations for, and stages in, spoken language occur during this particular period.

If language is learned from individuals who helpfully adapt their utterances to infants' levels of comprehension and expression (cf. Snow, 1995), it is logical that severe parental neglect would slow the rate of language development. Christopoulos *et al.* (1988) found that neglectful mothers produce less language and fewer accepting comments than normal parents. One would not be surprised to find additional differences in, say, contingent responding, a process whose facilitative effects have been documented (Hardy-Brown *et al.* 1981; Veneziano, 1988). And the evidence suggests that parentally neglected infants do develop language more slowly than those receiving normal levels of attention (Provence & Lipton 1962; Fox *et al.* 1988; Culp *et al.* 1991).

What about infants who have been *abused*? Eibl-Eibesfeldt (1989) has suggested that punishment and other forms of maltreatment may increase emotional bonds between children and their mothers, pointing to a finding that chicks, ducklings, and young rhesus monkeys tend to follow their mothers even if they are punished for doing so (Rajecki *et al.* 1978). Among rhesus monkeys there is evidence for stronger attachment relations among punished than unpunished animals (Seay *et al.* 1964). Scott (1962) and other researchers have found that dogs punished by their trainers become more strongly bonded to them than dogs that are handled in a normal fashion. He concluded, as others have, that fear – like other strong emotions such as hunger, pain, or loneliness – speeds up and intensifies socialization. Recently, Kraemer (1992, p. 495) concluded from a review of the literature that "in a variety of species, exaggerated attachment behavior ('anxious attachment') is maintained in the face of parental maltreatment of the infant".

There is evidence of this in humans, too. Egeland & Sroufe (1981) conducted a prospective study of 200 mother–infant dyads from birth to two years. The mothers were young, uneducated, poor, and unfamiliar with standard caregiving practices, and it was expected that they would be neglectful and/or abusive. Although little was made of it in the report, the infants who were physically abused were more likely than control children to be securely attached. At 12 months, 25% of the abused children and 18% of the control children were securely attached; six months later the gap had widened to 33% and 16% respectively.

Fox *et al.* (1988) evaluated the receptive language of three groups of children who had been intentionally mistreated by their parents. Children in the abused group had been physically injured. Severely neglected children's health had been compromised because of denied access to nutrition or medical care, or needed food, clothing, and shelter. Generally neglected children had been similarly neglected, but without physical injury. Only the children who had been severely neglected were significantly delayed in the development of receptive language, relative to normally treated children in a control group. Children who were generally neglected or abused were not significantly delayed.

The literature suggests that abused children do not necessarily develop language more closely than children who have never been abused. Allen & Oliver (1982) evaluated the receptive and expressive language of three groups of 2.5- to 5-year-old mistreated preschool children: abused, neglected, and abused and neglected, and compared their performance to that of a control group of non-mistreated children. Multiple regression analyses indicated that language could be predicted from neglect but not from abuse or a combination of neglect and abuse. The children who were abused had *higher* language scores than children who were neglected, and the children who were neglected and abused achieved slightly better scores than children who were only neglected.

Culp *et al.* (1991) evaluated the language of 74 preschool children who had been abused, neglected, or both abused and neglected. Abuse was defined as intentional physical injury. Neglect was defined as failure to provide support, education, or health and other care necessary for normal well-being. Standardized measures that were analyzed retrospectively included the receptive and expressive sections of the Preschool Language Scale (PLS), the language and cognitive sections of the Early Intervention Development Profile and the Goldman–Fristoe Test of Articulation. The three groups did not differ on the articulation test.

The neglected children and the abused and neglected children scored well below their chronological age levels. Statistical analyses revealed that the neglected group, but not the abused and neglected group, scored significantly below the abused group on all language measures, i.e., the receptive and expressive sections of the PLS and the language section of the Profile. There were no differences between groups on the cognitive section of the Profile. These analyses aside, it is of considerable interest that while both groups experiencing some amount of neglect scored six to nine months below chronological age level, abused children scored *at or slightly above* age level. In an attempt to explain their findings, Culp *et al.* suggested that abuse ought to be less injurious to linguistic development than neglect because the former might occur only once before the case came to light, whereas neglect might occur continuously over a protracted period of time. However, it is also possible that abuse may have marginally enhanced language development, since abused children, as a group, displayed (a) language scores that were above those of neglected children, and at or slightly above chronological age level, and (b) cognitive scores that were below age level, and not above the cognitive scores achieved by the other two groups.

We thus see developmental linguistic differences between abusive and neglectful parents. One might well ask why physically damaging levels of parental abuse fail to delay the development of language. A clue may have been suggested by Blager (1979), who found that children may respond to parental abuse with *precocious adaptations.* "Some," Blager said, "were very stylized in voice inflection and manner, like a 'Little Mr. Big Man' or a 'Perfect Little Child.'" (*ibid.*, p. 991) Eibl-Eibesfeldt (1989, p. 77) has suggested that fear "induces a childlike readiness to accept and learn." He pointed out that dictators take advantage of this connection, as do military captors (Watson 1978), and this is presumably what underlies the so-called Stockholm syndrome (cf. Ochberg 1978). One might therefore speculate that parental abuse induces fear and adaptive accommodations to parental vocal behaviors. In this context it is useful to recall that accommodation to the speech of another person is typically performed in order to gain that person's approval (Giles & Coupland 1991).

Humans continue to show vocal learning through adulthood

Much attention has been directed to the failures of older language learners, precisely in the domain of vocal learn-

ing, e.g., acquiring a native-sounding accent in a second language. These failures are widely attested for perception of segments (Takagi & Mann, 1996) and of word boundaries (DeJean de la Batie & Bradley 1995) as well as for production of phonetic detail (Flege *et al.* 1995) and developing target-language-like phonological processes. What remains unclear from the catalogs of failure is precisely *why* adults typically fail to achieve perfect foreign accents, whether this reveals an age-related reduction of susceptibility to social effects on vocal learning, and if so whether this reduction is biological or social in character. One hypothesis, related to the findings reviewed above concerning attachment, is that the loss of self-identity associated with being too convincing a speaker of a foreign language is in fact threatening to learners, and brakes their acquisition short of perfection. Additional complexities, though, are introduced by the observation that later language learners are more likely to retain high levels of proficiency and many opportunities for use of their first language, and thus are subject to the psycholinguistic constraints of keeping two, partially conflicting systems "up and running." Even young children learning two languages simultaneously display some merging and overlap of the phonological systems of their two languages (e.g., Pearson *et al.* 1994) – indication that one needs to deal with the issues of bilingualism in order to approach an assessment of "native-like" production in second language speakers.

Support for the notion that less-than-perfect vocal learning in adulthood does not reflect a biologically induced incapacity for new vocal learning comes from studies that have manipulated conditions of access to information about the second language. These studies are based on the notion that optimal vocal learning in L2 should replicate successful L1 vocal learning, i.e., allow for a period of perceptual learning prior to any demand for "phonological" output. Neufeld (1978) taught English speakers Thai by subjecting them to only about 20 hours of listening before allowing them to speak – considerably less "ear training" than the average child second language learner receives; a relatively high percentage of the learners were subsequently able to repeat modelled Thai sentences with accents that were judged as native like. It is important to note that these were pure vocal learners – they had not acquired any Thai lexicon or grammar, but could be considered to be good Thai babblers or prelinguistic vocal turn-takers.

Young children's relatively good success at acquiring accents may reflect their tendency to observe a mute period at first exposure to a second language (Saville-Troike 1988). Although they presumably do not understand the value of the strategy, children's shyness or reluctance to speak up until they control a target language prevents them from distorting the target language through the first language of phonology. Adults in second language learning situations typically do not have the luxury of several months of muteness, and in foreign language classrooms in fact are enjoined to start talking immediately.

Vocal accommodation

Vocal accommodation, evidently a universal, largely unconscious process, has been demonstrated to occur early in infancy (see above), but also is present – and represents a support for the continued existence of vocal learning capacities – throughout adulthood. Vocal accommodation refers to the tendency to adapt aspects of one's vocal production (including aspects as varied as pitch of voice, vowel production, intonation, accent, lexical choice, and grammatical features) to match or approach those of the conversational partner, particularly when the interaction is proceeding in a way characterized by positive affect and/or the partner has more power. If vocal accommodation reflects a natural human tendency that accounts for young children's speech approaching that of their adult interlocutors, it at the same time creates a social context in which young speakers can influence the speech they hear from interlocutors. The process of accommodation clearly involves short-term vocal learning.

Vocal accommodation is a useful social process precisely because so many aspects of vocal productions serve as social markers. It is important to emphasize here that we are talking about purely *vocal* phenomena – obviously linguistic phenomena can also serve as social markers (e.g., saying "I am a member of the Communist Party" constitutes social marking), but even without conveying any content through speech we convey considerable content through ways of speaking. Speech has been identified as socially marked for the personal features gender, age, geographic origin, social status, and sexual preference, as well as for situational features such as status of interlocutor and formality of setting (for a review, see Scherer & Giles 1979). All these features which constitute markers of iden-

tity thus becomes as well targets for accommodation from interlocutors (Giles & Coupland 1991; Giles *et al.* 1991). Accommodation typically is not absolute; just as foreign language learners normally retain some trace of their first language in their accents, conversational accommodators approach but do not match the vocal characteristics of their interlocutors.

More complete long-term accommodation can occur, though, when children or adolescents move to new dialect areas and accommodate to dialect features of their companions. Consider the case, for example, of children moving with their Boston- or Chicago-born parents to Philadelphia: will they or won't they acquire the Philadelphia variety of American English? Extensive work by Payne (1980) on precisely this question showed that children younger than eight acquired regular, widespread Philadelphia features, but that more lexically specific and complex features were not typically acquired even by very young in-migrants; furthermore, even Philadelphia-born children did not acquire these reliably unless their parents were also Philadelphia natives!

These findings somewhat modify our understanding of the importance of the peer group, which had been widely assumed to be definitive in inducing language change. It seems that phonologial and phonetic details of parental speech have a lasting effect on children's vocal language. Children don't just learn their parents' language, they learn to speak it with a degree of articulatory match to the model which is quite impressive. They can, however, shift certain aspects of their language production quite strikingly, shifts which are only likely to be seen under conditions of dialect mixing. Kerswill (1994) has been studying the emergence of a local dialect in a southeastern English new town with residents from a variety of different dialect areas. He has found evidence for the influence of both parental and local norms in children's speech. In one striking case, a child who at four had replicated his father's Scottish vowel system by six had shifted radically to the very different local southeastern English vowels. This is one example of how children during the age period five to 15 years can shift their accents to match those of their desired social companions, adopting a regional variety that deviates from that of their parents or even participating in the development of new dialect forms. Interestingly, accent is the aspect of a foreign or second language acquired after childhood that is the least likely to be fully achieved – a striking failure of normally effortless vocal learning that deserves further attention.

This suggests the need to reconsider a subject on which each of us has written something in the past – a sensitive period for language (Snow 1987; Locke 1993, 1994a) – but here we look only at the vocal learning phase of the larger developmental process.

SENSITIVE PERIOD FOR VOCAL LEARNING

There is less documentation than might be supposed for a critical or sensitive period for language learning, and much of the evidence that does exist is at the phonetic or phonological level of language (Snow 1987; Flege 1991). These facts concur with a point made by Bateson (1979) about critical periods – if there is reduced performance in later life, this may reflect little more than a reduction in "motivation" to use learning mechanisms that were activated at an earlier time. Reduced motivation may, in effect, imply reduced role of emotionality and dependency, not in some abnormal sense, but as they apply in the typically developing infant who is nurturantly treated by its loving parents (Locke 1993). These infants, too, are dependent, frequently frightened, and presumably in perceived need of parental rescue.

In the debates about a critical period for language, there has been little or no mention of the possibility that there might be an *optimum period for vocal learning in the human infant*, or a circumscribed period in which vocal learning mechanisms may be activated and stabilized. Such a prospect is reasonable given evidence of a circumscribed period of phonetic or phonological learning for one's native language (Hurford 1991; Shriberg *et al.* 1994), as well as a second language (cf. Asher & Garcia 1969; Flege 1991; Flege *et al.* 1995), that may exist apart from any comparable period for the learning of vocabulary, morphology or syntax (Snow & Hoefnagel-Hohle 1978; Snow 1987). A recent review of the literature suggests that this period occurs at about five to seven or eight years of age (cf. Locke 1993).

A sensitive period for vocal learning also makes biological sense inasmuch as our species potential to learn spoken language may rest on – and phylogenetically may have piggy-backed on – functions associated with kin recognition and social regulation. After all, in the languages that

arose and are spoken today the vocal cues that identify individuals and their emotional state are intermixed with cues that carry linguistic meaning. And it is reasonable to suppose that the vocal component of language learning uses a system already evolved for indexical and affective communication, a system whose functions in infancy may conceivably be extended by emotionality and fear.

We think there may be an optimal period for the vocal learning component of language, but such a period would exist not because it is advantageous for the developing members of our species to lose this *linguistic* capability, but because language takes advantage of functions that happen to adaptively weaken with maturation of the individual. The offset of sensitivity may be caused by the pre-emption of structure by specific experience, as Bateson's (1987) model of filial learning suggests.

What advantage might accrue to the individual that can carry out vocal learning only during a circumscribed interval occurring early in life? If animals have a *voice recognition system* that operates more efficiently early in life than later, they might tend to associate with, and respond preferentially to, individuals who are likely to protect them. There would be an advantage to learning early on who your mother is, and for the contents of that system not to be overwritten by information that is admitted later on. This scenario represents an application to humans of the competitive exclusion hypothesis set forth by Bateson (1987) based mainly on work with chicks.

With maternal attachment there is, by definition, diminished responsiveness to strangers, suggesting that up to some point the infant's vocal learning and utterance storage capabilities operate preferentially on the vocal behaviors of the mother. If other languages are learned to some level of proficiency, their prosody would be expected to be nonnative-like, and this is what the evidence shows (Oyama 1976; Snow & Hoefnagel-Hohle 1977; Snow 1987; Flege & Fletcher 1992).

We have suggested that vocal learning is, to some degree, associated with feeling states that accompany helplessness and dependency. It would not be surprising, then, to find that independence tends to rise as vocal and phonetic abilities decline. It is interesting, in this regard, that from five to seven years children from widely divergent cultures are thought to be able to assume responsibility and to express capabilities associated with responsible conduct (Super 1991). In Western societies, children of about this age tend to "leave home" to begin school. There are also some cross-cultural behavioral changes that occur during the five to seven year interval, including increased resistance to the verbal transformation effect (Super 1991) and susceptibility to the disruptive effects of delayed auditory feedback (White 1970).

Whatever decline in vocal learning occurs in midchildhood, it is clear that adults retain the ability to modify the sound-making practices first revealed in early childhood. Research on sound change (Labov 1974) indicates that phonological change frequently occurs within generations; that is, within particular speakers. Labov's studies of New York City vernacular suggests that sound change is not merely possible in adult speakers, but a necessary means of signalling group membership and achieving intimacy with listeners.

NEUROBIOLOGY OF VOCAL LEARNING

The ability to infer dispositions and intentions is a highly evolved social capability. In primates, such socially cognitive acts depend on neural specializations and modular properties that stand somewhat apart from other forms of cognition (Brothers 1990; Brothers & Ring 1992; Karmiloff-Smith *et al.* 1995). If we wish to explore the potential for vocal learning in primates, and the biological foundations of spoken language in human primates, there are several things we must know about these specializations.

One of the first things we must learn is how the brain responds to faces and facial movements, and we must appreciate the neurophysiology of demonstrated functional links between face perception and social vocalization. In humans, behavioral and physiological studies reveal a greater involvement of the right cerebral hemisphere during the display and interpretation of facial emotion (cf. Etcoff 1989). This effect also has been obtained in monkeys (Hamilton & Vermeire 1988; Hauser 1993). Moreover, in nonhuman primates, single unit recordings from temporal cortex have revealed cell ensembles with differential reactions to faces and other visually complex objects and cells that respond more vigorously to upright and intact faces than to inverted or rearranged faces (cf. Perrett & Mistlin 1990; Desimone 1991). There also are cells that appear to be sensitive to the directionality of facial view, line of gaze of the stimulus face, facial

expression, facial orientation, specific facial features, expressions, and a variety of movements. Interestingly, cells that code for facial identity are distinct from cells coding for facial expression.

We also need to learn what goes on in the brain when listeners attempt to extract from the human *voice* information that is relevant to the speaker's identity, intentions in speaking, and attitude toward his or her addressees. There is evidence that, when human listeners analyze such cues, they activate processing machinery in the nondominant, right cerebral hemisphere (cf. Locke 1992, 1993). Evidence suggests that with similar cues, nonhuman primates also may favor the right hemisphere, although they, too, may prefer the left hemisphere when processing conspecific calls (Petersen *et al.*, 1978, 1984; Hauser & Andersson 1994; Heffner & Heffner 1995)

As one might suppose, there is support for a functional linkage between face and voice processing. At present, much of it comes from clinical studies showing that, in adult patients a lesion in the right hemisphere tends to impair interpretation and execution of both facial expressivity *and* vocal prosody and emotion (Ross 1981). Among the developmentally disabled, there is now a small literature indicating that autistic children characteristically show aversion to or disinterest in the human face, atypical face-scanning patterns, and deficits in the interpretation of facial affect (Langdell 1978; Hobson *et al.* 1988). Malformations associated with autism have been found in the cerebellum and limbic system, including the amygdala (Bauman & Kemper 1985), a structure that has come to be identified with affective processing (Kling & Brothers 1992).

Research on rhesus monkeys suggests that the desire by infant primates to affiliate with their mother is regulated by distinct brain systems. Monkeys begin to react to maternal absences in a differentiated way – adaptively modulating defensive behavior – when they reach the age of about 9 to 12 weeks (corresponding to 7 to 12 months in humans). At this age, if left alone, infant monkeys tend to coo. If confronted by a person not looking at them, they tend to freeze. If confronted by a person who looks directly into their eyes, they tend to bark.

If involved in the circumstances above, the hypothalamic–pituitary–adrenal system (h-p-a) is likely to become involved. The prefrontal cortex assesses the situation, and reports that there is reason for defensive behavior. The hypothalamus secretes corticotropin-releasing hormone, which triggers the pituitary gland to secrete adrenocorticotropic hormone, which, in turn, activates the adrenal gland to release cortisol, which prepares the body to defend itself. The h-p-a system becomes mature in the monkey by 9 to 12 weeks. Drugs that suppress functions of the h-p-a system suppress cooing, freezing, and barking. When an infant monkey is separated from its mother, opiate-releasing and opiate-sensitive neurons are inhibited, causing the infant to miss its mother and to feel vulnerable. Reduction of activity in the opiate-sensitive pathways enables motor systems in the brain to produce cooing. When a predator appears, there is suppression of neurons that produce endogenous benzodiazepines, leading to elevated anxiety and fear responses.

It should be clear from the above that we need a better understanding of the effect of fear on mechanisms that contribute to vocal learning in primates.

CONCLUSION

The work reviewed here seems to us to generate two conclusions: that vocal learning can usefully be viewed as separate from language learning, and that issues of personal and social identity are expressed vocally precisely because the vocal system retains considerable flexibility and susceptibility to social influences throughout life.

There are many ways in which human primates seem well adapted for vocal learning. These adaptations clearly include biological capacities (brain specialization and lateralization, activity-dependent brain development, motor control particularly in the vocal tract, auditory discrimination abilities, and so forth) that create quite spectacular learning abilities (e.g., speed and accuracy of articulation, categorical perception of phonemes, rapid word-learing based on referential principles and concept formation), which, however, also vary enormously in their speed and efficiency as function of social environment. Fortunately, human primates have also developed universal solutions to the challenges of infant survival and socialization that make available to almost all children rich social support for vocal and verbal learning.

In addition, the processes of personal identity formation and of formation of groups and group identities recruit the vocal capacities of the human learner, selecting accent, prosody, voice quality, and lexical components as identifying features even for speakers of a single language. These vocal processes could not be recruited for use in such identity procedures if they were not subject to rapid learning and considerable flexibility even at later ages.

REFERENCES

Allen, R. E. & Oliver, J. M. (1982). The effect of child maltreatment on language development. *Child Abuse and Neglect* 6: 299–305.

Anthoney, T. R. (1968). The ontogeny of greeting, grooming and sexual motor patterns in captive baboons (superspecies *Papio cynocephalus*). *Behaviour* 31: 358–72.

Asher, J. & Garcia, R. (1969). The optimal age to learn a foreign language. *Modern Language Journal* 38: 334–41.

Bates, E., Marchman, V., Thal, D., Fenson, L., Dale, P., Reznick, J. S., Reilly, J. & Hartung, J. (1994). Developmental and stylistic variation in the composition of early vocabulary. *Journal of Child Language* 21: 85–123.

Bateson, P. (1979). How do sensitive periods arise and what are they for? *Animal Behaviour* 27: 470–86.

Bateson, P. (1987). Imprinting as a process of competitive exclusion. In *Imprinting and Cortical Plasticity: Comparative Aspects of Sensitive Periods*, ed. J. P. Rauschecker & P. Marler, pp. 151–68. Wiley, New York.

Bauman, M. & Kemper, T. L. (1985). Histoanatomic observations of the brain in early infantile autism. *Neurology* 35: 866–74.

Beals, D. & Tabors, P. (1995). Arboretum, bureaucratic, and carbohydrates: Preschoolers' exposure to rare vocabulary at home. *First Language* 15: 57–76.

Bell, M. A. & Fox, N. A. (1992). The relations between frontal brain electrical activity and cognitive development during infancy. *Child Development* 63: 1142–63.

Berntson, G. G. & Boysen, S. T. (1989). Specificity of the cardiac response to conspecific vocalizations in chimpanzees. *Behavioral Neuroscience* 103: 235–45.

Berntson, G. G., Boysen, S. T., Bauer, H. R. & Torello, M. S. (1990). Conspecific screams and laughter: Cardiac and behavioral reactions of infant chimpanzees. *Developmental Psychobiology* 22: 771–87.

Biben, M., Symmes, D. & Masataka, N. (1986). Temporal and structural analysis of affiliative vocal exchanges in squirrel monkeys (*Saimiri sciureus*). *Behaviour* 98: 259–73.

Blager, F. B. (1979). The effect of intervention on the speech and language of abused children. *Child Abuse and Neglect* 5: 991–6.

Bloom, K., Russell, A. & Wassenberg, K. (1987). Turn taking affects the quality of infant vocalizations. *Journal of Child Language* 14: 211–27.

Bolhuis, J. J. (1991). Mechanisms of avian imprinting: A review. *Biological Review* **66**: 303–45.

Bornstein, M. H. & Tamis-LeMonda, C. S. (1990). Activities and interactions of mothers and their firstborn infants in the first six months of life: Covariation, stability, continuity, correspondence, and prediction. *Child Development* **61**: 1206–17.

Bornstein, M. H., Tamis-LeMonda, C. S., Tal, J., Ludemann, P., Toda, S., Rahn, C. W., Pecheux, M-G., Azuma, H. & Vardi, D. (1992). Maternal responsiveness to infants in three societies: The United States, France, and Japan. *Child Development* **63**: 808–21.

Bowlby, J. (1969). *Attachment*, vol. 1, *Attachment and Loss*. Basic Books, New York.

Bretherton, I., Bates, E., Benigni, L., Camaioni, L. & Volterra, V. (1979). Relationships between cognition, communication and quality of attachment. In *The Emergence of Symbols: Cognition and Communication in Infancy*, ed. E. Bates, L. Benigni, I. Bretherton, L. Camaioni & V. Volterra, pp. 223–69. Academic Press, New York.

Brothers, L. (1990). The social brain: A project for integrating primate behavior and neurophysiology in a new domain. *Concepts in Neuroscience* **1**: 27–51.

Brothers, L. & Ring, B. A. (1992). Neuroethological framework for the representation of minds. *Journal of Cognitive Neuroscience* **4**: 107–18.

Bushnell, I. W. R., Sai, F. & Mullin, J. T. (1989). Neonatal recognition of the mother's face. *British Journal of Developmental Psychology* **7**: 3–15.

Carpenter, R. L., Mastergeorge, A. M. & Coggins, T. E. (1983). The acquisition of communicative intentions in infants eight to fifteen months of age. *Language and Speech* **26**: 101–16.

Cheney, D. L. & Seyfarth, R. M. (1980). Vocal recognition in free-ranging vervet monkeys. *Animal Behaviour* **28**: 362–7.

Cheney, D. L. & Seyfarth, R. M. (1990). *How Monkeys See the World: Inside the Mind of Another Species*. University of Chicago Press, Chicago.

Christopoulos, C., Bonvillian, J. D. & Crittenden, P. M. (1988). Maternal language input and child maltreatment. *Infant Mental Health* **9**: 272–86.

Chugani, H. T. & Phelps, M. E. (1986). Maturational changes in cerebral function in infants determined by [18]FDG positron emission tomography. *Science* **231**: 840–3.

Chugani, H. T., Phelps, M. E. & Mazziota, J. C. (1987). Positron emission tomography study of human brain functional development. *Annals of Neurology* **22**: 487–97.

Cohen, A. & Starkweather, J. (1961). Vocal cues to language identification. *American Journal of Psychology* **74**: 90–3.

Culp, R. E., Watkins, R. V., Lawrence, H., Letts, D., Kelly, D. J. & Rice, M. L. (1991). Maltreated children's language and speech development: abused, neglected, and abused and neglected. *First Language* **11**: 377–89.

de Boysson-Bardies, B., Halle, P., Sagart, L. & Durand, C. (1989). A crosslinguistic investigation of vowel formants in babbling. *Journal of Child Language* **16**: 1–18.

de Boysson-Bardies, B., Vihman, M. M., Roug-Hellichius, L., Durand, C., Landberg, I. & Arao, F. (1992). Material evidence of infant selection from the target language: a cross-linguistic phonetic study. In *Phonological Development: Models, Research, Implications*, ed. C. Ferguson, L. Menn, & C. Stoel-Gammon, pp. 369–91. York Press, Timonium, MD.

DeCasper, A. & Fifer, W. P. (1980). On human bonding: Newborns prefer their mothers' voices. *Science* **208**: 1174–6.

DeCasper, A., Lecanuet, J. P., Busnel, M.-C., Granier-Deferre, C. & Maugeais, R. (1994). Fetal reactions to recurrent maternal speech. *Infant Behavior and Development* **17**: 159–64.

DeCasper, A. & Spence, M. (1986). Prenatal maternal speech influences newborns' perception of speech sounds. *Infant Behavior and Development* **9**: 133–50.

DeJean de la Batie, B. & Bradley, D. C. (1995). Resolving word boundaries in spoken French: Native and non-native strategies. *Applied Psycholinguistics* **16**: 59–82.

Dennis, M. & Kohn, B. (1975). Comprehension of syntax in infantile hemiplegics after cerebral hemidecortication: Left-hemisphere superiority. *Brain and Language* **2**: 472–82.

Desimone, R. (1991). Face-selective cells in the temporal cortex of monkeys. *Journal of Cognitive Neuroscience* **3**: 1–8.

Diamond, A. (1990). The development and neural bases of memory functions as indexed by the AB and delayed response tasks in human infants and infant monkeys. In *The Development and Neural bases of Higher Cognitive Functions*, ed. A. Diamond. *Annals of the New York Academy of Science* **608**: 267–317.

Dodd, B. (1972). Effects of social and vocal stimulation on infant babbling. *Developmental Psychology* **7**: 80–3.

Egeland, B. & Sroufe, A. (1981). Developmental sequelae of maltreatment in infancy. *New Directions for Child Development* **11**: 77–92.

Eibl-Eibesfeldt, I. (1989). *Human Ethology*. Aldine de Gruyter, New York.

Etcoff, N. L. (1989). Asymmetries in recognition of emotion. In *Handbook of Neuropsychology*, vol. 3, ed. F. Boller & J. Grafman, pp. 363–82. Elsevier, New York.

Falk, D. (1978). External neuroanatomy of Old World monkeys

(Cercopithecoidea). *Contributions to Primatology* **15**. Karger, Basel.

Fernald, A. (1992a). Human maternal vocalisations to infants as biologically relevant signals: An evolutionary perspective. In *The Adapted Mind: Evolutionary Psychology and the Generation of Culture*, ed. J. H. Barkow, L. Cosmides & J. Tooby, pp. 391–428. Oxford University Press, New York.

Fernald, A. (1992b). Meaningful melodies in mothers' speech to infants. In *Nonverbal Vocal Communication: Comparative and Developmental Approaches*, ed. H. Papoušek, U. Jürgens & M. Papoušek, pp. 262–82. Cambridge University Press, Cambridge.

Fernald, A. & McRoberts, G. (1994). Prosodic bootstrapping: A critical analysis of the argument and the evidence. In *Signal to Syntax: Bootstrapping from Speech to Syntax in Early Acquisition*, ed. J. L. Morgan & K. Demuth, pp. 110–38. Erlbaum, Hillsdale, NJ.

Field, T. M., Cohen, D., Garcia, R. & Greenberg, R. (1984). Mother–stranger face discrimination by the newborn. *Infant Behavior and Development* **7**: 19–25.

Flege, J. E. (1991). Age of learning affects the authenticity of voice-onset time VOT in stop consonants produced in a second language. *Journal of the Acoustical Society of America* **89**: 395–411.

Flege, J. E. & Fletcher, K. L. (1992). Talker and listener effects on the perception of degree of foreign accent. *Journal of the Acoustical Society of America* **91**: 370–89.

Flege, J. E., Munro, M. J. & MacKay, I. R. A. (1995). Effects of age of second-language learning on the production of English consonants. *Speech Communication* **16**: 1–26.

Fox, L., Long, S. H. & Langlois, A. (1988). Patterns of language comprehension deficit in abused and neglected children. *Journal of Speech and Hearing Disorders* **53**: 239–44.

Giles, H. & Coupland, N. (1991). *Language: Contexts and Consequences*. Brooks/Cole Publishing Company, Pacific Grove, CA.

Giles, H., Coupland, N. & Coupland, J. (1991). Accommodation theory: Communication, context, and consequence. In *Context of Accommodation: Developments in Applied Sociolinguistics*, ed. H. Giles, J. Coupland & N. Coupland, pp. 1–61. Cambridge University Press, Cambridge.

Ginsburg, G. P. & Kilbourne, B. K. (1988). Emergence of vocal alternation in mother–infant interchanges. *Journal of Child Language* **15**: 221–35.

Gleitman, L. (1990). The structural sources of word meaning. *Language Acquisition* **1**: 3–55.

Goldman-Rakic, P. S. (1987). Development of cortical circuitry and cognitive function. *Child Development* **58**: 601–22.

Goodsit, J. V., Morgan, J. L. & Kuhl, P. K. (1993). Perceptual

strategies in prelingual speech segmentation. *Journal of Child Language* **20**: 229–52.

Gottlieb, G. (1991a). Experiential canalization of behavioral development: Theory. *Developmental Psychology* **27**: 4–13.

Gottlieb, G. (1991b). Experiential canalization of behavioral development: Results. *Developmental Psychology* **27**: 35–9.

Haith, M. M., Bergman, T. & Moore, M. J. (1977). Eye contact and face scanning in early infancy. *Science* **198**: 853–5.

Hamilton, C. R. & Vermeire, B. A. (1988). Complementary hemisphere specialization in monkeys. *Science* **242**: 1691–4.

Hardy-Brown, K., Plomin, R. & DeFries, J. C. (1981). Genetic and environmental influences on the rate of communicative development in the first year of life. *Developmental Psychology* **17**: 704–17.

Hart, B. (1991). Input frequency and children's first words. *First Language* **11**: 289–300.

Hauser, M. D. (1992a). A mechanism guiding conversational turn-taking in vervet monkeys and rhesus macaques. *Topics in Primatology* vol. 1, *Human Origins*, pp. 235–48. Tokyo University Press, Tokyo.

Hauser, M. D. (1992b). Articulatory and social factors influence the acoustic structure of rhesus monkey vocalizations: A learned mode of production? *Journal of the Acoustical Society of America* **91**: 2175–9.

Hauser, M. D. (1993). Right hemisphere dominance for the production of facial expression in monkeys. *Science* **261**: 475–7.

Hauser, M. D. & Andersson, K. (1994). Left hemisphere dominance for processing vocalizations in adult, but not infant rhesus monkeys: Field experiments. *Proceedings of the National Academy of Sciences* **91**: 3946–8.

Heffner, H. & Heffner, R. (1995). Lesioning studies and their role in the perception of communicative vocalizations in nonhuman primate with special emphasis on macaques. In *Current Topics in Primate Vocal Communication*, ed. E. Zimmerman, J. Newman & U. Jürgens, pp. 207–19. Plenum, New York.

Hirsh-Pasek, K., Kemler Nelson, D. G., Jusczyk, P. W., Cassidy, K. W. Druss, B. & Kennedy, L. (1987). Clauses are perceptual units for young infants. *Cognition* **26**: 269–86.

Hobson, R. P., Ouston, J. & Lee, A. (1988). What's in a face? The case of autism. *British Journal of Psychology* **79**: 441–53.

Hurford, J. R. (1991). The evolution of the critical period for language acquisition. *Cognition* **40**: 159–201.

Huttenlocher, J., Haight, W., Bryk, A., Selzer, M. & Lyons, T. (1991). Early vocabulary growth: Relation to language input and gender. *Developmental Psychology* **27**: 236–48.

Jakobson, R. (1968). *Child Language, Aphasia, and Phonological*

Universals. Mouton, The Hague (originally published, 1941).

Jürgens, U. (1995). Neuronal control of vocal production in nonhuman and human primates. In *Current Topics in Primate Vocal Communication*, ed. E. Zimmerman, J. Newman & U. Jürgens, pp. 199–206. Plenum, New York.

Jusczyk, P. W., Cutler, A. & Redanz, N. J. (1993a). Infants' preference for the predominant stress patterns of English words. *Child Development* **64**: 675–87.

Jusczyk, P. W., Friederici, A. D., Wessels, J. M. I., Svenkerud, V. Y. & Jusczyk, A. M. (1993b). Infants' sensitivity to the sound patterns of native language words. *Journal of Memory and Language* **32**: 402–20.

Kalin, N. H. & Shelton, S. E. (1989). Defensive behaviors in infant rhesus monkeys: Environmental cues and neurochemical regulation. *Science* **243**: 1718–21.

Kalin, N. H., Shelton, S. E. & Takahashi, L. K. (1991). Defensive behaviors in infant rhesus monkeys: Ontogeny and context-dependent selective expression. *Child Development* **62**: 1175–83.

Karmiloff-Smith, A., Bellugi, U., Klima, E., Grant, J. & Baron-Cohen, S. (1995). Is there a social module? Language, face processing, and theory of mind in individuals with Williams syndrome. *Journal of Cognitive Neuroscience* **7**: 196–208.

Keating, C. F. & Keating, E. G. (1982). Visual scan patterns of rhesus monkeys viewing faces. *Perception* **11**: 211–19.

Keating, C. F. & Keating, E. G. (1993). Monkeys and mug shots: Cues used by rhesus monkeys (*Macaca mulatta*) to recognize a human face. *Journal of Comparative Psychology* **107**: 131–9.

Kenney, M. D., Mason, W. A. & Hill, S. D. (1979). Effects of age, objects, and visual experience on affective responses of rhesus monkeys to strangers. *Developmental Psychology* **15**: 176–84.

Kerswill, P. (1994). Babel in Buckinghamshire? Pre-school children acquiring accent features in the new town of Milton Keynes. In *Nonstandard Varieties of Language*, ed. G. Melchers, pp. 64–84. Acta Universitatis Stockholmiensis. Almqvist & Wiksell, Stockholm.

Kling, A. S. & Brothers, L. A. (1992). The amygdala and social behavior. In *The Amygdala: Neurobiological Aspects of Emotion, Memory, and Mental Disorders*, ed. J. Aggleton, pp. 353–77. John Wiley & Sons, New York.

Koopmans-van Beinum, F. J. (1993). Cyclic effects of infant speech perception, early sound production, and maternal speech. *Institute of Phonetic Sciences* **17**: 65–78.

Kraemer, G. W. (1992). A psychobiological theory of attachment. *Behavioral and Brain Sciences* **15**: 493–541.

Kugiumutzakis, G. (1993). Intersubjective vocal imitation in early mother–infant interaction. In *New Perspectives in Early Communicative Development*, ed. J. Nadel & L. Camaioni, pp. 23–47. Routledge, New York.

Kuhl, P. K. & Meltzoff, A. N. (1982). The bimodal perception of speech in infancy. *Science* **218**: 1138–41.

Kuhl, P. K. & Meltzoff, A. N. (1988). Speech as an intermodal object of perception. In *Perceptual Development in Infancy*, ed. A. Yonas, pp. 235–66. Minnesota Symposia on Child Psychology. Erlbaum, Hillsdale, NJ.

Kuhl, P, K., Williams, K. A., Lacerda, F., Stevens, K. N. & Lindblom, B. (1992). Linguistic experience alters phonetic perception in infants by 6 months of age. *Science* **255**: 606–8.

Labov, W. (1974). Linguistic change as a form of communication. In *Human Communication: Theoretical Explorations*, ed. A. Silverstein, pp. 221–56. Erlbaum, Hillsdale, NJ.

Langdell, T. (1978). Recognition of faces: An approach to the study of autism. *Journal of Child Psychology and Psychiatry* **19**: 255–68.

Larson, C. R. (1988). Brain mechanisms involved in the control of vocalization. *Journal of Voice* **2**: 301–11.

Larson, C. R. & Kistler, M. K. (1986). The relationship of periaqueductal gray neurons to vocalization and laryngeal EMG in the behaving monkey. *Experimental Brain Research* **63**: 596–606.

LeMay, M. & Geschwind, N. (1975). Hemispheric differences in the brains of great apes. *Brain and Behavior Evolution* **11**: 48–52.

Lieberman, P. (1984). *The Biology and Evolution of Language*. Harvard University Press, Cambridge, MA.

Locke, J. L. (1992). Neural specializations for language: A developmental perspective. *Seminars in Neuroscience* **4**: 425–31.

Locke, J. L. (1993). *The Child's Path to Spoken Language*. Harvard University Press, Cambridge, MA.

Locke, J. L. (1994a). Phases in the development of linguistic capacity. In *Evolution and Neurology of Language*, ed. D. C. Gajdusek, G. M. McKhann & C. L. Bolis, pp. 26–34. Discussions in Neuroscience **10**. Elsevier, Amsterdam.

Locke, J. L. (1994b). Phases in the child's development of language. *American Scientist* **82**: 436–45.

Locke, J. L. (1996). Why do infants begin to speak? Language as an unintended consequence. *Journal of Child Language*, in press.

Locke, J. L. & Pearson, D. M. (1992). Vocal learning and the emergence of phonological capacity: A neurobiological approach. In *Phonological Development: Models, Research, Implications*, ed. C. Ferguson, L. Menn, & C. Stoel-Gammon, pp. 91–130. York Press, Timonium, MD.

MacKain, K. S. (1984). Speaking without a tongue. *Journal of the National Student Speech Language Hearing Association* **12**: 46–71.

Mandel, D. R., Jusczyk, P. W. & Pisoni, D. B. (1995). Infants' recognition of the sound patterns of their own names. *Psychological Science* **6**: 314–17.

Marler, P. & Peters, S. (1982). Long-term storage of learned birdsongs prior to production. *Animal Behaviour* **30**: 479–82.

Marler, P. & Tenaza, R. (1977). Signaling behavior of apes with special reference to vocalization. In *How Animals Communicate*, ed. T. A. Sebeok, pp. 965–1033. Indiana University Press, Bloomington, IN.

Masataka, N. (1992). Pitch characteristics of Japanese maternal speech to infants. *Journal of Child Language* **19**: 213–23.

Maurus, M., Barclay, D. & Striet, K.-M. (1988). Acoustic patterns common to human communication and communication between monkeys. *Language and Communication* **8**: 87–94.

Maurus, M., Kühlmorgen, B., Wiesner, E., Barclay, D. & Streit, K.-M. (1987). Interrelations between structure and function in the vocal repertoire of Saimiri: Asking the monkeys themselves where to split and where to lump. *European Archives of Psychiatric and Neurological Sciences* **236**: 35–9.

Mehler, J., Jusczyk, P., Lambertz, G., Halsted, N., Bertoncini, J. & Amiel-Tison, C. (1988). A precursor of language acquisition in young infants. *Cognition* **29**: 143–78.

Mendelson, M. J., Haith, M. M. & Goldman-Rakic, P. S. (1982). Face scanning and responsiveness to social cues in infant rhesus monkeys. *Developmental Psychology* **18**: 222–8.

Moon, C., Cooper, R. P. & Fifer, W. P. (1993). Two-day olds prefer their native language. *Infant Behavior and Development* **16**: 495–500.

Moran, G., Krupka, A., Tutton, A. & Symons, D. (1987). Patterns of maternal and infant imitation during play. *Infant Behavior and Development* **10**: 477–91.

Neufeld, G. (1978). On the acquisition of prosodic and articulatory features in adult second language learning. *Canadian Modern Language Review* **34**: 163–74.

Ninio, A. (1992) The relation of children's single word utterances to single word utterances in the input. *Journal of Child Language* **19**: 87–110.

Ninio, A. (1993). On the fringes of the system: Children's acquisition of syntactically isolated forms at the onset of speech. *First Language* **13**: 291–313.

Ninio, A. & Snow, C. E. (1996). *Pragmatic Development*. Westview, New York, in press.

Ochberg, F. M. (1978). The victim of terrorism. *The Practitioner* **1**: 293–302.

Oller, D. K. & Eilers, R. E. (1988). The role of audition in infant babbling. *Child Development* **59**: 441–9.

Oyama, S. (1976). A sensitive period for the acquisition of a non-native phonological system. *Journal of Psycholinguistic Research* **5**: 261–85.

Papoušek, M. (1994). *Vom ersten Schrei zum ersten Wort: Anfänge der Sprachentwicklung in der vorsprachlichen Kommunikation.* Verlag Hans Huber, Bern.

Papoušek, M. & Papoušek, H. (1989). Forms and functions of vocal matching in interactions between mothers and their precanonical infants. *First Language* **9**: 137–58.

Payne, A. (1980). Factors controlling the acquisition of the Philadelphia dialect by out-of-state children. In *Locating Language in Time and Space*, ed. W. Labov. Academic Press, New York.

Pearson, B., Navarro, A. & Gathercole, V. (1994). Assessment of phonetic differentiation in bilingual learning infants. Paper presented to Boston University Child Language Conference, November.

Perrett, D. I. & Mistlin, A. J. (1990). Perception of facial characteristics by monkeys. In *Comparative Perception: Complex Signals*, vol. 2, ed. W. C. Stebbins, & M. A. Berkley, pp. 187–216. Wiley, New York.

Petersen, M. R., Beecher, M. D., Zoloth, S. R., Moody, D. B. & Stebbins, W. C. (1978). Neural lateralization of species-specific vocalizations by Japanese macaques (*Macaca fuscata*). *Science* **202**: 324–7.

Petersen, M. R., Zoloth, S. R., Beecher, M. D., Green, S., Marler, P. R., Moody, D. B. & Stebbins, W. C. (1984). Neural lateralization of vocalizations by Japanese macaques: communicative significance is more important than acoustic structure. *Behavioral Neuroscience* **98**: 779–90.

Provence, S. & Lipton, R. C. (1962). *Infants in Institutions: A Comparison of Their Development with Family-Reared Infants During the First Year of Life*. International Universities Press, New York.

Querleu, D., Renard, X. & Versyp, F. (1985). Vie sensorielle du foetus. In *L'environnement de la naissance*, ed. G. Levy & M. Tournaire, pp. 114–36. Vigot, Paris.

Rajecki, D. W., Lamb, M. E. & Obmascher, P. (1978). Toward a general theory of infantile attachment: A comparative review of aspects of the social bond. *Behavioral and Brain Sciences* **3**: 417–64.

Redican, W. F. (1975). Facial expressions in nonhuman primates. In *Primate Behavior: Developments in Field and Laboratory Research*, ed. L. A. Rosenblum, vol. 4, pp. 104–94. Academic Press, New York.

Ross, E. D. (1981). The aprosodias: Functional-anatomic

organization of the affective components of language in the right hemisphere. *Archives of Neurology* **38**: 561–9.

Ross, M., Duffy, R. J., Cooker, H. S. & Sargeant, R. L. (1973). Contribution of the lower audible frequencies to the recognition of emotions. *American Annals of the Deaf* **118**: 37–42.

Sackett, G. P. & Ruppenthal, G. C. (1974). Some factors influencing the attraction of adult female macaque monkeys to neonates. In *The Effect of the Infant on its Caregiver*, ed. M. Lewis & L. A. Rosenblum, pp. 163–85. Wiley, New York.

Savage-Rumbaugh, E. S., Murphy, J., Sevcik, R. A., Brakke, K. E., Williams, S. L. & Rumbaugh, D. M. (1993). Language comprehension in ape and child. *Monographs of the Society for Research and Child Development* **58**.

Saville-Troike, M. (1988). Private speech: Evidence for second language strategies during the "silent period." *Journal of Child Language* **15**: 567–90.

Scherer, K. R. & Giles, H. (eds.). (1979). *Social Markers in Speech*. Cambridge University Press, Cambridge.

Scherer, K. R., Koivumaki, J. & Rosenthal, R. (1972). Minimal cues in the vocal communication of affect: Judging emotions from content-masked speech. *Journal of Psycholinguistic Research* **1**: 269–85.

Scott, J. P. (1962). Critical periods in behavioral development. *Science* **138**: 949–58.

Seay, B., Alexander, B. K. & Harlow, H. F. (1964). Maternal behavior of socially deprived rhesus monkeys. *Journal of Abnormal Social Psychology* **69**: 345–54.

Shriberg, L. D., Gruber, F. A. & Kwiatkowski, J. (1994). Developmental phonological disorders. III. Long-term speech-sound normalization. *Journal of Speech and Hearing Research* **37**: 1151–77.

Snow, C. (1987). Relevance of the notion of a critical period to language acquisition. In *Sensitive Periods in Development: Interdisciplinary Perspectives*, ed. M. H. Bornstein, pp. 183–209. Erlbaum, Hillsdale, NJ.

Snow, C. (1995). Issues in the study of input: Fine-tuning, universality, individual and developmental differences, and necessary causes. In *NETwerken: Bijdragen van het vijfde NET Symposium*, ed. B. MacWhinney & P. Fletcher, Antwerp Papers in Linguistics **74**, pp. 5–17. University of Antwerp, Antwerp.

Snow, C. E. & Hoefnagel-Hohle, M. (1977). Age differences in the pronunciation of foreign sounds. *Language and Speech* **20**: 357–65.

Snow, C. E. & Hoefnagel-Hohle, M. (1978). Critical period for language acquisition: Evidence from second language learning. *Child Development* **49**: 1263–79.

Snowdon, C. T., Coe, C. L. & Hodun, A. (1985). Population recognition of infant isolation peeps in the squirrel monkey. *Animal Behaviour* **33**: 1145–51.

Stoel-Gammon, C. (1988). Prelinguistic vocalizations of hearing-impaired and normally hearing subjects: A comparison of consonantal inventories. *Journal of Speech and Hearing Disorders* **53**: 302–15.

Stoel-Gammon, C. & Otomo, K. (1986). Babbling development of hearing-impaired and normally hearing subjects. *Journal of Speech and Hearing Disorders* **51**: 33–41.

Super, C. M. (1991). Developmental transitions of cognitive functioning in rural Kenya and metropolitan America. In *Brain Maturation and Cognitive Development: Comparative and Cross-cultural Perspectives*, ed. K. R. Gibson & A. C. Petersen, pp. 225–51. Aldine de Gruyter, New York.

Sutton, D., Larson, C., Taylor, E. M. & Lindeman, R. C. (1973). Vocalization in rhesus monkeys: Conditionability. *Brain Research* **52**: 225–31.

Sutton, D., Samson, H. H. & Larson, C. R. (1978). Brain mechanisms in learned phonation of *Macaca mulatta*. In *Recent Advances in Primatology*, ed. D. J. Chivers & J. Herbert, pp. 769–84. Academic Press, London.

Sutton, D., Trachy, R. E. & Lindeman, R. C. (1981). Primate phonation: Unilateral and bilateral cingulate lesion effects. *Behavioral and Brain Research* **3**: 99–114.

Symmes, D. & Biben, M. (1985). Maternal recognition of individual infant squirrel monkeys from isolation call playbacks. *American Journal of Primatology* **9**: 39–46.

Takagi, N. & Mann, V. (1996). The limits of extended naturalistic exposure on the perceptual mastery of English /r/ and /l/ by adult Japanese learners of English. *Applied Psycholinguistics*, in press.

Tomasello, M. & Barton, M. (1994). Learning words in nonostensive contexts. *Developmental Psychology* **30**: 639–650.

Tomasello, M. & Farrar, J. (1986). Joint attention and early language. *Child Development* **57**: 1454–63.

Tomasello, M., Savage-Rumbaugh, S. & Kruger, A. C. (1993). Imitative learning of actions on objects by children, chimpanzees, and enculturated chimpanzees. *Child Development* **64**: 1688–705.

Trachy, R. E., Sutton, D. & Lindeman, R. C. (1981). Primate phonation: Anterior cingulate lesion effects on response rate and acoustical structure. *American Journal of Primatology* **1**: 43–55.

Trevarthen, C. (1979). Instincts for human understanding and for cultural cooperation: Their development in infancy. In *Human Ethology: Claims and Limits of a New Discipline*, ed. M. von Cranach, K. Foppa, W. Lepenies & D. Ploog, pp. 530–71. Cambridge University Press, Cambridge.

Veneziano, E. (1988). Vocal–verbal interaction and the construction of early lexical knowledge. In *The Emergent Lexicon: The Child's Development of a Linguistic Vocabulary*, ed. M. D. Smith & J. L. Locke, pp. 109–42. Academic Press, New York.

Waser, P. M. (1977). Individual recognition, intragroup cohesion and intergroup spacing: Evidence from sound playback to forest monkeys. *Behaviour* **60**: 28–74.

Watson, P. (1978). *War on the Mind*. Hutchinson, London.

Werker, J. F. & Pegg, J. E. (1992). Infant speech perception and phonological acquisition. In *Phonological Development: Models, Research, Implications*, ed. C. A. Ferguson, L. Menn & C. Stoel-Gammon, pp. 285–311. York Press, Timonium, MD.

Werker, J. F. & Polka, L. (1993). Developmental changes in speech perception: New challenges and new directions. *Journal of Phonetics* **21**: 83–101.

White, S. (1970). Some general outlines of the matrix of developmental changes between five and seven years. *Bulletin of the Orton Society* **20**: 41–57.

15 The resilience of language in humans

SUSAN GOLDIN-MEADOW

INTRODUCTION

In 1974, Ernst Mayr published a now classic paper on the distinction between innate and acquired characteristics. In this work, Mayr broke away from some of the more confining features of traditional accounts of innateness, and proposed a dimension along which behaviors might be expected to vary with respect to innateness. Mayr proposed a distinction between "closed" and "open" programs[1] – a program that does not allow appreciable modifications during the lifespan of its owner is a "closed" program, while a program that does allow for the effects of additional input is "open." Since it seems unlikely that any developmental program can be completely closed, Wimsatt (1986) has suggested that Mayr's notion of a closed program may be most fruitfully viewed as a relative one – "relative to the period of time of development under investigation, and the class of inputs being investigated, and probably also to the environment and the prior state of the developing phenotype" (Wimsatt 1986, p. 203). In Wimsatt's terms, a closed developmental program is one which is canalized with respect to the relevant inputs.

Mayr's classificatory schemes distinguished two types of behavior: A behavior is considered *communicative* if it is directed toward a recipient who is capable of responding with behavior of its own, and *noncommunicative* if it is directed toward a "recipient" that is passive and does not itself react (e.g., behaviors involved in selecting a habitat or seeking food). Mayr further subdivided communicative behavior into behavior directed toward a member of one's own species (intraspecies behavior) and behavior directed toward an individual of another species (interspecies behavior). He then suggested that these three types of behavior vary with respect to innateness – intraspecies communicative behavior is more likely to be governed by a relatively closed program than is interspecies communicative behavior, which, in turn, is more likely to be governed by a relatively closed program than is noncommunicative behavior. According to Mayr, a program is likely to be relatively closed if it governs the formal signals and appropriate responses to those signals that are used by individuals within a species; a relatively closed program guarantees the development of these intraspecies signals, which are most effective when they are predictable. In contrast, a program is likely to be relatively open if it governs behavior with respect to natural resources, which are, in large part, unpredictable; a relatively open program thus permits an opportunistic adjustment to changes in the environment.

Mayr's illustrations of intraspecies communicative behavior were derived from nonhuman species providing three rather stereotyped categories of actions – courtship displays between males and females, interactions between parents and offspring, and threat displays between males of the same species. My purpose here is to deal with the question of open versus closed intraspecies behavior on quite another level altogether. The question I address is whether communicative behavior among humans – in particular, human language – is governed by a relatively closed or open program.

At first glance, it seems obvious that human language is very open to environmental input. A child who is exposed to Swahili learns Swahili, while a child who is exposed to German learns German. Perhaps even more strikingly, a child who is exposed to language processed by the hand

---

[1] Mayr (1974) used the term "genetic" program. Despite the terminology, genetic programs (either open or closed) are not parts of the genome for Mayr (1974). They are developmental programs and as such are properties of the phenotype; for a discussion, see Wimsatt 1986, p. 202).

and eye rather than the mouth and ear (e.g., a conventional sign language such as American Sign Language) will learn that language as effortlessly and naturally as a child learns any spoken language (Newport & Meier 1985). Thus, contrary to what Mayr might have predicted, human language appears to be relatively open, not only with respect to the particular language learned but also with respect to the modality in which the language is processed.

Nevertheless, some aspects of human language (and human language learning) may not be entirely environment sensitive. Despite wide variety in cultural environments, all human cultures (no matter how isolated) have a language. Moreover, although progress has been made in teaching aspects of human language to other animals (for examples, see Roitblat et al. 1993), language learning comes naturally to humans in a way that it does not to other animals. Thus, there may be aspects of language whose development is almost inevitable among humans, despite the fact that these aspects are not inevitable in communication in general. The development of these aspects of human language can be thought of as governed by a relatively closed program. How might we begin to discover these particular aspects of language?

One approach often taken by ethologists to distinguish behaviors along the open/closed dimension is to determine whether a behavior will be developed even when an organism is removed from its typical environment, the environment in which the organism and its developmental program evolved. If so, the behavior can be said to be "resilient" across these atypical environments and thus "buffered," canalized, or relatively closed against certain kinds of experience (cf. Goldin-Meadow 1982; Alcock 1988).[2] The notion that a behavior is invariant across different experiential histories has been called "developmental fixity" by Wimsatt (1986), who considers the idea central to the innate/acquired distinction. The term developmental fixity is not used by Wimsatt to refer to the fact that there may be a point in development after which the behavior cannot be altered (in other words, the notion

is not synonymous with the term "crystallization" as used in the literature on bird song and monkey calls, cf. Snowdon et al. Chapter 12). Rather, the term implies that there is an inevitability to the development of this behavior in members of the species – the behavior is overdetermined in the organism so that one is unlikely to find a viable member of the species who does not exhibit it. On this view, observing language development across environments that vary with respect to the learning conditions children typically experience should provide a technique for exploring the developmental fixity of language and for isolating those aspects of human language that are resilient.

At one end of the continuum, language development has been explored in circumstances in which children have been raised by animals with no human contact at all (Brown 1958) or by humans who have treated children inhumanely, depriving them of physical, social, and linguistic stimulation (Skuse 1988). Perhaps not surprisingly, children do not develop human language under such extreme developmental conditions. Thus, there appear to be some limits on the resilience of language. Human contact seems to be essential for language development in children – and those humans, at a minimum, must be humane. But contact with humans involves many factors, not the least of which is exposure to a model for language. My purpose here is to explore whether the humans who interact with a child must present a model for language in order for that child to develop language.

This question is a difficult one to address simply because most children if treated humanely are also exposed naturally to the language used in the social world surrounding them. There are, however, circumstances in which children are raised in supportive households but are unable to make use of the language model that surrounds them – deaf children whose hearing losses are so severe that they cannot naturally make use of input from a spoken language, and who are born to hearing parents who know no sign language and have not yet placed their children in an environment where they would be exposed to a signed language. Such children are, to all intents and purposes, deprived of a usable model for language – although, importantly, they are not deprived of other aspects of human social interaction. The question is whether these deaf children, despite this particular and rather extreme variation from the language-learning conditions children

---

[2] I follow Wimsatt (1986) in identifying resilience, or developmental fixity in Wimsatt's terms, as a criterion for innateness, whether or not such fixity is genetically determined. In other words, what is crucial is that a developmental outcome be invariant over a wide range of environments – the mechanism for such fixity is left open for the moment and is *not* assumed to be genetic.

typically experience (i.e., the lack of a usable linguistic model), develop a communication system that resembles human language. Such an outcome is consistent with the view that language development is buffered against many variations in environmental conditions and, in this sense, is governed by a relatively closed program.

My colleagues and I have studied ten deaf children of hearing parents in two American cities, six in the Philadelphia area and four in the Chicago area. Not surprisingly, these children, whose hearing losses were large, did not use spoken language to communicate. However, each of the ten children did communicate with the hearing individuals in their worlds and did so using gesture (Feldman *et al.* 1978; Goldin-Meadow & Mylander 1984, 1990a). Moreover, the gestures that each of the children used resembled many – but not all – aspects of the spoken and signed language systems developed by children learning language from conventional linguistic models. It is these aspects of language that are resilient in the face of at least one relatively extreme variation in language-learning circumstances, and can be considered relatively closed along Mayr's continuum.

I begin by describing the properties of language identified as resilient on the basis of our observations of the deaf children. To better understand the nature of this resilience, I then explore the environmental conditions under which the deaf children's gesture systems were developed. I focus first on the types of gesture that the deaf children's hearing parents produced as they spoke to their children to determine whether the parents' gestures might have provided a model (albeit an idiosyncratic model rather than a conventional one) for the language-like structure found in the deaf children's gestures.

I then focus on two other aspects of social interaction that might have influenced the deaf children's gesture systems. (a) I explore how the hearing parents responded to their deaf children's gestures to determine whether the parents might have shaped the language-like structure in their children's gestures by responding differentially to those gestures. (b) I explore how the hearing parents structured their interactions with their deaf children to determine whether the interactions themselves might have contributed to the structure of the deaf children's gesture systems. To do so, I also explore the development of gesture systems by deaf children in another culture (China) where hearing parents have been reported to

structure their interactions with their children quite differently from American parents. To the extent that the gesture systems of the deaf children in the two cultures contain similar properties despite differences in the social interactions the children experience with their parents, an increasingly powerful argument can be made for the resilience of these properties of language.

Finally, I briefly review a number of other studies that have taken advantage of variations that occur naturally in language-learning conditions – variations from the norm in terms of environmental conditions, as well as organic conditions – in order to investigate the boundary conditions under which language development is possible. These studies, in conjunction with the findings presented here, suggest that language development is resilient across a wide range of both environmental and organic conditions.

BACKGROUND ON DEAFNESS AND THE CHILDREN IN OUR STUDIES

Deaf children born to deaf parents and exposed from birth to a conventional sign language such as American Sign Language (ASL) acquire that language naturally; that is, these children progress through stages in acquiring sign language similar to those of hearing children acquiring a spoken language (Newport & Meier 1985). However, 90% of deaf children are not born to deaf parents who could provide early exposure to a conventional sign language. Rather, they are born to hearing parents who, quite naturally, tend to expose their children to speech (Hoffmeister & Wilbur 1980). Unfortunately, it is extremely uncommon for deaf children with severe to profound hearing losses to acquire the spoken language of their hearing parents naturally; that is, without intensive and specialized instruction. Even with instruction, deaf children's acquisition of speech is markedly delayed when compared either to the acquisition of speech by hearing children of hearing parents, or to the acquisition of sign by deaf children of deaf parents. By age five or six years, and despite intensive early training programs, the average profoundly deaf child has limited linguistic skills in speech (Meadow 1968; Conrad 1979; Mayberry 1992). Moreover, although many hearing parents of deaf children send their children to schools in which one of the manually coded systems of English is taught, some hearing parents send their deaf children to "oral" schools in which sign systems are neither taught nor encouraged; thus, these deaf children are not likely to receive input in a conventional sign system.

The children in our studies are severely (70–90 dB bilateral hearing loss) to profoundly (>90 dB bilateral hearing loss) deaf, and their hearing parents chose to educate them using an oral method. At the time of our observations, the children ranged in age from 14 months to 4 years 10 months and had made little progress in oral language, occasionally producing single words but never combining those words into sentences. In addition, at the time of our observations, the children had not been exposed to ASL or to a manual code of English. As preschoolers in oral schools for the deaf, the children spent very little time with the older deaf children in the school who might have had some knowledge of a conventional sign system (i.e., the preschoolers attended school only for a few hours a day and were not on the playground at the same time as the older children). In addition, the children's families knew no deaf adults socially and interacted only with other hearing families, typically those with hearing children.[3] Under such inopportune circumstances, these deaf children might be expected to fail to communicate at all, or perhaps to communicate only in nonsymbolic ways. This turns out not to be the case.

Studies of deaf children of hearing parents in general have shown that these children spontaneously use gestures (referred to as "home signs") to communicate, even if they are not exposed to a conventional sign language model (Tervoort 1961; Lenneberg 1964; Fant 1972; Moores 1974). Given a home environment in which family members communicate with each other through many different channels, one might expect that the deaf child would exploit the accessible modality (the manual modality) for the purposes of communication. The question is whether the gestures the deaf child uses to communicate are structured in language-like ways. In the next section, I

describe the properties of the deaf child's gestures that can be considered language-like; that is, the properties of language that appear to be resilient and therefore governed by a relatively closed or canalized developmental program.

It is important to stress that the behaviors upon which studies of the deaf children focus are grounded in a communicative context. Indeed, in order for a manual movement to be considered a gesture and included in our database, it must be produced with the intent to communicate. The difficulty lies in discriminating acts that communicate indirectly (e.g., pushing a plate away, which indicates that the eater has had enough) – acts that we do not want to include in our study – from those acts whose sole purpose is to communicate symbolically (e.g., a "stop-like" movement of the hands produced in order to suggest to the host that another helping is not necessary). Lacking a generally accepted behavioral index of deliberate or intentional communication, we decided that a communicative gesture must meet the following two criteria (Feldman et al. 1978; Goldin-Meadow & Mylander 1984). First, the movement must be directed to another individual. This criterion is satisfied if the child attempts to establish eye contact with the communication partner. Since manual communication cannot be received unless the partner is looking, checking for a partner's visual attention is a good sign that the child intended the movement to be seen. Second, the movement must not be a direct act on the other person or relevant object. As an example, if the child attempts to twist open a jar, that act is *not* considered a gesture for "open," even if the act does inform others that help is needed in opening the jar. If, however, the child makes a twisting motion in the air, with eyes first on the other person to establish contact, the movement is considered a communicative gesture. Thus, as West *et al.* (Chapter 4) advocate for the study of bird communication (as opposed to bird song), we examine the deaf child's behaviors within a communicative context. Indeed, behaviors are included in the study only if they are produced with the intent to communicate.

THE RESILIENT PROPERTIES OF LANGUAGE

The self-styled gesture systems of the deaf children in our studies were indexical and iconic systems of representation. The "lexicon" of the gesture systems contained both

---

[3] One of the primary reasons we were convinced that the children in our studies had had no exposure to a conventional sign system at the time of our observations was that they did not know even the most common lexical items of ASL or Signed English (i.e., when a native deaf signer reviewed our tapes, she found no evidence of any conventional signs; moreover, when we informally presented to the children common signs such as those for mother, father, boy, girl, dog, we found that they neither recognized nor understood any of these signs).

pointing gestures and characterizing gestures. Pointing gestures were used to index or indicate objects, people, places, and the like in the surroundings. Characterizing gestures were stylized pantomimes whose iconic forms varied with the intended meaning of each gesture (e.g., a C-hand twisted in the air to indicate that someone was twisting open a jar). Gestures of this sort, particularly pointing gestures but also some characterizing gestures, are produced by hearing children (Acredolo & Goodwyn 1988; Butcher *et al.* 1991). However, the deaf children's use of these getures is unique in that their gestures fit into a structured *system* (see below), while hearing children's gestures do not (Goldin-Meadow & Morford 1985; Morford & Goldin-Meadow 1992).

All natural languages, be they signed or spoken, have structure at more than one linguistic level (e.g., at the level of the sentence, word, and phoneme) and the deaf children's gesture systems are no exception. The gestures of the ten children whose systems were examined for sentence-level structure were found to have structure across gestures within a string (Goldin-Meadow & Feldman 1977; Feldman *et al.* 1978; Goldin-Meadow & Mylander 1984), and the gestures of the four children whose systems were examined for word-level structure were found to have structure within a gesture as well as across them (Goldin-Meadow *et al.* 1995). Thus, in this important respect, the deaf children's gesture systems resembled natural language. This finding suggests that more than one level of structure is a resilient or relatively canalized property of language. I describe, in turn, sentence-level and word-level structure in the deaf child's gesture system.

Combining words into sentences: Syntactic structure

The first important point to note is that each of the ten children combined their gestures into strings. Unlike the signals produced by other animals, which tend to be used as single units not combined with others signals (cf. Seyfarth & Cheney, Chapter 13), the deaf children's gestures were frequently combined with one another to create new meanings. For example, a deaf child might combine a point at a grape with an EAT gesture to comment on the fact that grapes can be eaten, and then later combine the EAT gesture with a point at the experimenter to invite the experimenter to have lunch with the family.

Moreover, and equally important, the gesture strings that the deaf children produced functioned in a number of respects like the sentences of early child language. On this basis, these strings warrant the label "sentence." In particular, the children produced gesture strings characterized by two types of *surface regularity*: (a) regularity in terms of which elements were produced and deleted in a string, and (b) regularity in terms of where in the string those elements were produced.

As an example of the first type of surface regularity, although the children did not produce gestures for all of the possible thematic roles that could be conveyed within a sentence, they were not haphazard in their selection of which roles to convey in gesture. The children were more likely to produce a gesture for the patient (e.g., the cheese, in a sentence about eating) than to produce a gesture for the actor (the mouse). In addition, the children produced gestures for the intransitive actor (e.g., the mouse in a sentence describing a mouse running to his hole) as often as they produced gestures for the patient (e.g., the cheese in a sentence describing a mouse eating cheese), and far more often than they produced gestures for the transitive actor (e.g., the mouse in a sentence describing a mouse eating cheese). In this way, the likelihood of production served to systematically distinguish among thematic roles and thus mark those roles – an important function of grammatical devices. It is also worth noting that the particular pattern found in the deaf children's gestures – patients and intransitive actors marked in the same way and both different from transitive actors – is an analog of a structural case-marking pattern found in naturally occurring human languages (in particular, ergative languages, cf. Silverstein 1976; Dixon 1979).

As an example of the second type of surface regularity, the children also distinguished among the thematic roles they did express by placing the gesture for a given role in a particular position in a gesture string; that is, the gestures the children produced within their strings were not produced in haphazard order but rather appeared to follow a small set of gesture order regularities (Goldin-Meadow 1979; Goldin-Meadow & Mylander 1984). Gesture order regularities describe where the gesture for a particular thematic role is likely to appear in a gesture string, e.g., in a string used to comment on the fact that the child intended to throw a toy grape, the child first produced a gesture for the grape, the *patient* (typically a pointing gesture at the grape but, at times, a characterizing gesture for the grape) before producing a gesture for the *act*, throw (a characterizing gesture). In general, the gesture for the object playing a patient role tended to *precede* the gesture for the act. As a second example of a gesture order, gestures for an object playing the role of recipient or goal tended to *follow* gestures for the act; e.g., in a string used to request that an object be moved to a puzzle, the child produced a gesture for the *act*, transfer (a characterizing gesture), before producing a gesture for the *recipient*, puzzle (a pointing gesture).

In addition to these regularities at the surface level, the children's gesture strings were also organized at *underlying levels*. Each string expressed one or more frames composed of a predicate and one, two, or three arguments (Feldman *et al.* 1978, p. 385–8; Goldin-Meadow 1985, pp. 215–19). For example, all of the children produced "transfer" or "give" predicates with an inferred frame containing three arguments – the actor, patient, and recipient[4] (e.g., *you/sister* give *duck* to *her/Susan*). The children also produced two types of two-argument predicates: transitive predicates such as "drink" with a frame containing the actor and patient (e.g., *you/Susan* drink *coffee*), and intransitive predicates such as "go" with a frame containing the actor and recipient (e.g., *I/child* go *downstairs*). Finally, the children produced predicates such as "sleep" or "dance" with a one-argument frame containing only the actor (e.g., *he/father* sleep).

The children frequently concatenated more than one predicate frame within the bounds of a single sentence – that is, they produced complex as opposed to simple sentences, thus demonstrating the important property of recursion in their gesture systems (e.g., a "climb" gesture, followed by a "sleep" gesture, followed by a point at a horse, to comment on the fact that the horse in a picture climbed up the house and then slept). Recursion gives language its generative capacity and is found in all natural language systems. Importantly, when the children concatenated more than one predicate frame within a single sentence, they did so systematically, allocating one

"slot" in underlying structure to the arguments and predicates playing a role in both frames (e.g., "he/horse" is assigned only one slot in underlying structure in the above sentence in which the horse played the actor role in both of the predicates of the concatenated frames, *he/horse* climbs and sleeps"; Goldin-Meadow 1982).

Thus, the deaf children conjoined the gestures they produced into strings characterized by surface regularities (regularities in likelihood of production and deletion and in gesture order) as well as regularities at underlying levels (the predicate frames underlying each simple and complex gesture sentence). The gesture strings could therefore be said to conform to a syntax, albeit a simple one.

Decomposing the word: Morphological structure

The deaf children's gestures not only formed parts of longer sentence-units but they themselves were made up of smaller parts. For example, to request the experimenter to lay a penny down flat on a toy, one deaf child produced a downward motion with his hand shaped like an O. In itself this could be a global gesture presenting the shape and trajectory as an unanalyzed whole. The experimenter pretended not to understand and, after several repetitions, the child factored the gesture into its components: first he statically held up the gesture for a round object (the O handshape) and then, quite deliberately and with his hand no longer in the O shape but exhibiting a flat palm, made the trajectory for downward movement. The original gesture was thus decomposed into two elements. This example implies the presence of a system of linguistic segments in which the complex meaning of "round-thing-moving-downward" is broken into components and the components combined into a gesture. Although the experimenter's feigned lack of understanding was undoubtedly important in getting the child to decompose his gesture at that particular moment, the point I want to stress here is that when the child did break his gesture into parts, those parts were elements of a wider system – one that accounted for virtually all of the gestures that this child produced (Goldin-Meadow & Mylander 1990b).

Thus, this child had devised a morphological system in which each gesture was a complex of simpler gesture elements (Goldin-Meadow & Mylander 1990b). Systematic compositionality of gestures within a system

---

[4] I use the term "recipient" to refer to the destination of predicates such as "go" or "put" whether that destination is animate ("go to Mother") or inanimate ("go to the table"). The children did not appear to distinguish animate from inanimate recipients in their gesture systems (i.e., both tended to occupy the same position in a two-gesture sentence).

of contrasts is crucial evidence of segmentation and combination. As an example of how this child's gestures formed a system of contrasts, a CMedium handshape (the hand shaped in a C with the fingers 1–3 inches (2–8 cm) from the thumb) meant "handle an object 2–3 inches (5–8 cm) wide," and a Revolve motion meant "rotate around an axis." When combined, these two components created a gesture whose meaning was a composite of the two meanings – "rotate an object 2–3 inches wide" (e.g., twist a jar lid). When the same CMedium handshape was combined with a different motion, a Short Arc (meaning "reposition"), the resulting combination had a predictably different meaning – "change the position of an object 2–3 inches wide" (e.g., tilt a cup). As a result, the child's gestures can be said to conform to a framework or system of contrasts.

The gesture systems of four children have been analyzed at this level (Goldin-Meadow *et al.* 1995), and all four children were found to produce gestures that could be characterized by paradigms of handshape and motion combinations. Thus, each child was found:

1. to use a limited set of discrete handshape and motion forms, i.e., the forms were categorical rather than continuous;

2. to consistently associate each handshape or motion form with a particular meaning (or set of meanings) throughout the corpus, i.e., each form was meaningful (and, in this sense, the children's handshape and motion units formed part of a morphological, as opposed to a phonological, system);

3. to produce most of the handshapes with more than one motion, and most of the motions with more than one handshape, i.e., each handshape and motion was an independent and meaningful morpheme that could combine with other morphemes in the system to create larger meaningful units – the system was combinatorial.

Although similar in many respects, the gesture systems produced by these four children were sufficiently different to suggest that the children had introduced relatively arbitrary – albeit still iconic – distinctions into their systems. For example, in contrast to the first child (and one other) who used the CMedium handshape to represent objects 2–3 inches in width (e.g., a cup or a box), the two other children used the same CMedium handshape to represent objects that were slightly smaller, 1–2 inches in width (e.g., a banana or a toy soldier; Goldin-Meadow *et al.* 1995). The fact that there were differences in the ways the children defined a particular morpheme suggests that there were choices to be made (although all of the choices still were transparent with respect to their referents). Moreover, the choices that a given child made could not be determined without knowing that child's individual system. In other words, one cannot predict the precise boundaries of a child's morphemes without knowing that child's individual system. It is in this sense that the deaf children's gesture systems can be said to be *arbitrary*.

In addition to suggesting that they were able to introduce arbitrariness into their gesture systems, the differences across the deaf children's systems suggest that the children had different *standards of well-formedness* within their individual systems. The children's gestures not only were adequate representations of objects and movements in the world, but they also conformed to an internally consistent system and, in this sense, each system had standards of form. Further evidence that the deaf child's gesture system is characterized by standards of form comes from the fact that, in an experimental test of the morphological system of one of the deaf children conducted several years later when he was 9 years 5 months, the child spontaneously corrected some of his hearing sister's gestures that did not conform to his gestural system (the sister used a handshape to convey a meaning that had to be conveyed by a different handshape in the deaf child's system, Singleton *et al.* 1993). Although it is not necessary for a language-user to correct another's "mispronunciation" in order to suggest that the user adheres to standards of form (such corrections imply a certain level of consciousness that a user need not have), corrections of another's performance can provide further evidence of a standard. Thus, the deaf child appeared to have a well-developed and articulated sense of what counts as an acceptable gesture, and he was not shy about informing others of his standards. In addition, in an analysis of the spontaneous gestures that this same deaf child used over a two-year period, Goldin-Meadow *et al.* (1994) found that the child tended to use precisely the same gestural form for the same meaning throughout this relatively long period; that is, he appeared to have a stable lexicon of gestures at his disposal. Taken together, these findings suggest that, even without a conventional language model, children are able to introduce relatively arbitrary standards of form into their communication systems (although these standards may vary from child to child).

Classes of word: Grammatical categories

The pointing gesture in the deaf child's gesture system serves an important discourse function played by the noun in conventional languages: it serves to single out an entity, which can then be commented upon. However, the pointing gesture does not fulfill all of the functions served by the noun, and in fact appears to function more like a pronoun than a noun. In particular, while the pointing gesture can indicate which object is the focus of attention, it does not categorize that object as one of a type; that is, it does not classify an entity in terms of its relationship to other entities of the world (cf. Stachowiak 1976). Nouns do serve this function in conventional languages. Moreover, even though the deaf child was able to use the pointing gesture to refer to objects that are not present in the room (e.g., pointing at the empty bubble jar to refer to the full bubble jar which was not in the room; Butcher *et al.* 1991), this function is not easily filled by the point, and its effectiveness depends crucially on the communication partner's willingness to interpret the present object as a symbol for the nonpresent object. In contrast, nouns allow the communicator easily to make reference to nonpresent objects.

In addition to the pointing gesture, however, the deaf child also produced characterizing gestures that, by virtue of their iconic form, have the potential to serve a categorizing function and thus call nonpresent objects to the attention of a communication partner. The question is whether characterizing gestures, in addition to serving predicate functions of verbs and adjectives in the deaf child's gesture system, also served some of the functions typically filled by nouns in conventional languages. In other words, does the deaf child have the category "noun" and, if so, does it contrast grammatically with the categories "verb" and "adjective"?

In a study of the characterizing gestures of one deaf child, Goldin-Meadow *et al.* (1994) identified characterizing gestures used to focus attention on the discourse topic as *nouns*, and characterizing gestures used to comment on that topic as predicates – *verbs* if the particular comment described an action, *adjectives* if it described

an attribute. They found that gestures playing noun-like roles were distinguished from those playing verb-like roles in two ways: by the form of the gesture (akin to a morphological marking), and by its position in a gesture sentence (akin to a syntactic marking). The distinction between nouns and verbs is most strikingly seen in gestures used in both roles. For example, if the child used a "twist" gesture to focus attention on a jar as the discourse topic (i.e., as a noun), the gesture was likely to be abbreviated in form (one twist of the hand rather than several – an alteration internal to the gesture and therefore a morphological marking) and not inflected, and was likely to *precede* a deictic pointing gesture at the jar (a relationship across gestures and therefore a syntactic marking). If, on another occasion, that same stem "twist" was used to say something about the jar (i.e., as verb), the gesture was likely to be inflected in form (produced in a space near the jar, the patient of this particular predicate, rather than in neutral space – a morphological marking) and not abbreviated, and was likely to *follow* a deictic pointing gesture at the jar (a syntactic marking).[5]

While all languages appear to distinguish nouns and verbs, only certain languages make a further distinction between nouns and verbs and a third class, the class of adjectives (Schachter 1985). Goldin-Meadow *et al.* (1994) found that, in the deaf child's system, gestures used as adjectives were treated like nouns with respect to morphology (i.e., adjectives tended to be abbreviated but not inflected), but like verbs with respect to syntax (i.e., adjectives tended to follow pointing gestures rather than precede them). The deaf child's adjective gestures consequently appear to behave as adjectives do in natural languages – sharing some morphosyntactic properties with nouns and others with verbs (cf. Thompson 1988). The distinction between nouns and verbs thus appears to be a resilient or relatively canalized property of language, as is the fact that adjectives comprise a category intermediate between the two.

---

[5] As described in the preceding sections, each gesture was coded in terms of the action the child was conveying and the object involved in that action. For example, the "twist" gesture stem was coded in terms of the action "twist" and the object "jar." When used as a noun to identify the jar, the object information in the stem is highlighted (e.g., a twistable-jar) and when used as a verb to request that the jar be twisted, the action information is highlighted (e.g., to jar-twist).

Language as a tool

The deaf children used their gestures for a wide variety of functions typically served by language – to convey information about current, past, and future events, and to manipulate the world around them. Like children learning conventional languages, the deaf children requested objects and actions from others and did so using their gestures. For example, one child produced a "round" gesture, a pointing gesture at a penny, and a pointing gesture at his own chest, to request that the experimenter give him the round penny. On another occasion, the child produced a pointing gesture at a duck followed by a pointing gesture at the experimenter, to request that his sister give the duck to the experimenter. Finally, a child produced a pointing gesture at a pile of toys on the floor followed by a "push-away" gesture, to request that an adult push the toys to the side.

Moreover, like children learning conventional languages, the deaf children commented on the actions of objects, people, and themselves, both in the immediate past (e.g., a pointing gesture at a bubble jar followed by an "expand" gesture produced in the context of having just blown a very big bubble that popped) as well as in the immediate future (e.g., a finger held straight up in the air (the child's marker for immediate future, see Morford & Goldin-Meadow 1996), a pointing gesture at a toy followed by a "hop" gesture, produced just before the child made the toy hop along the path displayed in the gesture). Gestures were also used to recount events that happened some time ago. For example, to describe a visit to Santa Claus, one of the deaf children first pointed at himself, indicated Santa via a "laugh" gesture and a "moustache" gesture, pointed at his own knee to indicate that he had sat on Santa's lap, produced a "firetruck" gesture to indicate that he had requested this toy from Santa, produced an "eat" gesture to indicate that he had had a snack (a pretzel), and then finished off the sequence with a palm hand arcing away from his body (his nonpresent marker, see Morford & Goldin-Meadow 1996) and a final point at himself.

In addition to the major function of communicating with others, some of the deaf children used their gestures for other functions that language typically serves. For example, the children used their gestures when they thought no one was paying attention, as though "talking" to themselves. Once when one of the deaf children was trying to copy a configuration of blocks off of a model, the child made an "arced" gesture in the air thus indicating the block he needed next; when the experimenter offered a block that fit this description, the child ignored her, making it clear that his gesture was not directed at her but was for his use only.[6] As an example of another important function of language, one deaf child used gesture to refer to his own gestures. To request a Donald Duck toy that the experimenter held behind her back, the child pursued his lips to imitate Donald Duck's bill, then pointed at his own pursed lips and pointed toward the Donald Duck toy. When offered a Mickey Mouse toy, the child shook his head, pursed his lips and pointed at his own pursed lips (Goldin-Meadow 1993). The point at the lips is roughly comparable to the words "I say," as in "I say 'Donald Duck bill.'" It therefore represents a communicative act in which gesture is used to refer to a particular act of gesturing and, in this sense, is reminiscent of a young hearing child's quoted speech (cf. Miller & Hoogstra 1989). The deaf child was able to distance himself from his own gestures and treat them as objects to be reflected on and referred to, thus exhibiting in his self-styled gesture system the very beginnings of the reflexive capacity that is found in all languages and that underlies much of the power of language (cf. Lucy 1993).

In sum, the deaf children were able to use their gestures for many of the major functions filled by hearing children's words and deaf children's signs, suggesting that these functions, as well as the structural properties described above, are resilient or relatively canalized properties of language.

THE CONDITIONS UNDER WHICH THE DEAF CHILD'S GESTURE SYSTEM IS DEVELOPED: INPUT FROM THE GESTURES OF HEARING INDIVIDUALS

The deaf children in these studies were found to elaborate gestural communication systems characterized by language-like properties without the benefit of a conventional

---

[6] These intended-for-self gestures were not included in the database on which all of our analyses were done because they did not meet our strict criteria for a gesture – in particular, they were not intended for communication with another (see earlier in text). However, even though gestures of this sort were self-directed, they cannot really be considered babbling because the contexts in which they were produced suggest that they were meaningful rather than meaningless forms.

language model. They therefore did not learn their gestural systems in the traditional sense of the word. Nevertheless, the children were exposed to the spontaneous gestures their hearing parents used when speaking to them (as are hearing children of hearing parents, cf. Shatz 1982; Bekken 1989). These gestures could conceivably have served as input to the children's gestural systems and, therefore, must be the background against which their gestural accomplishments are evaluated.

Syntactic structure

In an analysis of parental input to sentence-level structure in six of the deaf children, Goldin-Meadow & Mylander (1983, 1984) found that, although the deaf children's hearing mothers did indeed gesture, they produced relatively few gesture strings. Moreover, the few gesture strings the mothers did produce did not show the same structural regularities as those of their children. The mothers showed no reliable gesture order patterns in their gesture strings. Moreover, the production probability patterns in the mothers' gesture strings were different from the production probability patterns in the children's strings. Finally, the mothers began conveying complex gesture sentences later in the study than their children, and produced proportionately fewer gesture sentences with conjoined predicate frames than their children (Goldin-Meadow & Mylander 1983, 1984). Thus, the mothers' gestures did not provide a good model for the syntactic structure found in the deaf children's gesture systems.

Morphological structure

To determine whether the mothers' gestures offered a model for the morphological structure found in the deaf children's gestures, the gestures that each mother produced were coded within the framework of the morphological system developed by her child (Goldin-Meadow & Mylander 1990b; Goldin-Meadow et al. 1995). Each mother was found to use her gestures in a more restricted way than her child, omitting many of the morphemes that the child produced (or using the ones she did produce more narrowly than did the child), and omitting completely many of the handshape-motion combinations that the child produced. In addition, while there was good evidence that the gestures of each deaf child could be

characterized in terms of handshape and motion components which mapped onto a variety of related objects and a variety of related actions, respectively, there was no evidence that the mothers ever went beyond mapping gestures as wholes onto entire events; that is, the mothers' gestures did not appear to be organized in relation to one another to form a system of contrasts. Finally, when the mothers' gestures were analyzed with the same procedures used to analyze the children's gestures (that is, when a mother's gestures were treated as a system unto itself), the resulting systems for each mother did not capture her child's gestures well at all. Most importantly, the arbitrary differences that were found across the children's systems could not be traced to the mothers' gestures, but seemed instead to be shaped by the early gestures that the children themselves created. In other words, the differences could be traced to the gestural input that the children provided for themselves rather than to gestural input that their mothers provided for them (Goldin-Meadow et al. 1995).

Grammatical categories

In terms of providing a model for grammatical categories, the mother of the deaf child who distinguished among nouns, verbs, and adjectives did not use the same morphological and syntactic devices in her gestures that her child used in his to make these distinctions (Goldin-Meadow et al. 1994). Indeed, certain of the devices that the child used to distinguish these categories (abbreviation and gesture order) either were not used at all or were not used distinctively by mother. These devices were therefore likely to have been initiated by the child. The child's third device (inflection) was used by the mother; however, the child's inflections patterned systematically with the predicate structure of the verb and consistently marked entities playing particular thematic roles in those predicates – that is, they functioned as part of a system – while mother's did not. Thus, while the child may have used the gestural input his mother provided as a starting point for part of his system, he went well beyond that input, fashioning it into an integral component of the system and grammaticizing it as he did so.

Language as a tool

Finally, examining the functions of language that gesture assumed in mother and child, we found that all of the mothers used their gestures to make requests and comments about the here and now. However, the mothers were far less likely than their children to use gesture to describe events that were spatially or temporally removed from the present (Morford 1993; Morford & Goldin-Meadow 1996). For example, the deaf child frequently used the pointing gesture to indicate objects that were not present; one child pointed at the chair his father typically occupied at the dining room table in order to refer to the father, and then produced a "sleep" gesture to comment on the fact that father was asleep (in another room of the house). The deaf child's mother rarely used her pointing gestures for this purpose (Butcher *et al.* 1991). In addition, the deaf children's mothers were never observed to use gesture to refer to their own gestures; that is, they did not use gesture metalinguistically. Thus, while gesture assumed a large number of the functions language typically serves for the deaf child, it served a very restricted range of functions for the hearing mother – not surprising, given that the mother had a perfectly good spoken system that she used for these functions.

OTHER ASPECTS OF SOCIAL INTERACTION THAT MIGHT INFLUENCE THE DEAF CHILD'S GESTURE SYSTEM

In our studies of the deaf children, we have focused on isolating the properties of language whose development can withstand wide variations in learning conditions – properties of language that we have termed "resilient." The findings described above suggest that certain properties of language are so fundamental to human communication that they can be developed by a child who does not have access to a conventional language model. The fact that the structural properties of the deaf children's gesture systems were not patterned after the spontaneous gestures that their hearing parents used with them suggests that the children themselves may have played a large role in creating those structural properties.

Parental responsiveness to the deaf child's gestures

It is nevertheless possible that the structure in the children's gesture systems came not from the child but from other nonlinguistic aspects of the child's environment. For example, it is possible that, by responding with either

comprehension or noncomprehension to their children's gestures, the hearing parents of these deaf children might have (perhaps inadvertently) shaped the structure of those gestures. To explore this possibility, we divided each child's gesture strings into strings that followed the child's own preferred orders and strings that were exceptions to those orders. We then determined how often the child's hearing mother gave relevant and comprehending reactions to both types of string. We found that the mothers responded with comprehension to approximately half of each child's gesture strings, whether or not those strings followed the child's preferred orders. In other words, the mothers were just as likely to understand the children's ill-formed strings as their well-formed strings, suggesting that these particular patterns of parental responsivity did not shape the orders that the children developed in their gesture systems (see Goldin-Meadow & Mylander 1983, 1984, who also present similar results for the production probability patterns found in the deaf children's gestures).

There seems little doubt, however, that comprehensibility determined the form of the deaf children's gestures at a general level, for the children's gestures were iconic, with gesture forms transparently related to the intended meanings. Indeed, the overall iconicity of the children's gestures may have made it that much more likely that variations in gesture order would have little effect on the parents' comprehension – a mother could easily figure out that her child was describing apple-eating whether the child pointed at the apple before producing an "eat" gesture, or produced the "eat" gesture before pointing at the apple. Thus, although the children's gestures were quite comprehensible to the hearing individuals around them, there was no evidence that the structural details of each child's gesture system were shaped by the way in which the mothers responded to those gestures.

Parent–child interaction and its effect on the deaf child's gestures: A look across cultures

There may, however, be other, more subtle ways in which parent–child interaction affects child communication. For example, Bruner (1974/75) has suggested that the structure of joint activity between mother and child exerts a powerful influence on the structure of the child's communication. In order to determine the extent to which the structure in the deaf children's gestures is a product of the way in which mothers and children jointly interact in their culture – and in so doing, develop a more stringent test of the resilience of the deaf children's gesture systems – we have studied deaf children of hearing parents in a second culture, a Chinese culture.

We have chosen Chinese culture as a second culture in which to explore the spontaneous communication systems of deaf children because literature on socialization (Young 1972; Miller *et al.* 1991), on task-oriented activities (Smith & Freedman 1982) and on academic achievement (Chen & Uttal 1988; Stevenson *et al.* 1990) suggests that patterns of mother–child interaction in Chinese culture differ greatly from those in American culture, particularly those in white, middle-class American culture. In addition, our own studies of the interaction between hearing mothers and their deaf children in Chinese and American families replicate these differences (Wang 1992; Wang *et al.* 1995). In particular, we have found, first, that Chinese mothers are very active in *initiating* interactions with their deaf children, whereas American mothers tend to wait for their children to initiate interactions. Second, Chinese mothers tend to offer *directives* to their deaf children before they have the opportunity to try out the task. In contrast, American mothers tend to offer directives to their deaf children only after their children fail to accomplish the task or after their children request help. Finally, when *commenting on pictures or toys*, Chinese mothers tend not only to label the pictures or toys but also to supply additional information. For example, in reaction to a picture of a house, a Chinese mother might say "house, the house is very high; the roof is pointed; there are houses over there." In contrast, the American mothers tend merely to label pictures or toys, supplying no extra information. For example, in response to the same picture, an American mother might say only "house, that's a house."

The salient differences in Chinese and American maternal interaction patterns have provided us with an excellent opportunity to examine the role that mother–child interaction plays in the development of the gestural communication systems of deaf children. If we find similarities between the spontaneous gestural systems developed by deaf children in Chinese culture and deaf children in American culture, an increasingly powerful argument can be made for the noneffects of mother–child interaction patterns on the development of these gestural systems – that is, we will have increasingly compelling evi-

dence for the resilience of the linguistic properties found in the deaf children's gestural systems. Conversely, to the extent that the gestural systems of the Chinese deaf children are consistently different from the American deaf children's gestural systems, an equally compelling argument can be made for the effects of cultural variation – as instantiated in mother–child interaction patterns – on the spontaneous gestural systems of deaf children.

We have at this point only preliminary data comparing the gesture systems of Chinese and American deaf children and their hearing parents (Wang *et al.* 1995). We first explored the gestures that the hearing parents in the two cultures used when communicating with their children and found that the Chinese mothers were far more likely to produce gestures when interacting with their deaf children than were the American mothers. When we examined the types of gesture produced by the Chinese and American mothers, we found that there were very few similarities between them. In other words, the Chinese and American mothers provided their deaf children with very different gestural models. Despite the difference in the gestural models they received, the Chinese and American deaf children produced gestures that were comparable in many respects – in how frequently gestures were combined into strings and in the distribution of characterizing versus pointing gestures in their communications (we have not yet examined the more detailed aspects of syntactic and morphologic structure in the children's gestures). The children differed primarily in their use of culturally bound emblematic gestures (e.g., a finger brushed across the cheek used by the Chinese mothers to mean "shame"), essentially mirroring the types of emblem the mothers in each culture produced.

Given the differences in the worlds that the deaf children in these two cultures experienced, the similarities found in the spontaneous gesture systems developed by these children provide preliminary evidence for the noneffects of mother–child interaction patterns on the development of these linguistic properties; that is, we have evidence for the resilience of these aspects of language in the face of cultural variation. Also, the differences we have found in the conventional emblems used by the Chinese and American deaf children provide evidence for the impact of cultural variation on the development of certain aspects of the deaf children's gesture systems (particularly in portions of their gestural lexicons; an unsurprising result given that children easily learn the particular lexicons to which they are exposed).

RESILIENCE IN THE FACE OF EXTERNAL AND INTERNAL VARIABILITY

For obvious ethical reasons, researchers cannot deliberately manipulate the conditions under which language is acquired. One can, however, take advantage of variations that occur naturally in language-learning conditions in order to explore the boundary conditions under which language development is possible. Such "experiments of nature" explore particular deviations from typical language-learning circumstances and their effects (or non-effects) on the development of language. Deviations can involve environmental variations from the norm, variations that alter the quantity and/or quality of the linguistic input a child receives, or they can involve organic variations from the norm, variations that alter the way the child processes whatever input he or she receives. As described earlier, there are conditions that are incompatible with the development of language in humans (e.g., conditions of extreme deprivation under which children do not develop language during their periods of deprivation; Brown 1958; Skuse 1988). However, aside from these limiting cases, language development appears to be remarkably resilient across a wide range of both environmental and organic conditions.

One of the clearest instances of resilience in language development in the face of environmental variation comes from the children I have described here – deaf children whose hearing parents have not yet exposed them to the linguistic input they could effortlessly profit from (a conventional sign language). These children, deprived of an accessible conventional language model but otherwise experiencing normal home environments, use gesture to communicate and their gestures are structured in many respects as are the early communication systems of children learning conventional languages, signed or spoken. A less extreme example of environmental variation is found in hearing children born to deaf parents who themselves are not fluent speakers and who provide their children with limited access to spoken language input. A minimal amount of exposure to hearing speakers – five to ten hours per week – in an otherwise normal family situation is typically sufficient to allow acquisition of spoken language to

proceed normally (Schiff-Myers 1988), illustrating once again the resilience of language in humans.

As an example of resilience in language development in the face of organic variation, the acquisition of grammar in the earliest stages has been found to proceed in a relatively normal manner and at a normal rate even in the face of unilateral brain injury (Feldman 1994). As a second example, children with Down's syndrome have numerous intrinsic deficiencies that complicate the process of language acquisition; nevertheless, most Down's syndrome children acquire some basic language reflecting the fundamental grammatical organization of the language they are exposed to (the amount of language that is acquired is in general proportion to their cognitive capabilities; Rondal, 1988; see also Fowler et al. 1994). Finally, and strikingly given the social impairments that are at the core of the syndrome, autistic children do not appear to be impaired in their grammatical development, either in syntax or in morphology (Tager-Flusberg 1994; although they do often have deficits in the communicative, pragmatic, and functional aspects of their language).

Thus, language development can proceed in humans over a wide range of environments and a wide range of organic states, suggesting that the process of language development, if not developed by a completely closed program, may at least be buffered against a large number of both environmental and organic variations. No one factor seems to be ultimately responsible for the course and outcome of language development in humans, a not so surprising result given the complexity of human language. Indeed, what is striking about the deaf children we have described here is not that they are creating a language with little environmental support but that they are able to make use of the minimal support they do have to fashion a communication system that looks like human language. The deaf children in our studies are surrounded by a perfectly adequate spoken language model but, because of their hearing losses, they are unable to take advantage of that model. The only input they have is the spontaneous gesture that their hearing parents produce as they speak. The gestures the parents produce are no different from the gestures that any hearing individual uses along with speech (Goldin-Meadow et al. 1996) and thus are global and synthetic in form, with structure quite different from the structure of natural language (McNeill 1992). If the deaf children are using their parents' gestures as input, they must be taking the gestures they see and transforming them into a system with linguistic properties. Unlike their hearing parents' gestures, the deaf children's gestures are linear and segmented, having intergesture structure akin to syntax and intragesture structure akin to morphology, as well as the grammatical categories basic to human language. The surprising result is that the children's gestures are structured so much like natural language, given that their parents' gestures, which may serve as input to those gestures, are not.

Thus, there appears to be a form that human language naturally assumes, and that form can be reached through a wide variety of develomental paths. In other words, language development in humans is characterized by "equifinality" – a term coined by the embryologist Driesch (1908, cited by Gottlieb 1996) to describe a process by which a system reaches the same outcome despite widely differing input conditions. A system characterized by equifinality must rely either on a single developmental mechanism that can make effective use of a wide range of inputs, or on multiple developmental mechanisms each activated by different conditions but constrained in some way to lead to the same end product (cf. Miller *et al.* 1990).

Not surprisingly given the minimal input they have at their disposal, the deaf children do not produce in their gesture systems all of the properties found in natural human languages. Indeed, the absence of a particular linguistic property in the deaf child's gesture system could be taken as (indirect) evidence that exposure to a conventional linguistic system is necessary for the development of that property to take place (e.g., a bias in the direction that redundancy is reduced in complex sentences; Goldin-Meadow 1987) – evidence that the property is relatively fragile with respect to linguistic input, needing a more specified and particular set of environmental circumstances within which to develop than do the resilient properties of language.

One set of circumstances that might be necessary to develop the more fragile properties of language is a community of speakers or signers or, at the least, a willing communication partner. As described earlier, the deaf children's families chose to educate them through an oral method, and their emphasis was on their children's (minimal) verbal abilities. The families did not treat the children's gesture as though it were a language. In other words, they were not partners in the gestural communica-

tion that the children used. Thus, in order to be understood, the deaf children's gestures needed to be relatively iconic; that is, transparently related to their referents. An interesting question to pose is how far a deaf child can move toward arbitrariness and a more complex system without a conventional language as a model but *with* a willing communication partner who could enter into and share an arbitrary system with the child. To date, the circumstances that would allow us to address this question – two deaf children inventing a gestural system with no input from a conventional sign language – have not been identified. However, Kegl *et al.* (1996) have described a group of home signers in Nicaragua who were brought together to form a community in 1980. Over the course of a decade, a sign language appearing to have much of the grammatical complexity of well-established sign languages such as ASL evolved out of this set of home sign systems. This newly emergent language is referred to as Lengua de Signos Nicaraguense, Nicaraguan Sign Language, and it is far more complex than the home sign systems out of which it was formed – and far more complex than the gesture systems of the deaf children we have described here. The considerable distance between the deaf children's gesture systems and the newly formed Nicaraguan Sign Language highlights the importance of a community of signers in constructing a full-blown linguistic system.[7]

Nevertheless, the linguistic accomplishments of the deaf children in our studies should not be underestimated,

---

[7] A similar phenomenon has been reported in birds. Young chaffinches (*Fringilla coelebs*), when taken from the nest at five days and reared by hand in auditory isolation, each develops its own song and each of those songs lacks the normal features of chaffinch song (Thorpe 1957). However, if the young chaffinches are reared together as a group, albeit still in auditory isolation, the group develops a uniform community song – one that is unlike anything recorded in the wild but one that does have a tendency toward the division into phrases characteristic of typical chaffinch song (Thorpe 1957). In other words, developing song without a model but in a group led to a more structured output than did developing song without a model and on one's own. In this sense, the deaf child's relative performance was much closer to the norm than that of the isolated chaffinch, having developed a gesture system that has many structural properties characteristic of human language even without the benefit of a group. However, unlike the isolated chaffinch, the deaf child is not deprived of all social stimulation; although lacking a partner who might willingly enter into a joint gesture system, the deaf child produces gestures to communicate with others and thus, in this important sense, is isolated from a language model but not the social world.

Table 15.1. *Properties of the deaf child's gesture system: The resilient properties of language*

Sentence structure
Consistent production and deletion of semantic elements within a gesture sentence
Consistent ordering of semantic elements within a gesture sentence
Predicate frames underlying a gesture sentence
Complex gesture sentences reflecting recursion

Word structure
A limited set of handshape and motion forms
Each form is associated with a particular meaning
Each form-meaning pair combines freely with other form-meaning pairs

Grammatical categories
Noun gestures distinguished from verb gestures via morphologic form
Noun gestures distinguished from verb gestures via syntactic ordering within sentences

Language use
Gestures used to make requests
Gestures used to make comments about the present, past, and future
Gestures used to "talk" to oneself
Gestures used to refer to one's own gestures

particularly when compared with the linguistic accomplishments of chimpanzees (*Pan troglodytes*) exposed to models of human language. Although researchers have reportedly been successful in getting chimpanzees to respond to and produce aspects of human language (e.g., word order; Savage-Rumbaugh & Rumbaugh 1993), it is important to keep in mind that these chimpanzees are supported in extremely rich environments from a language-learning point of view – far richer, at least in terms of a usable language model, than the environments in which the deaf children find themselves. The chimpanzees have been extensively exposed to a model of a human language yet, despite this exposure, have developed relatively few properties of language found in the deaf children's gesture systems developed without a language model. For example, the chimps' language does not appear to have grammatical properties or structure at more than one level, nor do the chimps use whatever language they do have for the full range of linguistic functions – most of the chimps' productions are requests in the here and now, not comments on the present or the nonpresent or on talk itself (cf. Greenfield & Savage-Rumbaugh 1991).[8] Whatever

gains chimps make in learning language, they appear to do so at great cost and with much effort. In contrast, language comes naturally to human children – even when lacking an adequate model for language. The point here is that the most primitive of human language systems (like the deaf children's gesture systems) is still much richer than the most complex chimpanzee system, despite massive attempts at enriching the environment of one and the unfortunate impoverishment of the environment of the other.

SUMMARY

I have identified a set of linguistic properties (summarized in Table 15.1) that are found in the deaf child's gesture system and thus are, by my definition, resilient and likely to be developed by a relatively closed program – structure at the sentence level, structure at the word level, grammatical categories, and various aspects of language use. Note that I have not identified a particular structure as resilient but rather a class of structures. For example, I consider the child's tendency to break the words (or gestures in the case of the deaf child) that comprise his or her sentences into parts to be a resilient property of language. However, the particular units into which those words are decomposed vary across children and are likely to be influenced by a

[8] Even Kanzi, a pygmy chimpanzee who is the star of the language-learning chimps, used statements in only 4% of his linguistic productions (Greenfield & Savage-Rumbaugh 1991, p. 243).

number of factors (some of which may be serendipitous). What appears to be constant across all children is that they decompose their words and develop a morphological system, not the particular morphological system itself. A resilient property of language need not be completely impervious to input. Although the appearance of a resilient property in a child's communications is relatively independent of the particular input that the child receives (within limits, cf. children raised under inhumane conditions), the particular instantiation that the resilient property takes in that child's communication system will undoubtedly be influenced by the vagaries of his or her environment.

The fact that the resilient properties of language can be developed in an environment that varies quite substantially from the language-learning environment children typically experience suggests that children themselves may be constrained to interpret the environment in particular ways. Constraints of this sort may serve to narrow the range of possible outcomes in language development simply because they guide the child's search through the environment for relevant data. Although this sort of narrowing, or canalization, is often attributed to genetic causes (cf. Waddington 1957), there is evidence that canalization can be caused by the environment as well (Gottlieb 1991a). For example, Gottlieb (1991b) has shown that exposure to a particular stimulus at one point in development not only makes the organism particularly susceptible to that stimulus at later points in development but also makes the organism *less* susceptible to other stimuli; that is, it buffers the organism against those stimuli, thereby narrowing the range of possibilities open to the organism. Two points about the canalizing role of the environment are worth noting. First, when the environment plays a canalizing role, that role is often not easily categorizable as "learning" in the sense that output is neatly mappable onto characteristics of the environmental input. Second, in order for acquisition to be universal when the environment is playing a canalizing role, the relevant aspect of the environment must be reliably present in the world of each member of the species. In a sense, the environment must be considered as much a part of the species as its genes.

In this regard, it is important to acknowledge the aspects of environment that the deaf child has in common with all children. Although isolated from a language model, the deaf children were not isolated from the social world. Indeed, their gestures were produced in a naturalistic social situation and with the intent to communicate. It seems unlikely in the extreme that a child would invent a gestural communication system if there was no one to share it with.[9] The question, however, is whether the structural properties that turn out to characterize such an invented gesture system can be derived from properties of the child's social world. The answer will not be as simple as mothers reinforcing gestures that they find comprehensible (the children's gestures are just too easy to understand for such a mechanism to work). However, it is possible that more subtle aspects of the communication task itself may inexorably lead to certain types of structure (see Goldin-Meadow *et al.* 1996, who hypothesize that the resilient properties of segmentation and combination may arise to solve the problem generated by sharing information across two minds).

The studies I have described here identify properties of human language that are resilient in the face of one extreme variation from the conditions children typically experience when they learn language. These properties of language appear to be canalized with respect to this range of environmental inputs, suggesting that there are constraints on the development of these properties in humans. However, the studies described here do not identify (nor do they assume) a particular cause for the constraints that appear to guide human language development. Rather, the studies provide an empirical process by which such constraints can be identified; the findings can then serve as a framework within which causes can be explored. In this way, the findings inform the search for the biological and cultural foundations of the language-learning process.

ACKNOWLEDGEMENTS

This work was supported by grant no. BNS 8810769 from the National Science Foundation and grant no. RO1 DC00491 from NIH. I thank Martha McClintock

---

[9] As described above, we did find that the deaf child, at times, used gestures when no one was paying attention as though "talking" to himself or herself. Thus, the child was able to exploit a system that had been developed to communicate with another person in order to self-communicate. However, it is not likely that the child would have invented such a system in the first place if there had been no-one to receive it.

and Bill Wimsatt for their invaluable insights into innateness that, over a large number of years of teaching together, they have shared with me. I also thank them, Gilbert Gottlieb, and Susan Levine for their helpful comments on the manuscript itself.

REFERENCES

Acredolo, L. P. & Goodwyn, S. W. (1988). Symbolic gesturing in normal infants. *Child Development* **59**: 450–66.

Alcock, J. (1988). Singing down a blind alley. *Behavioral and Brain Sciences* **11**: 630–1.

Bekken, K. (1989). Is there "Motherese" in gesture? Doctoral dissertation, University of Chicago.

Brown, R. (1958). *Words and Things*. The Free Press, New York.

Bruner, J. (1974/5). From communication to language: A psychological perspective. *Cognition* **3**: 255–87.

Butcher, C., Mylander, C. & Goldin-Meadow, S. (1991). Displaced communication in a self-styled gesture system: Pointing at the non-present. *Cognitive Development* **6**: 315–42.

Chen, C. & Uttal, D. H. (1988). Cultural values, parents' beliefs, and children's achievement in the United States and China. *Human Development* **31**: 351–8.

Conrad, R. (1979). *The Deaf Child*. Harper & Row, London

Dixon, R. M. W. (1979). Ergativity. *Language* **55**: 59–138.

Fant, L. J. (1972). *Ameslan: An Introduction to American Sign Language*. National Association of the Deaf, Silver Springs, MD.

Feldman, H. M. (1994). Language development after early unilateral brain injury: A replication study. In *Constraints on Language Acquisition: Studies of Atypical Children*, ed. H. Tager-Flusberg, pp. 75–90. Erlbaum, Hillsdale, NJ.

Feldman, H., Goldin-Meadow, S. & Gleitman, L. (1978). Beyond Herodotus: The creation of language by linguistically deprived deaf children. In *Action, Symbol, and Gesture: The Emergence of Language*, ed. A. Lock, pp. 351–414. Academic Press, New York.

Fowler, A. E., Gelman, R. & Gleitman, L. R. (1994). The course of language learning in children with Down Syndrome: Longitudinal and language level comparisons with young normally developing children. In *Constraints on Language Acquisition: Studies of Atypical Children*, ed. H. Tager-Flusberg, pp. 91–140. Erlbaum, Hillsdale, NJ.

Goldin-Meadow, S. (1979). Structure in a manual communication system developed without a conventional language model: Language without a helping hand. In *Studies in Neurolinguistics*, vol. 4, ed. H. Whitaker & H. A. Whitaker, pp. 125–209. Academic Press, New York.

Goldin-Meadow, S. (1982). The resilience of recursion: A study of a communication system developed without a conventional language model. In *Language Acquisition: The State of the Art*, ed. E. Wanner & L. R. Gleitman, pp. 51–77. Cambridge University Press, New York.

Goldin-Meadow, S. (1985). Language development under atypical learning conditions: Replication and implications of a study of deaf children of hearing parents. In *Children's Language* vol. 5, ed. K. Nelson, pp. 197–245. Erlbaum, Hillsdale, NJ.

Goldin-Meadow, S. (1987). Underlying redundancy and its reduction in a language developed without a language model: The importance of conventional linguistic input. In *Studies in the Acquisition of Anaphora: Applying the Constraints*, vol. II, ed. B. Lust, pp. 105–33. D. Reidel Publishing Company, Boston, MA.

Goldin-Meadow, S. (1993). When does gesture become language? A study of gesture used as a primary communication system by deaf children of hearing parents. In *Tools, Language and Cognition in Human Evolution*, ed. K. R. Gibson & T. Ingold, pp. 63–85. Cambridge University Press, New York.

Goldin-Meadow, S., Butcher, C., Mylander, C. & Dodge, M. (1994). Nouns and verbs in a self-styled gesture system: What's in a name? *Cognitive Psychology* **27**: 259–319.

Goldin-Meadow, S. & Feldman, H. (1977). The development of language-like communication without a language model. *Science* **197**: 401–3.

Goldin-Meadow, S., McNeill, D. & Singleton, J. (1996). Silence is liberating: Removing the handcuffs on grammatical expression in the manual modality. *Psychological Review* **103**: 34–55.

Goldin-Meadow, S. & Morford, M. (1985). Gesture in early child language: Studies of deaf and hearing children. *Merrill-Palmer Quarterly* **31**: 145–76.

Goldin-Meadow, S. & Mylander, C. (1983). Gestural communication in deaf children: The non-effects of parental input on language development. *Science* **221**: 372–4.

Goldin-Meadow, S. & Mylander, C. (1984). Gestural communication in deaf children: The effects and non-effects of parental input on early language development. *Monographs of the Society for Research in Child Development* **49**: 1–121.

Goldin-Meadow, S. & Mylander, C. (1990a). Beyond the input given: The child's role in the acquisition of language. *Language* **66**: 323–55.

Goldin-Meadow, S. & Mylander, C. (1990b). The role of a language model in the development of a morphological system. *Journal of Child Language* **17**: 527–63.

Goldin-Meadow, S., Mylander, C. & Butcher, C. (1995). The resilience of combinatorial structure at the word level: Morphology in self-styled gesture systems. *Cognition* **56**: 195–262.

Gottlieb, G. (1991a). Experiential canalization of behavioral development: Theory. *Developmental Psychology* **27**: 4–13.

Gottlieb G. (1991b). Experiential canalization of behavioral development: Results. *Developmental Psychology* **27**: 35–9.

Gottlieb, G. (1996). A systems view of psychobiological development. In *Individual Development over the Lifespan: Biological and Psychosocial Perspectives*, ed. D. Magnusson. Cambridge University Press, New York, in press.

Greenfield, P. M. & Savage-Rumbaugh, E. S. (1991). Imitation, grammatical development, and the invention of protogrammar by an ape. In *Biological and Behavioral Determinants of Language Development*, ed. N. A. Krasnegor, D. M. Rumbaugh, R. L. Schiefelbusch & M. Studdert-Kennedy, pp. 235–62. Erlbaum, Hillsdale, NJ.

Hoffmeister, R. & Wilbur, R. (1980). Developmental: The acquisition of sign language. In *Recent Perspectives on American Sign Language*, ed. H. Lane & F. Grosjean, pp. 61–78. Erlbaum, Hillsdale, NJ.

Kegl, J., Senghas, A. & Coppola, M. (1996). Creation through contact: Sign language emergence and sign language change in Nicaragua. In *Comparative Grammatical Change: The Intersection of Language Acquisition, Creole Genesis, and Diachronic Syntax*, ed. M. DeGraff. MIT Press, Cambridge, MA, in press.

Lenneberg, E. H. (1964). Capacity for language acquisition. In *The Structure of Language: Readings in the Philosophy of Language*, ed. J. A. Fodor & J. J. Katz, pp. 579–603. Prentice-Hall, Englewood Cliffs, NJ.

Lucy, J. A. (1993). Reflexive language and the human disciplines. In *Reflexive Language: Reported Speech and Metapragmatics* ed. J. Lucy, pp. 9–32. Cambridge University Press, New York.

Mayberry, R. I. (1992). The cognitive development of deaf children: Recent insights. In *Child Neuropsychology*, vol. 7, *Handbook of Neuropsychology*, ed. S. Segalowitz & I. Rapin, pp. 51–68. Elsevier, Amsterdam.

Mayr, E. (1974). Behavior programs and evolutionary strategies. *American Scientist* **62**: 650–9.

McNeill, D. (1992). *Hand and Mind: What Gestures Reveal about Thought*. University of Chicago Press, Chicago.

Meadow, K. (1968). Early manual communication in relation to the deaf child's intellectual, social, and communicative functioning. *American Annals of the Deaf* **113**: 29–41.

Miller, D. B., Hicinbothom, G. & Blaich, C. F. (1990). Alarm call responsivity of mallard ducklings: Multiple pathways in behavioral development. *Animal Behavior* **39**: 1207–12.

Miller, P. J., Mintz, J. & Fung, H. (1991). Creating children's selves: An American and Chinese comparison of mothers' stories about their toddlers. Paper presented at the biennial meeting of the Society for Psychological Anthropology, October.

Miller, P. J. & Hoogstra, L. (1989). How to represent the native child's point of view: Methodological problems in language socialization. Paper presented at the Annual Meeting of the American Anthropological Association, Washington DC.

Moores, D. F. (1974). Nonvocal systems of verbal behavior. In *Language Perspectives: Acquisition, Retardation, and Intervention*, ed. R. L. Schiefelbusch & L. L. Lloyd, pp. 377–417. University Park Press, Baltimore, MD.

Morford, J. P. (1993). Creating the language of thought: The development of displaced reference in child-generated language. Doctoral dissertation, University of Chicago.

Morford, J. P. & Goldin-Meadow, S. (1996). From here and now to there and then: The development of displaced reference in Homesign and English. *Child Development*, in press.

Morford, M. & Goldin-Meadow, S. (1992). Comprehension and production of gesture in combination with speech in one-word speakers. *Journal of Child Language* **9**: 559–80.

Newport, E. L. & Meier, R. P. (1985). The acquisition of American Sign Language. In *The Cross-linguistic Study of Language Acquisition*, vol. 1, *The Data*, ed. D. I. Slobin, pp. 881–938. Erlbaum, Hillsdale, NJ.

Roitblat, H. L., Herman, L. M. & Nachtigall, P. E. (1993). *Language and Communication: Comparative Perspectives*. Erlbaum, Hillsdale, NJ.

Rondal, J. A. (1988). Down's Syndrome. In *Language Development in Exceptional Circumstances*, ed. D. Bishop & K. Mogford, pp. 165–76. Churchill Livingstone, New York.

Savage-Rumbaugh, E. S. & Rumbaugh, D. M. (1993). The emergence of language. In *Tools, Language and Cognition in Human Evolution*, ed. K. R. Gibson & T. Ingold, pp. 86–108. Cambridge University Press, New York.

Schachter, P. (1985). Parts-of-speech systems. In *Language Typology and Syntactic Description: Clause Structure*, vol. 1, ed. T. Shopen, pp. 3–61. Cambridge University Press, New York.

Schiff-Myers, N. (1988). Hearing children of deaf parents. In *Language Development in Exceptional Circumstances*, ed. D. Bishop & K. Mogford, pp. 47–61. Churchill Livingstone, New York.

Shatz, M. (1982). On mechanisms of language acquisition: Can features of the communicative environment account for development? In *Language Acquisition: The State of the Art*, ed. E. Wanner & L. R. Gleitman, pp. 102–27. Cambridge University Press, New York.

Silverstein, M. (1976). Hierarchy of features and ergativity. In *Grammatical Categories in Australian Languages*, ed. R. M. W. Dixon, pp. 112–71. Australian Institute of Aboriginal Studies, Canberra.

Singleton, J. L., Morford, J. P. & Goldin-Meadow, S. (1993). Once is not enough: Standards of well-formedness in manual communication created over three different timespans. *Language* **69**: 683–715.

Skuse, D. H. (1988). Extreme deprivation in early childhood. In *Language Development in Exceptional Circumstances*, ed. D. Bishop & K. Mogford, pp. 29–46. Churchill Livingstone, New York.

Smith, S. & Freedman, D. (1982). Mother–toddler interaction and maternal perception of child development in two ethnic groups: Chinese-American and European-American. Paper presented at the annual meeting of the Society for Research in Child Development, Detroit.

Stachowiak, F. J. (1976). Some universal aspects of naming as a language activity. In *Language universals: Papers from the Conference held at Gummersbach/Cologne, Germany*, ed. H. Seiler, pp. 207–28. Gunter Narr Verlag, Tübingen.

Stevenson, H. W., Lee, S.-L., Chen, C., Stigler, J. W., Hsu, C.-C. & Kitamura, S. (1990). Contexts of achievement. *Monographs of the Society for Research in Child Development* **55**.

Tager-Flusberg, H. (1994). Dissociations in form and function in the acquisition of language by autistic children. In *Constraints on Language Acquisition: Studies of Atypical Children*, ed. H. Tager-Flusberg, pp. 175–94. Erlbaum, Hillsdale, NJ.

Tervoort, B. T. (1961). Esoteric symbolism in the communication behavior of young deaf children. *American Annals of the Deaf* **106**: 436–80.

Thompson, S. A. (1988). A discourse approach to the cross-linguistic category "adjective". In *Explaining Language Universals*, ed. J. A. Hawkins, pp. 167–85. Basil Blackwell, Cambridge, MA.

Thorpe, W. H. (1957). The learning of song patterns by birds, with especial reference to the song of the chaffinch *Fringilla coelebs. Ibis* **100**: 535–70.

Waddington, C. H. (1957). *The Strategy of the Genes*. Allen & Unwin, London.

Wang, X.-l. (1992). Resilience and fragility in language acquisition: A comparative study of the gestural communication systems of Chinese and American deaf children. PhD dissertation, University of Chicago.

Wang, X.-l., Mylander, C. & Goldin-Meadow, S. (1995). The resilience of language: Mother–child interaction and its effects on the gesture systems of Chinese and American deaf children. In *Language, Gesture, and Space*, ed. K. Emmorey & J. Reilly, pp. 411–33. Erlbaum, Hillsdale, NJ.

Wimsatt, W. (1986). Developmental constraints, generative entrenchment, and the innate-acquired distinction. In *Integrating Scientific Disciplines*, ed. W. Bechtel, pp. 185–208. Martinus-Nijhoff, Dordrecht.

Young, N. F. (1972). Socialization patterns among the Chinese of Hawaii. *Amerasia Journal* **1**(4): 31–51.

16 Reciprocal interactions and the development of communication and language between parents and children

ANNICK JOUANJEAN-L'ANTOËNE

INTRODUCTION

The first studies of language acquisition by the child described mainly the developmental stages of this specific human ability. From Piaget (1923) to Brown (1973), authors were interested mostly in the different formal aspects of the acquisition: for example, the age of onset, total amount of language at any age, mean length of utterance, and emergence of grammar. These studies considered the abilities of each child as representative of the general linguistic abilities of the human species at this ontogenetic stage.

More recently, new trends have appeared, where language is studied in a more pragmatic way: it is considered as a means, at each developmental stage, for a child to elicit real communicative interactions. Thus, while admitting that the general stages of language development are alike in any child (Locke & Snow, Chapter 14) such an approach to the development of communication implies integrating various aspects that are usually considered separately by different researchers.

On the one hand, to consider the emerging linguistic skill as part of the larger phenomenon of communication implies integrating the analysis of the linguistic competence of a child at a given stage with that of previous stages, in particular with babbling. It implies also the integration of other communicative behaviors: for example, approaches, emotional addresses, and object exchanges. It can be hypothesized that, when a child is communicating, it is both acquiring the human language and developing its personal communicative style with human beings.

It is also necessary to consider the role of the social environment and of social actors in the development of linguistic abilities. In parent–child interactions, parents obviously act as tutors. But up to now, most studies done

in the family context have concentrated on the maternal influence only. This may lead one to underestimate the roles of the other family members, in particular that of the father.

Finally, when considering the family members as interacting actors, we have to consider the child itself. Because children of the same family are different (Plomin & Daniels 1987), individual differences in the family may be important and have to be taken into account in a study on language development. Moreover the acquisition of language and communication skills must be seen as a reciprocal process between parents and children.

The three main areas of our study (language and prelinguistic stages, language in the frame of communication, language and social actors) are not new issues, but it is new and original to integrate these areas in the same study. It is a necessary step if we want to investigate the social constructivist view (Mehan 1982), suggesting that all the acquisitions of a human being are socially scaffolded.

THEORETICAL AND METHODOLOGICAL CHOICES

A longitudinal naturalistic approach

It should be easier to understand the emergence of communication in individuals in social contexts by doing studies based on naturalistic observations in naturalistic settings rather than in experimentally controlled situations (See Goodwin 1990, and Chapter 17).

Most psychologists prefer "semi-naturalistic" situations that are, in fact, carefully controlled (e.g., control of the context, of the number of toys, of the activity proposed to the protagonists). Naturalistic situations are generally considered to be too complicated, which explains the choice of controlled settings.

We well know that experimentation is "the" method in science (Bernard 1965), but in the actual life of human beings "nature has been experimenting since the beginning of time . . . [our] mission is to observe and organize the data from the nature's experiments" (Cronbach 1957, p. 672).

Experimental psychologists also choose the experimental approach because during infancy the spontaneous occurrences of "important" behaviors are rather rare. Appropriate structured settings can facilitate emergence of these behaviors: semi-naturalistic situations are then comparable to an experimental approach because in arranging settings the researcher is also selecting the behavior to be studied.

The risk with studies based on an *a priori* selection of behaviors joined with control of the settings is that it leads to "empty cells" (Moerk 1977), and ignores the actual importance and role of the behaviors selected. For example, an experimental situation would lead to the underestimation of some behaviors (see our results on vocal-verbal imitation, p. 321) and overestimation of other behaviors.

Moreover the naturalistic approach (emphasized by ethologists) enables us to broach a knowledge of the "full pattern of activities" (Schneirla 1966). Observations in natural conditions are a necessary first step before designing appropriate experiments. Our choice of a naturalistic approach is not a revival of a naive empiricism (Clark 1994), but the asserted necessity to study speech and communication in ecologically valid situations (Hausberger 1993).

Furthermore, experimental studies tend to deal more with inter group differences ("the representative child") than with interindividual differences, which have been, up to now, sources of trouble rather than aims of study (Plomin *et al.* 1988). The naturalistic approach is useful in approaching individual differences and their possible origin.

Finally, such an approach to the development of communication also implies longitudinal studies. In most cases, experimental studies are designed for a limited number of ages (e.g., studies of language acquisition, age of onset), which means studying the result of the development rather than the developmental course itself. Thus, longitudinal studies seemed necessary for us to understand the developmental movement and not only the result of this movement (Vygotski 1985). For all these reasons, our study is a longitudinal naturalistic approach and deals with only two children in order to take into account individuality as a factor in the development of communication.

Language and prelinguistic stages

Bernicot (1992) described four stages in pscyholinguistics, from "structural psycholinguistics" to "psycholinguistics of communication." Based on the species-specific and universal characteristics of human vocal-verbal apparatus (Lenneberg 1967; Locke 1983; Lieberman 1984) a whole trend of studies initiated by Chomsky (1965) dealt first with proving the universality of structures of species-specific human linguistic skills and the universality of the acquisition mechanisms.

This led, up to the 1970s, to the neglect of the communicative function of language (except in philosophy; see e.g., Searle 1972) and to ignoring the relation between early vocal behavior, babbling, and later language development. The prevailing view, initiated by Jakobson (1963, 1968), was of a discontinuity between babbling and language. The methodological consequence was that vocalizations and babbling were simply excluded from language analysis for a long time. New questions have, however, arisen from recent studies.

A direct relation has been shown between the sounds present in the child's late babbling (after seven months) and the target-language (de Boysson-Bardies 1984; de Boysson-Bardies *et al.* 1984, 1985). Moreover the observation of progressive integration of adult language sounds into the child babbling (Oller 1980, 1986; Locke 1983; Bauer & Stasny 1987) and the statement of a "productive pairing of sound and meaning" before the mastering of language (Vihman *et al.* 1985; Blake & Fink 1987) have led authors to wonder about the possible relations between babbling and easy or rapid language development (Speidel & Nelson 1989).

Although species-specificity appears through the cross-cultural convergence that may be shown in the emerging repertoire of consonant sounds (MacNeilage 1980; Locke 1983), some differences may also be observed in infants from the same linguistic community (Ferguson & Farwell 1975; Vihman *et al.* 1985). Precise individual studies are therefore needed to comprehend these individual differences and individual relations between babbling and language. Here, the longitudinal naturalistic approach again can be fruitful. This kind of approach leads us to consider vocalizations and language in actual exchanges between the child and its protagonists, especially the parents.

Imitation and vocal sharing

The study of the emergence of language in naturalistic settings and in relation to previous vocal behaviors throws new light on behaviors such as imitation and leads to a better understanding of their contribution to language development. The importance and exact role of imitation in language acquisition is controversial (Snow 1981; Speidel & Nelson 1989), although it is clear that there are imitative children and nonimitative children (Bloom *et al.* 1974). Many authors are now convinced of the necessity to take imitation into account and go further with this "fuzzy" phenomenon (Moerk 1989).

Imitation will be considered here as a process playing a role in the mastering of interaction, and is expressed both by parents and children engaged in common exchanges. At this level, imitation seems to be very important and to have several functions that are different in children and in adults (Uzgiris *et al.* 1989). For the parents, imitation is part of a tutoring process that will enable the child to acquire new sounds or words, whereas for the child it is a learning strategy, imitative children expressing in this way their special rapid interest in language mastery. Moreover imitation in communicative exchanges is a proof of the reciprocity of interactions and may be considered as sharing (Uzgiris *et al.* 1989). Vocal or verbal sharing will have a third function, a social function in scaffolding relationships and attesting to favored relationships. These different viewpoints on imitation integrate the linguistic level and the social level of analysis, which can also lead to parallels with studies in nonhuman animals on vocal sharing and social relationships (see e.g., Hausberger, Chapter 8).

According to Bloom *et al.* (1974) and Snow (1981), a distinction was made on the one hand between *exact, expanded* and *reduced imitations*, and, on the other hand, between *immediate* and *deferred imitations*. I consider here the exact and expanded imitations and restrict the study to immediate imitations because it seems difficult to have a precise definition of deferred imitation, and also because of the theoretical orientation in which I consider imitation as a vocal sharing or sharing of the same linguistic topic.

Finally, I want to make a clear distinction between *imitation* of a protagonist and what is called *self-repetition*, i.e., the repetition by a protagonist of what he/she just previously said. The exact functions of the two kinds of imitation are not exactly similar: imitation of the adult's speech by the child can be a way of acquiring new linguistic items as underlined above, but to repeat his or her previous own speech maintains (or tries to maintain) the listener's attention on the same topic. Thus, it appears more as an immediate conversational pattern, related to the sharing function, than an acquisition process, despite the fact that self-repetition may be a kind of auto-reinforcement for the learning child. This underlines, as in animal vocal communication, the double function of each vocal occurrence: the immediate communicative and adaptative function as well as the ultimate function (Snowdon 1987), here the scaffolding of the linguistic skill.

The question now arises: "what other behaviors are involved since much of communication is nonverbal?"

Multimodality of communication

The necessity of studying all the communicative behaviors comes from the observation that human communication is largely multimodal (Cosnier 1977; Cosnier & Brossard 1984), although language and speaking remain the species typical mode of communication.

For example, approach behavior is needed to synchronize face-to-face interactions, which are quite common in daily life (Goffman 1974, 1987; Goodwin 1990), whether the exchanges are with adults or infants, or in youngsters' communication. However, if an approach first shows the readiness for communication, it also reveals the preference of individuals toward each other and can be considered as a criterion of the emerging preferential relationships between individuals. It can also, like imitations, be considered as acting at two levels: the immediate mastering of the interaction and the "ultimate" building of relationships.

Moreover the approaching behavior tends to prove that a given individual is the *actor* of his/her own development through the *choices of receivers during spontaneous exchanges*. This assertion is to be applied to human children as well as to nonhuman animals, here again showing the relevance of searching for common parallels in developmental studies in animals and humans.

The naturalistic approach will also allow us to investigate other behaviors that can be involved in communication: physical contacts (positive or agonistic), emotional expressions (smiles, laughs, cries), looks, and communication with objects.

Social actors and individual construction

The last level of our theoretical orientation is about the actual influence of the environment, especially that of the social actors. According to Piaget's theory the social environment was not important for the development and was said to be the result of an endogenous process (Piaget 1967, 1974). Thus, the infant or the young child was supposed to build its abilities in an environment comprising nonsocial objects, and to have access to the social world only later. Nowadays the social competency of the infant, even if its precocity has been discussed (Pêcheux *et al.* 1985) is widely acknowledged (for a review, see Cosnier 1984). Thus Piaget's constructivist view of development becomes a social constructivist view (Mehan 1982), and child development takes place in a world where social actors are interacting together.

However, reference to the social environment is often made at a very general level. A given environment is supposed to be composed of specific and identifiable criteria: parents' education level, wealth or poverty. Such characterization, even if it draws interesting comparisons for instance in language construction in different environments (Snow *et al.* 1976; Adams & Ramey 1980), cannot account for the question: "why are the children in a same family so different from one another?" (Plomin & Daniels 1987). The description and understanding of the actual and precise influences of social actors on the development of individual communication implies a detailed characterization of a given environment. To distinguish the influence of what is shared by all the children of the same family and what is not shared seems of great interest (Plomin & Daniels 1987; Plomin *et al.* 1988). This leads us beyond the general concept of environment to the understanding of each child's own environment within the family and allows us to take individuality into account to improve the study of development.

Moreover the construction of the environment of each child is due to the child itself as actor of its own construction, as well as to the parents' influence who adapt their behavior to each child. The importance of favored social links (both in quality and quantity) for all acquisition processes has been shown (Bandura 1977).

However, it is noticeable that most studies still focus on the mother's importance, "forgetting" the father and probably correlatively overstating the mother's role. All the interactors of the family will be taken into account in this study.

In conclusion, it is as important to look at how interaction partners are chosen and what is shared with them, especially vocal and linguistic sharing, in the family context as it is in larger social contexts. These sharings can explain the cultural transmission between individuals in socially identified groups (Goodwin 1990). Moreover, it may reveal parallels with the acquisition of bird song, where the choice of partners is related to song sharing (Henry 1994; West *et al.* Chapter 4; Hausberger, Chapter 8).

THE STUDY

The individuals and the observations

Two dizygotic twin sisters, Julie and Laura, were video recorded in their home environment. They are the first children born in the family. Their parents are research colleagues, and we can document that their environment is of a particularly high quality from a social enrichment point of view.

Why this choice of twin sisters? Since the end of the nineteenth century (Galton 1875), a great number of large longitudinal studies of twin children were performed (for example, The Louisville Twin Study (Falkner 1957; Wilson 1983) and The MacArthur Longitudinal Twin Study (Plomin *et al.* 1988, 1993)). In most cases these studies aimed at separating the influence of genetic factors from that of environmental factors on developmental aspects in the child. But twin studies may also be considered differently, particularly in regard to our aim of defining shared and nonshared environment. To this purpose we found it particularly interesting to study two children of exactly the same age, genetically different, but brought up together in the same environment. Although delays and disturbances in language development (Zazzo 1960, 1984;

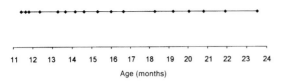

Fig. 16.1. Distribution of the number of observation during the year, and age of twins.

Bakker 1987) occur in 40% of twin children (Bakker 1987), mostly in homozygotic twins, we did not see any such effect in Julie and Laura, whose development was comparable to any other nontwin children coming from the same social environment.

The sessions lasted 45 minutes and took place in the evening, after the parents and their children had returned home between 6.00 and 7.30 p.m. The familial habits were maintained: during the observations, the two sisters played freely with their usual toys, while the father and mother had varied activities such as preparing the dinner or tidying. The parents could also play with the children. No modification was made in the environment and the observer merely asked the parents to behave as they did every day.

Data collection

The study started when the two children were 11 months old and stopped when they were four years old. This chapter presents one year of observation representing 17 sessions.

The observation frequency (Fig. 16.1) had to follow a regular pattern that was tolerated by the parents and took into account the rapid changes in, and the flexibility of, the children's behavior, during the first months (Emde 1988). Thus, two sessions per month were video recorded between ages 11 and 15 months and one session every month afterwards.

Ethological repertoire

The behaviors recorded on the video tapes were transcribed using codes corresponding to an ethological repertoire described earlier (Jouanjean-L'Antoëne 1982) and derived from the repertoire developed by Blurton Jones (Blurton Jones 1972; Blurton Jones & Woodson 1979). No selection was made *a priori* but all the individual behaviors

were described as much as possible in a precise and exhaustive way. Thus, the repertoire documents both nonverbal and vocal-verbal behaviors.

The repertoire of nonverbal behaviors is composed of "classical" ethological categories as, for example, approaches, physical contacts, facial expressions, and exchanges of objects. Each category includes a defined number of items.

The description of the repertoire of babbling and language is particularly complex, and raised different questions that required distinguishing between vocalizations and linguistic items.

It seemed simpler to identify linguistic occurrences, although the spoken language may be very different from the ideal model or the perfect adult expression free from any variation in pronunciation (Locke 1986).

Identifying vocal occurrences and babbling is much more complex and similar to the problems encountered by any ethologist studying unknown animal vocalizations. As in any other species there is a problem of natural categories (Marler 1982), since any infant vocalization can be seen as any kind of unknown sounds used in communication (Snowdon 1987), questioning the adultomorphic point of view that has prevailed up to now (Bauer & Stasny 1987).

Three principles have guided our methodology. First, we adopted an orthographic transcription following Poplack's (1993) directives, i.e., trying to retain the variability of production and being as regular as possible. Second, this transcription was done from direct audition of sounds, without (in this study) any sound spectrographic analysis. Finally, we did not restrict ourselves to the traditionally defined adult phonetic-linguistic categories, where the criteria of meaningful occurrence prevail (Bauer & Stasny 1987). Thus, following Bates (1976, 1979) and more recently Bloom (1993), each vocal occurrence uttered was considered in its pragmatic context, taking into account the behaviors of both the emitter and the receiver, their speech, the objects, and all kinds of information from the immediate environment.

We collected the entire corpus of vocal sounds included in exchanges between the protagonists of the family. No special problem (for the observer) occurred in the comprehension of the adults' vocal or verbal productions, except some verbal contractions very frequently used in spoken language (for example "y a" standing for "il y a", and "t'en as un aut'e" for "tu en as un autre"). The transcription of the children's utterances raised more difficulties. First, we had to distinguish precisely between vocalizations occurring during the babbling period and the first verbal expressions. It is a difficult task during the transition period between babbling and language. As I said above, we wanted to escape from the adultomorphic tendency of transcription, and we tried to combine the formal aspects of the sounds with the context of their production. In this way a vocal unit was considered as a word not only when it sounds like a word (adult form of words), but also when it was emitted in association with a pragmatic element which indicated that it stands for a word. For example, "wouwou" was considered as a linguistic production, despite its onomatopoeic form, when it was emitted as one child showed a furry toy dog to the mother from whom she obtained a response, or when a vocalization like "apu" (nothing left) was produced by a child when looking at an empty box.

Correlative to this method of classifying language, all the other productions were considered as vocalizations in which we made a distinction between monosyllabic and multisyllabic occurrences. Multisyllabic vocalizations could be formed by the duplication of a given sound (for example "aaa . . .") produced in any separated expression, or by the combination of several sounds or phonemes (for example "aeu . . ."). This distinction was not made *a priori* but chosen because it appeared to us that the parents were more sensitive to combined vocalizations than to monosyllabic ones. This methodological choice illustrates our will to elaborate the criteria from the data themselves (Bloom 1993) and to take into account the actual communicative function of the vocal production of the children.

Computer transcription

All of the observed behaviors were coded by three letters, and then transcribed on a computer. The first letters of the code stood for the general behavioral category (e.g., A for all kinds of approaches; F for facial expressions and gazes). The other two letters characterize the precise item inside the general category.

A computer program was devised specially (Rety 1994), allowing many types of analyses both at a general level of categories and on separate items or on any associated items chosen.

One of the major interests of this method was that the

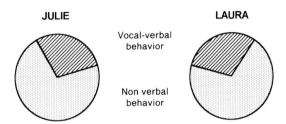

Fig. 16.2. Proportion of nonverbal and vocal–verbal behaviors expressed by Julie and Laura in communicative sequences with their parents.

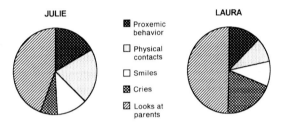

Fig. 16.3. Nonverbal behavior expressed by Julie and Laura in communicative sequences with their parents.

data selection was not performed *a priori* (see above) but after collecting and describing most of the communicative behaviors. Only then did we focus on particularly significant types of behavior.

RESULTS

Here I give first a general description of the communicative behavior of the two sisters, Julie and Laura, toward their parents, and then relate the observed interindividual differences both to their individual choices and styles in communicative situations and to the parents' behavior toward each child.

Communication and language in the two children

First, if we consider the total number of behaviors expressed by each child during the 17 sessions in the year, comparing nonverbal behaviors to vocal-linguistic behavior, the differences are not significant (Fig. 16.2).

But if we examine in detail some of the categories within the whole "nonverbal behavior" the two sisters' behaviors were very different (Fig. 16.3). Julie expressed

Table 16.1. *Frequencies of vocal and linguistic occurrences in the total amount of speech addressed by the twin sisters towards their parents*

	Julie	Laura
Vocal occurrences	573	516
Linguistic occurrences	480	634
Total amount of speech	1053	1150

Notes:
$[X]^2 = 20.04$, df = 1, $p < 0.001$.

Table 16.2. *Total amount of monsyllabic and multisyllabic vocalizations emitted by the twin sisters*

	Julie	Laura
Monosyllabic vocalizations	292	213
Multisyllabic vocalizations	281	303
Total	573	516

Notes:
$[X]^2 = 10.23$, df = 1, $p < 0.01$.

more proxemic behaviors, positive contacts, kisses, smiles, and laughs towards her parents than her sister did. On the contrary, Laura who approached her parents less often than Julie, looked and cried toward her parents more often than did her sister. The two sisters were also very different in their tendency to use objects as means of communication: Julie offered, showed, and exchanged, more objects with her parents than Laura did.

Julie and Laura are also different in their vocal and linguistic mode of communication: Julie vocalized more than Laura, who spoke to the parents more than Julie did (Table 16.1). Differences also appeared between both children in the temporal changes of these two modes of communication (Fig. 16.4).

For Laura, the maximum frequency of vocalizations occurred with the first (small) increase of linguistic occurrences at 15.09 months. After that, a decrease of both vocal and verbal occurrences was observed before the "explosion" of language at 18 months. This sudden increase was accompanied by an important amount of vocalizations. But after the nineteenth month of age the number of vocalizations decreased rapidly and they disappeared at the twenty-second month.

For Julie, whose linguistic ability emerged a little more slowly, we can see that vocalizations were emitted more regularly between 11 and 16 months. When the first words emerged (16 months), the number of vocalizations decreased, but when the verbal skill appeared the vocalizations were once again numerous. Vocalizations were still observed at the end of the year of study.

These results, which clearly differentiated the two children, suggest a more complex relationship between vocalizations and language than had been expected at first. We tried further therefore to explain the sisters' differences by analyzing more precisely their vocal behavior. For this purpose we distinguished between simple and complex vocalizations, as described in our methodology (Table 16.2).

Julie emitted more monosyllabic vocalizations than did her sister, and in contrast Laura emitted more multisyllabic vocalizations than Julie did. These results support the general hypothesis of the preparing function of complex babbling, but also suggest there is not necessarily a relation between the amount of simple vocalizations and further complex babbling. It is noticeable that Julie emitted vocalizations using a low voice, sometimes even with her lips closed. Moreover she frequently emitted short interjections such as "hein?" or "heu?" with an interrogative intonation. This kind of vocalizing behavior appeared more like a means of attracting the parents' attention than like an actual preparation to linguistic mastery. This suggests a different usage of vocal behavior by the two sisters: Julie using vocalizations to keep contact with the parents, Laura more interested in language itself, and trying to conform to the linguistic model of her parents.

In conclusion, beyond the general similarity, which was noticeable and perhaps reflected the general cultural level of this family, the twin sisters behaved very differently. Laura, who exhibited more looks, fewer approaches, fewer physical contacts and more negative emotions toward her parents than her sister did, mastered the linguistic code more rapidly than Julie.

Julie's linguistic skill developed more slowly than that of her sister. The numerous interjections in her vocalizations suggested her expressed will to be "in touch" with her parents, which is reminiscent of her general tendency to approach and have more contacts with her parents than her sister. The communication using objects was also different in that Julie offered, showed, and exchanged more objects with her parents than Laura did.

Thus, the differences observed give evidence of a per-

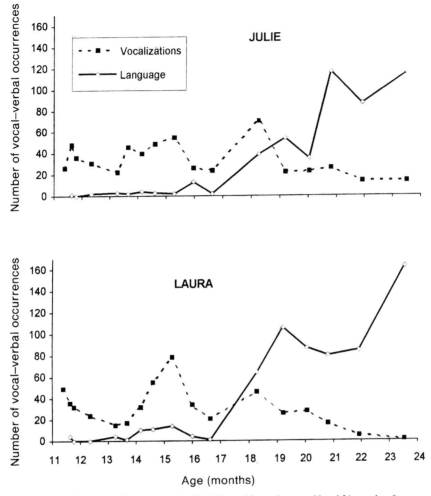

Fig. 16.4. Vocalizations and language emitted by Julie and Laura between 11 and 24 months of age.

sonal behavioral and emotional style of each child: we can say that the more "distant", Laura, is more interested in rapid language mastery, but Julie, who preferred direct contact and exchange objects, is more progressive in her language acquisition.

I now move toward the search for causality to explain these differences particularly by examining parents' behaviors.

Parents' presence and the question of shared/nonshared environment

I want here to look at the parents' behavior toward the children. The first noticeable fact is the great difference in the amount of time each parent was present during the sessions. The mother was present at each of the 17 sessions, but the father was absent (for professional reasons) during six sessions when the children were between 13.20 months and 16.17 months. Moreover, when both parents were present at home during a session, the timing of their presence was not equal. The mother was present during 82% and the father 23% of the total observation time. The actual influence expressed by the amount of time and number of behaviors addressed toward the children by each parent was different. But the comparison of the mean number of behaviors per minute expressed by each parent appeared to be quite similar (Table 16.3). At this global level, parents, and also children as previously mentioned,

Table 16.3. *Behaviors emitted by the parents*

	Father	Mother
Total amount of behaviors	1029	3947
Presence time (min)	167.13	586.65
Mean number of behaviors/min	6.15	6.72

Table 16.4. *Vocal–verbal occurrences observed in dyads between parents and children*

Dyads mother/children			
Julie → mother	695	Laura → mother	765
mother → Julie	1151	mother → Laura	1265
Dyads father/children			
Julie → father	214	Laura → father	186
father → Julie	289	father → Laura	221

are comparable, which may be related to the general cultural characteristics of this particular social environment. These characteristics are part of the shared environment for children of a same family (Plomin & Daniels 1987; Plomin *et al.* 1988).

Particularly because our subjects are twin sisters we further investigated what proportion of behaviors the mother and the father addressed toward each of the two sisters at the same time. Some authors hypothesize that one reason language is, in some cases, acquired more slowly by twins could be that parents of twins do not individualize their behaviors and especially their language toward each child. On the contrary, parents of twins are supposed to address themselves to both children at once. Surprisingly this proportion of common addresses in our study was very low: only 8.64% for the father and 12.82% for the mother. If these parents are comparable at a general level of behavior, they rarely communicated with both children at the same time and conversely each parent was probably adapting her or his behavior according to each child's individuality.

This result supports the idea that the influence of the nonshared environment may be very important, and can explain differences between children in the same family. This led us to look at reciprocal behaviors in the dyads formed by parents and their children when they communicated with each other.

Reciprocity in parents' and children's communicative behavior

Here I consider the total number of behavior each protagonist addressed to the others, first in dyads with the mother, when each sister was the emitter and sometimes receiver, and then in dyads with the father (Fig. 16.5).

Laura addressed more behaviors to her mother than Julie did, and the mother expressed the same tendency

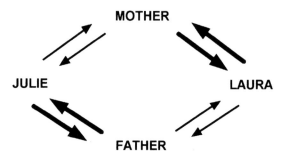

Fig. 16.5. Behaviors observed in dyads between parents and children. Thicker arrows indicated more interactions.

toward Laura. In contrast, in dyads with the father Julie emitted and received more behaviors than Laura did.

Thus, despite the great difference in the time each parent was present, the reciprocity in the choices of these parents and their twin children is very interesting and led us to wonder about the influence of each parent on each child, with regard to language development in particular.

It is remarkable that, in all cases, the total number of behaviors addressed by children to their parents was more important than the number of behaviors they received from their parents. This asymmetry suggested that children are active constructors of their own environment. Despite the fact that the twins shared the same environment with the same parents and moreover spent a long time together (the sisters were never separated during this period), individual and reciprocal choices between parents and children were clearly observed.

We then wanted to analyze possible reciprocity in vocal–verbal exchanges. We found the same reciprocal tendencies (Table 16.4): the mother exchanging more with Laura than with Julie, the father exchanging more with Julie than with Laura. An asymmetry was therefore observed: the children received more vocal–verbal exchanges from the parents than they produced toward them. One reason is certainly that at the beginning of the observations the twins' language was not yet developed and emerged during the year of observation, but also the parents were stimulating their children to create a scaffold for their linguistic acquisition.

These differences were not significant and perhaps reflected the parents' will to be egalitarian with their children, as they told us several times during spontaneous conversations about their educational practices. They also were aware of possible psychological problems in a twin situation in which children should be particularly sensitive to differences. However, the parents' apparent egalitarian desires did not prevent the emergence of differences.

Vocal–verbal reciprocal imitations or vocal–verbal shared patterns in parents and children

Imitation will be considered here, not only as a process acting on the linguistic acquisition, but also as reflecting shared topics in our naturalistic setting. Because these imitations are emitted in naturalistic exchanges and conversations, they can also be considered as sharing (Uzgiris et al. 1989), i.e., reflecting vocal–verbal spontaneous shared topics between the interacting protagonists. If differences appear in interactions between the two sisters and parents, this behavior will show more precisely how social influences are acting in this family via the speech exchanges.

The first result must be strongly emphasized: a large proportion of speech consisted in imitations (at the most global level of analysis) during interactions (Fig. 16.6). The global proportions are far more important than the proportions published in the literature, for children and for parents (see e.g., Bloom et al. 1974; Snow 1981).

At a global level, Julie is the most imitative child but only in the proportion of self-repeated speech. The father was more imitative than the mother. However, it must be remembered that these results are calculated from a limited amount of speech emitted by the father as compared to the mother, due to the limited duration of the presence of the father. The proportion of self-repeated speech was almost identical for both parents, but lower than that observed for the children.

In dyads with the father, the total amount of imitative

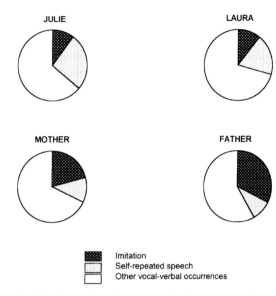

Fig. 16.6. Proportion of imitations and self-repeated speech in the total amount of vocal–verbal occurrences.

speech addressed by the father to each child was almost identical and reciprocal (Table 16.5). To be more precise, Laura imitated her father's language more than did Julie, who on the contrary repeated her own speech more often than did her sister. The father imitated more vocalizations emitted by Julie and more language occurrences of Laura, as expected because Julie generally emitted more vocalizations than her sister and Laura more language. But it is noticeable that the father tended to address more expanded imitations towards Julie than toward Laura, which suggests that he reinforced his teaching role more toward Julie than he did toward Laura.

Dyads with the mother appeared to be very different (Table 16.6). Julie addressed more and received fewer imitative occurrences from the mother than Laura did. These differences are significant ($\chi^2=15.51$, df$=1$, $p<0.001$). Laura's vocalizations in particular were far more reinforced than Julie's from the first stages of acquisition onward. Thus, the developmental course of expanded imitations showed that the mother addressed a greater number of these kinds of imitation to Laura, particularly between 13 and 15 months of age, and that the burst observed toward Laura occurred a month before the same phenomenon appeared toward Julie (Fig. 16.7).

Table 16.5. *Imitations in the dyads between the twin sisters and their father*

Dyads children → father	Julie	Laura
Imitated		
Vocalizations	0	1
Language	11	22
Self-repeated		
Vocalizations	14	2
Language	35	27
Total	60	52

Dyads father → children	Julie	Laura
Imitated		
Vocalizations	8	3
Language	15	25
Expanded imitations	67	59
Total	90	87

Total 16.6. *Imitations in dyads between the twin sisters and their mother*

Dyads children → mother	Julie	Laura
Imitated		
Vocalizations	23	9
Language	38	64
Self-repeated		
Vocalizations	145	90
Language	57	74
Total	263	237

Dyads mother → children	Julie	Laura
Imitated		
Vocalizations	37	54
Language	24	44
Expanded imitations	162	230
Total	223	328

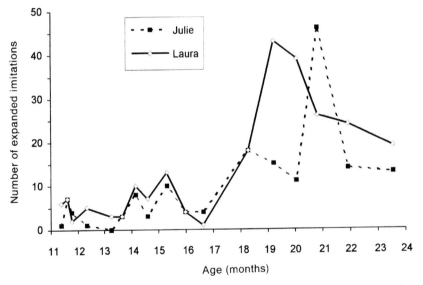

Fig. 16.7. Expanded imitations emitted by the mother toward Julie and Laura between 11 and 22 months of age.

DISCUSSION AND CONCLUSIONS

Our choice of studying two twin sisters was a risk because the twin situation raises special educational difficulties that can lead to developmental psychological troubles for the children. There is a paradox in the parents' educational practices in that they must be egalitarian with two children who are at the same age and probably expressing similar needs at the same time, but they must also differentiate between the children to allow them to construct their individuality (Robin & Casati 1994a,b). Some parents of twins tend to be too egalitarian and the children receive the same toys, are dressed with the same clothes, share all the everyday activities of the family, and are rarely separated (Robin & Casati 1994a,b). Their common life, or shared environment, is sometimes so important that the twins are behaving like a couple living in cryptophasia (Zazzo 1960, 1984) and developing glossophasia (Bakker 1987). This means that twins, mainly homozygotic twins, sometimes tend to speak a secret personal language that cannot be understood by other people, even by their parents.

The family we studied showed many aspects of twin families, but beyond the general similarity that reflected the general cultural level of this family, and despite the great importance of the shared environment of the children, these twin sisters never expressed any psychological trouble that can be considered as cryptophasia or glossophasia. Moreover the results showed that the parents succeeded in managing both egalitarian and differential educational practices.

Thus, the father and the mother were *similar at the more global level of their behavior* (for example, considering the total amount of communicative behaviors addressed to each child or the amount of vocal and verbal exchanges) and this can be considered as showing the egalitarian educational level that was consciously wanted by these parents for their children. This result also might represent the level of the shared environment of children in the same family and might represent the social characteristics of a given environment.

In the twin situation, the parents are also aware that it is necessary to differentiate each child in order to prevent psychological difficulties. In our study the parents' behaviors were rarely addressed to both children at the same time (only about 10%). On the contrary, the parents were addressing one child at a time and for this reason they also succeeded in *taking into account the individuality* of each child. Correlated with the general egalitarian attitude and the general similarity of the parents' behaviors, the results showed a great resemblance between the sisters at this behavioral level, and this probably reflects the influence of that particular social shared environment on the children.

The detailed analysis of the different categories of behavior expressed by each child showed their individual differences, both in their nonverbal behaviors, as for example approaches, emotional expressions, object exchanges and in their language development. These results led us to hypothesize the existence of *an actual relationship between the nonverbal typology of each child and her own language rhythm of acquisition.* Thus, Laura, who was

more "distant", using fewer approaches, fewer physical contacts, and fewer object exchanges than her sister, developed the language skills earlier than did Julie. Conversely Julie, who seemed to prefer nonverbal ways of communication developed language a little slower than did her sister. Beyond these individual differences, the two sisters showed a normal language development, and Julie, who was a little slower than her sister, cannot be considered as having a delayed acquisition.

Returning to the fact that the parents' behaviors were rarely addressed to the two children but were on the contrary clearly individualized, we noticed that the mother tended to address more behaviors to Laura than to Julie and that the father chose Julie a little more often than Laura. These *choices of favored protagonists were reciprocal.* Because the amount of the behaviors emitted by the children was greater than the amount expressed by the parents it appears that the children were making these choices themselves. These children appeared as *active constructors of their own development.*

These results acknowledge the social constructivist view (Mehan 1982), and we must consider that the social construction of children begins in the family. Moreover, the data illustrate that the distinction made between the shared and nonshared environment (Plomin *et al.* 1988) is fruitful and might be taken into account to explain the differences between the children reared in a same family (Plomin & Daniels 1987). Beyond the general level of similarity of a given environment there are more subtle causalities explaining individual differences that can only be revealed by *longitudinal studies* integrating various aspects of behavioral development expressed in actual communicative exchanges.

The same remarks hold in part for *language development.* Like the other behaviors, the language occurrences were clearly individualized and the protagonists showed the same tendencies in their individual choices. This is especially important because some authors hypothesized that the delay sometimes noticed in language acquisition by the twins might be due to common language addresses by parents to both children simultaneously (Bakker 1987). On the contrary, our twin sisters developed differentially but normally. However, it is more difficult to evaluate the exact influence of each parent on the linguistic development of each child. The mother exchanged (addressed and received) significantly more language with Laura than with Julie, and this can explain the more rapid mastering of language by Laura. But the role of the father on the language development of the two sisters raised more questions. We must emphasize that the father was always present far less often than the mother, and his prolonged absence from home took place when the children were between 14 and 17 months old, i.e., during the transition between the babbling period and the actual emergence of the language. This absence may have been of crucial importance particularly for Julie, who chose the father as her favored protagonist. Moreover, the father appeared more egalitarian than the mother in the amount of vocal and verbal occurrences addressed to each child, but his tendency to choose Julie a little more often could not "compensate" for the greater amount of reciprocal choices between the mother and Laura. We must also re-address here the childrens' own behaviors in order to have a more integrative view of their linguistic development.

Our distinction in language between the vocal and verbal occurrences was made at first to examine the contribution of each modality to the linguistic skill of each child. We wanted to see how the vocalizations of each child were preparing their individual access to language mastery. The results showed that *a simple and direct link between babbling and language must be questioned.* Julie, who emitted the greater total amount of vocalizations, was more progressive in language acquisition than Laura. But we noticed that Julie emitted more monosyllabic vocalizations and Laura, conversely, preferred using more complex vocal productions. Laura's complex vocal behavior may explain her rapid language mastery, and Julie seemed to use vocal expressions more in an interactive way than in efforts to master language rapidly. Perhaps this different use can explain the difference in the developmental course in language acquisition by the two sisters. These results give indications of why some children are more progressive in language acquisition whereas other children are "explosive" (Nelson 1973, 1981).

Because our study focuses on both parents and children, it also gives an opportunity to go beyond "the direct maternal influence model" (Yoder & Kaiser 1989, p. 142) toward an interactional view of linguistic development, taking into account bidirectional influences and the influence of the child itself on the mother's behavior. We could therefore suggest that the more rapid mastery of language by Laura, compared to Julie, is the result of her personal tendency to express more complex vocalizations that

better stimulate the mother, especially in the transition period between vocalizations and language when the father was absent.

Our results on language development in these twin sisters underline the necessity of integrative naturalistic studies. It is well known that parents are providing a scaffold for the vocal and verbal acquisitions of the infant and the child, but the actual effects of the *mother's influence* on the rhythm of acquisition has been discussed only recently (Hampson & Nelson 1993) because it is not clear how this influence is acting. What is the kind of input acting on the emerging linguistic skill in the child? Is it specifically the amount and form of language addressed to the child or are the opportunities to communicate given by the parents more important (Hoof-Ginsburg 1990)? And how can the *father's influence* be taken into account? In a recent review, Le Chanu & Marcos (1994) stressed that, if fathers resemble mothers in several aspects of their language, they are also different, particularly in the vocabulary addressed to children and in functional and conversational aspects of the model they are providing to their children. A final question is: "are *all children responding the same way* to identical stimulation provided by the parents?" Nelson (1973) showed that children have different strategies when acquiring the linguistic code: some of them called referential children have an analytic way of acquisition, whereas others have a more holistic access to language. It is an interesting possible development of our study to go in search of further causalities to explain the differences between the two sisters.

Finally there are common points between our research and the nonhuman animal studies. We stressed the necessity of studying individuals in their spontaneous communicative expressions. The main interest is that naturalistic studies lead to a better, integrative view of the development of communication. For example, in the context of social relationships animals tend to express behaviors that are not observed in controlled situations or in captive conditions (West *et al.*, Chapter 4). This is in agreement with our results on the greater amount of imitations compared to previous other results in experimental settings. Such naturalistic studies also give us an opportunity to the actual determinants of the observed behaviors. That is striking, and supporting parallels between several species is *the asserted evidence that the vocal means of communication and the language in the human species are related to favored*

social interactions and relationships. In the European starling (*Sturnus vulgaris*) (Henry 1994; Hausberger 1993, and
see Chapter 8) or in Australian magpies (Brown &
Farabaugh, Chapter 7) shared songs are observed in close
companions. In wild dolphins (*Tursiops truncatus*), imitation was shown between individuals who shared strong
social bonds (Tyack & Sayigh, Chapter 11). Tyach &
Sayigh also found that two calves had whistles that resembled those of the mother and that these two individuals
spent more time surfacing closely to their mother than did
the other observed animals.

In conclusion, our study and the others we referred to
showed the importance, both in several nonhuman species
and in the human species, of studying the emergence of
communication in the context of naturalistic exchanges
between individuals. Because every child (or every young
animal of the species cited) learned from its parents (and
conspecifics) behaving as active tutors, and moreover
because the young is himself or herself an active learner
developing his or her own learning strategy, the learning
processes must be studied in actual social contexts trying
to integrate the behaviors of the different social actors. We
showed the importance of favored reciprocal relations
between each parent and each child, and tried to relate
these individual choices to the development of the linguistic skills. We were also convinced that all the protagonists
in the human family have to be studied. The father, even if
he is not always present at home, appeared very important,
perhaps more qualitatively than quantitatively.

ACKNOWLEDGEMENTS

This research was supported by the CNRS. I would like
to thank the family, the parents and their twins, who participated in the longitudinal study for more than three
years. I am very grateful to all colleagues with whom
fruitful discussions have occurred during the preparation
of my Thèse d'Etat, from which these results are
extracted: J. Cosnier, R. Tremblay, J. Y. Gautier, J.
Bernicot, M. Robin, C. Garitte, and J. S. Pierre. I am also
grateful to M. Hausberger and C. Snowdon, who offered
me the opportunity to write a chapter for this book and
made helpful comments on the first draft of this chapter.
I acknowledge H. Schuelke-Grillou for her assistance all
through the research process and especially in preparing
this manuscript.

REFERENCES

Adams, J. & Ramey, C. (1980). Structural aspects of maternal
speech to infants reared in poverty. *Child Development* **51**:
1280–4.

Bakker, P. (1987). Autonomous languages of twins. *Acta
Geneticae Medicae et Gemellologiae* **36**: 233–8.

Bandura, A. (1977). *Social Modeling Theory*. Aldine-Atherton,
Chicago.

Bates, E. (1976). *Language and Context. The Acquisition of
Pragmatics*. Academic Press, New York.

Bates, E. (1979). *The Emergence of Symbols. Cognition and
Communication in Infancy*. Academic Press, New York.

Bauer, H. R. & Stasny, E. A. (1987). An ethologic model of
phonetic development: Nonrandom asymmetries in
prelinguistic and linguistic utterances. *Technical Report* no.
379, Department of Statistics, Ohio State University, pp.
1–25.

Bernard, C. (1965). *Introduction à l'étude de la médecine
expérimentale*, second edition. Flammarion, Paris. (First
edition, 1865).

Bernicot, J. (1992). *Les actes de langage chez l'enfant*. PUF, Paris.

Blake, J. & Fink, R. (1987). Sound-meaning correspondences in
babbling. *Journal of Child Language* **14**: 229–53.

Bloom, L. (1993). Transcription and coding for child language
research: The parts are more than the whole. In *Talking
Data: Transcription and Coding in Discourse Research*, ed. J.
Edwards & M. Lampert, pp. 149–66. Erlbaum, Hillsdale,
NJ.

Bloom, L., Hood, L. & Lightbown, P. (1974). Imitation in
language development: If, when and why. *Cognitive
Psychology* **6**: 380–420.

Blurton Jones, N. (1972). *Ethological Studies of Child Behaviour*.
Cambridge University Press, London.

Blurton Jones, N. G. & Woodson, R. H. (1979). Describing
behavior: The ethologists' perspective. In *Social Interaction
Analysis: Methodological Issues*, ed. M. E. Lamb, S. J.
Suomi & G. R. Stephenson, pp. 99–118. University of
Wisconsin Press, Madison.

Brown, R. (1973). *A First Language. The Early Stages*. George
Allen & Unwin Ltd, London.

Chomsky, N. (1965). *Aspects of the Theory of Syntax*. MIT Press,
Cambridge, MA.

Clark, G. A. (1994). Origine de l'homme: Le dialogue de sourds.
Recherche (Paris) **25**: 316–21.

Cosnier, J. (1977). Communication non verbale et langage.
Psychologie Médicale **9**: 2033–49.

Cosnier, J. (1984). Observation directe des interactions précoces
ou les bases de l'épigenèse interactionnelle. *Psychiatrie de
l'Enfant* **27**: 107–26.

Cosnier, J. & Brossard, A. (1984). *La communication non verbale*.

Textes de base en psychologie. Delachaux & Niestlé, Neuchâtel.

Cronbach, L. J. (1957). The two disciplines of scientific psychology. *American Psychologist* **12**: 671–84.

de Boysson-Bardies, B. (1984). L'influence des langues-cibles sur le développement de la parole: Etudes comparatives sur des enfants de 6 à 10 mois. In *Construction et actualisation du langage*, ed. M. Moscato & G. Pieraut-Le Bonniec, pp. 51–68. Presses Universitaires de Rouen, Rouen.

de Boysson-Bardies, B., Sagart, L. & Durand, C. (1984). Discernible differences in the babbling of infants according to target language. *Journal of Child Language* **11**: 1–15.

de Boysson-Bardies, B., Sagart, L., Halle, P. & Durand, C. (1985). Acoustic investigations of cross-linguistic variability in babbling. In *Precursors of Early Speech*, ed. B. Linblom & R. Zetterstrom, pp. 113–116. Macmillan Press, Basingstoke.

Emde, R. N. (1988). Development terminable and interminable. I. Innate and motivational factors from infancy. *International Journal of Psycho-Analysis* **69**: 23–42.

Falkner, F. (1957). An appraisal of the potential contribution of longitudinal twin studies. In *The Nature and Transmission of the Genetic and Cultural Characteristics of Human Populations*. Milbank Memorial Fund, New York.

Ferguson, C. A. & Farwell, C. B. (1975). Words and sounds in early language acquisition. *Language* **51**: 419–39.

Galton, F. (1875). The history of twins as a criterion of the relative powers of nature and nurture. *Journal of the Anthropological Institute* **6**: 391–406.

Goffman, E. (1974). *Les rites d'interaction*. Editions de Minuit, Paris.

Goffman, E. (1987). *Façons de parler*. Editions de Minuit, Paris.

Goodwin, M. H. (1990). *He Said, She Said. Talk as Social Organization among Black Children*. Indiana University Press, Bloomington, IN.

Hampson, J. & Nelson, K. (1993). The relation of maternal language to variation in rate and style of language acquisition. *Journal of Child Language* **20**: 313–42.

Hausberger, M. (1993). How studies on vocal communication in birds contribute to a comparative approach of cognition. *Etología* **3**: 171–85.

Henry, L. (1994). Influences du contexte sur le comportement vocal et socio-sexuel de la femelle étourneau (*Sturnus vulgaris*). Thèse, Université de Rennes.

Hoof-Ginsberg, E. (1990). Maternal speech and the child development of syntax: A further look. *Journal of Child Language* **17**: 85–99.

Jakobson, R. (1963). *Essais de linguistique générale*. Editions de Minuit, Paris.

Jakobson, R. (1968). *Child Language, Aphasia and Phonological Universals*. Mouton, The Hague. (Originally published 1941).

Jouanjean-L'Antoëne, A. (1982). Etude préliminaire de la communication gestuelle et verbale chez 16 enfants de 2 à 3 ans observés dans une crèche de Rennes: Cas particulier du geste de pointer du doigt. Thèse 3ème Cycle, Université de Rennes I.

LeChanu, M. & Marcos, H. (1994). Father–child and mother–child speech: A perspective on parental roles. *Europopean Journal of Psychology of Education* **9**: 3–13.

Lenneberg, E. H. (1967). *Biological Foundations of Language*. Wiley, London.

Lieberman, P. (1984). *The Biology and Evolution of Language*. Harvard University Press, Cambridge, MA.

Locke, J. L. (1983). *Phonological Acquisition and Change*. Academic Press, London.

Locke, J. L. (1986). Speech perception and the emergent lexicon: An ethological approach. In *Language Acquisition*, ed. P. Fletcher & M. Garman, pp. 240–50. Cambridge University Press, Cambridge.

MacNeilage, P. F. (1980). The control of speech production. In *Child Phonology*, ed. G. H. Yeni-Komshian, J. F. Kavanagh & C. A. Ferguson, pp. 9–21. Academic Press, New York.

Marler, P. (1982). Avian and primate communication: The problem of natural categories. *Neuroscience and Biobehavioral Reviews* **6**: 84–94.

Mehan, H. (1982). Le constructivisme social en psychologie et en sociologie. *Sociologie et Sociétés* **14**: 77–96.

Moerk, E. L. (1977). Processes and products of imitation: Additional evidence that imitation is progressive. *Journal of Psycholinguistic Research* **6**: 187–202.

Moerk, E. L. (1989). The fuzzy set called "imitation". In *The Many Faces of Imitation in Language Learning*, ed. G. E. Speidel & K. E. Nelson, pp. 277–300. Springer-Verlag, New York.

Nelson, K. (1973). Structure and strategy in learning to talk. *Society for Research in Child Development Monographs* **38** (1–2), Serial no. 149.

Nelson, K. (1981). Acquisition of words by first-language learners. *Annals of the New York Academy of Sciences* **379**: 148–59.

Oller, D. K. (1980). The emergence of the sounds of speech in infancy. In *Child Phonology*, vol. 1, *Production*, ed. G. Yeni-Komshian, J. Kavanagh & C. Ferguson, pp. 93–112. Academic Press, New York.

Oller, D. K. (1986). Metaphonology and infant vocalizations. In *Precursors of Early Speech*, ed. F. Lindbloom & R. Zetterstrom, pp. 21–35. Macmillan, Basingstoke.

Pêcheux, M. G., Streri, A. & Deleau, M. (1985). Discussion autour de la notion de compétence chez le nouveau né.

Psychologie Française **30**: 153–62.

Piaget, J. (1923). *Le Langage et la pensée chez l'enfant*. Delachaux et Niestlé, Neuchâtel.

Piaget, J. (1967). *Biologie et connaissance: Essai sur les relations entre régulations organiques et processus cognitifs*. Gallimard, Paris.

Piaget, J. (1974). *Adaptation vitale et psychologie de l'intelligence*. Hermann, Paris.

Plomin, R. & Daniels, D. (1987). Why are children in the same family so different from one another? *Behavioral and Brain Sciences* **10**: 1–60.

Plomin, R., De Fries, J. C. & Fulker, D. W. (1988). *Nature and Nurture during Infancy and Early Childhood*. Cambridge University Press, Cambridge.

Plomin, R., Emde, R. N., Braungart, J. L., Campos, J., Corley, R., Fulker, D. W., Kagan, J., Reznick, J. S., Robinson, J., Zahn-Waxler, C. & De Fries, J. C. (1993). Genetic change and continuity from fourteen to twenty months: The MacArthur Longitudinal Twin Study. *Child Development* **64**: 1354–76.

Poplack, S. (1993). Variation theory and language contact. *American Dialect Research* 252–86.

Rety, B. (1994). Logiciel en Pascal pour l'analyse de données comportementales sur ordinateur PC. Unpublished computer program.

Robin, M. & Casati, I. (1994a). Are twins different from singletons during early childhood? *Early Development and Parenting* **3**: 211–21.

Robin, M. & Casati, I. (1994b). Existe-t-il une spécificité du développement psychologique précoce des jumeaux? *Neuropsychiatrie de l'Enfance* **42**: 180–9.

Schneirla, T. C. (1966). Behavioral development and comparative psychology. *Quarterly Review of Biology* **41**: 283–302.

Searle, J. R. (1972). *Les actes de langage*, second edition. Hermann, Paris. (First edition, 1969).

Snow, C. E. (1981). The uses of imitation. *Journal of Child Language* **8**: 205–12.

Snow, C. E., Arlnan-Rupp, A., Hassing, Y., Jobse, J., Joosten, J. & Vorster, J. (1976) Mothers' speech in three social classes. *Journal of Psycholinguistic Research* **5**: 1–20.

Snowdon, C. T. (1987). A comparative approach to vocal communication. *Nebraska Symposium on Motivation* **35**: 145–99.

Speidel, G. E. & Nelson, K. E. (1989). *The Many Faces of Imitation in Language Learning*. Springer-Verlag, New York.

Uzgiris, C., Broom, S. & Kruper, J. C. (1989). Imitation in mother–child conversations: A focus on the mother. In *The Many Faces of Imitation in Language Learning*, ed. G. E. Speidel & K. E. Nelson, pp. 91–120. Springer-Verlag, New York.

Vihman, M. M., Macken, M. A., Miller, R., Simmons, H. & Miller, J. (1985). From babbling to speech: A reassessment of the continuity issue. *Language* **61**: 397–445.

Vygotski, L. S. (1985). *Pensée et langage*, second edition. Editions Sociales, Paris. (First edition, 1934).

Wilson, R. S. (1983). The Louisville Twin Study: Developmental synchronies in behavior. *Child Development* **54**: 298–316.

Yoder, P. J. & Kaiser, A. P. (1989). Alternative explanations for the relationship between maternal verbal interaction style and child language development. *Journal of Child Language* **16**: 141–60.

Zazzo, R. (1960). *Les jumeaux, le couple et la personne* vol. II, *L'individuation psychologique*. PUF, Paris.

Zazzo, R. (1984). *Le paradoxe des jumeaux*. Stock, Paris.

17 Crafting activities: Building social organization through language in girls' and boys' groups

MARJORIE HARNESS GOODWIN

THE SOCIAL FUNCTION OF LANGUAGE

This chapter analyzes alternative types of conversational action used to build social organization among girls and boys in an African-American working class neighborhood in Philadelphia. Participants work together to generate distinctive definitions of the situation appropriate to the task at hand, and the same individuals articulate talk and gender differently as they move from one activity to another. Making use of the same language system, children select alternative ways of putting these forms to use, constructing a range of diverse activites and social arrangements that can highlight either affiliation or competition.

From the perspective of ethology, Cullen (1972, p. 101) has argued that "all social life in animals depends on the coordination of interactions between them." To achieve collaborative activity humans need to display to one another culturally meaningful behavior – articulating for their recipients what they are up to and how they expect others to respond. Sociologist Georg Simmel (1950, pp. 21–2) has stated that "if society is concerned as interaction among individuals, the description of the forms of this interaction is the task of the science of society in its strictest and most essential sense." In that language provides the tool through which humans coordinate their behavior, then what is required for an adequate understanding of social organization is close attention to talk itself.

Conversation analysis, an approach to the study of naturally occurring interaction developed within sociology by the late Harvey Sacks and his colleagues,[1] provides a

methodology for rigorously describing the procedures used by participants in conducting talk in interaction. This discipline proceeds under the assumption (one very akin to ethology) that order is observable in conversational materials because participants must display order *to one another*:

If the materials (records of natural conversations) were orderly, they were so because they had been methodically produced by members of the society for one another, and it was a feature of the conversation that we treated as data that they were produced so as to allow the display by the co-participants to each other of their orderliness, and to allow the participants to display to each other their analysis, appreciation, and use of that orderliness. (Schegloff & Sacks 1973, p. 290)

FIELDWORK AND THEORETICAL APPROACH

The present study is based on fieldwork conducted over 18 months among a group of children, in an African-American working class neighborhood of southwest Philadelphia, whom I encountered during a walk around my neighborhood. Seeing a group of girls jumping over a rope (skipping), I asked if I could watch them, and told them that I was interested in making observations of their play over a long period of time, because I was doing a study of the everyday activities of children. I informed them that I wanted as accurate a record of what went on as possible

---

[1] Important collections of research in conversation analysis can be found in books by Schenkein (1978), Atkinson & Heritage (1984), Button & Lee (1987) the special double issue of *Sociological Inquiry* edited by Zimmerman & West (1980), and a special issue of *Human Studies* (1986) edited by Button *et al.* Turn-taking has been most extensively analyzed by Sacks *et al.* (1974). For an analysis of both basic ideas in ethnomethodology and work in conversation analysis that grows from it see the book by Heritage (1984).

and therefore would bring a tape-recorder after two months. At first I thought I would study only girls' activities; this plan changed when the boys insisted that their activities were equally as interesting and asked if I could record them as well. The parents of each child were visited and told about my purposes on the street.

In contrast to many studies of children's language, rather than being based on interviews, laboratory experiments, or idealized versions of interaction obtained through elicited scripts (Nelson 1981), puppet enactment (Andersen 1977) or role play (Brenneis & Lein 1977; Lein & Brenneis 1978), this study is based on field observation of naturally occurring interaction. I travelled with the children as they went about their activities on the street while I had a Sony TC 110 cassette recorder with an internal microphone over my shoulder. Because I used only the internal microphone, I never had to actively point something at the children to record them and could get good records of their conversations simply by staying with them. I attempted to minimize my interaction with the children so as to disturb as little as possible the activities under study. My role was different from that of most anthropologists, and other ethnographers of children (Corsaro 1981, 1985) in that I was more an observer of the people I was studying than a participant in their activities.

My study also differed from other studies of children's behavior in the social sciences. Very seldom do psychologists and sociologists take as their focus the study of talk itself. Instead investigators study phenomena such as *relationship formation* or *friendship development*, often focusing on the content rather than the process of building such relationships through talk. Weaknesses of the experimental paradigm have been noted by developmental psychologists such as Damon (1983, p. 61): "The more we structure a setting for the purposes of systematic observation, the more we risk losing the richness, complexity, and spontaneity of natural children's interactions." Although psychologists realize the importance of studying social and cognitive processes naturalistically, they nonetheless are generally reluctant to depart from their controlled experimental paradigms.[2]

In my fieldwork I did not elicit particular speech genres; instead I recorded whatever talk children were engaged in as the children played on the street after school, on weekends and during the summer months. Although it is common for researchers working within the tradition of the ethnography of speaking to elicit "events that are culturally encoded as lexemes of the language" (Agar 1975, p. 43),[3] this was not my focus. Instead, the relevant unit of analysis was focused gatherings or "situated activity systems" (Goffman 1961, p. 96), units of focused social encounters.[4] Although I agree with Goffman that face-to-face interaction provides the central locus for the organization of social life, and have selected the "situated activity system" as my unit of analysis, I utilized a different methodology in my fieldwork. I made use of actual instances of naturally occurring talk (possible because I made tape-recordings) rather than idealized versions of them in my work. The methodology I used was ethnographic and designed to capture as accurately as possible the structure of events in the children's world *as they unfolded in the ordinary settings where they habitually occurred*. The tapes I collected preserved a detailed record of the children's activities, including the way in which their talk emerged through time – in all, over 200 hours of transcribed talk form the corpus of this study. By capturing the actual sequences of interaction as they unfolded, a record is available for other researchers to conduct comparative work.

THE GROUPS

Although there were gangs within a few blocks of Maple Street,[5] none of the children were members of gangs, and mothers were active on local boards to keep gangs out of

---

[2] Exceptions include recent research by psychologists and activity theorists (e.g., Lave 1988, 1991; Caiklin & Lave 1993; Scribner 1984) that employs ethnographic methods in natural settings to study "everyday cognition."

[3] The liabilities in such an approach have been noted by Gumperz (1981, p. 12), who noted "the very labels we use are often quite different from what we really intend to do." In addition he stated that everyday conversation does not have the bounded features that ritual performances, formal lectures, courtroom scenes, etc. have.

[4] According to Goffman (1961, p. 17) these are "a type of social arrangement that occurs when persons are in one another's immediate physical presence," involving such things as "a single visual and cognitive focus of attention, a heightened mutual relevance of acts, and an eye-to-eye ecological huddle that maximizes each participant's opportunity to perceive the other participants' monitoring of him" (Goffman 1961, p. 18).

[5] All names of people and places are fictitious.

their neighborhood. Education was highly valued and all of the children on Maple Street completed high school. Most of the fathers worked during the day at jobs such as hardware store manager, public transportation driver, policeman, preacher, and independent handyman, while women worked in jobs such as hospital cashier, dry cleaner's store attendant, factory seamstress, health aide, traffic guard, school bus attendant or school cafeteria worker. The language spoken by them was Black English Vernacular; analyzing my transcripts of the talk of the Maple Street children, Labov (1972, p. 184) found it similar to that of black speakers in Harlem.

I observed the children (whom I will call the Maple Street group) for a year and a half (1970–1971) in their neighborhood, focusing on how the children used language within interaction to organize their everyday activities.[6] The children ranged in age from 4 through 14 years and spent much of their time in four same age/sex groups:

	Ages (yrs)	No. of children
Younger girls	4–10	8
Younger boys	5–6	3
Older girls	10–13	15
Older boys	9–14	23

Here I will be concerned principally with older children, ages 9–14.[6]

Age boundaries clearly delimited peer groups. Children younger than four played inside the house and children older than 14 years regularly interacted in couples, and not necessarily with children from the neighborhood. Although relatives or friends came to visit, most of the children's playmates lived within a block's distance of each other. Boys ventured away from the neighborhood to play sports and work on jobs such as paper routes; girls, by way of contrast, were more frequently near their homes, obligated to care for siblings and perform chores. The children generally played on the shaded steps of their row houses, though the girls' activity of jump rope, and boys' activities such as riding go carts and bikes, skating, and standing on hands took place on the street.

The subgroups differed in their play references. Older

---

[6] For a more complete description of this fieldwork, see Goodwin (1990).

girls participated in activities requiring a wider range of types of social organization and language skill. They not only competed in jump rope, but also participated in dramatic play such as "house" and "school." They planned elaborate club meetings, practiced original dance steps, made things to sell (such as cake, pizza, and water ice). Girls spent more time talking than they did in any type of play activity. Because they often cared for younger children, their playgroups had greater age heterogeneity.

This contrasted with the older boys, who spent most of their time playing games. They participated in an extensive repertoire of pastimes. All year round they played organized sports such as football and basketball. Other activities – such as playing with yoyos or tops, walking on hands, coolie or skelley (a game which for a successful "win" involves the moving of a token made of a bottle cap or glass bottle rim filled with tar through squares of a grid drawn in chalk on the street), half ball, pitching pennies, flying kites, and marbles – were played in two- or three-week cycles.

Boys chose to organize their activities in a form of hierarchical social organization. Boys discussed ranking in terms of skill displayed in games and contests. Bragging was explicit in the midst of a game such as coolie or skelley.[7]

1   Malcolm: WOO! In the pocket. AH::: HA HA *h
                     ((*Smiles*)). I beat three games.

In fact many individual pastimes were transformed into contests. For example, boys compared who could walk

---

[7] Data are transcribed according to a modified version of the system developed by Jefferson and described by Sacks *et al.* (1974, pp. 731–3). Double obliques (//) indicate the point at which a current speakers' talk overlaps the talk of another. An alternative system is to place a left bracket ([) at the point of overlapping talk. Punctuation marks are not used as grammatical symbols, but mark intonation. A period indicates falling intonation. A comma is used for falling rising intonation. Question marks are used for rising intonation. Numbers in parentheses (0.5) indicate elapsed time in tenths of a second. A colon indicates that the sound preceding the colon has been lengthened. When one strip of talk follows another with noticeable quickness an equals sign is placed between the two utterances. Arrows are used to call attention to particular lines within a longer transcript. Bold face italic type indicates stressing and may involve pitch and/or volume. Uppercase indicates increased volume. Materials in double parentheses indicate features of the audio materials other than actual verbalizations. A degree sign (°) indicates that talk it precedes is low in volume.

on their hands the farthest without falling down or who could do the most elaborate yoyo tricks:

2	William:	I could walk on my hands better than anybody out here. Except him. And Freddie. Thomas can't walk.
3	Carl:	((*Playing with yoyos.*)) I do it experience! I do it better than Ossie. Watch. I'll win again!

Literature on girls emphasizes cooperative and non-competitive features of their play (Lever 1976; Savasta & Sutton Smith 1979; Sutton-Smith 1979). Even though girls do not make explicit rankings among one another as boys do, within their play they can make distinctions among group members. For example, in jump rope, girls count by tens to see how many successful jumps a girl can make, preferring this chant to other rhymed verses. Within another activity, playing house, girls created extended family relationships with hierarchical relationships; those in the role of mother had the most power in making decisions and considerable negotiation occurred among these players.

By far the most common types of comparison, however, were with respect to relationships girls argued they maintained with boys and adults. Girls who openly brag about a relationship with someone can be subject to criticism for acting as if they are superior to others:[8]

4	Kerry	Julia going around tellin everybody that that Bea- that Bea mother like her more than anybody else. She said she think she so big just because um, Miss Smith let her work in the kitchen for her one time.

As other researchers of girls' social organization (Best 1983, p. 93; Berentzen 1984, p. 108; Eder 1985) have found, girls sanction friends who attempt to put themselves above others. Not unlike principles of social organization found in societies with an egalitarian ethos, such as the !Kung Bushmen (Lee 1986), girls' interactions display a similarity among group members, as opposed to the explicit ranking found among the boys.

[8] A similar finding has been noted by Eder (1985), who reported that working class sixth-through-eight grade white girls describe self-congratulation as being "conceited" or "stuck-up."

However, despite the nonhierarchical organization, there is considerable conflict within the girls' group as girls attempt to work out their alliances. As in other non-hierarchically organized groups (Barth 1959), conflict develops as group members vie with each other for the position of "best friends" and delineate who is excluded from friendship arrangements. Differences in notions about how to structure social relationships in girls' and boys' groups affect how language is used within same-sex groups.

DIRECTIVES ACROSS ACTIVITIES

The ways in which speakers shape directives, actions that request that others do something and recipients reply to them, provide for a range of alternative types of social arrangement between participants. Some directive/response sequences display an orientation toward a differentiation between participants, and result in asymmetrical forms of relationships. Directives may be constructed in a variety of different ways. For example they can be phrased as imperatives (i.e., "Do X!"), a format that is frequently heard as face-threatening to its addressee (Brown & Levinson 1978, pp. 100–1) and thus constituting an "aggravated" speech act (Labov & Fanshel 1977, pp. 84–6). Alternatively, directives may be softened or "mitigated" (Labov & Fanshel 1977, pp. 84–6), for example phrased as requests ("Could you please do X?"). In this section I compare how girls and boys participate in a similar activity, organizing a task, comparing girls' ring making with boys' sling shot making.

Directives in girls' task activity

When organizing their own task activities, making objects, girls opt for directives that include the speaker as well as the hearer within the scope of the proposed action. In the following girls are making rings from glass bottle rims. To procure necessary supplies they go to a corner store and ask for glass bottles:

5		((*Girls are searching for bottles to make rings.*))
	Kerry:	Well let's go- let's go around the corner- Let's- let's go around the corner where whatchacallem.

6		((*Girls are looking for bottles.*))
	Martha:	Let's go around Subs and Suds.
	Bea:	Let's ask her "Do you have any bottles."

7		((*Girls are looking for bottles.*))
	Kerry:	Let's go.
		There may be some more on Sixty Ninth Street.
	Martha:	Come on.
		Let's turn back y'all so we can safe keep em.
		Come on. Let's go get some.

Girls' directives frequently include accounts that attend to the needs of the situation or the well being of the recipient:

8		((*Regarding rings made from glass bottle tops.*))
	Martha:	We gotta clean em first.
	Bea:	⌈You know
		⌊I know.
	Martha:	⌈Cuz they got germs.
		⌊Wash em and stuff cuz just in case they got germs on em.

In organizing activities girls participate jointly in decision-making with minimal negotiation of status. Girls' directives are constructed as suggestions for action in the future. The verb used by the girls – "let's" – signals a proposal rather than either a command or a request and, as such, neither shows special deference toward the other party (as a request does) nor makes claims about special rights of control over the other (as a command does). Thus, through the way in which they format their directives the girls make visible an undifferentiated "egalitarian" relationship between speaker and addressee(s). The structure of directives used to organize making rings is not different from the types of action that are used to direct other girls' activities involving a joint task, such as playing jacks (i.e., "Let's play some more jacks."), jumping rope (i.e., "Let's play 'one two three footsies'."), or hunting for turtles (i.e., "Let's look around. See what we can find.")

Directive use in boys' task activity

By way of contrast, when boys on Maple Street organize a task activity, they construct hierarchies, with those assuming a position of authority utilizing direct imperative forms. During two hours of sling shot making only two instances of directives using "let's" were used – when the leader wanted to direct participants into another play area, his back yard. In the following, play takes place at the home of Tony and Malcolm; the place where play occurs is significant, and Malcolm and Tony frequently remind the others that because play occurs on their property, they could be asked to leave. Throughout the activity Tony and Malcolm, who assume the position of team leaders of two sling shot making groups, order the others what to do, and frequently boys comply with these orders.

9		((*Regarding a hanger Ossie is cutting with pliers.*))
	Tony:	Go downstairs. I don't care what you say you aren't- you ain't no good so go downstairs.
	Dave:	((*Moves down the steps.*))

10		((*Instructing how to use pliers to cut wire.*))
	Malcolm:	Put it on the ground and jump on it.
	Ossie:	Aye::: ((*Jumping on pliers.*))

11	Malcolm:	PL:IERS. I WANT THE PLIERS!
		Man y'all gonna have to get y'all **own** wire cutters if this the way y'll gonna be.
	Pete:	Okay. Okay. ((*Giving pliers to Malcolm.*))

Syntactically, the forms utilized by the boys generally differentiate between speaker and hearer. A boy playing leader orders another to do something. In addition pejorative character depictions (such as "You ain't no good") may be included in the directive. Explanations for why particular actions should be undertaken are framed in terms of individual wants of the speaker (example 6) rather than benefits for group members, as was characteristic of girls' directives. In response to these directives the boy issuing the order generally gets compliance when addressing a subordinate person.

Other forms of sequencing are possible. Rather than complying with the directive, a participant may reject it (examples 12 and 13), or counter it (examples 14 and 15), constructing a more symmetrical relationship with regard to speaker:

12		((*Chopper moves up the steps to where Tony is seated.*))
	Tony:	Get off my steps.
	Chopper:	No. You get on *my* steps.
		I get on yours.

13   Malcolm: Get *out* of here Tony.
       Tony: I'm not gettin out of *no*where.

14   ((*Requesting pliers.*))
       Tony: Gimme the *things*.
       Chopper: You shut up you big lips.

15   Malcolm: Man back out.
                   I don't need ya'll *in* here I keep tellin
                   ya,
       Chopper: You better shut up. I'll tell you that.

In examples 9–11, participants display to each other that they have asymmetrical rights and duties with respect to one another, and the recipient ratifies the speaker's claim to be able to do something ritually offensive. By way of contrast, examples 12–15 show participants opposing speaker's claims, displaying more symmetrical relationships. These data sets show that directives provide the tools for building alternative forms of relationships among group members.

Directives do not all take imperative forms. Different affective alignments can be displayed with other types of response. Rather than demanding that the addressee do something, the speaker can make requests for permission (Ervin-Tripp 1982, p. 35) or information, that might or might not be granted. In contrast to imperatives Gordon & Ervin-Tripp (1984, p. 308) contrasted these forms as follows: "conventional polite requests, with few exceptions, are interrogatives that appear to offer the hearer options in responding . . . Conventional polite forms . . . avoid the appearance of trying to control or impose on another." In the following when boys make requests to Ossie, a boy playing the role of second in command, he provides reasons for rejections to the request.

16   ((*Requesting use of the pliers.*))
       Chopper: Could I hold yours now Ossie?
       Ossie: No. I didn't even get five yet.

17   Pete: My brother got my hanger to make a
                   slingshot.
                   Ossie can I have one?
       Ossie: I gotta come make some more slings.

Of course other responses are possible. In the following, when subordinate boys use modal constructions (through the use of "could" and "can") and interrogatives, in response Malcolm, the party in authority, answers with negation (18) and imperatives (19 and 20).

18   Tommy: Could Pete be on your side and I be on
                   Malcolm's side?
                   Malcolm could I be on your side?
       Malcolm: Heck no!

19   ((*Tokay takes a hanger.*))
       Tokay: Can I have some hangers?
       Malcolm: Put that thing back!

20   Tokay: Anybody wanna buy any rubber bands?
       Malcolm: Put em in your pocket. Cuz you gonna
                   pop em.

Both symmetrical (equal) (16 and 17) and asymmetrical (hierarchically based) (18–20) relationships are possible in the organization of directive/response sequences. A boy in the position of leader (such as Malcolm) uses an aggravated action to display his power relative to the addressee through issuing imperatives and rejecting requests: reciprocally, those in the position of subordinate display their position by making requests and complying with imperatives.

Variation in girls' directive use

Whereas imperatives are commonplace among the boys, during same-sex play in the girls' group such actions are seldom used; they are reserved for sanctioning girls who are deviant. For example, one girl who was being ostracized was issued the following imperative.

21   ((*Annette, who is being ostracized, is
                   walking up her steps.*))
       Bea: GIRL YOU MADE- BETTER GO
                   BACK-
                   BETTER GO BACK IN THE *HOUSE!*

While girls have access to the same aggravated forms as the boys, they generally utilize them with each other only to mark particular stages in a relationship. Though unremarkable among boys, in the girls' group using imperative forms is considered inappropriate, insulting talk, referred to as "basing":

22   Kerry: GET OUTA MY STREET GIRL!
                   HEY GIRL GET OUTA MY STREET!
       Rhonda: → Now don't come basin at me!

In same-sex groups, girls rather than boys generally employ quite mitigated forms in their dealings with one another. Aggravated forms are used by both groups in cross-sex interaction. In countering boys, girls make use of imperatives as well as forms of opposition moves that boys employ in countering prior talk. For example, in response to infractions that boys commit, girls use imperatives in an attempt to extinguish behavior they deem unacceptable:

23   ((*Freddie starts drinking water from Kerry's spout.*))
    Kerry: You act so *greedy.*
             Go *home* if you want some water.

24   ((*Karl steps on Ruby's lawn.*))
    Ruby: Get out the way offa that-
             Get off that *lawn!*

25   ((*Tommy, Kerry's brother, is hitting Kerry.*))
    Kerry: Tommy, stop playin around. STOP.
             Quit playin around. STOP!

In cross-sex interaction, girls respond to imperatives not with compliance but with strong indications of opposition. Using expressions of polarity, they signal opposition immediately, in much the same way as the boys did in examples 12 and 18 above:

26 Chopper: Get outa here you wench!
              You better get outa here.
    Bea: No! You don't tell *me* to get out!

27   ((*Talking about Martha's hair.*))
    Billy: Wet it!
    Martha: No. I don't *wanna* wet it.

Girls make use of directive forms that are equally as forceful as those of boys in a number of other contexts as well. The following occur while girls are caring for younger siblings:

28   ((*Jolyn, Kerry's sister, puts down the hood of her jacket on a windy day.*))
    Kerry: *Don't* put that down.
              Put that back *up!*
              It's sup*posed* to be that way.

29 Martha: Stay out the street now man.
              Come on punk. Hurry up Glen.

Here clearly the caregiver is in a superior position with regard to her charge; a hierarchy similar to that of the boys is established through use of these actions.

The use of such bald imperatives to sanction wrong-doers is learned at an early age. Girls as young as age four can use bald imperatives effectively. In the following Jolyn (aged four) uses imperatives and polarity markers to rebuke a boy two years her senior:

30   ((*Larry begins to paint Jolyn' table.*))
    Jolyn: *Don't* paint that um table!
    Larry: Let me just paint.
    Jolyn: *No. Don't* paint that table.
              Don't put all that paint on that thing.
              That chair.

Authority in issuing directives changes across different activities. As was seen with the boy's group, participants *who control the space* where play takes place can dictate the rules for playing. In examples 31 and 32, play occurs on the adjacent porches of Martha and Patrice, girls who though older than the girls playing the role of their children do not normally issue controlling actions to others. In enacting the role of mother, girls address directives to their "children" that are very similar in structure to the bald imperatives their own mothers use in talk to them. Their imperatives are spoken loudly (as indicated by capitals) and with emphatic stress (indicated by italics):

31 Martha: BRING THOSE CARDS BACK,
              BRING THAT BOOK IN THE
              HOUSE AND C:OME HOME!
              Don't *climb* over that way.
              You climb over the *right* way.

In the next example, both girls playing mother, Patrice and Martha, as overseers of the unfolding drama, monitor the actions of participants and control how roles are played:

32 Patrice: HEY BRENDA YOU OUGHTA // be sleep!
    Kerry: I can't even get her in the bed.
    Patrice: I know.
    Patrice: SHE'S NOT // EVEN PLAYIN RIGHT.
              SHE NOT EVEN *PLAY*IN RIGHT.
    Martha: BRENDA PLAY RIGHT.
              THAT'S WHY NOBODY WANT YOU FOR A CHILD.
    Patrice: GET IN THERE AND GO TO SLEEP!

Girls in the position of "mother" can thus dictate for others the dimensions of play *outside* the frame of fantasy as well as within it. They can control not only who can play what roles but also how they play them. The frame of dramatic play thus closely resembles the frame of task activity among the boys, where participants use imperatives to display differences in social positionings among players.

Girls and boys make use of a similar speech form – directives – for establishing social relationships among group members. Both boys and girls use imperative forms to propose differentiation; and recipients of such actions may either accept or challenge such proposals, arguing either for symmetrical or asymmetrical relationships with respect to the action on the floor. In cross-sex groups during argumentative exchanges aggravated imperative forms rather than migitated directives are used. Inclusive forms such as "Let's" are used only by the girls, whereas boys use either imperative or request forms, depending upon the social position of the speaker. Girls have access to the same expanded repertoire of forms as boys, but they choose to use the more inclusive forms in conducting their same-sex social organization. Girls demonstrate considerable versatility in their use of directive formats; within other activity frames such as playing house, sanctioning younger children in their charge, and responding to offensive actions of others, girls use aggravated forms. They select a form from a repertoire of speech forms depending on its appropriateness for the current situation.

ARGUMENT FORMS IN GIRLS' HE-SAID SHE-SAID DISPUTES

The fact that girls presume equality in their dealings with one another has implications for the ways in which they carry out disputes and political processes. Although arguments among the boys on Maple Street almost always ended quickly (and rarely if ever led to physical fights), their disputes were frequent, direct, and constructed through use of aggravated speech forms. Boys issued direct accusations, followed by a series of denials:

33   ((*Making sling boards.*))
    William: You usin my *nails.*
    Carl: I ain't use *none* a your nails.
    William: Did *so.*
    Carl: Did *not.*
    William: Where are they // now.
    Carl: This is *none* a your nails.

    William: Where are they // then.
    Carl: They *Vin*cent nails.
    William: They *my* nails // egghead.
    Carl: ((*Falsetto.*)) Eh!
    William: → Ain't they Vincent.
    Vincent: Yeah, they are William's nails.
    William: *Now:.*

While similar forms of action/counter-action sequences (did so/did not) occur in cross-sex disputes as well (see Appendix 1), the most important form of girls' disputes called "he-said–she-said" was built with much more complicated linguistic and participation structures than those of the boys, had a much longer duration in time, encompassed many more participants, and had far more serious consequences. Girls' accusations within this activity were far more indirect than those of boys, and concerned a specific offense: talking about someone in her absence. Much like the strict forms of action used to initiate a case in the Anglo-American legal system, he-said–she-said accusations provide a history of how the accuser learned that her addressee had talked about her behind her back, which establishes the grounds for the charge. The utterances show that an implicit alliance exists between two parties who hold that the defendant said something about an absent party in the past. For example:

34   Barb to Bea: They say y'all say I wrote everything over there.

35   Ann to Benita: And *Sis*ter said that *you* said that I was showin off just because I had that bl:ouse on.

36   Bea to Ann: Kerry said *you* said that I wasn't gonna go around *Pop*lar no more.

The girls' accusations with their embedded reports thus differ from the boys' in critical ways. They do not directly report an event for which the addressee is held responsible, as do the boys, but instead frame the accusation as something learned through an intermediary party. In essence they constitute an implicit coalition of "two against one" against the defendant – a form of argument that may be stated explicitly by the accuser during a confrontation. The following occurs in response to a defendant's denial:

37   Ruby: Well I'm a get it straight with the people. What **Kerry**, (1.4)

It's between Kerry, and you, (1.0)

→   See **two** (0.5) two against one.

Who wins? The one is two.=Right? (0.5)

And that's Joycie and Kerry. (0.5)

They both say that you said it.

And you say that you didn't say it.

Who you got the **proof** that say that

you **did**n't say it.

Girls' accusations provide for a much more elaborated dispute than occurs among the boys. While the accuser may argue that an alliance exists against the hearer, usually the third party named in the accusation is absent from the confrontation. The addressee can thus answer that the intermediary party is making up the event in question or lying, and argument can continue indefinitely without clear resolution.

Though traditionally gossip has been analyzed in terms of women's powerlessness, and considered a speech form that is situated within the private sphere (Gal 1990), on Maple Street gossip constitutes a major public political event through which girls display their willingness to engage in what Goffman (1961, pp. 237–58) called *character contests*. In that the girls' social organization consists largely in shifting coalitions in triads, rather than a hierarchical organization, gossip can be used to rearrange the social organization of the moment. It functions to control aspiring individuals, girls who in various ways "think they are cute" or are perceived as trying to show that they are "better than" others. In less serious cases the defendant endures the deluge of accusations that are hurled against her and practices avoidance behavior for a few days. However, in the most serious he-said–she-said confrontation, a girl was ostracized for a period of a month and a half, and during this time she was subject to the taunts of others, ridiculing her siblings and mother, or composing songs about her physical traits, and pranks such as ringing her doorbell and running away.

Within the he-said–she-said event girls react with righteous indignation when they learn their character has been maligned. They display an intense interest in initiating and elaborating disputes about their rights. Nothing of the complexity of he-said–she-said embedded accusation statements nor the scale of the girls' he-said–she-said political event was observed among the boys. While boys used ridicule, "sounding" on the traits of a boy or his

mother through contests of ritual insult, boys who were targets of such taunting were not ostracized from the group. Through the constant participation in rounds of different games, boys knew their relative ranking, and argued openly about it. Thus, differences in cultural practices for making comparisons lead to differences in ways in which language is used to structure social relationships in same-sex groups. Though girls and boys used similar types of directive and counter in cross-sex disputes, the girls' inclusive directives and indirect he-said–she-said accusations were distinctive of their same-sex group organization.

ALTERNATIVE DISPUTE STRATEGIES IN GIRLS' GAMES

Psychologists often reify cultural differences in girls' and boys' groups, stating, for example, that while boys are "more concerned with and more forceful in pursuing their own agendas . . . girls are more concerned with maintaining interpersonal harmony" (Miller *et al.* 1986) (see also Leaper (1991, p. 798). This has been the view in the literature on girls' socialization, based on white middle class girls (Maccoby 1990). It has been argued that girls avoid direct competition and are little interested in negotiational involvements (Gilligan 1982; Lever 1978; Sutton-Smith 1979). Such positions resemble the "two cultures" view of language differences between males and females postulated by Maltz & Borker (1982) and the collaborative model of women's talk in work of sociolinguists such as Falk (1980), Troemel-Ploetz (1992), and Coates (1994).

Close examination of the directives and gossip activities of Maple Street girls defies such stereotypes. In addition, girls' interaction within the frame of games such as jump rope (Goodwin 1985), and four square (Hughes 1988) shows that girls can animatedly dispute, resist and probe the boundaries of rules, as referees and players together build the game event. In ongoing fieldwork I am conducting of girls playing the game of hopscotch, I have found different ways of expressing disagreement (either highlighting opposition or making it nonsalient) in the midst of games (Goodwin 1995, 1996). Two examples are presented below to show the contrasts.

The following provides an example of argument in which girls highlight opposition in the midst of hopscotch.

The data are of working class African-American girls, children of migrant workers in the rural South.

38	1	Lucianda:	((*Takes turn jumping twice in square two and possibly putting her foot on the line of square 1.*))
	2	Joy:	You out.
	3	Lucianda:	No I'm **not**. ((*Shaking head no.*))
	4	Joy:	You hit the line.
	5	Crystal:	**Yes** you did.
	6		You hit the line. ((*With hand pointing at line.*))
	7	Joy:	You hit the line.
	8	Lucianda:	I AIN'T HIT NO LINE! ((*Leaning her chest towards Crystal in a challenge posture.*))
	9	Alisha:	**Yes** you did.
	10	Crystal:	((*Smiling, shaking head, goes to the spot.*)) °You did. You s-
	11	Lucianda:	**No** I didn't.
	12	Alisha:	Yes you did.
	13	Crystal:	Didn't she go like this.
	14	Lucianda:	((*Does a challenge hit towards Alisha.*))
	15	Alisha:	You hit me.
	16	Crystal:	You did like this. ((*Stepping on the line as she replays the jump.*))
	17	Lucianda:	Shut up with your old fashioned clothes. ((*To Alisha.*))
	18	Crystal:	You did like that.
	19	Joy:	Yeah you hit that line right there honey. ((*As she goes up and uses her foot to index it, tapping it twice.*))
	20	Lucianda:	((*Throws the rock and it lands outside.*))
	21		My feet.
	22	Vanessa:	Y- you out now!

In this game of hopscotch, referees state unequivocally "You out" (line 2), followed by an explanation ("You hit the line.") (lines 4, 6, 7). Polarity markers such as "No," (lines 3, 11) and "Yes" (lines 5, 9, 12) preface opposition moves. The foul call – "You hit the line." – is emphatically opposed by the player with "I AIN'T HIT NO LINE!" and accompanied by a strong body stance – a challenge position in which the player extends her chest towards one of the judges.

In the above example girls work to forcefully construct salient opposition while holding each other accountable for deviations from rules. This occurred in examples from working class Latina girls' games as well (Goodwin 1995). However, by making other language choices it is possible to construct actors, events and social organization in a very different way. A group that did not make use of the participation possibilities of hopscotch to enact forceful position was middle class white girls from Columbia, South Carolina. The following provide some ways in which opposition was expressed in their games.

39	1	Linsey:	((*Throws stone and hits line.*))
	2	Liz:	Oh! Good job Linsey!
	3		You got it all the way on the seven.
	4	Kindrick:	((*Shaking head.*)) That's-
	5		I think that's sort of on the line though.
	6	Liz:	Uh- your foot's in the wr(hhh)ong-
	7		sp(hh)ot.
	8	Kendrick:	Sorry.
	9		That was a good try.
40	1	Linsey:	((*Throws token.*))
	2	Cathleen:	°You **did** it!=
	3	Linsey:	Yes! ((*Falsetto.*))
	4	Linsey:	((*Jumps on line.*))
	5	Cathleen:	Wh-
	6	Kendrick:	You- accidentally jumped on that.
	7		But that's okay (hh).

In contrast to definitively categorizing moves as fouls, here the girls acting as judges use a variety of language structures to *mitigate* their foul calls. Hedges such as "I think" (example 39, line 5) and "sort of" (example 39, line 5) display uncertainty about the accuracy of the call. The force of a fault-finding word such as "wrong" is undercut by embedding laugh tokens within it (example 39, line 6). Agency is removed from the offender's action through use of terms such as "accidentally" (example 40, line 6), and divorcing the foot that lands off the line from the actor controlling that foot. Moreover, committing a foul may have no real consequences. It's "okay" (example 40, line 7) if someone "accidentally" (example 40, line 6) jumps on the line. Finally, even a failed attempt is praised as "a good try" (example 39, line 9). Completely missing from the way these girls play the game is any articulation of strong stances or accountability for one's actions. While the working class girls fiercely hold one another accountable for moves they make that break the rules, middle class girls prefer to ignore fouls and violations and avoid conflict at all costs, even indulging in group hugs when a player does successfully complete a difficult turn.

REPERTOIRES OF LANGUAGE PRACTICES

Within the social sciences a great deal of research has recently been directed toward investigating disparate accounts for gender differences in language. Disputes rage over whether women speak through a "different voice" (Gilligan 1982) and, if so, whether it should be attributed to a model of "difference" or "dominance."[9] The model of "difference" is based on the belief that psychological differences in personality traits, skills beliefs, attitudes, or goals are the result of socialization, and are located within individuals. According to this perspective men and women speak differently because they, like members of distinct ethnic minorities, are socialized into different cultures from childhood. By way of contrast the dominance model holds that women speak the way they do because men keep women in their place; social structure is viewed as controlling human behavior, and power differentials alone account for gender differences in language use.

While providing differing explanations for the relationship of language to gender, both the difference and dominance models are alike in accepting rather simplistic views of the relationship of language and gender. In both paradigms, based on middle class American models of female roles, females are viewed as tentative, deferential, and imprecise. Both the explanations of dominance and difference fail to take into account that females' relationships with males change throughout the life course (i.e., when girls play in their same-sex playgroups they are not dominated by boys) or that societies differ widely in how assertively females speak. For example, ethnographic studies of women's language use find that the women of Malagasy (Keenan 1974) or Gapun, Papua New Guinea (Kulick 1992, 1993)) are anything but tentative or indirect in their speech use, and among Mayan women of Tenejapa, Chiapas, directness takes precedence over indirection within the context of the court (Brown 1990).

Both the difference and dominance models ignore the situated, emergent nature of everyday speech – something Ochs (1992, p. 345) has argued is desperately needed if we are to understand the relationship of language and gender.

In this Chapter I have focused on a repertoire of speech forms available to girls across different activities that defy stereotypes of women as powerless speakers. Girls selectively build alternative types of social organization in a range of speech activities: (a) directives may be formatted both to include the speaker in the scope of the action in question or create a differentiation between speaker and hearer; (b) accusations in the he-said–she-said, which provide an exact past history of events through time and generate extended disputes about character. In contrast to many studies that find female speech to be mitigated or apolitical, among the girls I studied power is clearly evident in the actions of girls initiating a he-said–she-said dispute, in the many situations in which girls use aggravated directives, as well as in the ways in which they can dispute a player's move in the midst of a game such as hopscotch.

Analysis of the processes observed here would have been impossible without long-term ethnographic study across a range of diverse activities. A situated view of language allows us to examine the repertoire of language forms human beings make use of across diverse activities to achieve their social organization. Through different language games (Wittgenstein 1958) children craft diverse forms of social order, including ones proposing equality as well as hierarchy, and exclusive alliances as well as cooperatively functioning task forces within a girls' group. When children's social organization is investigated, we find a group exhibiting full female agency. If research on gender and language is to move forward, beyond the recycling of old canons of gender differences, we need to enlarge the scope of our inquiry to include detailed ethnographic research of situated language use, in a variety of ethnic communities (Henley 1996), without neglecting important stages of the life course (Thorne 1987).

APPENDIX 1

In response to statements made by a first speaker, the recipient may refute the prior action and provide an account or explanation for the position taken up (marked with "#" in the transcript). In the following Martha and

---

[9] See for example Freed (1992) and Crawford (1995).

Kerry undercut the grounds for a position taken by William.

((*While children skate, William bumps into Kerry.*))

	Kerry:	Get off William. Get off!
	Martha:	Now Kerry just aim at William butt and let's see if we could knock him down.
	Kerry:	Oy yeah you- you be // you better
	William:	Y'all better *not* knock me *down!*=
	Kerry:	Yeah?
#	Martha:	If we do that's what we // playin'.
#	Kerry:	Play and you gonna get knock down.
	William:	Nuh *uh:!*
	Kerry:	Mm *hm!*
#	William:	Nuh *uh* y'all. I ain't playin'.
	Kerry:	Yes you *are* playin'.
	William:	I can't af // fird
#	Kerry:	If you- if you put a skate on you playin'.
	William:	*No* it ain't.
	Kerry:	*Yes* it is.
	William:	I ain't playin' // nuttin!
	Kerry:	Is you playin' Martha,
#	Martha:	If you-
		if you put that skate on // you are.
	Kerry:	Yep. If you put the skate on you playin'.

REFERENCES

Agar, M. (1975). Cognition and events. In *Sociocultural Dimensions of Language Use*, ed. M. Sanches & B. G. Blount, pp. 41–56. Academic Press, New York.

Andersen, E. S. (1977). *Learning to Speak with Style: A Study of the Sociolinguistic Skills of Children*. Routledge, London.

Atkinson, J. M. & Heritage, J. (eds.) (1984). *Structures of Social Action*. Cambridge University Press, Cambridge.

Barth, F. (1959). *Political Leadership among Swat Pathans*. Athlone Press, New York.

Berentzen, S. (1984). *Children Constructing Their Social World: An Analysis of Gender Contrast in Children's Interaction in a Nursery School*. Bergen Occasional Papers in Social Anthropology, no. 36. Department of Social Anthropology, University of Bergen, Bergen.

Best, R. (1983). *We've All Got Scars*. Indiana University Press, Bloomington, IN.

Brenneis, D. & Lein, L. (1977). "You Fruithead": A sociolinguistic approach to children's disputes. In *Child Discourse*, ed. S. Ervin-Tripp & C. Mitchell-Kernan, pp. 49–66. Academic Press, New York.

Brown, P. (1990). Gender, politeness and confrontation in Tenejapa. *Discourse Processes* 13: 123–41.

Brown, P. & Levinson, S. C. (1978). Universals of language usage: Politeness phenomena. In *Questions and Politeness Strategies in Social Interaction*, ed. E. N. Goody, pp. 56–311. Cambridge University Press, Cambridge.

Button, G., Drew, P. & Heritage, J. (1986). *Human Studies: Special Issue*. Martinus Nijhoff, Dordrecht.

Button, G. & Lee, J. R. (eds.) (1987). *Talk and Social Organisation*. Multilingual Matters, Clevedon.

Caiklin, S. & Lave, J. (eds.) (1993). *Understanding Practice: Perspectives on Activity and Context*. Cambridge University Press, Cambridge.

Coates, J. (1994). The language of the professions: Discourse and career. In *Women and Career: Themes and Issues in Advanced Industrial Societies*, ed. J. Evetts, pp. 72–86. Longman, London.

Corsaro, W. A. (1981). Friendship in the nursery school: Social organization in a peer environment. In *The Development of Children's Friendships*, ed. S. R. Asher & J. M. Gottman, pp. 207–41. Cambridge University Press, Cambridge.

Corsaro, W. A. (1985). *Friendship and Peer Culture in the Early Years*. Ablex, Norwood, NJ.

Crawford, M. (1995). *Talking Difference: On Gender and Language*. Sage, London.

Cullen, J. M. (1972). Some principles of animal communication. In *Non-verbal Communication*, ed. R. A. Hinde, pp. 101–25. Cambridge University Press, Cambridge.

Damon, W. (1983). The nature of social-cognitive change in the developing child. In *The Relationship between Social and Cognitive Development*, ed. W. F. Overton, pp. 103–41. Erlbaum, Hillsdale, NJ.

Eder, D. (1985). The cycle of popularity: Interpersonal relations among female adolescents. *Sociology of Education* 58: 154–65.

Ervin-Tripp, S. (1982). Structures of control. In *Communicating in the Classroom*, ed. L. C. Wilkinson, pp. 27–47. Academic Press, New York.

Falk, J. (1980). The conversational duet. *Proceedings of the 6th Annual Meeting of the Berkeley Linguistics Society* 6: 507–14.

Freed, A. F. (1992). We understand perfectly: A critique of Tannen's view of cross-sex communication. In *Locating Power: Proceedings of the Second Berkeley Women and Language Conference*, ed. K. Hall, M. Bucholtz & B. Moonwomon, pp. 144–52. Berkeley Women and Language Group, Department of Linguistics, University of California, Berkeley.

Gal, S. (1990). Between speech and silence: The problematics of research on language and gender. In *Gender at the Crossroads of Knowledge: Feminist Anthropology in the Postmodern Era*, ed. M. di Leonardo, pp. 175–203. University of California Press, Berkeley, CA.

Gilligan, C. (1982). *In a Different Voice: Psychological Theory and Women's Development*. Harvard University Press, Cambridge, MA.

Goffman, E. (1961). *Encounters: Two Studies in the Sociology of Interaction*. Bobbs-Merrill, Indianapolis.

Goodwin, M. H. (1985). The serious side of jump rope: Conversational practices and social organization in the frame of play. *Journal of American Folklore* **98**: 315–30.

Goodwin, M. H. (1990). *He-Said–She-Said: Talk as Social Organization among Black Children*. Indiana University Press, Bloomington, IN.

Goodwin, M. H. (1995). Co-construction in girls' hopscotch. *Research on Language and Social Interaction* (Special issue of *Co-Construction*, ed. S. Jacoby & E. Ochs), **28**: 261–82.

Goodwin, M. H. (1996). Games of stance. In *Language Practices of Older Children*, ed. S. Hoyle & C. T. Adger, Oxford University Press, New York, in press.

Gordon, D. & Ervin-Tripp, S. (1984). The structure of children's requests. In *The Acquisition of Communicative Competence*, ed. R. L. Schiefelbusch & J. Pickar, pp. 298–321. University Park Press, Baltimore, MD.

Gumperz, J. (1981). Conversational inference and classroom learning. In *Ethnography and Language in Educational Settings*, J. Green & C. Wallat, pp. 3–23. Ablex, Norwood, NJ.

Henley, N. M. (1996). Ethnicity and gender issues in language. In *Handbook of Cultural Diversity in Feminist Psychology*, ed. H. Landrine, in press.

Heritage, J. (1984). *Garfinkel and Ethnomethodology*. Polity Press, Cambridge, MA.

Hughes, L. (1988). "But that's not really mean": Competing in a cooperative mode. *Sex Roles* **19**: 669–87.

Keenan, E. Ochs (1974). Norm-makers, norm-breakers: Uses of speech by men and women in a Malagasy community. In *Explorations in the Ethnography of Speaking*, ed. R. Bauman & J. Sherzer, pp. 125–43. Cambridge University Press, Cambridge.

Kulick, D. (1992). *Language Shift and Cultural Reproduction: Socialization, Self, and Syncretism in a Papua New Guinean Village*. Cambridge University Press, Cambridge.

Kulick, D. (1993). Speaking as a woman: Structure and gender in domestic arguments in a New Guinea village. *Cultural Anthropology* **8**: 510–41.

Labov, W. (1972). *Language in the Inner City: Studies in the Black English Vernacular*. University of Pennsylvania Press, Philadelphia.

Labov, W. & Fanshel, D. (1977). *Therapeutic Discourse: Psychotherapy as Conversation*. Academic Press, New York.

Lave, J. (1988). *Cognition in Practice*. Cambridge University Press, Cambridge.

Lave, J. (1991). Situating learning in communities of practice. In *Perspectives on Socially Shared Cognition*, ed. L. Resnick, J. M. Levine & S. D. Teasley, pp. 63–84. American Psychological Association, Washington, DC.

Leaper, C. (1991). Influence and involvement in children's discourse: Age, gender and partner effects. *Child Development* **62**: 797–811.

Lee, R. B. (1986). Eating Christmas in the Kalahari. In *Anthropology 86/87: Annual Editions*, ed. E. Angeloni, pp. 17–20. Dushkin, Guilford, CN. (Reprinted from *Natural History*, 1969).

Lein, L. & Brenneis, D. (1978). Children's disputes in three speech communities. *Language in Society* **7**: 299–323.

Lever, J. R. (1976). Sex differences in the games children play. *Social Problems* **23**: 478–87.

Lever, J. R. (1978). Sex differences in the complexity of children's play and games. *American Sociological Review* **43**: 471–83.

Maccoby, E. E. (1990). Gender and relationships: A developmental account. *American Psychologist* **45**: 513–20.

Maltz, D. N. & Borker, R. A. (1982). A cultural approach to male–female miscommunication. In *Communication, Language and Social Identity*, ed. J. J. Gumperz, pp. 196–216. Cambridge University Press, Cambridge.

Miller, P. M., Danaher, D. L. & Forbes, D. (1986). Sex-related strategies for coping with interpersonal conflict in children aged five and seven. *Developmental Psychology* **22**: 543–8.

Nelson, K. (1981). Social cognition in script framework. In *Social Cognitive Development: Frontiers and Possible Futures*, ed. J. H. Flavell & L. Ross, pp. 97–118. Cambridge University Press, Cambridge.

Ochs, E. (1992). Indexing gender. In *Rethinking Context*, ed. A. Duranti & C. Goodwin, pp. 335–58. Cambridge University Press, Cambridge.

Sacks, H., Schegloff, E. A. & Jefferson, G. (1974). A simplest systematics for the organization of turn-taking for conversation. *Language* **50**: 696–735.

Savasta, M. L. & Sutton-Smith, B. (1979). Sex differences in play and power. In *Die Dialektk des Spiels*, ed. B. Sutton-Smith, pp. 143–50. Holtman, Schorndoff.

Schegloff, E. A. & Sacks, H. (1973). Opening up closings. *Semiotica* **8**: 289–327.

Schenkein, J. (1978). *Studies in the Organization of Conversational Interaction*. Academic Press, New York.

Scribner, S. (ed.) (1984). Cognitive studies of work. *Quarterly Newsletter of the Laboratory of Comparative Human Cognition* **6**: (1–2): 1–4.

Simmel, G. (1950). *The Sociology of Georg Simmel*. Transl. K. Wolff. Free Press, Glencoe, IL.

Sutton-Smith, B. (1979). The play of girls. In *Becoming Female*, ed. C. B. Kopp & M. Kirkpatrick, pp. 229–57. Plenum, New York.

Thorne, B. (1987). Re-visioning women and social change: Where are the children? *Gender and Society* 1: 85–109.

Troemel-Ploetz, S. (1992). The construction of conversational equality by women. In *Locating Power: Proceedings of the Second Berkeley Women and Language Group*, ed. K. Hall, M. Bucholtz & B. Moonwomon, pp. 581–9. Berkeley Women and Language Group, Department of Linguistics, University of California, Berkeley, CA.

Wittgenstein, L. (1958). *Philosophical Investigations*, ed. G. E. M. Anscombe & R. Rhees, transl. G. E. M. Anscombe, 2nd edn. Blackwell, Oxford.

Zimmerman, D. H. & West, C. (eds.) (1980). Sociological Inquiry. Special Double Issue on Language and Social Interaction.

Index